D0163004

NEUROETHOLOGY
Nerve Cells and the Natural Behavior of Animals

Jeffrey M. Camhi
HEBREW UNIVERSITY, JERUSALEM

NEUROETHOLOGY

Nerve Cells and the Natural Behavior of Animals

SINAUER ASSOCIATES INC.
PUBLISHERS
Sunderland, Massachusetts

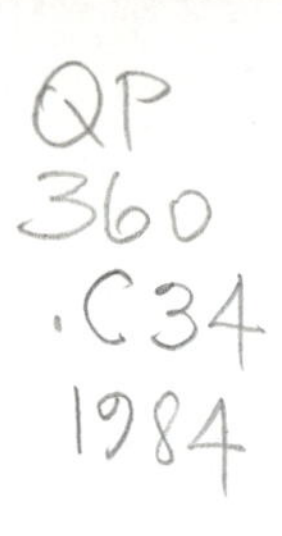

NEUROETHOLOGY:
Nerve Cells and the Natural Behavior of Animals

Cover photo courtesy of M. Konishi.

Library of Congress Cataloging in Publication Data

Camhi, Jeffrey M., 1941–
Neuroethology: nerve cells and the natural behavior of animals.

Bibliography: p.
Includes index.
1. Neuropsychology. 2. Animal behavior. I. Title.
[DNLM: 1. Behavior, Animal. 2. Neurons. QL 751 C182n]
QP360.C34 1983 591.5′1 83-14957
ISBN 0-87893-075-2

Printed in U.S.A.

9 8 7 6 5 4 3 2 1

To Jane

Contents

Preface

The aim of this book is to introduce readers to the major concepts and research strategies in the rapidly developing field of neuroethology. Rather than presenting a broad, comprehensive treatment covering a great many of the animals that have been studied, I have chosen to illustrate each selected principle with just one, or at most a few, animal subjects. Each of these is discussed in depth, including the framing of research questions, the research approaches and techniques, the findings themselves, problems of interpretation, and the prospects for future work. In order to show the findings in a broader biological perspective, most such discussions are followed by a brief comparative treatment. The hope is that even a reader with little or no background in animal behavior or neurobiology will be led quickly to appreciate neuroethological issues as they are viewed by researchers working in this field today. As a teaching aid, each chapter is followed by a set of questions for open-ended discussion and by an annotated list of suggested readings.

This book, then, is designed for university undergraduates with little or no prior course work in animal behavior or neurobiology. The necessary concepts from both these fields are presented in Chapters 2 and 3. However, owing to the emphasis I have placed on current research on selected topics, it is hoped that the book will also find use in more advanced courses, including graduate courses.

In the past, there has been a tendency among many researchers, especially in neurobiology, to specialize narrowly on either vertebrate or invertebrate animals. The backbone somehow became a fence dividing this field of research in half. Increasingly, however, workers are realizing the artificiality of this division. It is clear that all animals face similar behavioral problems, such as obtaining mates, food, and shelter and avoiding predation. Moreover, the neuronal signals in the brain that produce behavior appear to be nearly identical in all animals. These considerations make any species an appropriate candidate for investigation. An appropriate research strategy, therefore, is first to ask a general question of neuroethological interest and then to select the specific animal that best allows one to answer this question. For some questions, an invertebrate animal will fit the bill, whereas for others it will be a vertebrate. In this book, I have followed this logic and have given roughly equal billing to the spineless and the spined in illustrating the various issues discussed. If the former group is featured slightly more heavily, this reflects my own greater research experience with invertebrate animals.

The book is divided into three parts. Part I is introductory; it presents the kinds of issues neuroethologists address (Chapter 1), background material in behavior (Chapter 2) and in neurobiology (Chapter 3), and a case study in neuroethology (Chapter 4). Part II concerns the reception of sensory information from the environment. For instance, Chapter 5 discusses visual information and visual reception, and Chapter 6 discusses auditory information and reception. Chapter 7 treats the processing by the brain of visual and, to a lesser degree, auditory signals. Part III deals with behavioral outputs from the nervous system. For instance, in Chapter 8, the neural organization underlying brief, individual behavioral acts is discussed. Chapter 9 turns to the neural organization for temporally patterned behaviors. And Chapter 10 discusses the role of sensory feedback in producing motor outputs. The book concludes with a brief epilog.

There has been less than uniform agreement as to the meaning of the word *neuroethology*. Some researchers who might appear to be strict neurophysiologists, and others who seem to be pure field behaviorists, regard themselves as subsumed under the neuroethological rubric. Yet others, whose work might appear more neuroethological, seem to avoid this designation like a plague. Among all these workers, there seems to be little agreement in defining the important questions. A second aim of this book, then, is to suggest an outlook, a set of issues and research questions, and a broad set of experimental strategies that represent at least one coherent delineation of the field of neuroethology. I have made little effort to stifle my enthusiasm, shared by many of my colleagues, for this field of endeavor—an enthusiasm based upon the belief that neuroethology confronts some of the greatest unsolved mysteries of the animal world.

Jeffrey M. Camhi
Jerusalem, Israel

Acknowledgments

It was my great good fortune to be a faculty member for 15 years in Cornell University's Section of Neurobiology and Behavior. The colleagues and students with whom I worked were my teachers, without whom this book could not have been written. During the period of preparation of the manuscript, I have pestered many of these friends with endless questions, which they have tirelessly answered. Most frequently pestered were Robert Capranica and Melvin Kreithen, to whom I owe a great debt. Ron Hoy also offered many stimulating ideas.

Many friends and colleagues were kind enough to read and comment on early drafts of the manuscript. The feedback received was extremely valuable. Those who read most or all of the manuscript were Christopher Comer, Darryl Daley, Donald Griffin, William Kristan, Paul Stein, Susan Volman, Jeffrey Wine, and the students of the neuroethology course I taught at Cornell in the spring of 1982. Others who read shorter sections were Robert Campenot, Marshall Devor, Thomas Eisner, Stephen Emlen, Ronald Harris-Warrick, Peter Hillman, Howard Howland, William Keeton, Melvin Kreithen, Ellis Loew, Randolf Menzel, William O'Neill, John Phillips, Elizabeth Sherman, James Simmons, Micha Spira, and Jurgen Tautz.

Barbara Seeley is to be thanked and congratulated for her perseverance and patience in typing the several drafts of the manuscript. I also thank the chairman, Thomas Podleski, and the administrative staff of Cornell's Section of Neurobiology and Behavior for easing many travails inevitably involved in a writing project of this size. My editor and publisher Andrew Sinauer provided at all times the necessary encouragement, as well as thoughtful planning and excellent judgment in all phases of preparation. And finally, the constant encouragement I received from my wife Jane was an indispensable ingredient in the completion of this work.

PART 1 INTRODUCTORY CONCEPTS

Many students of animal behavior have become so fascinated with its directedness, with the question "what for?" or "toward what end?" that they have quite forgotten to ask about its causal explanation. Yet the great question . . . "How?" [is] quite as fascinating as the question "What for?"—only they fascinate a different type of scientist. If wonder at the directedness of life is typical of the field student of nature, the quest for understanding of causation is typical of the laboratory worker. It is a regrettable symptom of the limitations inherent to the human mind that very few scientists are able to keep both questions in mind simultaneously.

(K. Lorenz, 1960)

The four chapters of this section set the stage for the detailed study of neuroethology that constitutes the remainder of this book. Chapter 1 sketches historically the disparate traditions of ethology and neurobiology and suggests a range of problems that can be solved best through a synthesis of these two disciplines. Chapters 2 and 3 then present some concepts from, respectively, ethology and neurobiology, which will be important for understanding the remainder of the book. Chapter 4, drawing on several of these concepts, presents a broad description of a neuroethological case study, the escape mechanism of the cockroach.

CHAPTER 1

Roots and Potential Growth of Neuroethology

The word *neuroethology* implies a blending of two scientific traditions that are about as different as one could imagine—that of the laboratory *neurobiologist* and that of the field *ethologist* (from the Greek *ethos*, meaning "manner," or "behavior"). Let us examine in turn each of these scientific traditions.

Neurobiologists work in laboratories equipped with electrical stimulators, amplifiers, oscilloscopes, possibly a computer, and a diversity of other sophisticated instruments. This work centers about an animal that has perhaps been anesthetized and highly dissected, actions enabling placement of recording electrodes on neural structures of particular interest. Alternatively, a fragment of nervous tissue may have been extracted altogether from the animal and placed for study in a dish filled with a saline solution. The animal under investigation probably has been selected from a limited number of species that are of common neurobiological interest. (This choice may be based upon factors such as the ease of obtaining and maintaining the animals, as in the case of domestic cats, dogs, and laboratory rats; or it may be based upon some special interest in animal species that are phyletically close to ourselves, such as monkeys; or upon some technical advantage offered by a given species, such as the presence of several especially large nerve cells, as in the mollusks squid and *Aplysia* and in the arthropods crayfish, cricket, and cockroach.) During the several hours of an experiment, the investigators may need to assist the animal's respiration and maintain its body temperature, and they may bathe its nervous system with a precisely formulated saline solution. Depending upon the purpose of the experiment, the animal's sensory receptors may be exposed to a carefully selected and highly controlled set of stimuli. At the end of the experiment the animal may be killed and samples of its neural tissue prepared for histological examination.

Quite a different picture describes ethological research. In this approach, spearheaded in Europe in the 1930s to 1950s by Konrad Lorenz, Niko Tinbergen, and Karl von Frisch, the study site was not the laboratory but the out-of-doors, in particular, the natural habitat of the species under investigation. Rather than imposing highly restrictive conditions upon an animal under study, these workers often viewed their subjects from a vantage point of complete concealment so as not to disturb at all the animals' natural activities. Some contemporary behaviorists have adopted an alternative strategy—obtaining a close-up view of their subjects' natural behaviors by actually allowing themselves to become accepted (often through years of patient effort) as a member of the animals' social group. Jane Goodall and Dian Fossey have been successful in this way with, respectively, chimpanzees and gorillas in the wild. In these studies, then, the animals and their social structure set the rules as to how the biologist behaves during the course of the study. This relationship between biologist and animal subject is diametrically opposed to that of the laboratory neurobiologist. Although an ethologist's careful description of his subject's behavior may lead, in a subsequent stage of the study, to his carrying out some experimental manipulations, these are generally performed either in the field or under simulated field conditions in a laboratory. At all times, the animal is maintained under conditions that are as natural as possible. Finally, ethologists, unlike neurobiologists, have tended to study a wide diversity of animal species. Many, however, have shown an intense aversion to the study of domestic species such as dogs, cats, and rats (among the animals most studied by neurobiologists) because these have long been removed from their natural habitats; thus, they may have learned, or even evolved, unnatural behavioral patterns.

The differences between ethologists and neurobiologists might be said to extend beyond matters of science to those of personal taste and life style. Most ethologists simply love being out-of-doors and looking at nature. Lorenz, for instance, would "become emotionally involved to the point of 'falling in love' " with the birds and other animals he studied. He recognized that he and other ethologists were "regarded as quite crazy by some scientists specializing in the [laboratory] way of thinking" (Lorenz, 1970). By contrast, some neurobiologists are more fascinated by the inner workings of machines, whether these be man-made instruments or living organisms. For instance, one famous senior physicist-turned-neurobiologist has described himself as being "influenced particularly by the grandeur of the Niagara plants and the Panama Canal locks" (Cole, 1975), a scientific frame of reference no doubt foreign to many ethologists. The laboratory of another famous neurobiologist, Sir Charles Sherrington of Oxford (often regarded as the founder of modern neurobiology) has been described as follows: "Each cubicle was a maze of wires so dense you had to stoop to enter," and the equipment in each cubicle "made a tremendous noise" (Young, 1975). Ethologist Lorenz probably would have been as uncomfortable in this laboratory setting as neurobiologist Sherrington would have been standing knee deep in mud while gazing for hours at a flock of geese.

Given this dichotomy, what hope is there for synthesizing these two disciplines into the topic of neuroethology? In fact, what's the point of even trying? The key point is this: the reason that nervous systems evolved in the first place was to produce behavior, and to do so in the out-of-doors, under the full blare of natures' physical forces. Therefore, a true understanding either of the nervous system or of animal behavior cannot be had without a combination of neurobiological and ethological approaches.

Let us look at a set of questions suitable for a neuroethological study. Imagine that it is spring and you are in a northern temperate pine forest, watching a male white-throated sparrow. The male bird, perched atop a pine tree, is broadcasting his song, which advertises his presence, species identity, and mating readiness to females of the same species. This advertisement competes with the calls of other males of the same species to attract the attention of females. A female white-throated sparrow, perhaps 50 meters away, hears this male's song. If the female is to take this opportunity to establish a bond with this male and ultimately produce offspring, she must first respond to the song by approaching the male. What are the problems faced by the female in doing so? First, her ears must be able to *detect* the physical features of the male's song. Next she must *recognize* this as the song of an adult male of her own species that is ready for courtship. In the process, she must *discriminate* this song from other songs broadcast simultaneously by nearby birds of the same and of different species, as well as from the calls of insects and frogs and from nonbiological sources of sound such as wind and the rustling of leaves. Furthermore, she must determine the *location* of the sound source. Then she must somehow make a *decision* to initiate the behavioral response and fly off toward this male. The act of flying requires *coordination* of the wings and body in a highly complex set of movements that function to keep the bird aloft and progressing on course. It also requires a complex set of *orientation* reactions to avoid obstacles while heading toward its target. Having arrived at the male, she engages in a set of *complex behavioral acts,* usually body gestures intended to win acceptance by the male. In this seemingly simple act of hearing and flying toward its potential mate, the female sparrow reveals to us some highly complex problems faced by its nervous system.

The neuroethologist would wish to know how the nervous system solves these problems. It might seem that the way to proceed experimentally would be to implant in the female's brain recording electrodes that can monitor the activity of her nerve cells while she flies toward, and courts, the male. However, the techniques for making such cellular recordings from totally unrestrained animals, out-of-doors, have not been developed and would be extremely difficult to achieve. Rather, neuroethologists must often resort to a study in tandem, analyzing behavior in the field and the nervous system in the laboratory. However, these two facets of the study should be clearly coordinated; the field observations may raise questions for the laboratory, and the answers to these may pose further questions for the field.

Some neuroethologists, then, move their study site repeatedly back and forth between field and lab. Indeed, if one restricts one's view to a single research site, one may overlook crucial aspects of the problem at hand. For instance, studies on the neuronal mechanisms of song recognition in birds show how an overly lab-bound perspective can lead an experimenter astray. Until recently, the auditory stimuli used in most such studies were not the natural songs of the species under investigation but rather much simpler, physically pure, sound stimuli. These stimuli, generated electronically, included pure tones resembling sustained musical notes, and brief click sounds. By recording from nerve cells at various locations within the auditory nerves and the brain, investigators searched for cells that would respond selectively either to pure tones of a particular pitch or to clicks. This approach was reasonable from a purely experimental point of view—one often wishes to use physically simple and well-defined stimuli. The problem is that, as anyone who has walked through the woods knows, most bird songs do not consist of sustained pure tones or clicks; rather, they can be extremely complex and can include changes of tone and intensity in a detailed temporal pattern. It was found that very few nerve cells in the higher auditory centers of birds respond at all strongly to pure tones or clicks. This seemed to signal the end of progress on understanding the mechanisms of song recognition in birds. However, when tape recordings of natural calls, recorded in the field, were played to these animals, many of their neurons gave intense responses (Leppelsack, 1978). One could thus begin to search in earnest for the particular brain structures and their neural mechanisms used to recognize songs. Similar findings have been made in the auditory systems of a number of animals tested with their own natural, species-specific sounds (Bullock, 1977; O'Neill and Suga, 1979, 1982).

The problems that are encountered by a female white-throated sparrow exposed to the male's song—signal detection, recognition, discrimination, localization, decision making, coordination, orientation, and the control of complex behavioral acts—are problems faced by most animals in numerous aspects of their daily lives. Understanding how the nervous system solves these problems is a major concern of neuroethology and the principal focus of this book.

Other aspects of the behavior of white-throated sparrows raise other important neuroethological questions that, for the sake of brevity, are touched upon only lightly in this book. One category of such questions relates to spans of time much greater than that occupied by a male's song and a female's flight and courtship. For instance, singing in sparrows occurs only in the spring, following their annual northward migration. What neuronal and hormonal mechanisms synchronize this behavior to occur just at this time of year? Also, by what developmental processes does a male white-throated sparrow grow up to sing its particular species-specific song, rather than some other song, and how does the female grow up to recognize this particular song? And finally, how did the particular song pattern and other behaviors of white-throated sparrows evolve from this species' phyletic precursors?

Given that neuroethology involves both behavioral field studies and neurobiological lab work, the modern neuroethologist needs to obtain training in both these types of research. So far, however, few scientists have been seriously trained in both areas. This book is intended to introduce the reader to some of the exciting problems available for research by those who pursue their studies from this broad, neuroethological perspective.

Summary

Neurobiologists and ethologists stem from very different research traditions; yet a blending of these two traditions into the discipline of neuroethology is both possible and desirable. In fact, it has already begun to achieve notable success. Neuroethology involves (1) observing an animal in its native habitat to discover the problems that its nervous system confronts in producing natural behaviors; and (2) carrying out neurobiological studies with the animal in as natural a state as possible. This work often involves a reciprocal process of behavioral experiments giving rise to neurobiological questions, followed by neurobiological experiments giving rise to behavioral questions. Researchers in this area need to be trained to work in both the field and the laboratory. Problems of interest to neuroethology include signal detection, recognition, discrimination, localization, decision making, coordination, orientation, and the control of complex acts. Further areas of interest are the neuronal and hormonal mechanisms underlying periodic changes in behavior, as well as the ontogeny and the evolution of behavior and its mechanisms.

CHAPTER 2

Behavioral Concepts

If one watches a flock of birds, a school of fish, or an ant hill, one observes a dazzlingly complex array of behaviors and behavioral interactions. Even if one simplifies the situation by watching just one isolated animal, describing accurately its movements is a task that can tax the powers of even the keenest observer. Yet animal watchers have made considerable progress over the years in comprehending these complex activities. This chapter presents a brief sketch of some of their findings—specifically, those that will be most useful in understanding the neuroethological material presented in the succeeding chapters.

The Mechanistic View of Behavior

We will begin by looking backward in time. The renaissance naturalists inherited from Greek thought the conception that all life activities, including behavior, were ultimately attributable to some form of nonphysical, or spiritual, force. The idea that such nonphysical forces play a role in living things is called VITALISM. The alternative view—that only physicochemical processes are involved in living things—is called MECHANISM. At various times during the last few centuries, a bitter controversy has raged between mechanists and vitalists.

The most noteworthy early mechanist was the seventeenth century French philosopher and scientist René Descartes. To Descartes, only human beings have a nonphysical soul. This soul directs our thoughts and some other higher functions. However, he regarded all other faculties of the human brain and behavior, together with *all* activities of nonhuman animals, as strictly mechanistic. Greatly impressed by the lifelike movements performed by the new renaissance machines being invented throughout Europe, Descartes argued that "we see clocks, artificial fountains, mills, and similar machines which . . . do not lack the power to move, of themselves, in various ways. . . . And truly one can well compare the nerves of [a living animal] to the tubes of the mechanisms of these fountains, its muscles and tendons to divers other engines and springs which serve to move these mechanisms, and its nerve activity to the water which drives them" (Descartes, 1972).

This Cartesian outlook remained quite firmly entrenched until threatened by important new discoveries in zoology and psychology in the late nineteenth and early twentieth centuries. Most notable was Darwin's *Origin of Species*, published in 1859. Darwin pointed to a phyletic relationship between human beings and other animals, which prompted some later workers to seek vitalistic forces (as had been imputed to the human mind) underlying the behavior of nonhuman species as well. Moths were said to be attracted to a flame by their "curiosity," and birds would sing "for the joy of it." In psychology, findings by Sigmund Freud and others, using the method of introspection (thinking about one's own mental experiences), as well as findings by a group called Gestalt psychologists (Chapter 7), added fuel to the vitalistic fire by uncovering new realms of human mental activity that seemed assured of defying any mechanistic explanation.

A reaction to these events by mechanistically oriented researchers came swiftly, around the beginning of this century. The principal figures were Edward Thorndike (1911), John Watson (1930), and Jacques Loeb (1912) in America, Ivan Pavlov (1927) in Russia, and Sir Charles Sherrington (1906) in England. Thorndike and Watson, both psychologists, reacted strongly against the use of the technique of introspection. They claimed that because only overtly observable movements of an animal or a human being are measurable and verifiable, only behavior itself, and not mental experience, is worthy of study. Both carried out extensive studies on animal and human learning. Thorndike's work helped to establish the field known today as comparative psychology, and Watson established the school of psychology known as behaviorism.

Sherrington and his colleagues explored mechanisms of movement through their study of the REFLEX—a simple, brief stereotyped movement carried out in automatic fashion in response to some sensory stimulus. A well-known example is the flexion reflex in which, for instance, a dog that steps on a thorn quickly lifts its foot in response. This and many other reflexes were shown to persist even after the spinal cord had been isolated from the brain by a spinal transection in the region of the neck. Such SPINAL REFLEXES were regarded as strong evidence for the strictly mechanistic control of behavior because it was thought highly unlikely that any soul that an animal might possess could be located anywhere except the brain.

Meanwhile, Pavlov extended the concept of the reflex by showing that a learned component could be incorporated into it. For instance, if a dog was presented with food, the odor of the food induced the reflexive flow of saliva. If each time the dog was fed, it heard the sound of a bell, soon the bell alone, without any food or its odor, would induce the flow of saliva. Thus, a stimulus that is unrelated to food (bell) has taken over the response-eliciting role of a stimulus that *is* related to food (odor). This learned salivary response was called a CONDITIONED REFLEX, and the process of learning this response is called CLASSICAL CONDITIONING or ASSOCIATIVE LEARNING.[1] Watson, working with human sub-

[1]Classical conditioning has provided an important experimental tool in studying those sensory cues that an animal is able to detect but to which it normally gives no observable response. One attempts to "condition" the animal to the stimulus in question. That is, one repeatedly presents this stimulus at about the same time as one presents a second stimulus of a different sort, to which the animal *does* respond. After many such paired stimuli, one now presents only the stimulus to which there initially was no observable response. If there now *is* a response, one has shown that the animal's nervous system is capable of

jects, described similar, simple, learned phenomena that he called CONDITIONED RESPONSES. The fact that these learned responses could be tied to reflexes, which were regarded by many as strictly mechanistic, suggested that learning itself might also occur as a strictly mechanistic process.

Although all these workers were devoted mechanists, it was Jacques Loeb who most forcefully championed this cause, through both scientific and popular writings. If Descartes had been impressed by the mechanical actions of clocks and fountains, Loeb was equally impressed by the seemingly automatic bending movements carried out by the stems of plants toward the sun or away from the earth's center of gravity. Loeb regarded these TROPISMS of plants as assuredly mechanistic, driven strictly by light and gravitational force; the time had long since passed when sceintists in general seriously posited a role for vitalistic forces in plant life. Loeb examined similar bending actions in some sessile lower invertebrates and also studied the orientation movements of insects and other animals toward or away from light sources or other stimulating agents. He termed all these reactions *animal tropisms* and developed a theory as to how they might operate. He then greatly elaborated this theory in a quite extravagant attempt to explain mechanistically all of animal and human behavior. The controversies that ensued produced more heat than light. Loeb's proposed theory for the mechanism of animal tropisms became discredited in the 1930s and 1940s (Fraenkel and Gunn, 1940), and those reactions in which the whole animal moved toward or away from the stimulating agent were renamed TAXES (singular, taxis). Taxes were categorized by Fraenkel and Gunn according to the stimulating agent [examples being phototaxes (light), geotaxes (gravity), and chemotaxes (airborne odorants)] and according to the direction of movement [such as positive phototaxes (a movement toward a light source) and negative chemotaxis (a movement away from a particular odorant)]. In spite of the renaming of the behaviors Loeb had studied and the formulation of new explanations for these behaviors, Loeb's work occupies an important historical position in the development of the mechanistic conception of animal behavior.

In the past 50 years, the mechanism/vitalism controversy in animal behavior has quieted considerably. Because it is certain that physicochemical factors play a role in behavior, most scientists have gone about the business of trying to understand these factors. Although it cannot be absolutely proved that such factors will *fully* explain *all* the behavior of animals, scientists simply have no adequate means to study any nonphysical forces that may exist. Therefore, most animal behaviorists have stopped trying.

Contributions from Ethology

Konrad Lorenz, Niko Tinbergen, and other ethologists carried out most of their studies on behaviors more complex than either reflexes or taxes. The be-

encoding this stimulus, even though it normally produces no noticeable effect. By such a procedure, it has been found that numerous forms of sensory stimuli previously thought to be undetectable by animals are, in fact, detected. Some of these are discussed in Chapters 5 and 6.

haviors that these ethologists explored most fully were called FIXED ACTION PATTERNS. Such behaviors are characterized by four properties.[2]

1. They are generally elicited by fairly specific stimuli (such as the sight of a *particular* object) rather than by highly generalized stimuli (such as the mere presence of light, which can lead to a phototaxis). We will see momentarily some examples of these more specific stimuli.

2. The stimulus intensity that is just sufficient to elicit a fixed action pattern, called the BEHAVIORAL THRESHOLD, may vary much more widely than is generally the case for reflexes. For instance, many animals, once having performed copulatory behavior, cannot be made for a long period to copulate again, except in response to the most intense sexual stimulation. Thus, copulation results in a profound elevation of the copulatory behavioral threshold. By contrast, if one prevents a fixed action pattern from occurring for an unnaturally long period of time, it may occur in the complete absence of any apparent triggering stimulus. Such behavioral occurrences, called VACUUM ACTIVITIES, can be thought of as an extreme lowering of the behavioral threshold to the point where no stimulus at all is required. An example is the performance of nest-building behavior by a weaver bird that is kept in a cage without grass or other of its needed nest materials. The normal nest-weaving movements of the head and body ensue, even though there are no materials with which to weave (Lorenz, 1981).

3. Although fixed action patterns are normally elicited by an environmental stimulus, an experimenter can remove this stimulus once the animal's movements have begun and the behavior will usually continue in its normal form until the act has been completed. Thus, a fixed action pattern, even though it may involve a sequence of activities, is triggered by just one environmental stimulus. Although such behaviors may be performed with varying degrees of vigor, the basic form and sequence of the movements, once initiated, is generally unalterable. A fixed action pattern, therefore, represents a functional unit of behavior that occurs either completely or not at all. This ALL-OR-NOTHING property contrasts with reflexes and taxes, which generally require continued maintenance of the eliciting stimulus in order to maintain the ongoing movements.

4. A given fixed action pattern is generally performed by all, or nearly all, members of a species of a given age and sex when presented with the appropriate stimulus under the appropriate conditions. Some fixed action patterns in fact occur in a somewhat larger taxonomic grouping, such as a genus, but not in *much* larger taxonomic groupings, such as a whole phylum. In fact, by studying the precise form of certain fixed action patterns as they vary among closely related species, ethologists have been able to deduce taxonomic relationships, just as though they were studying morphological, rather than behavioral, characteristics. By contrast, some reflexes, such as the flexion reflex, are found in most four-legged vertebrates; and the reflexive closure of the eye's pupil in response to light occurs in almost all vertebrate animals.

A few of the better-studied fixed action patterns are illustrated in Figure 1. In general, when beginning a study of the behavior of some species, the early

[2]A number of qualifications have been raised concerning these properties of fixed action patterns (Hinde, 1970; Barlow, 1977). Nevertheless, the four properties enumerated here represent a fair general description of these behaviors.

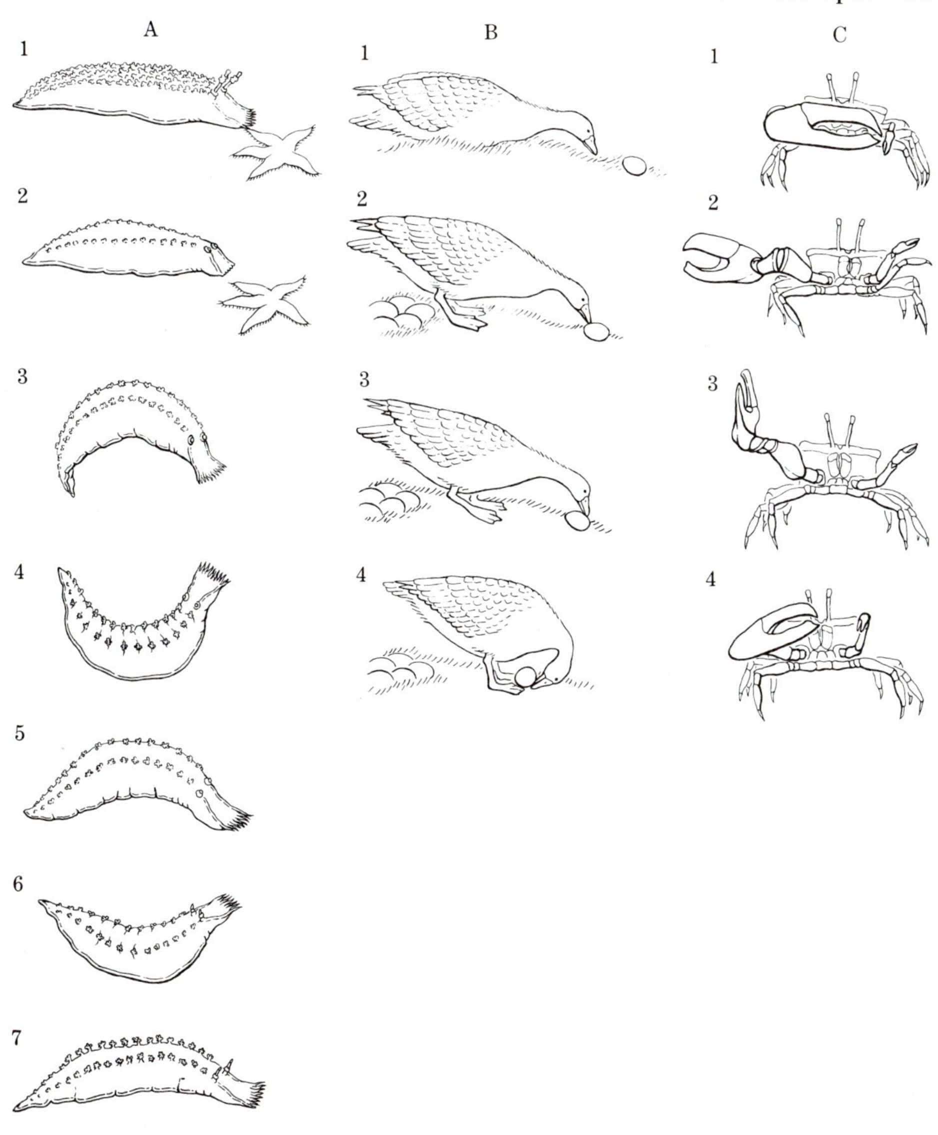

FIGURE 1. Some fixed action patterns. A. Escape response of the mollusk *Tritonia*. In 1, the cruising *Tritonia* contacts a starfish that preys upon this mollusk. Specific chemicals in the skin of the starfish release the retraction movement shown in 2 and the swimming sequence shown in 3 through 7. B. Egg retrieval by the greylag goose. In 1, the brooding goose detects an egg that has rolled out of its nest. It approaches and prods the egg (2), tucks the egg under its bill (3), and pulls it while walking backward (4). C. Courtship behavior in a male fiddler crab. The one enlarged claw engages in signaling to the female, by moving forward (1), then laterally (2), then upward (3), and again forward (4). (A after Willows; B after Lorenz and Tinbergen, 1938; C after Crane, 1957.)

ethologists would painstakingly accumulate a catalog of all of the fixed action patterns exhibited by the species in its natural habitat. This catalog was called an ETHOGRAM. In practice, not all of the behaviors entered on an ethogram necessarily exhibited all four defining properties of fixed action patterns just enumerated. In fact, some such behaviors exhibited properties intermediate between fixed action patterns and reflexes or taxes. The existence of such intermediate forms should serve as a caution that all schemes for categorizing animal behavior, though they may be heuristically useful, are to some extent artificial. Nevertheless, a wealth of information and insight has been garnered through ethological studies on behaviors, several of which do qualify as fixed action patterns. We shall now examine in greater detail a few key findings that will help to place our study of neuroethology in a broad behavioral perspective.

SIGN STIMULI

Many of the fixed action patterns first studied were stereotyped gestures that serve as communication signals among members of the same species of animal. Such gestures are involved primarily in courtship behavior, the rearing of offspring, and the aggressive maintenance of territorial boundaries. Many such gestures involve intricate body movements as well as the exposure of bright patches of color on the body surface. Such a communication signal by one individual may in turn evoke some other behavior by its social partner, and often complex behavioral interactions ensue.

A method was needed to sort out the salient stimulus features involved in such complex behavioral signaling systems. The ethologists' method of choice was the construction of dummy models of a signaling animal and the presentation of these dummies, one at a time, as stimuli to a recipient individual. The dummies, usually made of clay or cardboard, were each constructed and painted so as to resemble to a greater or lesser degree the shape and markings of the real signaling animal. The effectiveness of a given dummy in evoking the recipient's natural behavioral response was termed the RELEASING VALUE for that particular variant of dummy. By comparing the releasing values for all the dummy variants, investigators could determine the most effective stimulus properties.

One important early study of this type was carried out by Tinbergen on herring gulls. In the springtime, adult gulls fly out from their nests to feed. They then return home, where their nestlings await them. When a nestling sees its parent standing over it, it pecks at the tip of the parent's bill, an act that in turn induces the parent to regurgitate food into the nestling's mouth. What visual features of the parent evoke pecking at its bill? The adult herring gull's head and body are white and its bill, which is mostly yellow, bears a bright red spot near its tip. By comparing the pecking response of the nestlings at a range of dummy models of the head and bill, Tinbergen (1951) deduced the visual features having the greatest releasing value for pecking. For instance, he varied the color of the spot on the bill or removed the spot altogether, and thereby found that the red color was an important cue (Figure 2A). Then, by using dummies with a gray bill and spots of various intensities of neutral shades (ranging from white through gray to black), he found that not only the spot's red color but also its contrast

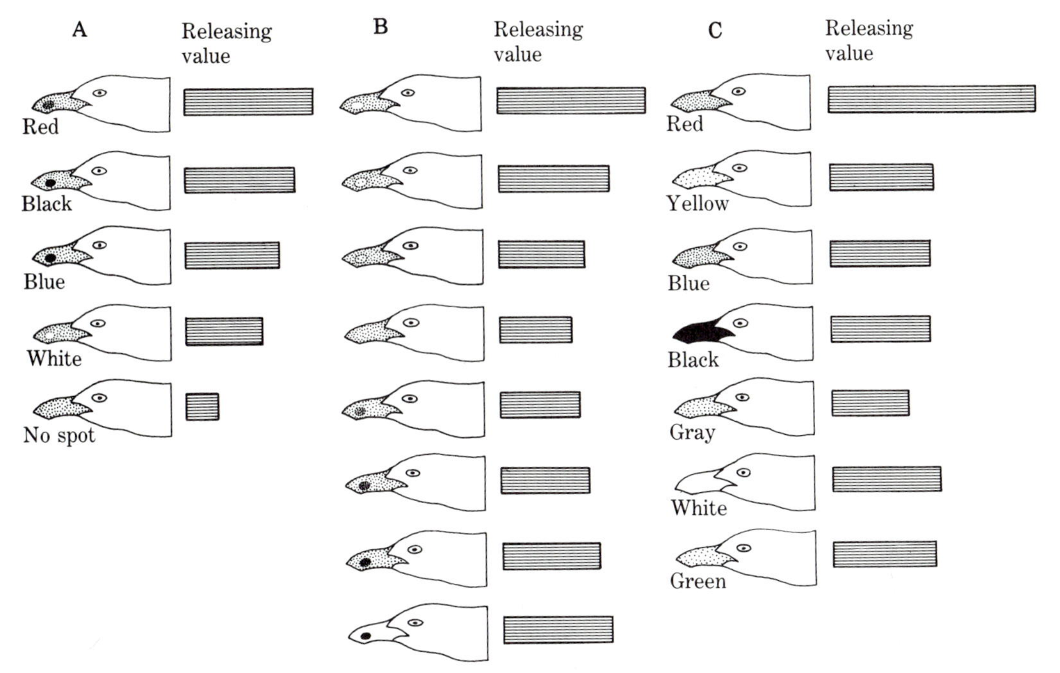

FIGURE 2. Differences in releasing value (shown by the lengths of the bars) of different models of a herring gull's head. A. Variations of the color of the bill's spot. B. Variations in the darkness of a spot (white through gray to black) on a gray bill. C. Variations in the color of the solidly painted bill. (From Tinbergen, 1951.)

with the bill is an important cue (Figure 2B). By omitting the spot from the bill of the dummy and painting the whole bill with a series of single colors, he showed that the yellow bill color was unimportant in releasing pecking (Figure 2C). He also showed that moving dummies had a higher releasing value than stationary ones. Finally, varying the color of the head and body had no effect on the pecking induced by the dummies. Thus, it is the spot's color and its contrast with the rest of the bill, together with the dummy's movement, that constitute the major stimuli for releasing pecking behavior. This interaction of several different aspects of a stimulus to produce a more effective sensory signal is called STIMULUS SUMMATION. Stimuli involving separate sensory modalities, such as vision, olfaction, and audition, may also summate to trigger behavior.

Because the red spot on the parent gull's bill constitutes only a small part of the nestling's view of its parent, this spot can be thought of as a symbol, or sign, for the parent bird. Such a highly specific stimulus that symbolizes some larger object or event of biologic importance is called a SIGN STIMULUS. Sign stimuli that are used in social communication within a species (as is the gull's spot) are also called RELEASERS, or SOCIAL RELEASERS. Sign stimuli have been studied in a large number of behaviors, especially in birds, fish, and arthropods.

Because only the herring gull's red spot and not other available visual features (such as the color of its bill, head, or body) constitute the chick's sign stimulus for pecking, we can conclude that information from these other potential stimuli is somehow filtered out in the chick's visual processing. In fact, the chick has perfectly good visual receptors in its eyes that are excited by yellow and white stimuli such as the bill, head, and body; but the information from these receptors apparently is not used to identify a good object at which to peck. Thus, there is a sequential reduction of the information encoded in the neuronal pathway leading to pecking, beginning at the level of the visual receptors (which as a group respond to many colors including red, yellow, and white), onward to that point in the pathway that activates the pecking behavior (which responds selectively to red). Such reductions, as well as other transformations, of encoded information are characteristic of sensory systems. We will encounter some well-studied examples in Chapter 7.

ANIMAL COMMUNICATION

Many of the behaviors studied by the classical ethologists were body gestures used by animals for communicating with their own species. Any communication system is characterized by the components shown in Figure 3. In this figure, the *sender* is an individual that is transmitting some message, such as, "I am a male of species X ready to defend my nest territory and accept a mate." Before the sender can transmit its message, it must first transform the message into a CODED SIGNAL, just as we do by encoding our messages into spoken or written language. The coded signal may be a particular set of vocalizations, body gestures, changes of skin color, release of odorants (called pheromones), or other signals. Depending upon the type of coded signal employed, one or another *transmission channel* will be used—sound waves for vocalizations, light transmission for body gestures, and so forth. The particular transmission channel selected must be suitable for the environmental conditions under which the communication is to occur. For instance, body gestures and color changes, which would be detected visually, would be practically useless at night or in the midst of extremely dense vegetation. Other factors affecting the choice of a given transmission channel are (1) the amount of information that needs to be transmitted (for instance, gestures and vocalizations generally can contain much more information than pheromones); (2) the required speed of transmission (visual and auditory signaling is much faster than pheromonal transmission); (3) whether the signal needs to persist or can be transient (the marking of objects in the environment by pheromones provides a persistent signal); (4) whether the location of the signal needs to be conveyed (in

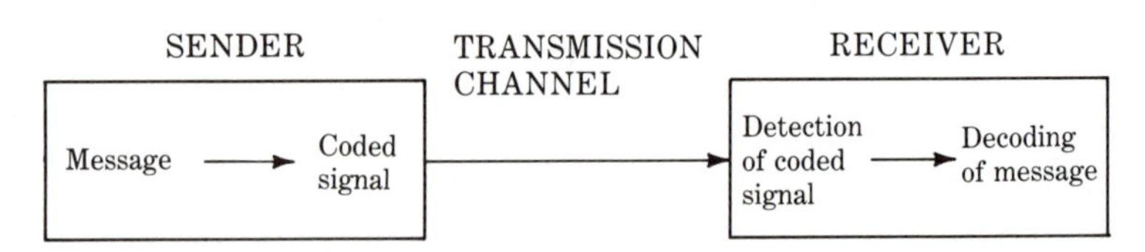

FIGURE 3. A system of animal communication. Explanation in text.

daylight in open terrain visual signals are easy to localize, whereas at night or in dense vegetation sounds are generally easier); (5) the energetic cost of producing a particular type of message (it may be energetically more costly to make continuous sounds than to deposit a small amount of pheromone).

An animal that detects the coded signal and decodes its message is called a *receiver*. Experimentally, the only way that one can verify the detection and decoding of a message by a receiver is to observe some meaningful behavioral response by this animal to the message, such as the sending of an appropriate return message.

The ethologists' experimental strategy of using dummy models to investigate animal signaling can readily be understood in terms of the communication scheme of Figure 3. The ethologist has simply replaced the sender and its real coded signal by a dummy that incorporates a code *suspected* of carrying part or all of the sender's message. For instance, a dummy in the shape of a herring gull's head and bill painted entirely gray, but with a black spot on the bill, is used to test whether spot-to-background contrast is part of the code. As we have seen, by observing the receiver's behavior (in this case pecking by a chick), investigators determined that this contrast is indeed part of the code (Figure 2B). In recent years, clay and cardboard dummies have largely been replaced by much more sophisticated models of suspected coded signals. One notable example, used in studies of auditory communication, is electronic or computer simulation of an animal's vocalizations. These instruments can be made to vary the sound signal systematically in a quantitatively controlled manner, a procedure that enables one to study the role in communication of particular components of the vocalization (Simmons, 1973; Pollack and Hoy, 1979; O'Neill and Suga, 1982).

One useful approach to the study of communication is that of INFORMATION THEORY, a method of studying quantitatively the amount of information contained in coded messages. The following questions are examples of problems that can be studied through this approach. (1) How much information is contained in a given coded message? (2) How much information is contained in the entire repertoire of coded messages that a given sender is able to transmit? (3) To what extent does the addition of complexity to a coded signal add to the information being transmitted? (4) To what extent does the frequency of repetition of a signal affect the overall amount of information transmitted? (5) Is more information transmitted by signals that are graded in form or by those that are discrete? Because of the highly quantitative nature of studies utilizing information theory, further consideration of this approach is beyond the scope of this book. However, interested readers can find useful reviews in Chapter 8 of Wilson (1975) and in Steinberg (1977).

One property of any effective communication system is that both the sender and the receiver must "agree" upon the meaning of the coded message. Because the coded messages employed by a given species have been molded by evolution, this agreement must be the result of coevolution of the sender's motor pattern that produces the coded signal and the receiver's sensory mechanism for recognizing the code. Evidence suggesting such coevolution has come from breeding experiments on crickets (Hoy and Paul, 1973; Hoy *et al.*, 1977).

In each of two species of crickets tested (*Teleogryllus commodus* and *Te-*

leogryllus oceanicus), the males produce calls that attract females of their own species. In these and other species of crickets, the selective response of a female to calls of her conspecific males is based upon the species-specific temporal pattern of sound pulses within the call (Figure 4A and B). Male hybrids from crosses of *T. commodus* and *T. oceanicus* make calls that have temporal patterns different from either parental species. Moreover, male hybrids from crosses of male *T. commodus* and female *T. oceanicus* (termed $c \times o$ males; Figure 4C) make calls that are slightly different from male hybrids from a male *T. oceanicus* and a female *T. commodus* (termed $o \times c$ males; Figure 4D). When females were tested quantitatively to determine their preferences among these various calls, the following observations were made. Female *T. commodus* were attracted selectively to the calls of male *T. commodus* and female *T. oceanicus* to the calls of male *T. oceanicus*. Female $c \times o$ hybrids were attracted more effectively to calls of their brothers (that is, $c \times o$ males) then to the calls of either parental species or of $o \times c$ males. Likewise, $o \times c$ females were attracted more often to calls of $o \times c$ males than to calls of $c \times o$ males or males of either parental species. Thus, not only did hybrid males of a given cross produce a new coded signal, but also the hybrid females of the same cross were selectively attracted to that signal. We can conclude that the production of the signal by the sender and the response by the receiver are under common genetic control, a conclusion that offers a mechanism for coevolution of signal and response.

A further property of effective communication is that the coded signal must be transmitted clearly and distinctly in order for its meaning to be unambiguously comprehended. This is particularly important where the signal has an especially great survival value for the communicating individuals or for their present or future offspring. The clearest examples are found among courtship signals, which often involve greatly exaggerated movements or extremely loud vocalizations. A courting individual must compete with others of its own species for a mate, and it is this competition that, through the process that Darwin called sexual selection, has led to the evolution of such distinct, exaggerated courtship signals.

INTERSPECIES SIGNALING

Some animals have evolved especially distinct gestures or body markings that they flaunt as messages intended not for their own but for different species. The intended receiver is often a potential predator. Many of these brightly marked animals possess body chemicals that render them distasteful, or they have the ability to sting, or they possess other close-range deterrents to predation. The bright coloration could then serve to advertise to predators that the prey is armed for a counterattack, and this could keep predators at bay. Displaying such WARNING COLORATION or, as it is often called, APOSEMATIC COLORATION, has in fact been shown experimentally to confer protection from predators (Edmunds, 1974; Smith, 1975, 1977). Some other species of animals, having no chemical or other close-range defenses, also flaunt bright colors. These color patterns and the movements by which they are displayed often show a striking resemblance to those of some species that *is* chemically or otherwise protected (Wickler, 1968).

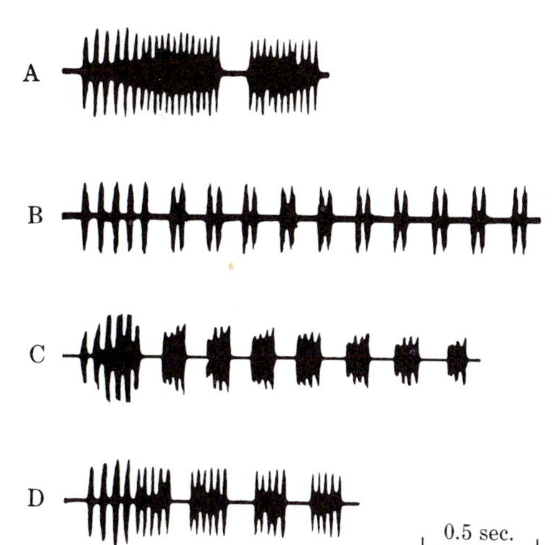

FIGURE 4.

Sound recordings of the calls of two cricket species of the genus *Teleogryllus*, and of hybrids between these two species. A. *T. oceanicus*. B. *T. commodus*. C. Hybrid of the cross male *commodus* × female *oceanicus* (*c* × *o*). D. Hybrids of the cross *o* × *c*. The temporal pattern of each call is shown by the sequence of vertical blips. (After Hoy *et al.*, 1977.)

The naturalist Henry Bates in 1862 suggested that this resemblance might confer protection upon the chemically defenseless species by its pretense of being chemically armed. In this association of different species, called BATESIAN MIMICRY, the chemically defenseless species is called the MIMIC and the defended species, the MODEL. Batesian mimicry can be shown in laboratory experiments to confer upon the mimic protection from predation (Brower *et al.*, 1960; Brower and Brower, 1962). There is also strong supportive evidence for such protection based upon experiments performed on artificially colored butterflies in natural habitats (Sternberg *et al.*, 1977; Jeffords *et al.*, 1979). The flaunting of the color signal by the mimic, then, constitutes a lie; it says, "There is danger to any that attack me," when no such danger in fact exists.

In some cases of mimicry, two different species have evolved nearly identical aposematic color patterns and movements for displaying them; and yet *both* species are armed with chemical or other close-range deterrents. That is, unlike Batesian mimicry, each species acts in a sense both as a model for and as a mimic of the other species. The selective advantage to any one individual partaking of this arrangement is that by associating itself with a larger number of similarly marked individuals than are contained in just its own species the individual increases the probability that a predator will learn not to attack him through its trial attacks on *other* individuals. This explanation is a variation of the first coherent interpretation of such mimetic groupings of self-defending species, which was offered originally by Fritz Müller in 1878. Such a mimetic association is called MÜLLERIAN MIMICRY. The more species joining such a mimetic group, the better for each individual member of any of these species. In fact, numerous multispecies assemblages of Müllerian mimics have been discovered, most notably in butterflies and snakes. Such large groups also offer ideal opportunities for other species, which do *not* possess defensive mechanisms, to receive protection through Batesian mimicry by also flaunting the same color patterns. Several such joint Müllerian–Batesian mimicry systems have been discovered (Figure 5).

Not only prey species, but also predators, have developed trickery in interspecies signaling. For instance, predatory angler fish have evolved small ap-

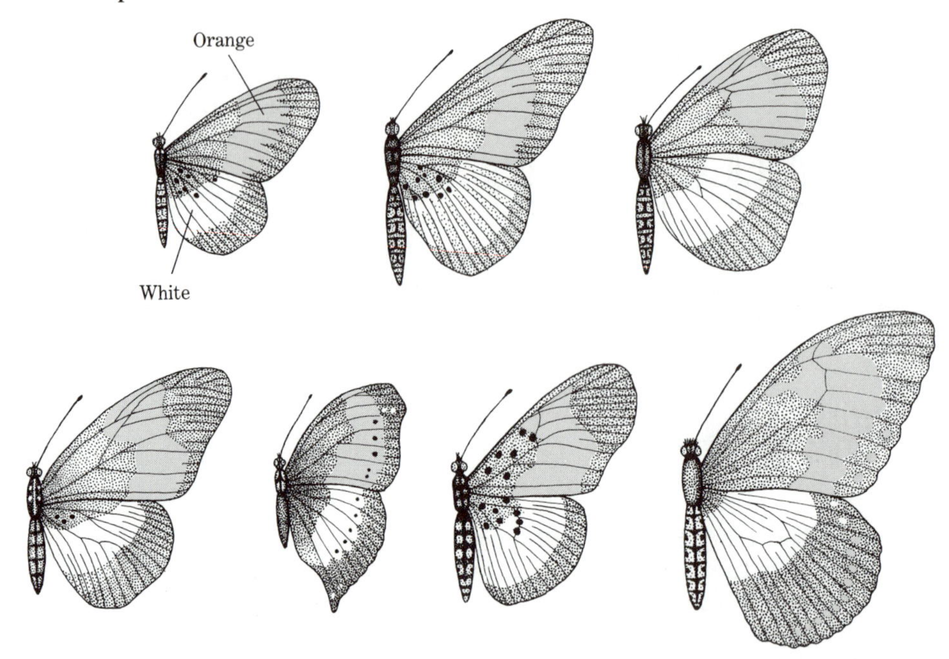

FIGURE 5. Müllerian and Batesian mimicry in a group of butterfly species. All seven species shown closely resemble one another. The background color of the wings is brown. Each forewing has an orange stripe and each hindwing a white stripe. The three butterfly species in the top row are all inedible and distasteful and thus may serve as Müllerian mimics for one another. The four butterflies in the bottom row are all edible and thus may be Batesian mimics of the three top species. (After Wickler, 1968.)

pendages that, when wriggled by the fish, bear an uncanny resemblance to worms. When small fish approach such a worm mimic for a meal, they often become the meal of the angler. Because the aggressor is the mimic in this case, this phenomenon is often called AGGRESSIVE MIMICRY.

Many species of animals, rather than sending distinct warning signals to potential predators, seem to attempt the opposite—a reduction to zero of any such signals, rendering them undetectable by predators. For instance, CRYPTIC COLORATION, or CRYPSIS, results in a blending in with the background by matching the body coloration and the reflectivity of light with that of the surrounding environment (Figure 6). Often particular positions and postures are maintained that afford an optimal matching with the background. Another strategy that would seem to serve in signal reduction is DISRUPTIVE COLORATION, that is, the presence of often fairly bold body markings that help to disguise key body features (such as the eyes or the body outline) that would otherwise render the animal easily recognizable (Figure 7).

FIGURE 6. The African hawkmoth *Xanthopan morgani*, seen in its normal resting position on a tree trunk. (From Cott, 1957.)

It is not sufficient for behaviorists merely to observe a color pattern or an adopted posture and to conclude that because the animal is difficult for *us* to see or to recognize this animal therefore derives protection from its natural predators. What is required is an experimental demonstration of this fact. One approach is to show that the same animal, if painted artificially so as to interfere with its signal-reducing strategy, becomes more readily attacked by predators. Some such experimental results have been obtained on cryptically colored prey (Edmunds,

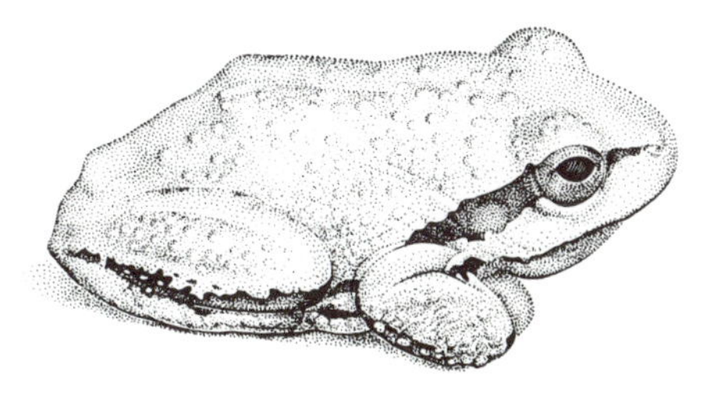

FIGURE 7.

Disruptive coloration in a tree frog. The stripe along the body, covering the eye and continuing over the legs and trunk, may render this animal less recognizable to its predators. A hint of this can be seen by observing this figure in dim light such as that of the early evening when this species comes out into the open. (From Alcock, 1979.)

1974; Pietrewicz and Kamil, 1977). As for disruptive coloration, however, the only experimental test to date carried out in a natural habitat showed that obliteration of the putatively disruptive markings of certain butterflies did not lead to increased predation (Silberglied *et al.*, 1980). Additional work is therefore needed to verify the role of disruptive coloration.

Given that signal-reducing strategies may help an animal to hide from its potential predators, the problem arises of how an animal so concealed manages to transmit necessary signals to members of its own species, such as its mates, offspring, and parents. One approach to this problem is the use of PRIVATE CHANNELS OF COMMUNICATION—the transmission of signals to ones own species through a form of energy that other species cannot detect. One example is the use of highly specific sex pheromones released by female silk moths to attract males (Schneider, 1974). Because the male is highly specialized to detect these particular molecules in the very low concentrations that occur far downwind of the sender, most other animals remain uninformed that this signal transmission is occurring. No known predator is able to detect the silk moth pheromone. However, in a bizarre example of aggressive mimicry,the bolas spider manufactures and releases the identical pheromone molecule and thus attracts male silk moths, which it then captures and ingests (Eberhard, 1977). Thus, there has occurred a coevolution of specific pheromone production by a silk moth and by its predator. Other aspects of strategic evolutionary battles between predators and prey will be touched upon in several chapters of this book.

NATURE AND NURTURE IN BEHAVIORAL DEVELOPMENT

The synonymous terms INSTINCTIVE and INNATE behavior have been used by many ethologists to describe the fixed action patterns that they study. According to the original definition, an instinctive or innate behavior was one whose form was determined only by the genetic makeup of the animal; environmental influences were thought to play no role in the ontogeny of such behaviors. That is, the animal's genetic *nature*, and not any *nurturing* from the environment, was regarded as the responsible agent. In particular, one form of interaction with the environment that was thought most definitely *not* to be involved in establishing an animal's instinctive behavior was LEARNING—a modification of behavior through the animal's own individual experience.

An example of a behavior that was thought to be instinctive according to this definition (requiring no learning or other forms of input from the environment) was the pecking of a herring gull chick at its parent's bill. The ethologists used as evidence of this behavior's innateness the fact that the pecking occurred in a fully functional form on the chick's very first trial, which usually took place on its first day after hatching. It appeared, then, that the chick did not need to learn through practice how to perform this pecking behavior; rather, the controlling neural circuits were presumed to be wired up under strictly genetic instructions.

One can raise a number of objections to this interpretation, however. First, although pecking at the parent's bill was functional on the first trial, it could be that prior to this trial the chick had practiced some components of the pecking movement in its parent's absence, for instance, during the act of hatching out of its shell. Second, it has been shown that even though the first pecking at the bill is adequate to induce the parent's regurgitation of food, the accuracy of the chick's pecking at the red spot on the bill does improve with time (Hailman, 1967, 1969). Some of this improvement may simply result from a strengthening of the muscles and from the normal growth of neural connections in the developing chick's brain—both processes of maturation rather than of learning through individual experience. However, learning may also be involved. This is suggested by the fact that a very young chick will direct its pecking at the bill of *any* adult herring gull that approaches it, but in time comes to peck only at the bill of it's own parents. Thus, learning to recognize its parents seems to be a component of the improved pecking behavior.

If the development of the behavior involves learning—a change in performance as a result of individual experience—this behavior cannot be entirely genetically determined. Moreover, other types of environmental influence aside from learning surely influence behavior. These include the ingestion of proper and adequate nutrients, the opportunity for adequate exercise, and the maintenance of an adequate body temperature.

Given this complex state of affairs, the terms *instinctive* and *innate* have now lost much of their original meaning and, thus, most of their usefulness. Most investigators now use these terms, if at all, to refer to behaviors that are carried out in a functional manner on their initial performance. This avoids totally the issues of whether any future improvement will occur and of whether the task as first performed was acquired through learning of individual task components, through genetic instruction, or through a combination of the two. In fact, it is now generally recognized that most or all behaviors involve a combination of nature and nurture, though the proportion of each of these two influences differs for different behaviors.

Close to one end of this nature–nurture spectrum are behaviors like the calling song of a male cricket, which is performed perfectly on its first trial and which shows no change from that time forward. This high degree of fixedness makes sense for communication signals for which, as we have seen, agreement between sender and receiver is of prime importance. Other behaviors develop gradually through individual learning, although the bounds of what any individual can learn are determined ultimately by its genetic endowment. For example, people can learn to drive a car, play the piano, or do mathematics, but probably not to run a mile in much less than four minutes. Finally, some behaviors have a large learned component, but this learning must occur during a restricted, genetically determined interval during the animal's ontogeny. After this time in the animal's life, little or no further learning of this particular task is possible. Such a restricted learning interval is called a CRITICAL PERIOD.

An example of a critical period is seen in the singing behavior of some species of song bird. For instance, in order for a male white-crowned sparrow to develop the ability to sing its species-specific song when it matures during the second spring season of its life, it must have heard during its *first* spring season the normal song of adults of its own species. In fact, this song must be heard during the period from 10 to 50 days after hatching (Konishi, 1965; Notobohm, 1970). If a white-crowned sparrow reared in the laboratory is exposed during this critical period to the song only of a *different* species, such as the song sparrow, it develops the next spring neither the song sparrow's song nor its own. Thus, the learning that is able to occur during this critical period is of a highly restricted type. The young white-crowned sparrow is tuned by its genetic instructions to hear, remember, and later to produce just its species-specific song.

A much briefer critical period is seen in a number of species of ducks and geese. During a period of just a few hours on the first day after hatching, these birds learn to recognize their mother. For the next several weeks, they follow her wherever she goes. If, during the chick's critical period, their mother is removed and replaced by a different species of bird, or even by a human being, they will instead follow this substituted individual. They even prefer this substitute to their own mother if she is returned to them immediately after the critical period has ended. Later, when these chicks mature, they court not with their own species but with the individual, or its likeness, that they had learned to follow during their critical period. This process of early learning of a long-term attachment to a specific individual is called IMPRINTING. Similar examples of imprinting have been reported in several species of bird and in some fish and mammals (Immelmann, 1980).

THE EVOLUTIONARY APPROACH TO BEHAVIOR

A major share of the work carried out by the classical ethologists was concerned with the evolution of behavior. Believing in general that fixed action patterns were under strictly genetic control, Lorenz, Tinbergen, and others emphasized that such behaviors must have evolved through Darwinian natural selection—the enhanced reproductive success conferred upon given animals by virtue of their possessing a particular heritable trait—in this case, a particular behavior. The evolution of behavior, it was argued, was no different in principle from the evolution of morphological features of a species. These arguments are not much diminished by more recent studies showing that learning and other forms of environmental interaction play a substantial role in some fixed action patterns. For as we have seen, the range of learning that is possible for a particular species is determined by its genetic endowment and thus is itself a heritable trait that can undergo natural selection.

Evolutionary thinking during these early years of ethology ran along lines such as these: animals will perform those behaviors that maximize the reproductive success of the species, as measured by the total number of viable offspring produced by that species. In the mid 1960s, however, evolutionary thinking began to focus on the reproductive success not of the species as a whole, but of the *individual* (Williams, 1966). In fact, current evolutionists point out that some behaviors, while serving the best interests of the individual, actually act *against* the best interests of the species. An extreme example is infanticide, which is practiced by males of some species against the offspring of conspecific males. This may help to assure adequate food and other resources for the infanticidal male's own offspring; but it does not appear to favor the reproductive success of the species as a whole (Alcock, 1979).

On the basis of this new emphasis upon the evolution of the traits of individuals, a new system of evolutionary thinking has developed (Wilson, 1975; Dawkins, 1977; Alcock, 1979). This line of thought runs as follows: (1) different individuals within a species differ somewhat in their behavior; (2) at least part of this individual difference is based upon genetic dissimilarities among the members of the species and thus is heritable; (3) at least some of these behavioral differences will affect an individual's reproductive success—the number of progeny that it produces; (4) each of the offspring produced will show some behavioral traits generally resembling those of its own parents; (5) therefore, the enhanced reproductive success associated with a particular behavior of a parent leads to evolution through the natural selection of this behavior, that is, a higher proportion of individuals in subsequent generations will come to have this behavioral trait; (6) if one individual performs a particular heritable behavior in a manner superior to that of another individual of the same species, this better performer is more likely to have its behavioral traits reappear in subsequent generations; (7) at any given moment, individuals of any species are competing behaviorally with one another, each to assure that its *own* offspring will be among those that are born and that survive and reproduce. It can be said that each individual animal behaves selfishly (though no conscious self-awareness is implied here), driven by its own genetic constitution to assure that its own genes make their way into subsequent generations. Animals behave selfishly because they have inherited from their own parents a set of genes that produces selfish, or competitive, behavior. Those individuals of past generations that did not receive such selfishness-producing genes by and large did not survive to produce progeny in the current generation. Though these principles will no doubt be modified by future work, they have been adopted provisionally by a great many current behaviorists.

Given this selfishness theory, it would seem difficult to explain numerous, well-known cases where an individual animal acts in an apparently altruistic manner, that is, instances in which an individual commits its finite energy resources to helping others of its species reproduce, at the expense of its own reproductive success. The help provided can take such forms as obtaining food for the offspring of other individuals, protecting those offsprings from predation, grooming them, or providing them with shelter. Apparently altruistic behavior has been explained in this way: an individual that aids *its own relatives* (its parents, its offspring, its siblings, its cousins, aunts, and uncles) to achieve a heightened reproductive success is, by this process, helping to pass on its *own* genes to the next generation. This is because each of these relatives shares with the helping individual some of its own genetic makeup (more so for closer relatives). Under certain ecological or social circumstances, this indirect means of passing on one's own genetic makeup may be more effective than producing one's own offspring. In support of this explanation, it is generally observed that altruistic behavior, when it occurs, is directed selectively at an individual's relatives and not at unrelated members of the species. Evolutionary selection for such helping of one's relatives at the expense of one's own personal reproduction is called KIN SELECTION.

Thus, there are two general forms of social behavior by which an individual can help to assure the passing on of its genetic constitution (and therefore of its own behavioral traits) to subsequent generations: (1) behaving competitively so as to heighten the individual's own reproductive success and thus the transmission of his or her own genes directly; and (2) assisting one's own relatives to survive and reproduce and, in the process, helping to pass along some genes that are identical to those of the helping individual. The sum total of an animal's success, using both these means, in passing its genetic constitution to the next generation is called that animal's INCLUSIVE FITNESS.

A major concern of behaviorists whose work involves this evolutionary approach is to discover in what way a particular behavioral act of an animal results in a heightened inclusive fitness for that individual animal. That is, why, within the explanatory umbrella of Darwin's theory of natural selection, does an animal do the things that it does? It is often argued, as Lorenz does in the quote at the beginning of this section of the book, that such *why* questions involve an approach to behavior quite distinct from, even irreconcilable with, the *how* questions of behaviorists interested in studying mechanisms. These two groups of behaviorists—those asking why and those asking how—have now formed largely separate and noninteracting behavioral disciplines.

This separation is both unnecessary and malproductive. If evolutionary and mechanistic behaviorists would interact more closely, important new research questions would emerge. For example, what differences are there among the nervous systems of individuals of a species that show behavioral differences? To take one example, male field crickets show two different mating strategies. Some individuals expend considerable energy emitting loud calls to attract females. Other males are silent and wait near the calling males, where they try to intercept approaching females and mate with them (Cade, 1979). Are these silent males forever mute, or can an individual act as a caller on one night (or on one part of a night) and as a silent mate-thief on another night (or other parts of the night)? If a mute is always a mute, does he lack the neural machinery for calling? Or is this machinery present but merely shut down? If it is shut down, does this occur through a hormonal signal, or by the action of some inhibiting nerve cells, or both? Are such hormonal signals or inhibiting nerve cells absent in calling males?

Research questions such as these probably would not occur to those who are interested in mechanisms but are not familiar with evolutionary studies; and those engaged in evolutionary work will miss the opportunity to have their findings pursued on the level of neural mechanisms if they remain insulated from neuroethologists. It seems clear that the present trend toward isolation is contrary to the individual interests of all concerned. What is needed, then, is a new species of researcher trained in both field and laboratory, in both evolution and mechanism, who can use this broad perspective to chart the future course of the emerging discipline of neuroethology.

Summary

The two opposing views of mechanism *vs* vitalism have been debated, on and off, especially since the time of Descartes. Categories of behavior that have been adduced in support of the mechanistic view are reflexes, including conditioned reflexes and taxes.

Fixed action patterns have been studied by the use of dummy models to reveal the sign stimuli involved in releasing behavior. Many sign stimuli are employed by animals for communicating among members of their own species. In addition, some animals have evolved especially exaggerated behaviors used for signaling other species, such as potential predators. Fixed action patterns and other behaviors develop within the individual through a combination of genetic instructions and environmental influences.

The evolutionary approach to animal behavior offers insights important for neuroethologists. One such insight is that individual differences and competition among members of a species are common and are the basis of the evolution of behavior. This should lead to a search for the mechanisms underlying individual differences and competition within a species.

Questions for Thought and Discussion

1. Are you a mechanist or a vitalist? Why?
2. The next time you are out-of-doors, make detailed observations of the behavior of some animal or group of animals. (Even if you live in an urban area, animals such as ants, pigeons, dogs, and cats should be available for observation.) Write two sets of research questions (that is, questions that one can hope to answer through further observations or experiments) that emerge from your observations. The first set of questions should concern the mechanisms of behavior, that is, these should be *how* questions. (One example: How does an ant find its way home from a distant food source?) The second set should concern the evoutionary approach to behavior, that is, these should be *why* questions. (One example: Why do some species of birds gather into flocks?) Outline briefly how you would go about answering each of your research questions.
3. Would you say that human beings have fixed action patterns? If so, list a few. Do human beings respond to sign stimuli; specifically, to social releasers? If so, list a few.
4. Spiders of the genus *Cupiennius* live on banana plants. An adult male, standing on one broad leaf, sends courtship signals to a potential mate, standing on another leaf of the same plant. The signals take the form of vibrations of the plant that the male produces by vibrating his legs. These vibrations are transmitted over considerable distances along the plant. In response to detecting this male's vibration signal, a female produces her own vibration signal. This helps to guide the male toward her for the purpose of mating. Both male and female continue signaling while he approaches her (Rovner and Barth, 1981). List several research questions that you think could help elucidate this particular animal communication system. Outline briefly how you would try to answer each of your questions experimentally.
5. Do you think human beings show *truly* altruistic behavior in which an individual's inclusive fitness is lowered in the course of his helping *unrelated* individuals? How do you relate your answer to the evolutionary approach to animal behavior?

Recommended Readings

BOOKS

Alcock, J. (1979) *Animal Behavior: An Evolutionary Approach,* 2nd Edition. Sinauer Associates, Sunderland, Massachusetts.
An excellent and highly readable account of the modern evolutionary approach to behavior.

Gould, J.L. (1982) *Ethology: The Mechanisms and Evolution of Behavior.* W. W. Norton, New York.
A highly readable general textbook of animal behavior written for the introductory level.

Griffin, D.R. (1981) *The Question of Animal Awareness: Evolutionary Continuity of Mental Experience.* The Rockefeller University Press, New York.
This thoughtful book, by a leading mechanistically oriented behaviorist, explores some interesting approaches to the mechanism–vitalism controversy.

Hinde, R.A. (1970) *Animal Behavior: A Synthesis of Ethology and Comparative Psychology.* McGraw-Hill, New York.
A well-written, thoughtful, and detailed account of the mechanisms of behavior. Though slightly dated, this book is still very useful.

Immelmann, K. (1980) *Introduction to Ethology.* Plenum Press, New York.
A good summary of classical ethology is presented in this text; however, it contains little that is new in the field of animal behavior.

Krebs, J.R. and Davies, N.B. (eds.). (1978) *Behavioural Ecology: An Evolutionary Approach.* Sinauer Associates, Sunderland, Massachusetts.
This book discusses a number of contemporary issues in the evolutionary approach to behavior.

Krebs, J.R. and Davies, N.B. (1981) *An Introduction to Behavioural Ecology.* Sinauer Associates, Sunderland, Massachusetts.
An excellent introductory presentation of the evolutionary approach to animal behavior.

Lorenz, K. (1981) *The Foundations of Ethology.* Springer-Verlag, New York.
This book, by the Nobel prize winner who helped found the field of ethology, summarizes his current view of this field.

Manning, A. (1979) *Introduction to Animal Behavior.* Addison-Wesley, Reading, Massachusetts.
An excellent, brief introduction to the subject, primarily from the point of view of mechanisms of behavior.

Marler, P. and Hamilton, W.J. III. (1966) *Mechanisms of Animal Behavior.* John Wiley & Sons, New York.
Though somewhat dated, this detailed account of behavioral mechanisms is still one of the finest books on the subject of animal behavior.

Sebeok, T.A. (ed.). (1977) *How Animals Communicate.* Indiana University Press, Bloomington, Indiana.
A huge compendium of review articles by many of the leaders in the field of animal communication.

Wilson, E.O. (1975) *Sociobiology: The New Synthesis.* Harvard University Press, Cambridge, Massachusetts.
This monumental tome, more than any other writing, helped to establish the evolutionary approach to the study of social behavior.

BOOKS OF HISTORICAL INTEREST

Several classic books on the subjects in this chapter have been reprinted in modern editions. The following list includes these more recent editions.

Descartes, R. (1972) *Treatise of Man.** Harvard University Press, Cambridge, Massachusetts.
Fraenkel, G.S. and Gunn, D.L. (1961) *The Orientation of Animals: Kineses, Taxes and Compass Reactions.* Dover Publications, New York.
Loeb, J. (1964) *The Mechanistic Conception of Life.* Harvard University Press, Cambridge, Massachusetts.
Lorenz, K. (1970) *Studies in Animal and Human Behavior.** Vols. I and II. Harvard University Press, Cambridge, Massachusetts.
Pavlov, I.P. *Conditional Reflexes: An Investigation of the Physiological Activity of the Cerebral Cortex.** Dover Publications, New York.
Sherrington, C. (1961) *The Integrative Action of the Nervous System.* Yale University Press, New Haven, Connecticut.
Tinbergen, N. (1951) *The Study of Instinct.* Oxford University Press, New York.
Watson, J.B. (1930) *Behaviorism.* W.W. Norton, New York.

*These are translations of works originally published in other languages.

CHAPTER 3

Concepts from Cellular Neurobiology

Since the goal of neuroethology is to understand animal behavior in terms of the structure and function of the nervous system, ideas from both behavior and neurobiology form the basis of this discipline. In the last chapter, we considered some key concepts from the field of animal behavior. We must now examine fundamental neurobiological concepts.

The functioning of a nervous system can be compared to that of an enormously sophisticated computer. A computer takes in information, processes it, and produces some form of output. Inside the computer, the information travels along wires from one component, or chip, to the next. Each chip changes the information that it receives into information of a different form. The nature of this change depends upon the particular type of chip. Thus, to understand the functioning of a computer, one would need to know how each chip operates and to know the wiring pattern by which different chips are connected together to produce a well-organized, or integrated, flow of information.

As with a computer, the nervous system takes in and processes information and produces outputs. The fundamental component that processes this information is the nerve cell, or NEURON. Like a computer chip, information enters a neuron, is transformed by it, and leaves it to exert an influence on other neurons. Different nerve cells carry out different types of transformation on the information they receive. Thus, a major task of the neurobiologist is to understand how individual neurons function and by what patterns of connectivity they are "wired" together.

It is important for a neuroethologist to have a good grasp of this subject. However, it is a subject of great complexity, both conceptually and technically. In this one chapter, then, it is possible to present only the most salient principles. The subjects found here are those that will assist the reader in understanding the material in the subsequent chapters of this book.[1]

[1]Given the elementary nature of this chapter, the material is presented here without reference citations. Those readers interested in pursuing this field further should consult the readings recommended at the end of this chapter.

The Structure of Neurons

Neurons share with all living cells several characteristic structural features; a plasma membrane, a nucleus, and a full range of other cytoplasmic organelles. A neuron's most striking structural difference from other cells of the body is its shape. Although there is a wide range of neuronal shapes, most nerve cells include a thickened region called the CELL BODY (also called the SOMA, or PERIKARYON) and one or more thin, tubular projections from the cell body, called NEURITES, or PROCESSES. Depending upon whether one, two, or several neurites emerge from the cell body, the neuron would be termed a MONOPOLAR, BIPOLAR, or MULTIPOLAR cell (Figure 1). The cell body contains the nucleus and the cell's major machinery for synthesizing needed molecules. The neurites, by contrast, are designed for conveying information signals. They can be very long (even several meters long in some neurons of large animals) and they can ramify profusely, like the branches of a tree.

In many neurons, most of the information enters the cell through a group of branching neurites, then travels along a single long neurite, at the end of which it is passed on to other neurons. In such a cell, the receiving neurites are called DENDRITES and the single long neurite is called an AXON. The branching pattern of the dendrites, which is often profuse, is called the DENDRITIC TREE, and the usually less profuse branches at the end of the axon are called AXON TERMINALS.

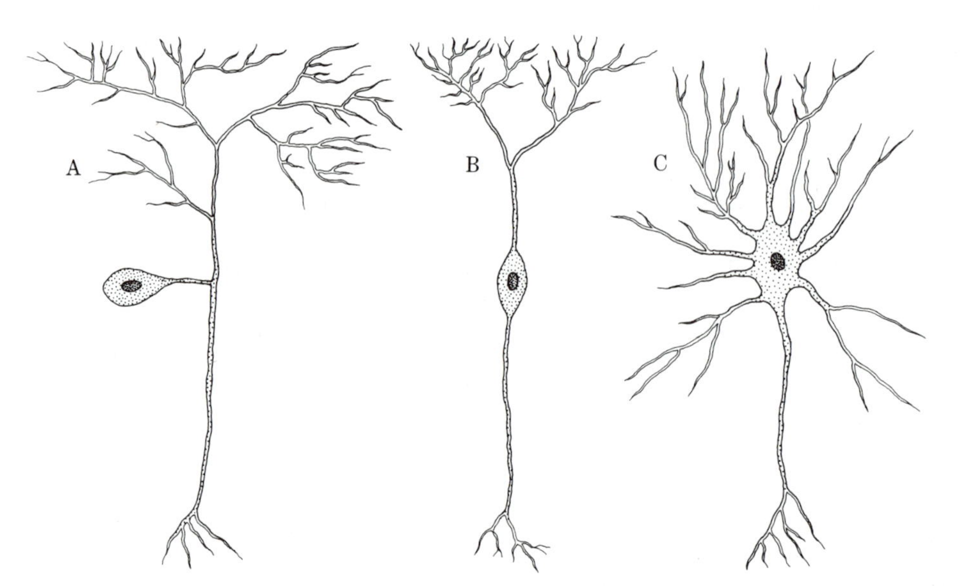

FIGURE 1. Three neurons, drawn schematically. A. Monopolar neuron. Most motor neurons and interneurons in higher invertebrates, and many sensory neurons in vertebrates, are of this type. B. Bipolar neuron. Cells of this type are found in several sensory systems among both vertebrates and invertebrates. C. Multipolar neuron. Most vertebrate interneurons and motor neurons are of this type. In general, dendritic branching is more elaborate than is shown here.

In practice, however, many neurons are not so neatly parceled into separate receiving and transmiting neurites. Rather, information can enter or leave such cells at many different locations. Thus, although the use of the terms DENDRITE and AXON is common, the specific direction of informational flow that these terms imply has not been proved in very many cases.

As we shall see later in this chapter, the shape of a neuron has an important influence on the way the cell transforms the information it receives. Therefore, there is great interest in studying neuronal shapes. In practice, however, one cannot simply examine a piece of neural tissue under the microscope and observe clearly the shapes of its component cells. Not only is each cell small and transparent, but its neurites are complexly intertwined with those of other cells. To help reveal a neuron's shape, a variety of histological methods have been, and continue to be, developed. Though most of these are beyond the scope of this chapter, one especially useful set of techniques deserves mention. These techniques involve placing directly inside an individual neuron a substance that is visible or can be made visible. The substances are usually applied by some form of microinjection. The injected material spreads throughout the neuron but cannot escape through the plasma membrane. The highly complex shape of the cell can thus be observed under the microscope, against the transparent background of other cells. Three major classes of visualizable substances are utilized in this way: fluorescent dyes such as LUCIFER YELLOW and PROCION YELLOW; metal precipitates such as COBALT SULFATE; and the enzyme HORSERADISH PEROXIDASE. An example of a neuron filled with cobalt sulfate is shown in Figure 2.

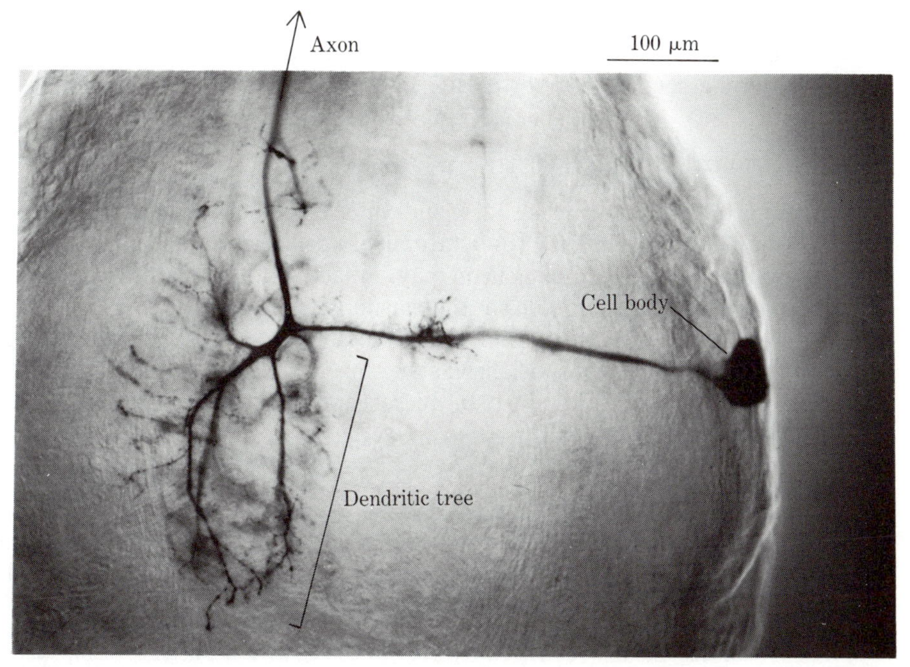

FIGURE 2. An interneuron of a cockroach, filled with cobalt sulfate. The axon extends far beyond the limits of this photograph. This particular cell, called Giant Interneuron 1, is discussed in Chapter 4. (Courtesy of D. Daley.)

ORGANIZATION OF NEURONS IN THE NERVOUS SYSTEM

Nerve cells can be classified into three functional groups: (1) SENSORY, or AFFERENT NEURONS, which convey information generally from the receptor organs in the periphery to the central nervous system, that is to the brain and nerve cord.[2] (In vertebrates, the nerve cord is called the spinal cord.) (2) INTERNEURONS, whose cell bodies and neurites are contained entirely within the central nervous system;[3] and (3) MOTOR, or EFFERENT NEURONS, whose axons control the action of muscles or glands. The long neurites of sensory and motor neurons, communicating between the central nervous system and the periphery, are bundled together into PERIPHERAL NERVES, or, as they are sometimes called near their points of emergence from the central nervous system, PERIPHERAL ROOTS. In some animals, some peripheral nerves (or branches of them) contain neurites only of sensory neurons and are thus called SENSORY NERVES. Others, containing neurites only of motor neurons, are called MOTOR NERVES. Still others containing both types of neurites are called MIXED NERVES. Within an animal's nerve cord, many interneurons that are excited by sensory cells, called SENSORY INTERNEURONS, send long neurites to the brain. In addition, many brain interneurons send long neurites down into the nerve cord. These are called, respectively, ascending and descending interneurons. The particular site at which long neurites of many cells of similar function terminate is called a PROJECTION AREA. As we shall see in subsequent chapters, particular projection areas are devoted to the analysis of particular aspects of the sensory environment or to particular motor tasks. Aside from interneurons whose neurites project over long distances, there are others that have shorter projections within the brain or the nerve cord, and even some that extend over only a few millimeters, called LOCAL INTERNEURONS. The organization of these several functional types of neurons is illustrated in Figure 3, which shows a section of a vertebrate's spinal cord.

THE STRUCTURE OF SYNAPSES

A major task of a neuron is to transmit information to other neurons. Therefore, we shall examine briefly the anatomical specializations that permit this transmission to occur. A single neuron may, in fact, communicate with hundreds or thousands of other cells. To do so, the plasma membrane covering its cell body or neurites generally must abut each of these many other cells. At each such point of membrane apposition is a specialized structure called a SYNAPSE, so small that it is visible only through the electron microscope. The intercellular transfer of

[2]In some sensory organs, such as the vertebrate eye, ear, nose, and tongue, the cells that actually detect the environmental signal, called RECEPTOR CELLS, have no neurites. These cells synapse, within their receptor organs, upon sensory neurons whose long neurites convey sensory signals into the central nervous system. For other sensations, such as touch, no separate class of receptor cells has evolved, and each sensory neuron both receives the environmental stimulus and transmits it centrally via its long neurite.

[3]Some neurons, though restricted entirely to the central nervous system, function as sensory cells, detecting such properties as the concentration of sugar or carbon dioxide in the blood.

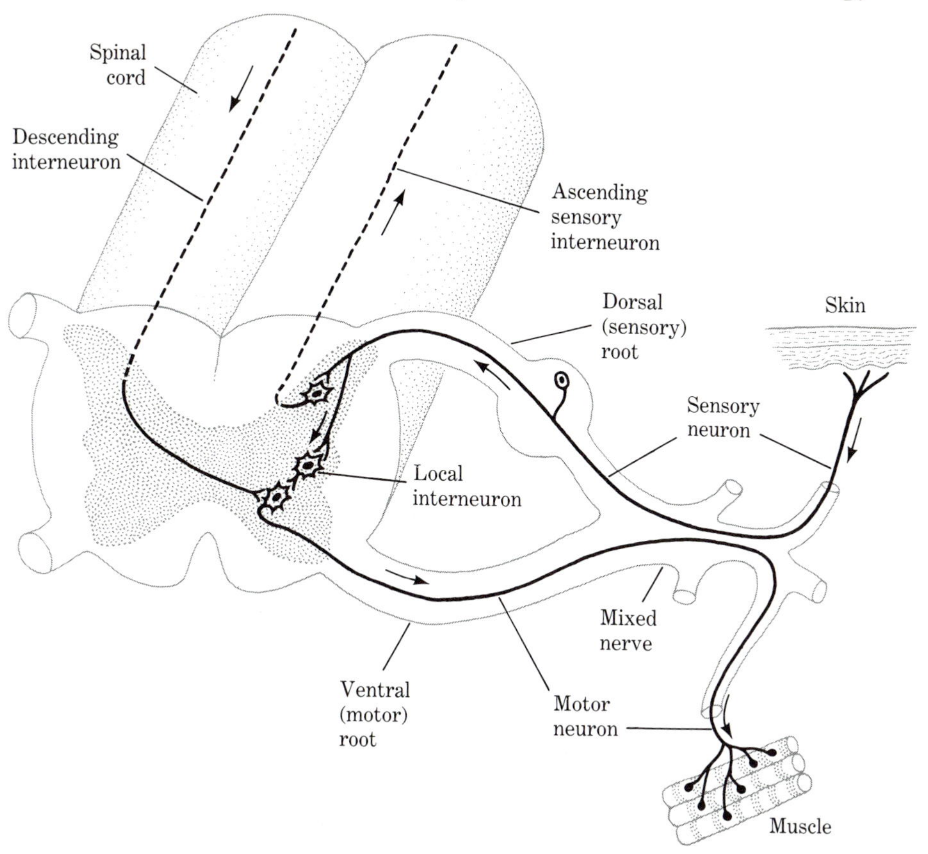

FIGURE 3. A fragment of the spinal cord from a vertebrate animal; several functional categories of neurons are shown. (See text for explanation.)

information, called SYNAPTIC TRANSMISSION, generally occurs unidirectionally, from the PRESYNAPTIC to the POSTSYNAPTIC neuron.[4]

There are two major anatomical categories of synapses. These correspond to two different physiological categories and will be discussed later in this chapter. They are called ELECTRICAL (or ELECTROTONIC) and CHEMICAL SYNAPSES. An electrical synapse (Figure 4A) is identical in structure to a gap junction, found among many cellular types in the body. At such a synapse, there is practically no space between the presynaptic and the postsynaptic neuron. Running between the cytoplasms of the two cells are tiny tubular structures that enable atoms or small molecules to pass directly from one neuron to the other. As we shall see,

[4]The designation of a neuron as presynaptic or postsynaptic refers to the transmission of information at a *specific* synapse. Though a given cell is presynaptic to a second cell, it will at the same time be postsynaptic to a third cell. In fact, one sometimes finds a pair of neurons with two synapses between them, one transmitting in one direction and the other in the opposite direction. In this arrangement, called reciprocal synapses, each of the two cells is both presynaptic and postsynaptic to the other.

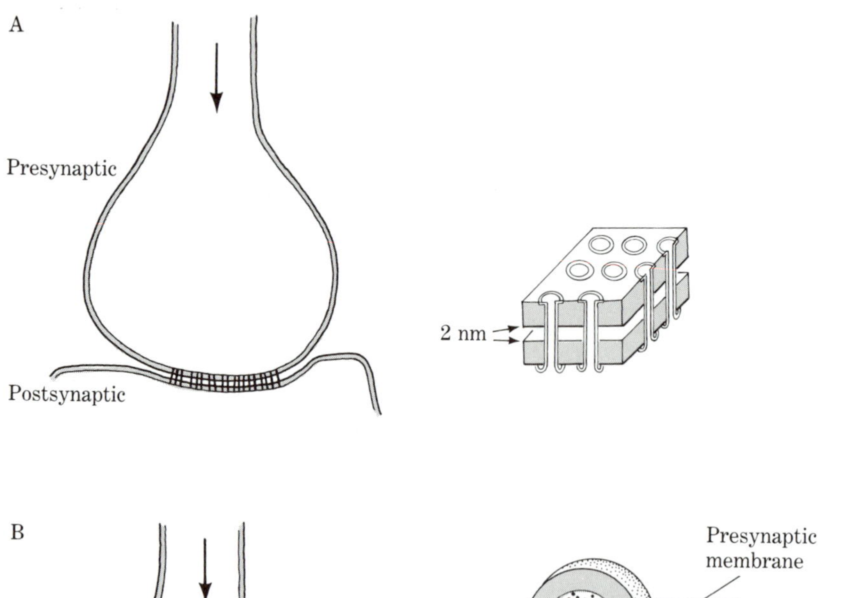

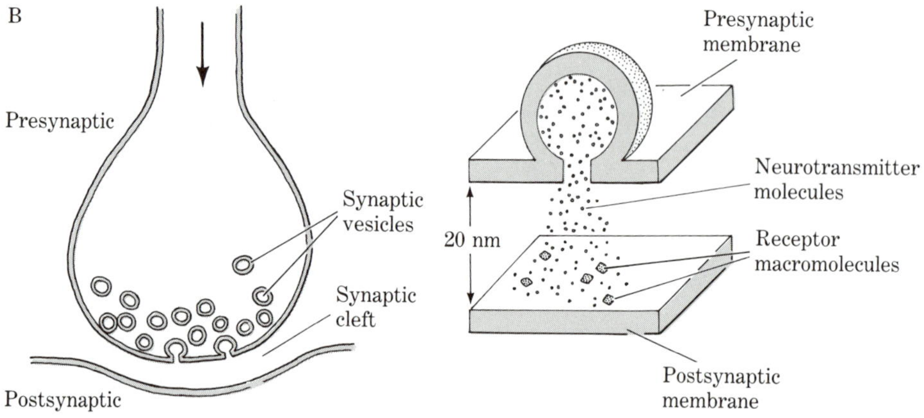

FIGURE 4. Synaptic structure. A. An electrical synapse. The inset shows an expanded view of the tiny tubular structures that connect the two cells. The space between the two cells is only 2 nanometers, or 2×10^{-9} meters. B. A chemical synapse. The inset shows a synaptic vesicle that has momentarily fused with the presynaptic plasma membrane and has opened to release its neurotransmitter molecules to the synaptic cleft. Some of this transmitter will bind temporarily to the postsynaptic receptor macromolecules inducing a postsynaptic signal. The synaptic cleft is 20 nanometers wide, 10 times the intercellular spacing at an electrical synapse.

these tubes provide the means by which neuronal signals pass directly across the electrical synapse, from one neuron to the other.

At a chemical synapse (Figure 4B), the space between the plasma membranes of the two neurons is considerably larger than at an electrical synapse. This space is called the SYNAPTIC CLEFT. There are no tubular structures connecting the cytoplasms of the two cells. Owing to these structural features, transmission at a chemical synapse is indirect and thus more complicated than at an electrical synapse. Chemical synaptic transmission occurs through the release by the presynaptic neuron, into the synaptic cleft, of a specific chemical agent called a NEUROTRANSMITTER. The molecules of neurotransmitters are apparently stored presynaptically in tiny, membrane-bound structures called SYNAPTIC VESICLES. During synaptic transmission, some of these vesicles fuse temporarily with the presynaptic plasma membrane and open, releasing their neurotransmitter into the synaptic cleft. Some of these neurotransmitter molecules bind to specific receptor macromolecules embedded in the plasma membrane of the postsynaptic cell, just across the cleft. This neurotransmitter–receptor bond results in the onset of a neuronal signal in the postsynaptic cell, through mechanisms that will be discussed later in this chapter.

The Communication Signals of Neurons

Most of the informational signals used by neurons can be understood on the basis of a single principle—the selective permeability of the neuron's plasma membrane to specific types of charged atoms, or IONS.[5] The ions appear to move through specific pores, or CHANNELS, in the membrane. Each channel is highly selective, allowing just one or a few types of ions to pass. However, the channel does not specify whether these ions will move into or out of the cell. Rather, the direction of ion flow is determined by other factors, which we will examine shortly. The channels are of two types: PASSIVE channels, which are always open and available for ions to pass through; and ACTIVE channels, which remain closed unless opened by a specific triggering signal. As we shall see, different types of active channels open in response to different triggering signals.

Whenever there is a net movement of either positively or negatively charged particles, this movement constitutes an electrical current. Thus, a net movement of positive or negative ions through channels in the plasma membrane constitutes such a current. These currents give rise to related electrical phenomena that constitute a neuron's informational signals. In the next several sections we shall explore the ionic movements that occur through these channels, the electrical consequences of these movements, and the way that nerve cells use these events as signals. We will begin by examining the electrical situation of a resting neuron—one that is not producing any signals.

THE RESTING NEURON

Studies of the ionic contents of neurons and their extracellular surroundings have shown that the concentration of most types of ions inside a neuron is very different from that outside. Of the most abundant ionic types, the concentrations of sodium (Na^+) and chloride (Cl^-) are much higher outside the cell than inside, whereas the concentration of potassium (K^+) and of a group of negatively charged organic molecules (A^-) are much higher inside the cell than out (Figure 5A). An important electrical effect results from this uneven ionic distribution, when considered together with the plasma membrane's ionic channels.

[5]Charged molecules, as well as charged atoms, are called ions. However, the plasma membrane is impermeable to most types of charged molecules.

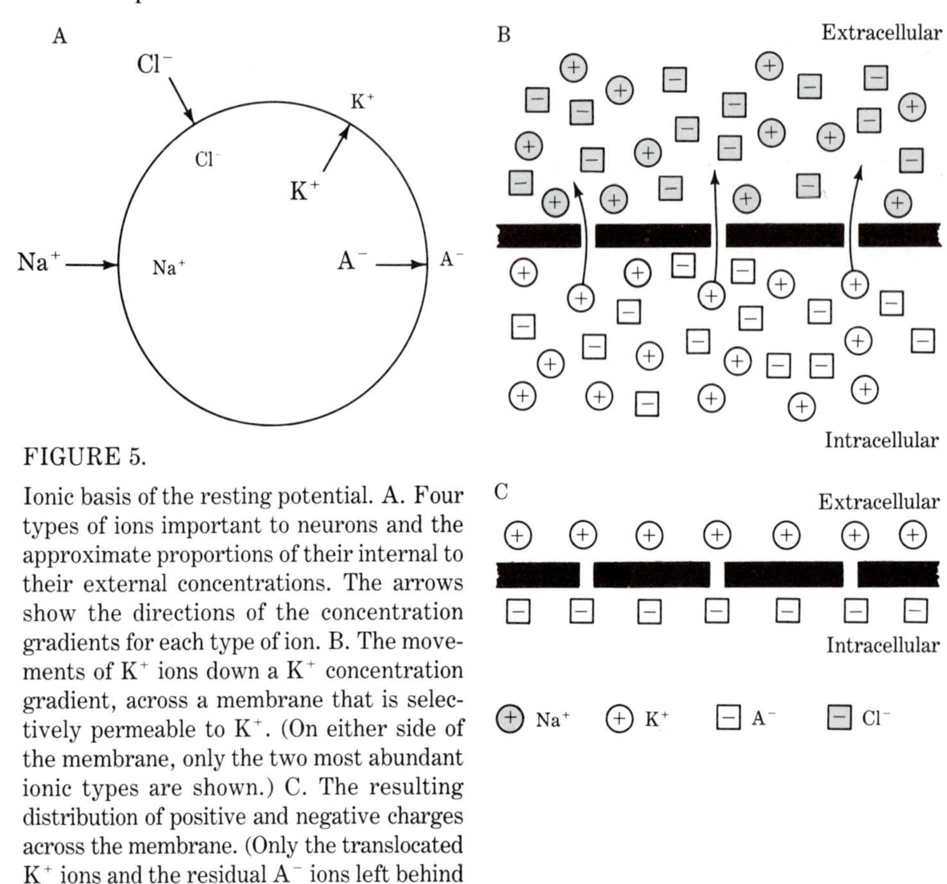

FIGURE 5.

Ionic basis of the resting potential. A. Four types of ions important to neurons and the approximate proportions of their internal to their external concentrations. The arrows show the directions of the concentration gradients for each type of ion. B. The movements of K^+ ions down a K^+ concentration gradient, across a membrane that is selectively permeable to K^+. (On either side of the membrane, only the two most abundant ionic types are shown.) C. The resulting distribution of positive and negative charges across the membrane. (Only the translocated K^+ ions and the residual A^- ions left behind are shown.)

When a neuron is at rest, most of its *active* channels remain closed. Of the *passive* channels, the great majority are specific for K^+ ions. Thus, the membrane of a resting neuron is much more permeable to K^+ than to any other type of ion. For simplicity, let us assume for the moment that the resting membrane is permeable *only* to K^+. As we have seen, the K^+ concentration inside the cell is different from that outside the cell, that is, there is a CONCENTRATION GRADIENT of K^+ across the plasma membrane. Whenever there is a concentration gradient of some type of ion across a membrane and the membrane is permeable to this ionic type, some of these ions will cross from the side of higher concentration to the side of lower concentration. The concentration gradient, then, can be thought of as exerting a force, pushing these ions in the direction of their lower concentration. Under the influence of this force, then, some K^+ ions will pass outward through the membrane's passive K^+ channel (Figure 5B, arrows).

As K^+ ions move outward, they contribute positive charges to the cell's extracellular fluid. Moreover, they leave behind an excess of negative charges (mostly A^-) inside the cell. These internal negative charges (A^-) and the external positive charges (K^+) attract one another electrostatically across the plasma membrane. This results in a thin cloud of negativity held just against the membrane's inner surface and positivity just against its outer surface (Figure 5C). The more K^+ ions leaving the cell, the greater these two charge clouds become.

The separation across the membrane of these two charge clouds of opposite sign is referred to as an ELECTRICAL GRADIENT. This gradient will exert a force on any ion moving through a membrane channel; a positively charged ion such as K^+, sitting inside a K^+ channel, will be repelled by the positive cloud outside the cell and will be attracted by the negative cloud inside. Thus, although it was the exit from the cell of K^+ ions that established these two charge clouds in the first place, these same clouds soon begin to oppose the exit of further K^+ ions. Ultimately, an equilibrium develops, the charge clouds having grown exactly strong enough to oppose any further exit of K^+ ions. The outward force on K^+ (resulting from its concentration gradient)[6] is now exactly balanced by an equal inward force on K^+ resulting from the neuron's electrical gradient. These are the only two forces that push K^+ through open membrane channels. Thus, there is now no further net movement of K^+ ions across the membrane.

The separation of an external cloud of positive charges from an internal negative cloud has an important electrical consequence. Like any separation of electrical charges, this constitutes an electrical potential difference, or voltage difference, across the membrane. The particular value of voltage, measured in millivolts (mV), at which the concentration and electrical forces on K^+ are equal and opposite is called the EQUILIBRIUM POTENTIAL for K^+. Physical chemists have found that the value of this equilibrium potential is proportional to the ratio of K^+ concentrations inside and outside the cell. Because it is possible to measure those concentrations, one can accurately calculate the K^+ equilibrium potential. For most neurons studied, its value is about -75 mV. (The minus sign can be taken to mean that the inside of the cell is negative, in agreement with the presence on the membrane's inner surface of the negative cloud charge.)

Recall that in reality a resting neuron is slightly permeable to ions other than K^+. Thus, the charge clouds on either side of the membrane will have some contribution from transmembrane movements of other ionic types. One might expect, therefore, that the actual transmembrane voltage will be somewhat different from the K^+ equilibrium potential of -75 mV. How can this real transmembrane voltage be measured?

To measure the voltage difference between *any* two points, one simply places the two sensing ends, or electrodes, from a voltmeter, one at each of these two places of interest. In the case of a neuron, the tip of one electrode must be placed inside the cell (that is, intracellularly) and the other outside (extracellularly) in the fluid bathing the cell. The intracellular tip, of course, must be tiny. For this purpose, one uses a MICROELECTRODE, usually a glass pipette whose tip has been drawn out to less than one micrometer (1/1000 millimeter) in diameter. The bore of the glass pipette is filled with an ionic solution, such as KCl, which can conduct electricity. With such a microelectrode, one can successfully impale a neuron's

[6]The actual number of K^+ ions that leave and establish the charge clouds is but a tiny fraction of the number originally contained within the neuron. Thus, the K^+ concentration within the cell has hardly changed at all; there remains a much higher internal than external K^+ concentration and thus there remains a strong outward K^+ concentration gradient.

cell body or neurites, as long as these are larger than a few micrometers in diameter (though if the diameter is less than perhaps 15 micrometers, considerable skill is required to avoid damaging the cell).

From both the microelectrode and the extracellular electrode, wires lead first to an amplifier and then to the voltmeter. The voltmeter of choice for most neurobiological studies is an oscilloscope, which can register accurately very rapid electrical events at the plasma membrane. (These events are the neuronal signals, which will be discussed in the next few sections.) By convention, the intracellular electrode is generally connected, via the amplifier, to the oscilloscope's positive input terminal; and the extracellular electrode is connected to the negative input terminal (Figure 6A). This means that the oscilloscope trace would deflect upward from its starting position if the inside of the neuron were to become positive relative to the outside, and it would deflect downward if the inside were to become negative relative to the outside. Because the oscilloscope trace moves across the

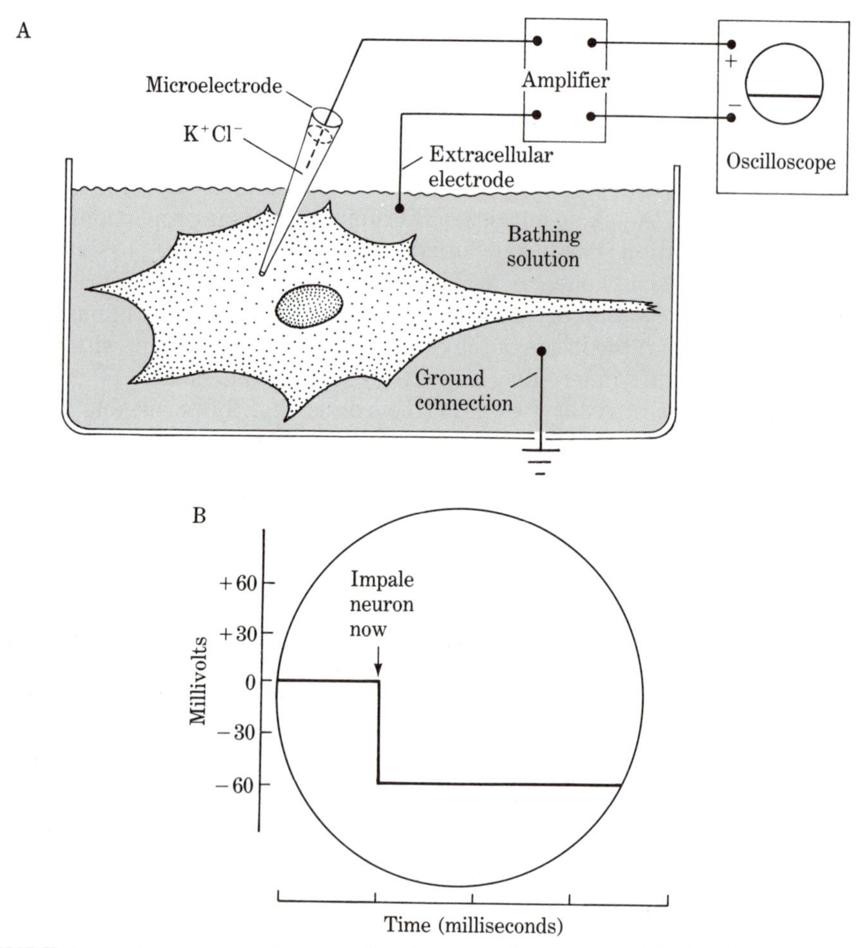

FIGURE 6. A. Arrangement for recording transmembrane potentials. B. The resting potential, recorded as soon as the microelectrode impales the neuron. (See text for explanation.)

screen at a constant rate (which is adjustable by the experimenter), the screen displays the way that the transmembrane voltage changes with time.

Figure 6B shows that, prior to impaling the neuron (that is, with both electrodes sitting outside a cell), no potential difference is registered. But the moment one impales the neuron with the microelectrode, one records a stable voltage, which, as expected, registers the inside as negative. Its value for many neurons is about −60 mV. This represents a significant deviation from the calculated value of the K^+ equilibrium potential: −75 mV. This difference indicates that at least one other type of ion also contributes to the transmembrane potential of a resting neuron, by crossing the membrane. This ion is Na^+, to which a neuron's plasma membrane has slight, but significant, permeability. Because the concentration of Na^+ ions is higher outside the neuron than inside, some Na^+ ions enter the neuron. The entry of these positive ions partially dilutes the internal negative charge cloud, leading to the measured reduction from −75 to −60 mV. This measured transmembrane potential of a resting neuron is called its RESTING POTENTIAL. Although its actual value varies somewhat among different neurons, it can be recorded from any cell body or neurite large enough to impale with a microelectrode.

For the purpose of this discussion, the major significance of the resting potential is that most information signals of neurons consist of transient changes of the transmembrane potential *away* from the resting potential. Most of these voltage signals result from the transient opening of one or more groups of active channels, each group being specific for one or a few types of ions. Thus, as we shall see, the opening of active Na^+ channels allows many more Na^+ ions to enter the cell than do so in a resting neuron. This makes the inside of the cell more positive (that is, less negative) than at rest. Such a transient decrease of the internal negativity (whether it results from Na^+ entering or from some other source) is called a DEPOLARIZATION. A depolarization shows up on the oscilloscope screen as an upward deflection. Alternatively, the opening of active K^+ channels allows even more K^+ ions to leave the cell than do so in a resting neuron.[7] This leaves the inside of the cell more negative than at rest. Such an increase in internal negativity is called a HYPERPOLARIZATION. It shows up on the oscilloscope screen as a downward deflection from the resting potential.

THE DRIVING FORCES ON IONS

We have just seen that the movement of K^+ ions through K^+ channels is influenced by two forces: that of the K^+ concentration gradient and that of the membrane's electrical gradient. Likewise, each other type of ion is influenced both by the concentration gradient of *its own* ionic type, and by the membrane's

[7]As we noted earlier, if the resting potential were equal to the potassium equilibrium potential, there would be no net force to push K^+ ions through the membrane. Thus, even the opening of active K^+ channels would result in no K^+ movement and, therefore, no voltage change. But because the resting potential is only −60 mV, there is a voltage of 15 mV that is not balanced out by the force of the K^+ concentration gradient. These excess 15 mV thus can push K^+ ions in the outward direction.

electrical gradient. For instance, because the Na^+ concentration is higher outside than inside the cell (Figure 5A), the Na^+ concentration gradient produces an inward force on Na^+ ions (Figure 7A). Moreover, because the resting neuron's electrical gradient is in the direction positive outside and negative inside (Figure 6B), the cell's electrical gradient also produces an inward force on the positive Na^+ ions (Figure 7B). This combination of the concentration force and the electrical force on an ion is called the DRIVING FORCE for that ion. The driving force on Na^+ ions, then, is inwardly directed. It is a strong force because, with both of the component forces inwardly directed, they have an additive effect (Figure 7C).

Given this constant force, or push, on Na^+ ions to enter a resting neuron, why doesn't the cell fill up with sodium? One reason is that the resting cell, as we have seen, has a low sodium permeability because most of the Na^+ channels are closed. However, a small amount of Na^+ does leak through. Moreover, if the Na^+ channels were to open (as they do in producing certain neuronal signals), a great many Na^+ ions would enter the cell. This excess internal Na^+ is extruded gradually by an energy-requiring mechanism called the sodium–potassium pump, which is located in the membrane.

Let us now consider the situation of calcium (Ca^{2+}) ions. Ca^{2+} has not been mentioned so far because its total concentration in the nervous system is much lower than that for any of the other ions mentioned. However, as we shall see, Ca^{2+} plays an important role in neuronal signaling. The forces on Ca^{2+} are similar to those on Na^+. As with Na^+, the Ca^{2+} concentration is much higher extracellularly than intracellularly, a situation producing an inward concentration force (Figure 7A). And because Ca^{2+} is positively charged, the membrane's electrical gradient produces an inward force on these ions (Figure 7B). Thus, because there is a strong inward driving force on Ca^{2+} (Figure 7C), these ions would stream into the cell if Ca^{2+} channels were to open.

By contrast, the driving force on Cl^- is very small. This is because the inward concentration force on Cl^- is about equal to the outward electrical force on these negatively charged ions (Figure 7). Thus, the opening of Cl^- channels would result in very little Cl^- movement across the membrane.

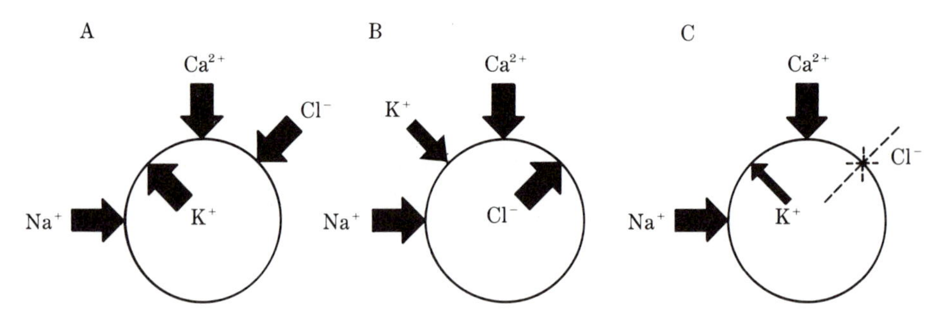

FIGURE 7. A. The forces produced on four types of ions, each by its own concentration gradient. B. The forces produced on each of the four ionic types by the cell's electrical gradient. C. The driving force on each of these four types of ions, derived by adding for each type its concentration (A) and its electrical (B) force. The strength of each parameter is represented by the thickness of the arrow and its direction by the arrowhead.

Returning now to K^+ ions, recall that there is a large outward force deriving from the K^+ concentration gradient and a large inward force deriving from the cell's electrical gradient. If the cell's resting potential were -75 mV, the electrical force on K^+ would exactly balance the concentration force, so that the driving force would be zero. But the -60 mV resting potential is insufficient to fully balance the concentration force. Thus, in the resting neuron, there is a slight outward driving force on K^+ (Figure 7). Therefore, if active K^+ channels were to open, some additional K^+ ions would move out of the cell. These are restored gradually to their intracellular location by the sodium–potassium pump.

What would be the effect on the transmembrane voltage of the opening of specific ion channels? If active Na^+ or Ca^{2+} channels were to open, these ions upon entering the cell would add their positive charges to the internal charge cloud and so would depolarize the cell. By contrast, if active K^+ channels were to open, the K^+ ions that would leave the cell would add their positive charges to the external charge cloud and would thus hyperpolarize the neuron. If active Cl^- channels were to open, there would be little or no movement of Cl^- ions, and, therefore, there would be little or no voltage change. These are the principles that we require to understand the production of information signals by neurons.

TWO TYPES OF SIGNALS

In a postsynaptic neuron, voltage signals arise at the synapses in response to information received from the presynaptic neurons. These induced voltage signals are called POSTSYNAPTIC POTENTIALS (PSPs) or JUNCTIONAL POTENTIALS (JPs). They are fairly small depolarizations or hyperpolarizations that remain local, traveling within the postsynaptic cell no more than a few millimeters from the synapse before fading out. Because some neurons, such as local interneurons, do not have neurites longer than these few millimeters, postsynaptic potentials can reach all parts of such a cell, including its axon terminals. Thus, the postsynaptic potentials themselves can give rise to synaptic transmission onto the next neurons in the chain.

However, many nerve cells have neurites that extend over considerable distances. In such cells, the postsynaptic potentials would die out long before reaching the axon terminals. These neurons (as well as some local interneurons) supplement the postsynaptic potentials with another type of voltage signal that *can* conduct over long distances. This signal is called the ACTION POTENTIAL, or nerve impulse. It is a *large* depolarization that is evoked by the smaller depolarizations of some types of PSPs. Once evoked, the action potential travels rapidly along the axon without a decrease in voltage. We turn now to a discussion of the action potential; we will consider postsynaptic potentials in a later section.

THE ACTION POTENTIAL

The action begins with the opening of active Na^+ channels in the plasma membrane. These Na^+ channels open in response to a depolarization of the membrane, that is, they are VOLTAGE-SENSITIVE channels. A patch of membrane that contains voltage-sensitive channels is said to be ELECTROGENIC. Although the voltage signals that normally open these active Na^+ channels are depolarizing postsynaptic potentials, one can produce experimentally the same channel opening by using an intracellular microelectrode to depolarize the neuron. One simply connects the wire at the back of this microelectrode to the positive terminal of a battery or to an electronic stimulator, and the extracellular electrode to the negative terminal. When the current flows, positive charges are delivered to the inside of the cell (Figure 8A).

Suppose that we depolarize a neuron only slightly—by passing a brief, weak current through such an intracellular microelectrode. For very small depolarizations, only a few of the active voltage-sensitive Na^+ channels open. Each springs open abruptly and closes again automatically within about half a millisecond. (A millisecond is 1/1000 second.) Na^+ ions, driven by their strong inward driving force (Figure 7C) will suddenly enter the cell through these open channels. However, because such a small number of channels has opened, only a small number

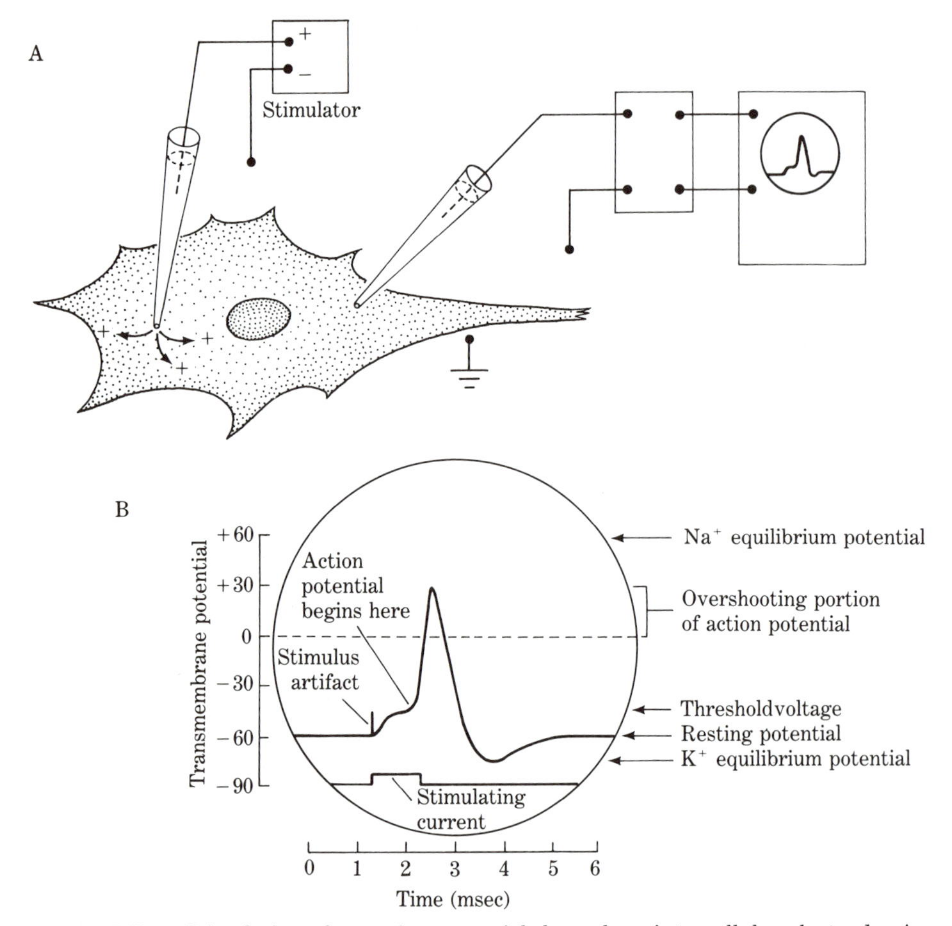

FIGURE 8. Stimulation of an action potential through an intracellular electrode. A. Arrangement of stimulating and recording apparatus. B. The action potential, shown with the time scale expanded to reveal the signal's shape in detail. (See text for explanation.)

of Na^+ ions enter. These incoming Na^+ ions produce a small, sudden depolarization. However, this depolarization ceases quickly because of compensating movements of K^+ ions. Because the depolarization has slightly increased the outward driving force on K^+ (Figure 7), these ions, free to move through passive K^+ channels, exit the cell. In so doing, they bring the transmembrane potential back toward the resting potential.

However, if we produce with our stimulating electrode a somewhat larger depolarization, then a somewhat greater proportion of the voltage-sensitive Na^+ channels in the region of the stimulating electrode suddenly spring open. When they do, a greater number of Na^+ ions suddenly enter the cell, causing a larger depolarization. Once again, K^+ ions respond by leaving the cell, tending to restore the transmembrane potential to the resting potential. But the K^+ exit, driven by a weak driving force, is slower than the abrupt Na^+ entry driven by the strong Na^+ driving force (Figure 7). If sufficient Na^+ has entered before the K^+ movements can reverse the depolarization, a second important effect sets in; the depolarization resulting from the Na^+ entry serves as a stimulus that opens additional voltage-sensitive Na^+ channels in the same region of the membrane. With these additional Na^+ channels opened, still more Na^+ ions enter the cell, producing a still greater depolarization and thus leading to the opening of still more Na^+ channels. This positive feedback process, in which depolarization leads to channel opening, which leads to more depolarization, keeps repeating itself until a very large depolarization has been achieved. This all happens very quickly, usually within less than 1 millisecond.

In order for this self-reinforcing depolarization to have been initiated, a specific amount of initial depolarization had to be produced by the microelectrode (or by any other source). The particular value of transmembrane voltage which, if achieved, initiates the self-reinforcing depolarization, is called the THRESHOLD voltage for this neuron. In different neurons, the threshold voltage varies between about 5 and 30 mV depolarized from the resting potential (thus it is typically at about -30 to -55 mV).

As soon as the self-reinforcing depolarization has reached its peak, the voltage begins to drop quickly toward the resting potential. This drop comes about for two reasons. First, each active Na^+ channel closes about 0.5 millisecond after it has opened and cannot reopen again until the membrane potential has returned to near the resting potential and until a few additional milliseconds have elapsed. Second, shortly after the Na^+ channels have begun to open (when the depolarization is near its peak), a group of voltage-sensitive K^+ channels open as well. Thus, K^+ ions stream out of the cell, thereby reversing the depolarization that had been caused by Na^+ entry. Actually, the K^+ ions are now being pushed by an outward driving force that is much greater than that of the resting neuron (Figure 7C). This is because, with the inside of the cell now charged much more positively than at rest, the cell's electrical gradient no longer produces a strong inward force on K^+ (Figure 7B). Thus, the outward force of the K^+ concentration gradient (Figure 7A), now much stronger than this reduced force of the electrical gradient, dominates the electrical gradient in determining the overall driving force on K^+ ions. Given the resulting strong outward driving force on K^+, when the voltage-sensitive K^+ channels open during the action potential, K^+ ions stream out of the cell and thus return the transmembrane potential to near the resting potential.[8]

The voltage changes just described constitute a single action potential. We can record this action potential by impaling the neuron with a second microelectrode, in addition to the microelectrode used for stimulation (Figure 8A). Actually, the first voltage change that is recorded when the stimulating current begins has nothing at all to do with the neuron's response. It is a brief blip that represents the nearly instantaneous spread of some current (literally at the speed of light) directly from the stimulating electrode to the recording electrode. This blip, called the STIMULUS ARTIFACT, can provide a useful time reference as to the moment when the electrical stimulus begins. As Figure 8B shows, when the depolarization produced by the stimulating electrode's current reaches the cell's threshold voltage, the action potential begins suddenly with a rapid increase in the rate of depolarization. This depolarization continues to well beyond zero transmembrane voltage, the inside of the cell actually becoming positive relative to the outside. For this reason, the action potential is said to be an OVERSHOOTING potential.[9] Following its peak, the voltage drops quickly toward the resting potential. The overall duration of the action potential in most neurons studied is only about a millisecond, though there are some neurons in which the duration is many times longer. The detailed shape of this signal (Figure 8B) should now be understandable on the basis of the underlying ionic events.

A special feature of the action potential is its ALL-OR-NOTHING property. That is, a depolarization that does not reach the threshold voltage produces no action potential at all; and a depolarization that exceeds threshold, even by a great amount, does not produce a larger than normal action potential. Rather, a depolarization to threshold or beyond (whether from a stimulating electrode or from a postsynaptic potential) serves only to initiate the self-reinforcing opening

[8]Actually, the transmembrane potential returns to near -75 mV (the K^+ equilibrium potential) and not to -60 mV (the resting potential). This is because there is now a greater abundance than at rest of open K^+ channels and almost all the active Na^+ channels have been closed. Therefore, the membrane's permeability is dominated, even more than at rest, by K^+. It is as though the membrane is now permeable *only* to K^+. As explained earlier in this chapter, under such conditions the resting potential would equal the K^+ equilibrium potential. After the voltage-sensitive K^+ channels have closed, the transmembrane potential returns to the resting potential of about -60 mV.

[9]One can readily understand why the action potential overshoots the value of zero transmembrane potential, on the basis of the principles already developed in this chapter. With Na^+ channels open, Na^+ ions enter the cell, driven by the strong Na^+ driving force. As they continue to enter, and thus as their accumulation on the inner surface of the membrane continues to depolarize the cell, the cell's electrical gradient is reduced toward zero. The Na^+ concentration gradient, however, remains essentially unchanged, because the actual number of ions moving through the membrane is but a small percentage of the total number present. It is only after the cell's electrical gradient has actually reversed, being positive inside and negative outside, that the concentration and electrical forces on Na^+ become equal and opposite. The voltage at which this occurs, called the Na^+ equilibrium potential, is about $+55$ mV. The driving force on Na^+ would then be zero. Only at this voltage would the net entry of Na^+ through the open channels cease. In fact, the action potential never reaches this voltage, because the Na^+ channels begin to close and the K^+ channels begin to open before $+55$ mV is achieved.

of Na^+ channels and subsequent opening of K^+ channels. Once this process has been initiated, it continues on its own to completion.

THE CONDUCTION OF ACTION POTENTIALS

In general, there is one special region, somewhere near a neuron's input synapses, where all of its action potentials are initiated. This region, called the TRIGGER ZONE, in many cells is the closest region to the synapses with enough voltage-sensitive channels to support an action potential. Once the action potential has begun there, it is conducted along the cell, usually all the way to the axon terminal.

How does the conduction of an action potential along a neurite occur? When Na^+ channels open to admit Na^+ ions, this produces locally a partial neutralization of the internal negative charge cloud. Thus, a potential difference, or voltage difference, occurs between this location inside the neurite and neighboring locations further down the neurite. As a result, positive ions will flow from this location to the neighboring locations (Figure 9, thin arrows). Most of this movement is carried out by K^+ ions, which are the most abundant intracellular positive ions. This delivery of positive charges to the neighboring region in the neurite causes this next region to become somewhat depolarized. For some small distance along

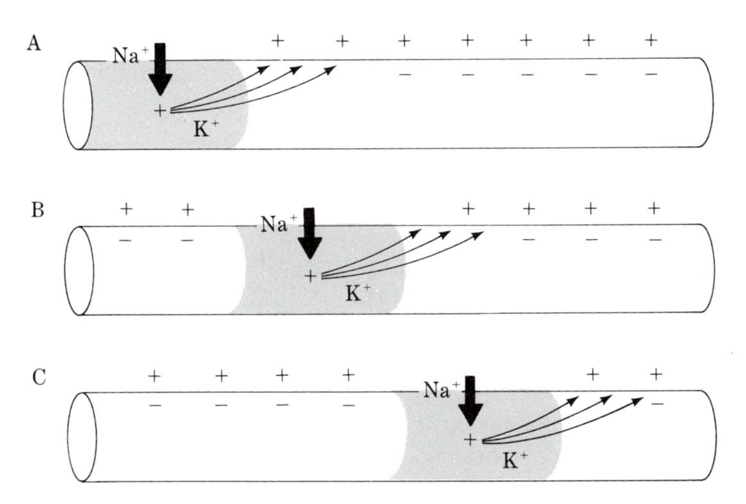

FIGURE 9. The conduction of an action potential along a neurite. A, B, and C are three successive moments in time. The location of the action potential at a given moment is indicated by the shaded portion. For simplicity, Na^+ entry is indicated as occurring only in the middle of this shaded region. Internal K^+ movements are drawn as though they were unidirectional. Actually, K^+ also moves to the left; however, here it would be ineffective in inducing an action potential because the Na^+ channels to the left of the action potential have just closed and cannot reopen for a few milliseconds, by which time these K^+ ions would have exited from the axon. Though not discussed in the text, K^+ ions actually pass outward through passive K^+ channels in the membrane, thus completing the electrical circuit that began with the inward movement of Na^+.

the neurite, this depolarization is sufficient to reach the threshold voltage. Therefore, the Na^+ channels in this neighboring region spring open and a full-blown action potential develops there (Figure 9B). The entry of Na^+ ions through the active Na^+ channels in this second location will cause a suprathreshold depolarization in the next region of membrane, yet farther down the neurite, leading to a further spread of the action potential (Figure 9C), and so on down the neurite. Although Figure 9, for simplicity's sake, represents the conduction of an action potential as a stepwise affair, it is actually a continuous moving wave (except in specialized cases to be described momentarily). Because a self-reinforcing opening of active Na^+ channels occurs at every point along the neurite, the action potential maintains a constant amplitude, overshooting to about +30 mV at each location.

The speed of conduction of action potentials ranges, in different axons, from about 0.1 to 120 meters per second. Two major structural features contribute to determining the conduction speed in a given neurite. First, the greater the diameter of the neurite, the faster the conduction because a larger diameter offers less resistance to the lateral spread of ions along the inside of the axon. This permits an action potential at a given location to depolarize to the threshold voltage a greater stretch of neurite. Thus, fewer total regenerations of the action potential (like the three shown in Figure 9) are required for this signal to travel all the way to the axon terminal. Because each such regeneration occupies a finite time, reducing the overall number of regenerations increases the conduction velocity. As we shall see in Chapters 4 and 8, many behaviors that need to be executed very quickly utilize neurons of very great axonal diameter, capitalizing on their rapid conduction of action potentials.

The second structural feature affecting the speed of action potential conduction is found almost exclusively in vertebrates. Around some vertebrate axons there is an insulating wrapping of multiple layers of cell membrane that belong to nonneural, satellite cells. These are called GLIAL CELLS in the central nervous system or SCHWANN CELLS in the peripheral nervous system. The wrapping that they form around axons is called MYELIN. Between the myelin wrappings formed by adjacent glial or Schwann cells, a small length of axon is exposed to the fluid environment. This region is called a NODE OF RANVIER.

Myelinated axons conduct action potentials much faster than unmyelinated axons of the same diameter. The reason is that the presence of the myelin wrappings affects the axon's electrical properties[10] in such a way as to facilitate the lateral flow of K^+ ions inside the axons that occurs during an action potential (Figure 9). In fact, it is only at the nodes of Ranvier that voltage-sensitive Na^+ and K^+ channels open to produce an action potential. The action potential, then, travels in jumps from node to node, a process called SALTATORY CONDUCTION. The presence of the myelin wrappings permits the intraaxonal K^+ flow to cover

[10] The myelin wrapping of a glial or Schwann cell around a segment of axonal membrane makes that membrane less able to hold an internal charge cloud. In electrical terms, the electrical capacitance of the membrane has been decreased, so charge storage is reduced. Thus, K^+ ions are free to move onto the next node of Ranvier. Also, whereas in unmyelinated axons some of the laterally spreading K^+ ions leak out of the membrane through passive K^+ channels, myelin wrappings severely limit this leakage. In electrical terms, the myelin wrapping between two nodes increases the electrical resistance of the underlying membrane. Thus, again more K^+ ions are available to move on to the next node.

the relatively great distance from one node to the next. Owing to this enhanced K^+ flow, fewer time-consuming regenerations of the action potential are required for a given length of axon. Rapidly executed behaviors in vertebrates often utilize neurons that have myelinated axons of great diameter, both the myelin and the axonal diameter contributing to the speed.

POSTSYNAPTIC POTENTIALS

We shall now examine postsynaptic potentials (PSPs)—the signals that can lead to the production of an action potential at a neuron's trigger zone. We will consider first the PSPs at electrical synapses which, though far less common than chemical synapses, are simpler to understand. When an action potential arrives at the presynaptic axon terminal, the entry of Na^+ ions into this terminal creates a potential difference between the insides of the presynaptic cell and the postsynaptic cell. As a result, positive ions, mostly K^+, will flow from the presynaptic cell to the postsynaptic cell through the tiny tubular structures that connect the two cells (Figures 10A and 4A). The K^+ ions that cross the synapse depolarize

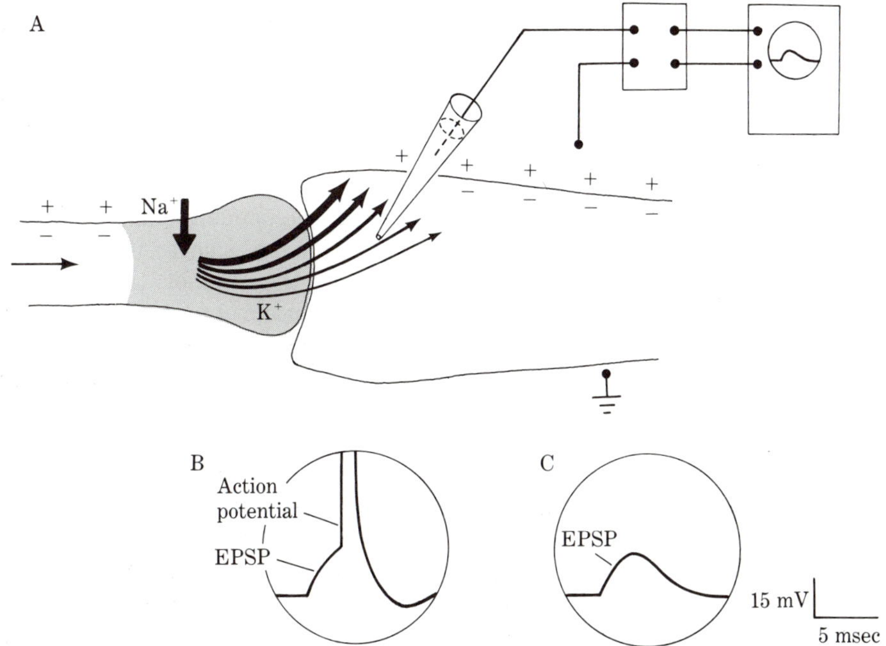

FIGURE 10. An electrical synapse. A. An action potential has reached the presynaptic terminal (shaded portion). Here Na^+ ions enter through active Na^+ channels. This results in the movement of K^+ ions into the postsynaptic neuron, which they depolarize. Though not discussed in the text, K^+ ions subsequently pass outward through passive K^+ channels in the membrane of the postsynaptic cell, thus completing the circuit that began with the presynaptic entry of Na^+. B. A recording showing a case in which the electrical EPSP has depolarized the trigger zone to threshold and evoked an action potential. C. Another case, in which the electrical EPSP has not reached threshold and thus can be seen in its entirety.

the postsynaptic neuron. If there are many tubular passages in the synapse, enough positive ions may cross over to depolarize the postsynaptic cell's trigger zone to threshold and thus to induce an action potential (Figure 10B). Alternatively, there may be insufficient ions reaching the trigger zone, in which case the depolarization would remain subthreshold (Figure 10C).

Even though such a subthreshold depolarization would not by itself evoke an action potential, it could assist other, similar subthreshold potentials, coming from other synapses, in doing so. Thus, because it can *contribute* to producing a threshold depolarization and an action potential, it is called an EXCITATORY POSTSYNAPTIC POTENTIAL, or EPSP. Specifically, it is an ELECTRICAL EPSP because it has occurred at an electrical synapse. The electrical EPSP in Figure 10C is seen undistorted by an action potential. By contrast, in Figure 10B, the electrical EPSP is visible only until threshold is reached; after that, its shape is obscured by the superimposed action potential.

The situation is considerably more complicated at a chemical synapse. Again, imagine that an action potential has arrived at the axon terminal of a presynaptic neuron (Figure 11). At the axon terminal of a chemical synapse, not only Na^+ channels, but also Ca^+ channels open. Thus, both Na^+ and Ca^{2+} enter the presynaptic axon terminal. The entry of Ca^{2+} into this terminal is essential for the synapse to function. The entering Ca^{2+} ions somehow induce the synaptic vesicles to fuse very briefly with the presynaptic plasma membrane adjacent to the synaptic cleft. The fused vesicles open and extrude their neurotransmitter molecules into the cleft. These transmitter molecules diffuse across the cleft, where they bind with the receptor macromolecules found locally on the postsynaptic membrane. In most chemical synapses, this bond results in the opening of a special class of active membrane channels. These channels, then, are said to be CHEMICALLY SENSITIVE. They are not voltage sensitive as are the channels that produce the action potential. In fact, the neuronal plasma membrane appears to be subdivided into separate areas: those that contain voltage-sensitive channels (for Na^+, for K^+, and, in some regions, for Ca^{2+}) and other areas, restricted primarily to the synaptic sites, that contain chemically sensitive channels.

Some chemical synapses are excitatory, as is the electrical synapse of Figure 10. Others inhibit, rather than excite, the postsynaptic neuron. We will consider first excitatory chemical synapses. Here, the postsynaptic channels that open in response to neurotransmitter molecules permit Na^+ and K^+ ions to pass. Recall that an action potential also results from the opening of channels for Na^+ and for K^+. However, whereas the action potential uses two separate groups of channels (Na^+ channels that open first and K^+ channels that open later), each channel that opens at an excitatory chemical synapse permits simultaneous passage of both Na^+ and K^+.

What is the consequence of the opening of these postsynaptic Na^+/K^+ channels? Na^+ ions will move inward and K^+ ions outward. However, because the driving force on Na^+ is much greater than that on K^+ (Figure 7C), much more Na^+ than K^+ will move through these channels. This overall ionic movement is called the SYNAPTIC CURRENT. Some of the Na^+ coming into the cell will enter the negative charge cloud along the inner surface of the plasma membrane and, by partly neutralizing it, will depolarize the cell. Within a few milliseconds, how-

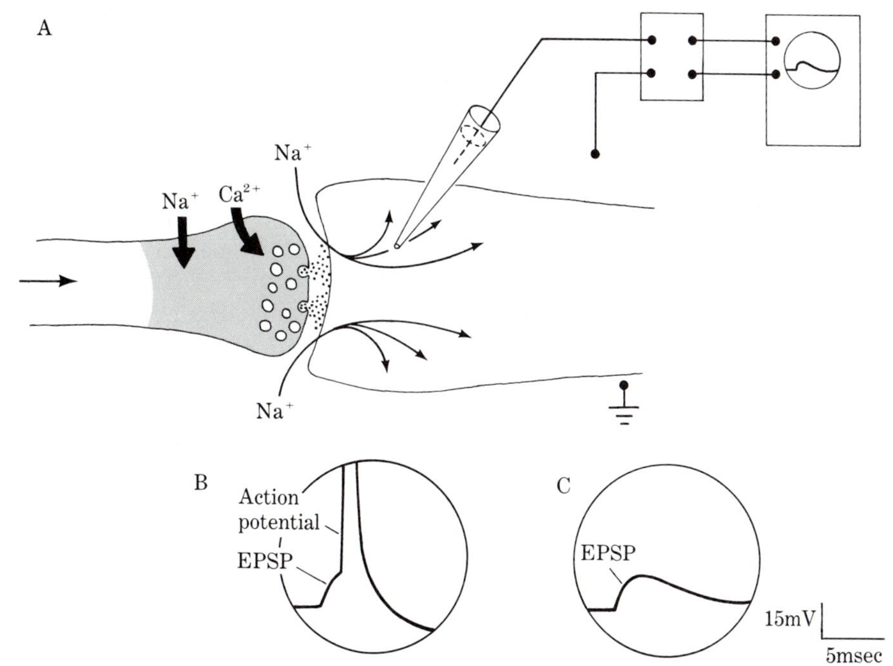

FIGURE 11. A chemical excitatory synapse. A. When an action potential arrives at the presynaptic terminal, primarily Na^+ moves through the subsynaptic channels into the postsynaptic cells, as shown. B. A suprathreshold EPSP evokes an action potential. As a result, only the initial part of the EPSP is visible. C. A subthreshold EPSP, visible in its entirety. The rising, or depolarizing, phase of a chemical EPSP is more rapid than the falling, or repolarizing, phase. The slower repolarization is due partly to the continuing action of some of the neurotransmitter, which continues to permit some Na^+ ions to enter the postsynaptic cell.

ever, the release of neurotransmitter ceases, the excess neurotransmitter molecules are removed by various means, the postsynaptic Na^+/K^+ channels close, and the Na^+ entry through these channels terminates. If during this interval the cell's trigger zone has been depolarized to threshold, an action potential results (Figure 11B). But if the transmembrane voltage at the trigger zone does not reach threshold, the depolarization soon falls off to zero without further effect (Figure 11C). This depolarization is called a CHEMICAL EPSP.

At the chemical synapses between motor neurons and muscles, called NEUROMUSCULAR JUNCTIONS, the EPSP has a special name, the END PLATE POTENTIAL, or EPP. This term derives from the flattened axonal terminals of some motor neurons, called end plates. In many muscle cells, an EPP depolarizes the plasma membrane to threshold and thus evokes an action potential just like that in a neuron. Some other muscle cells, especially in invertebrates and lower vertebrates, produce EPPs but no action potentials.

As has been said, some chemical synapses are not excitatory but rather are inhibitory. In most of these, the postsynaptic potential is not in the depolarizing but in the hyperpolarizing direction (Figure 12). That is, it moves the transmembrane potential further away from the threshold voltage. This makes it momentarily more difficult for other synapses that are excitatory to depolarize the postsynaptic neuron to threshold. Just as at a chemical excitatory synapse, neurotransmitter is released from the presynaptic cell, and this released neurotransmitter binds with postsynaptic receptor macromolecules. But here, the channels that open in response to this bond are not Na^+/K^+ channels but rather K^+ (or, more rarely, K^+/Cl^-) channels. We have already seen that in the resting cell there is an outward driving force on K^+ ions (Figure 7C). Thus, the opening of the K^+ channels leads to an exit of K^+ ions and thus to a hyperpolarization. The hyperpolarizing potential produced is called a CHEMICAL INHIBITORY POSTSYNAPTIC POTENTIAL, or chemical IPSP.

The situation can be slightly different if the channels that open are not K^+, but rather are Cl^- channels, as occurs at some inhibitory synapses. If, as occurs in some cells, the intracellular Cl^- concentration of the postsynaptic neuron is slightly elevated, this cell will have a slight outward driving force on Cl^- (Figure 7). Therefore, when Cl^- channels open, a limited number of Cl^- ions move outward. Such an outward flow of negative ions causes a depolarization. However, in this case it is only a slight depolarization because of the small driving force. This depolarization has the appearance of an EPSP (Figure 13A). However, one might guess that it is not actually an EPSP because the Cl^- ions will continue to move outward only until the cell has been depolarized to the Cl^- equilibrium potential (the voltage at which the driving force on Cl^- is zero), which is only very slightly depolarized from the resting potential. The opening of chloride channels, then, could never depolarize a neuron all the way to threshhold.

In fact, the opening of the Cl^- channels actually *resists* any attempt by excitatory synapses to depolarize the cell to threshold. To understand this, suppose that just prior to the opening of the Cl^- channels at the inhibitory synapse, the

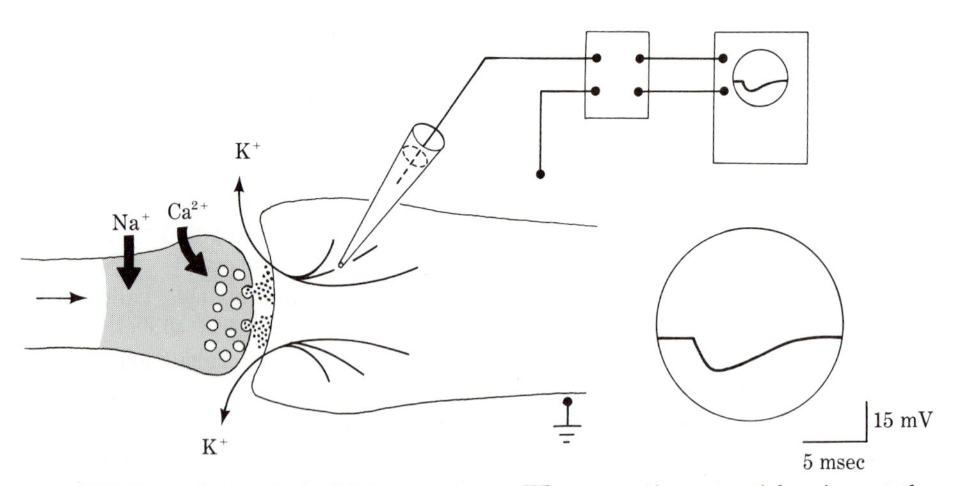

FIGURE 12. A chemical inhibitory synapse. When an action potential arrives at the presynaptic terminal, ionic events occur as shown. The resulting IPSP is shown enlarged on the right.

postsynaptic cell had received from some excitatory neuron a large EPSP (Figure 13B). During muchof the EPSP, the postsynaptic membrane voltage remains *more* depolarized than the Cl^- equilibrium potential E_{Cl^-}. During this time, the driving force on Cl^- is reversed; the outward force of the electrical gradient is

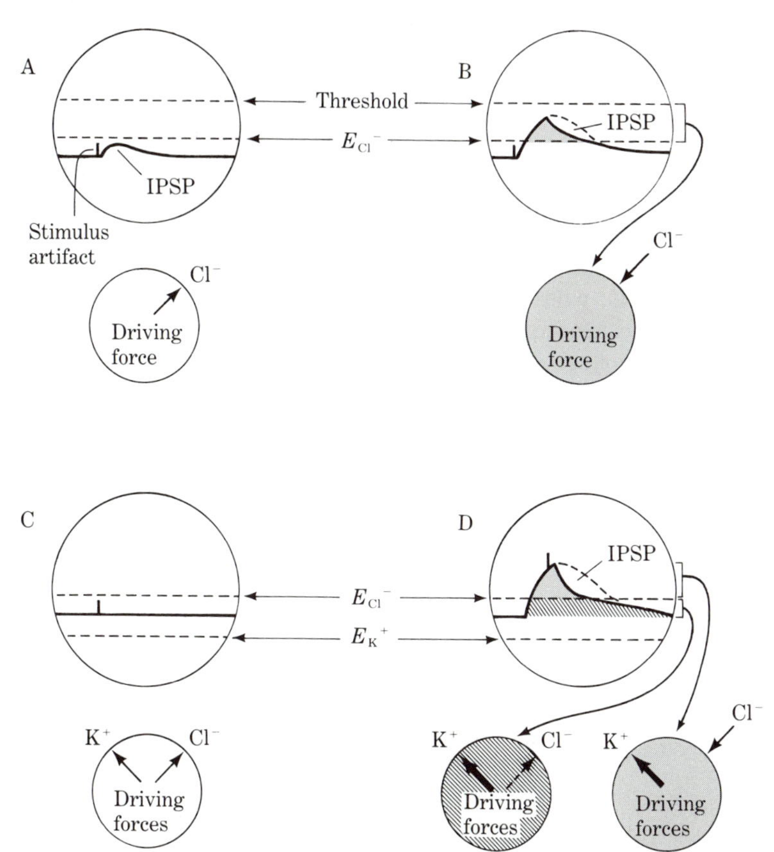

FIGURE 13. Synaptic inhibition of two unusual forms. A. A depolarizing IPSP, which can result from the opening of synaptic Cl^- channels. B. The fact that the depolarizing postsynaptic potential in A was an IPSP is revealed by the reversal of its direction when it occurs during a large EPSP. This reversal results from the reversal in the direction of the driving force when the postsynaptic potential is depolarized beyond the Cl^- equilibrium potential (lower circles in A and B). C. An oscilloscope trace showing no postsynaptic potential, even though a known presynaptic neuron was stimulated and gave an action potential. The synapse between these two neurons could utilize K^+/Cl^- channels. (In this example, the Cl^- driving force at rest is the same as in part A.) D. The existence of inhibition at the synapse tested in C is revealed by stimulating the presynaptic neuron during a large EPSP. The IPSP now appears because the depolarization of the EPSP has produced an increased outward driving force on K^+ and an inward driving force on Cl^- (lower circles). These directions of movement for these two ions produce a hyperpolarization. E_{Cl^-}, E_{K^+}—chloride and potassium equilibrium potentials.

now reduced and is less than the inward force of the Cl^- concentration gradient, so the Cl^- driving force is now directed inward. As a result, if Cl^- channels open during the peak of this large EPSP, Cl^- ions will flow inward, and will thus reduce the size of the EPSP. Consequently, although these Cl^- channels acting alone produced a depolarization, when acting simultaneously with a large depolarization, they decrease this depolarization and thus inhibit the cell. In view of this inhibition, the voltage signal of Figure 13A, even though it is a depolarization, is an IPSP. The IPSP can be seen inverted in Figure 13B, in the form of a reduction of a large EPSP.

As an extension of these ideas, if an action potential in a presynaptic neuron evokes in a postsynaptic cell *no* voltage signal—neither a depolarization nor a hyperpolarization (Figure 13C)—there could still be a very potent inhibitory effect. This could result, for instance, if the channels that open at this synapse are K^+/Cl^- channels, whose *net* effect in the resting neuron is an equal exit of K^+ and of Cl^-. Because equal numbers of positive and negative ions would leave the neuron, no voltage signal would result. The inhibitory effect of this channel opening

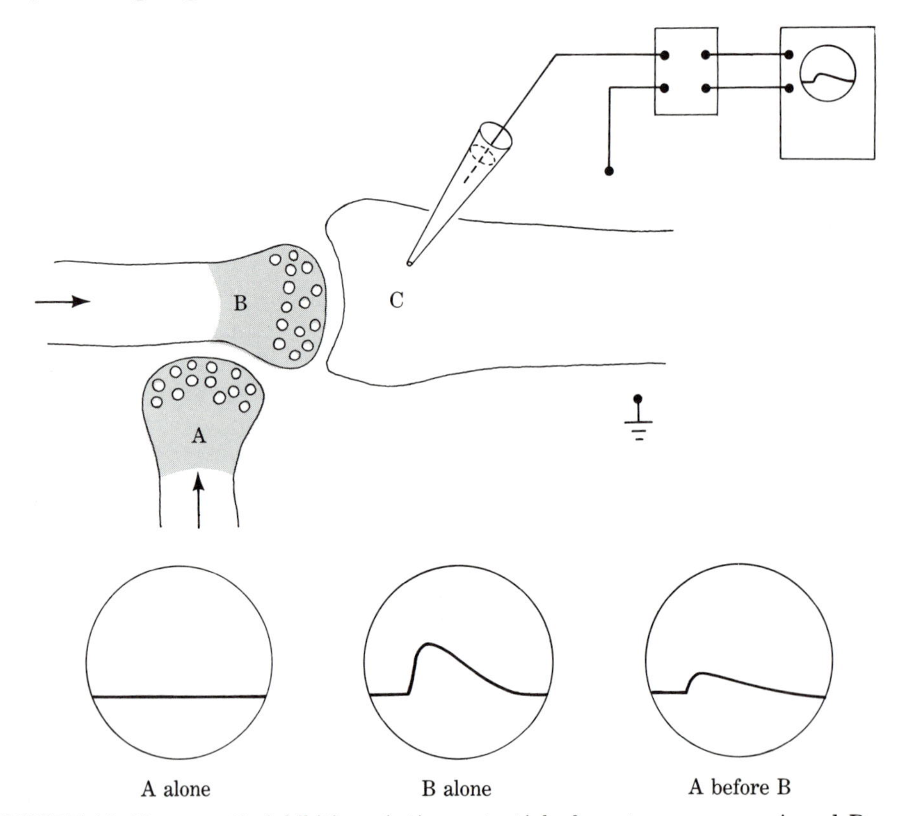

FIGURE 14. Presynaptic inhibition. Action potentials from two neurons, A and B, are shown along with a dendrite from a third neuron, C. Recordings from cell C show no PSP when cell A alone gives an action potential, a large PSP when cell B alone gives an action potential, and a smaller PSP when cell A gives an action potential just before that in cell B. (See text for further explanation.)

would be revealed only if it occurred during an EPSP or some other source of depolarization (Figure 13D).

As a final example of synaptic interactions, the following synaptic arrangement among a set of three neurons (A, B, and C) has been found in many nervous systems (Figure 14). Intracellular recordings are made from neuron C. An action potential in neuron A alone produces no depolarization or hyperpolarization in cell C. (This is the case even if tested while cell C is momentarily depolarized from its resting potential to near its threshold.) By contrast, an action potential in cell B alone produces in cell C a large EPSP. Now, if an action potential in cell A *precedes* that in cell B by a very brief interval (from roughly one-half to several milliseconds) this reduces or abolishes the EPSP in cell C. This reduction occurs because cell A synapses on the terminal of cell B and acts on this terminal so as to reduce the amount of neurotransmitter released at the B-to-C synapse. This comes about because of the opening of postsynaptic channels at the A-to-B synapse, which reduce the size of the action potential. (This action potential reduction occurs in fundamentally the same manner as the reduction of EPSP size by the opening of Cl^- channels as shown in Figure 13B and D.) Because the action potential is now a smaller voltage signal, the voltage-sensitive Ca^{2+} channels of the axon terminal in cell B receive less of a stimulus to open, so that fewer Ca^{2+} ions enter the axon terminal. Consequently, less neurotransmitter is released from cell B, resulting in the smaller postsynaptic potential in cell C. This mechanism, in which cell A exerts its effect on cell C only indirectly, by acting on a neuron presynaptic to cell C, is called PRESYNAPTIC INHIBITION.

CHEMICAL VERSUS ELECTRICAL SYNAPSES: A COMPARISON

In chemical, as compared to electrical synapses, several additional steps occur between the arrival of an action potential at the presynaptic axon terminal and the onset of the postsynaptic potential. These steps are the presynaptic entry of Ca^{2+} ions, the fusion of a synaptic vesicle with the presynaptic plasma membrane, the release of neurotransmitter, the diffusion of this neurotransmitter across the synaptic cleft, the binding of the neurotransmitter to postsynaptic receptor macromolecules, the opening of the chemically sensitive channels, and the resulting synaptic current. Taken together, these are time-consuming steps and cause a SYNAPTIC DELAY at a chemical synapse of more than 0.3 milliseconds. (A typical value is 0.5 milliseconds.) By contrast, electrical synapses show virtually no synaptic delay. We shall see in Chapter 8 one adaptive use of this time difference, namely, that some behaviors that must be executed very quickly employ neural circuits containing primarily electrical synapses.

Another distinction is that chemical synapses generally show a greater capacity than do electrical synapses for modifiability of function. For instance, in some chemical synapses, if the presynaptic neuron gives a rapid barrage of several action potentials, each action potential results in more transmitter release than did the one before. Thus, each successive postsynaptic potential is larger than the one before. This process is called SYNAPTIC FACILITATION (Figure 15A and B). Some other chemical synapses function in the opposite way; there is *less* transmitter release for each successive action potential. This results in progres-

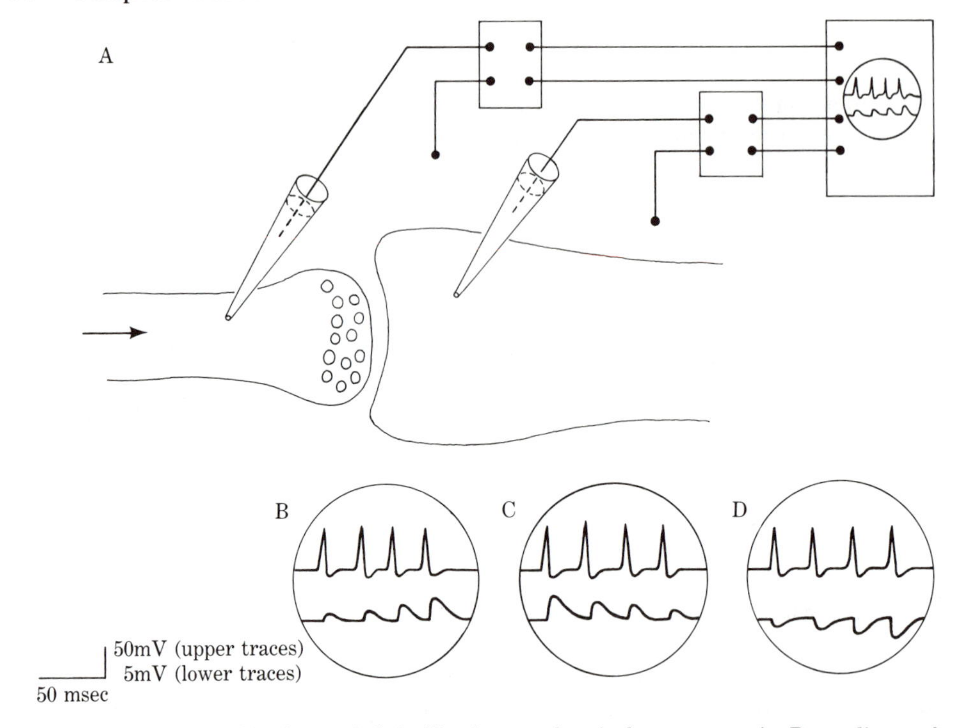

FIGURE 15. Facilitation and defacilitation at chemical synapses. A. Recordings of presynaptic action potentials and EPSPs (B and C) or IPSPs (D). (In reality, most axon terminals and dendrites are too small to impale with microelectrodes, so both recordings would most likely be made from the cell bodies.) B, C, and D are representative of three different types of chemical synapses: B, facilitating EPSPs; C, defacilitating EPSPs; D, facilitating IPSPs. (See text for further explanation.)

sively smaller PSPs, a process called SYNAPTIC DEFACILITATION or ANTIFACILITATION (Figure 15C). Some synapses recover from defacilitation in a matter of seconds, whereas others may require hours to recover. Synaptic facilitation and defacilitation can occur not only at excitatory but also at inhibitory synapses (Figure 15B and D). In either facilitating or defacilitating synapses, then, the size of the postsynaptic potential at any moment depends in part on the recent activity, or experience, of that synapse. No comparable experience-dependent properties are found in electrical synapses.

An extreme form of modifiability is seen at some chemical synapses that have a spatial arrangement like that of presynaptic inhibition (Figure 14). As in Figure 14, an action potential in neuron A would alter the number of Ca^{2+} channels that open in the axon terminal of cell B when this terminal is invaded by an action potential. But beyond this similarity, the situation here differs in several ways from that of presynaptic inhibition. First, the effect on Ca^{2+} channels is an *increase,* not a decrease, in the number of these channels that open. Second, this effect lasts not just for milliseconds, but rather for several minutes, hours, or even days. And third, this long-term enhancement of the readiness of Ca^{2+} channels to open is mediated by a complex set of intermediate biochemical steps, occurring within the cytoplasm of the axon terminal of neuron B itself. These steps are initiated when axon terminal B binds the neurotransmitter from neuron A. The end result of this process is a long-term enhancement of the size of the postsynaptic potential in neuron C. This overall process is called synaptic SENSITIZATION. Such long-term changes in synaptic function are generally not found in electrical synapses.

Given that electrical and chemical synapses can function in such different ways, a neuroethologist often wishes to determine which of these two synaptic types has produced a given recorded postsynaptic potential. But because electrical and chemical EPSPs (as well as depolarizing IPSPs) look very much alike on the oscilloscope, the mere appearance of such a signal does not help to identify the type of synapse. Rather, this identification requires special tests. These are described, along with other tests of synaptic function, in the Appendix at the end of this chapter.

NEUROTRANSMITTER MOLECULES AND THEIR POSTSYNAPTIC RECEPTOR MOLECULES

Different neurons release different types of neurotransmitter molecules. The most fully studied neurotransmitter, ACETYLCHOLINE, is used by all known vertebrate neuromuscular junctions at skeletal muscles and by many synapses in the central nervous system. It is also used by some synapses of invertebrates. A group of neurotransmitters that are chemically related to one another, the BIOGENIC AMINES, includes DOPAMINE, NOREPINEPHRINE, SEROTONIN, and OCTOPAMINE. Two other major neurotransmitter categories are the AMINO ACIDS and the PEPTIDES.

As we have seen, the formation of a chemical bond between neurotransmitter and receptor molecules results in the opening of ionic channels in the postsynaptic membrane. It is the particular type of RECEPTOR MOLECULE present and not the type of neurotransmitter molecule used by a synapse that determines which type of ionic channel will open. Actually, the receptor macromolecule is a complex of at least two components: a binding component that binds to the neurotransmitter and an ionophore component that regulates the channel opening. Differences in the ionophore component give rise to differences in the types of ions that will be admitted by the channel. For instance, acetylcholine released at some synapses binds with a class of receptor macromolecules whose ionophore opens Na^+/K^+ channels. This synapse thus produces an EPSP. At other synapses, the acetylcholine binds with a receptor macromolecule whose ionophore opens K^+ channels and thus leads to an IPSP. Moreover, a single postsynaptic neuron may contain both these types of receptor macromolecules for acetylcholine, either at different synapses or even mixed at the same synaptic site. This cell may also contain, at yet other synapses, receptors for several different neurotransmitter types.

This complex mosaic of different classes of receptor macromolecules on a postsynaptic neuron often confounds certain types of neurochemical experiments. For instance, some workers have attempted to inject a neurotransmitter into the bloodstream or into restricted regions of the brain and to search for a consistent behavioral outcome. (In some experiments, the substance injected is an AGONIST of a neurotransmitter: a chemically similar substance producing a similar biochemical effect; or an ANTAGONIST: an agent that blocks the ability of receptor macromolecules to respond to their neurotransmitter.) There are several potential pitfalls in such experiments. First, because a given type of neurotransmitter can open one type of channel at one synapse and another type at another synapse, a mixture of depolarizing and hyperpolarizing effects would result as the injected neurotransmitter contacts a mixed population of synapses. Moreover, cells that are functionally unrelated would be activated as this neurotransmitter spreads through the neural tissue. In spite of these pitfalls, a few such experiments have produced quite clear and interesting results, which we will examine in Chapter 8.

RECEPTOR POTENTIALS

For a neuron in the central nervous system, the principal source of excitatory input is synaptic transmission from other neurons. However, for a receptor cell residing in some sensory organ, the primary source of excitation is physical energy from the animal's external environment. Nevertheless, similar principles apply to the excitation of receptor cells as to postsynaptic neurons. For instance, at an olfactory chemoreceptor, one or more specific types of odor-producing molecules in the environment bind to specific macromolecules on the receptor cell's plasma membrane. This chemical bond results in the opening of Na^+/K^+ channels like those at a chemical excitatory synapse. Thus, a depolarizing RECEPTOR POTENTIAL is evoked that closely resembles an EPSP. The situation is the same in most other types of receptors, except that the agent initiating the opening of channels is not a molecule from the environment but rather some other stimulus. This general process of converting some form of environmental energy into a receptor potential is called SENSORY TRANSDUCTION.

As we have seen, some sensory organs contain receptor cells that have no neurite but rather synapse directly onto an afferent neuron whose axon projects to the central nervous system. Such receptor cells in general produce no action potentials. Rather, the receptor potential itself induces the release of neurotransmitter from the receptor cell, and this evokes EPSPs and consequent action potentials in the afferent neuron. One example of this arrangement, found in the vertebrate ear, is described in detail in Chapter 6. A similar, though more complex, situation is found in the vertebrate eye and will be discussed in Chapter 5.

SUMMARY OF NEURONAL SIGNALS

Let us now summarize the properties of the major types of signals used by neurons (Table 1):

1. *Action potentials* serve to conduct information from one end of the neuron to the other. They are brief, overshooting, all-or-nothing potentials that prop-

TABLE 1.

Signals	Signaling Role	Typical Duration	Size	All-or-Nothing *vs* Graded	Actively Propagated *vs* Local	Channel Opening	Channel Sensitivity
1. Action potential	Conduction along a neuron	1–2 milliseconds	About 100 millivolts; overshooting	All-or-nothing	Actively propagated, undecremented	First Na^+, then K^+; different channels	Voltage (depolarization)
2. Electrical EPSPs	Transmission between neurons	Several milliseconds	From less than 1 to more than 20 millivolts	Graded	Local	None	(No synaptic channels open)
3. Chemical EPSPs[a]	Transmission between neurons	10–100 milliseconds	From less than 1 to more than 20 millivolts	Graded	Local	Na^+/K^+	Chemical (neurotransmitter)
4. Chemical IPSPs	Transmission between neurons	10–100 milliseconds	Depolarizing or hyperpolarizing, from less than 1 to about 15 millivolts	Graded	Local	K^+ or Cl^- or K^+/Cl^-	Chemical (neurotransmitter)
5. Receptor potentials	Registering environmental energy	Depends in part on duration of stimulus	From less than 1 to more than 20 millivolts	Graded	Local	Na^+/K^+	Environmental energy, often through several intermediate steps of transduction

[a]The properties shown are those of the most common types of chemical EPSPs and IPSPs. Other types have also been recorded but are not discussed in this book.

agate, undecremented, along electrogenic membrane. They result from the sequential opening of separate voltage-sensitive Na^+, and then K^+, channels.

2. *Electrical EPSPs* are local potentials. They result from the direct flow of current produced by the presynaptic action potential across an electrical synapse. They do not involve the opening of channels in the postsynaptic membrane, as occurs at chemical synapses.
3. *Chemical EPSPs* (or, at neuromuscular junctions, EPPs) generally, have a somewhat longer duration than electrical EPSPs. They are local potentials, graded in size, that result from the opening of chemically sensitive Na^+/K^+ channels, rather than from the opening of two separate classes of Na^+ and K^+ channels, as occurs with the action potential.
4. *Chemical IPSPs* are hyperpolarizing or depolarizing graded, local potentials that result from the opening of K^+, Cl^-, or K^+/Cl^- channels.
5. *Receptor potentials* are also graded, local potentials that result generally from the opening of Na^+/K^+ channels. The channel opening results from the transduction of some form of environmental energy.

There are also other types of neuronal signals (many of which have only recently been discovered) that have properties different from those described here. For the purpose of this book, however, the five types of neuronal signals listed above provide a sufficient background. Because these five types of signals are found in nearly identical form in practically all invertebrate and vertebrate animals studied, they appear to represent the major informational signals of the nervous system.

Figure 16 shows how three of the five types of signals we have discussed are distributed in a simple neuronal pathway. Shown here are receptor potentials, action potentials, and chemical EPSPs (as well as EPPs in a muscle cell). The pathway consists of a sensory neuron, a motor neuron, and a muscle cell. The sensory neuron, a stretch-sensitive mechanoreceptor, is stimulated three times by successively larger stretch amplitudes. This produces receptor potentials of three different sizes. The largest of these exceeds the threshold for producing an action potential in the sensory neuron. The detailed appearances of the signals recorded in various regions along this neuronal chain should now be familiar, based upon the foregoing discussion.

Extracellular Recording and Stimulating of Neurons

As we have seen, in order to record accurately a neuron's resting potential or any of its voltage signals, one must place one electrode inside the cell and the other outside. The electrodes thus span the voltage-producing membrane. This is also the most effective electrode placement for passing current to depolarize a neuron or, by reversing the current direction, to hyperpolarize the neuron. However, a cell of small diameter may be difficult to penetrate without damage. Moreover, in an experiment where the animal is to be permitted some freedom of movement, an intracellular electrode may easily become dislodged from the neuron. It is fortunate, therefore, that some information about a neuron's signals can

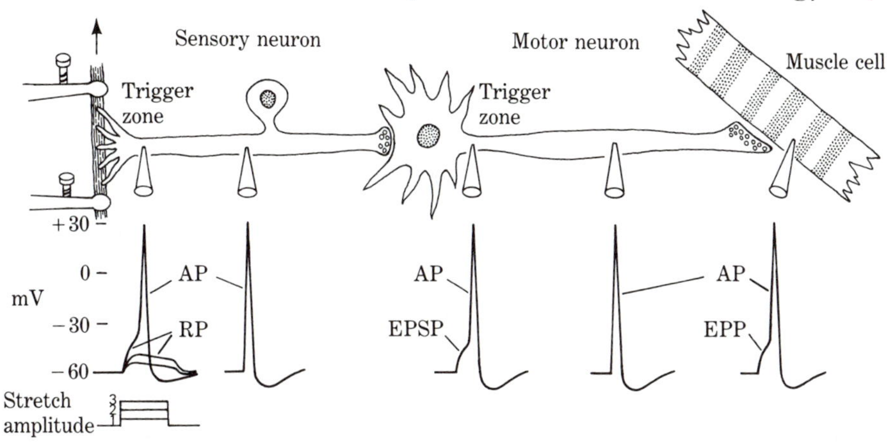

FIGURE 16. Neuronal signals in a simple synaptic pathway. The sensory neuron is stimulated by stretching its dendrites. On the largest of three stretches, the threshold voltage is reached and an action potential is evoked. The recorded signals further along the pathway are in response to the action potential produced by the largest stretch. This sensory neuron synapses on a motor neuron, which in turn synapses on a muscle cell. RP, Receptor potential; AP, action potential; EPSP, excitatory postsynaptic potential; EPP, end plate potential.

be obtained without intracellular impalement, simply by placing the electrodes extracellularly but near the cell. However, owing to the inefficiency of this technique, the only type of signal than can generally be recorded extracellularly is the action potential, which is many times greater in amplitude than other signal types (Table 1).

As shown in Figure 17, one can use as extracellular electrodes two wires hooked under an axon to record passing action potentials. At time 1 in the figure, the action potential has not yet arrived at the nearer electrode. Thus, with both electrodes sitting in the same extracellular cloud of positive charge, there is no potential difference between them, and thus no upward or downward deflection occurs on the oscilloscope. At time 2, the action potential has arrived at the left electrode. This electrode now sits in an extracellular charge cloud that, partially depleted of its Na^+ ions, is negative relative to the electrode on the right. This potential difference between the two electrodes is registered as a deflection on the oscilloscope. This is a downward deflection because the oscilloscope's positive terminal (to which the left electrode is connected) is sensing a negative voltage. At time 3, with the action potential between the two electrodes (and the resting potential restored at the site of the first electrode), there is again no potential difference between the electrodes. Thus, the oscilloscope trace returns to its original elevation. At time 4, the right electrode becomes negative relative to the left, so the oscilloscope trace deflects upward. And at time 5, with the action potential having passed both electrodes, the trace returns again to its original elevation.

Extracellularly recorded action potentials have voltages much smaller than those recorded intracellularly, where the two electrodes span the voltage-pro-

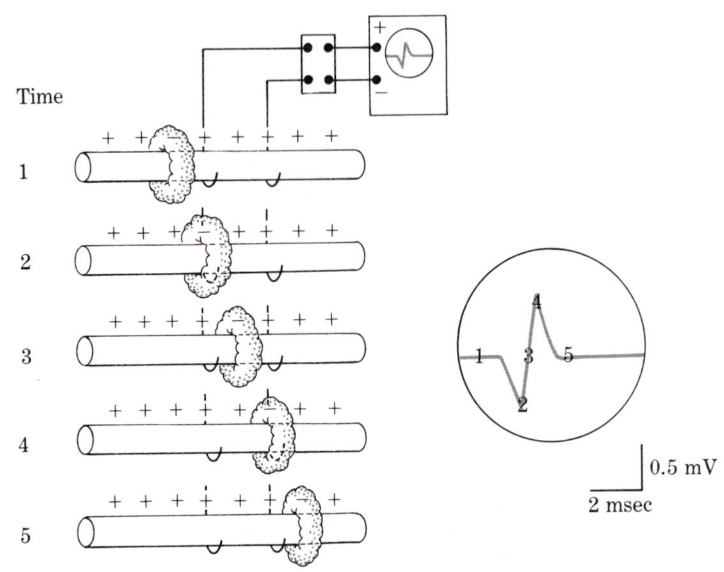

FIGURE 17. Extracellular recordings with hook electrodes of an action potential in a single axon. As the action potential passes, a small extracellular cloud of negative charge (ring) contacts first one electrode and then the other, producing on the oscilloscope successive deflections in opposite directions.

ducing membrane. Typical sizes are 0.1 to a few millivolts, only 1/10 to 1/100 the size of an intracellularly recorded action potential. Moreover, both the size and the shape of the action potential depend upon the exact placement of the extracellular electrodes. For instance, if there is a space between the axon and the electrodes, such that the wires do not sit directly in the extracellular charge cloud, the action potentials recorded will be much smaller. Also, in extracellular (but not in intracellular) recordings, the greater the diameter of the axon producing an action potential, the greater the voltage of the recorded signal. This effect of axon diameter on the size of the extracellular action potential has important experimental uses. For instance, if one records with hook electrodes from a small nerve containing relatively few axons, it is possible to ascribe the larger action potentials to the axons of larger diameter. In fact, where there are fewer than perhaps five axons in the nerve, as occurs in many small peripheral nerves of invertebrates, one can sometimes ascribe each recorded action potential to a particular axon, simply on the basis of relative sizes. Such identification at the level of individual neurons is generally impossible to achieve when recording from larger nerves contining many axons. The only exception is if one of these axons is of much greater diameter than the rest, giving it a uniquely identifiable large action potential.

In some experiments, one records not from axons in a nerve but from some point deep within the brain or the nerve cord. Here there can be hundreds or thousands of neurons packed densely in complex spatial patterns. But here too one can record extracellularly from single neurons, though it is generally impossible to specify from which particular cell one is recording. Such recordings are made with a microelectrode, typically one made of tungsten or some other metal, covered all along its length except at the tip with a thin coating of insulation. The tip diameter may be several micrometers, a size that, though larger than electrode tips used for intracellular impalement, is sufficiently small to permit the tip to rest just next to one neuron. An electrode so placed will record the action potentials of this nearby cell as much larger than those from other, more distant cells (Figure 18). The second electrode, connected to the opposite terminal of the oscilloscope, is placed at some distant point on the animal's body and thus is not affected by the voltage signals in any of the neurons. Rather, this second electrode serves as a reference against which the voltage changes detected by the microelectrode are registered.

One can use extracellular electrodes, not only to record action potentials, but also to stimulate neurons. Recall that to stimulate a cell, one must make positive charges accumulate on the inside of its plasma membrane. If one connects two extracellular electrodes to a battery, or to a similar voltage source such as an electronic stimulator, ions will pass between the tips of the two electrodes. Most of the ions that flow travel on a direct path between the electrodes and therefore remain outside the neuron. However, some lesser proportion of the ions will flow in less direct pathways, and some of these will enter and then leave a nearby neuron (Figure 19). Positive charges (mostly K^+) will enter the neuron near the tip of the positive electrode and will exit near the tip of the negative electrode. However, before actually passing out through the membrane, some of the positive ions pile up inside the membrane near the negative electrode. Such an internal accumulation of positive charges, as we have seen, contributes to the depolarization of the membrane. If this depolarization reaches threshold, an action potential will be evoked and thus will propagate along the axon.[11]

Neurons deep within the brain can also be stimulated electrically. For this purpose, one generally uses an extracellular microelectrode, whose tip is placed in the region of interest. The principle is the same as for stimulating axons in a nerve; positive charges enter a neuron at one point and then accumulate on the inside of the membrane prior to leaving the cell at another point. It is at this point of exit that the depolarization occurs.

In using extracellular electrodes to stimulate neurons, if one increases the strength of stimulating current, additional neurons that are further from the electrode tips will be depolarized. Therefore, a greater number of neurons may be induced to give action potentials. Thus, one can exercise some rough control over the number of cells excited by the stimulating current, simply by varying the strength of this current. In Chapters 5, 7, 8, and 9, we will encounter experiments that utilize these extracellular stimulating techniques.

INTEGRATION WITHIN A NEURON

As we have seen, an EPSP or IPSP originates in a postsynaptic neuron right at the location of a synapse (Figures 10, 11, and 12). As it spreads away

[11]Note that there is also a pileup of positive charges *outside* the membrane at the positive electrode, prior to their entry into the cell. This results in a hyperpolarization of the membrane in this region. However, if the two electrodes are spread sufficiently far apart, this hyperpolarization does not interfere with the production of an action potential near the negative electrode.

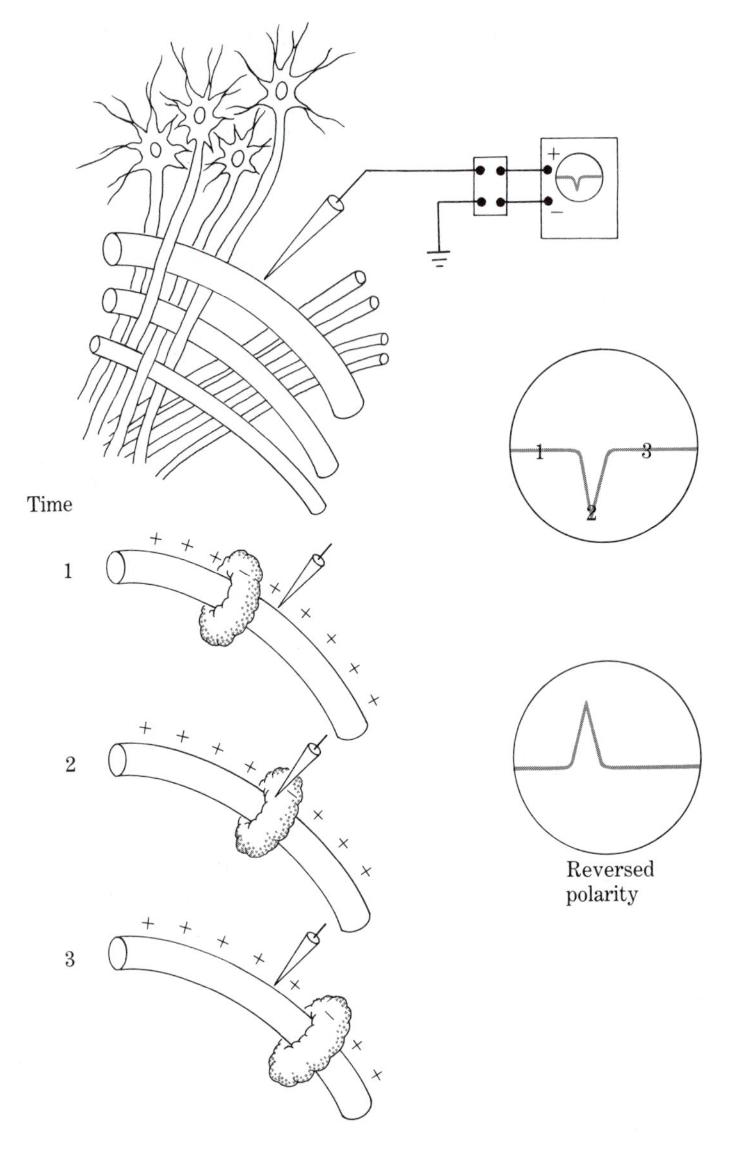

from the synapse, its voltage decreases, reaching zero within only a short distance. This spread is often called ELECTROTONIC, or PASSIVE DECAY. This is shown in Figure 20 by three hypothetical recordings of the same EPSP at three different locations along a dendrite. The electrotonic decay has an exponential shape. This relationship of EPSP size to distance along the neurite is quantified by a parameter called the LENGTH CONSTANT or SPACE CONSTANT. As derived mathematically, the length constant is the distance required for a postsynaptic potential (or any other subthreshold, passively decaying voltage) to drop to $1/e$, or 37%, of its height at the point of origin. In the example of Figure 20, the length constant is 0.5 millimeters (mm). That is, because the EPSP has a height of 10 mV right at the synapse, at 0.5mm the height is only 3.7 mV. And at 1mm from the synapse, it is only 37% of 3.7, or about 1.4 mV, and so forth. Thus, within roughly three

◀ FIGURE 18. Extracellular recording, by means of a microelectrode, of an action potential from an axon that courses through a meshwork of neurons. The action potential is depicted as a moving extracellular cloud of negative charge. The axon producing the action potential is drawn separately three times, representing three successive moments in time. At time 2, when the microelectrode senses the negative charge cloud, the oscilloscope trace deflects downward. At time 3, the trace returns to its original position. Because action potentials, by convention, are usually represented as upward deflections, one often chooses to interchange the wire connections from the two electrodes to the oscilloscope, thus reversing the action potential's polarity. If the tip of the extracellular microelectrode is sufficiently close to more than one neuron and falls within the moving charge clouds of each, it will record the action potentials of each. The size and detailed shape of the action potentials from each of these neurons may differ, depending upon the relative positions of the electrode tip and the cells. These differences in appearance can often be used to determine which of the action potentials belongs to the same cell and which to a different cell. However, there is often considerable uncertainty in this judgment.

length constants, which for this particular dendrite equals 1.5 mm, the EPSP has decayed almost to zero.

Not all neurites have the same length constant. One determining factor is the neurite's diameter: the greater the diameter, the greater the length constant, and so the farther a subthreshold signal is able to spread. The basis of this relationship is that a thicker neurite offers less resistance to the longitudinal movement of intracellular ions, enabling these ions to travel a greater distance down the neurite.

The electrotonic decay that an EPSP undergoes in spreading from its synapse of origin to the neuron's trigger zone limits the effectiveness of this EPSP in producing action potentials. For an EPSP to be maximally effective, then, its synapse should be located right at the trigger zone. As a next best alternative, the synapse should be close to the trigger zone, with the connecting neurite having a large diameter. The farther the synapse is from the trigger zone and the narrower the connecting neurite, the less effective the EPSP.[12]

[12]As Figure 20 shows, when a voltage signal such as an EPSP spreads electrotonically, it becomes not only smaller in amplitude, but also spread out in time. Thus, EPSPs that originate on distant dendrites produce small, slowly varying voltage changes at the neuron's trigger zone. Their effect is more to modulate than to drive directly the cell's impulse activity.

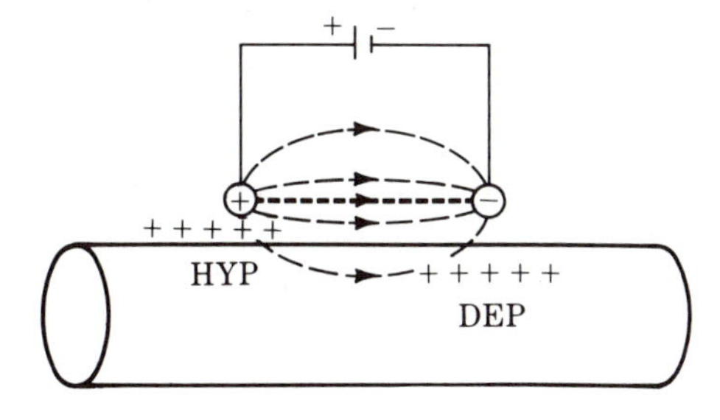

FIGURE 19.

Extracellular stimulation of an axon. Upon connecting a battery or stimulator (at top) to two electrodes, positive ions flow along the paths shown by the dashed lines. Positive ions entering the axon hyperpolarize it (HYP) and those leaving the axon depolarize it (DEP).

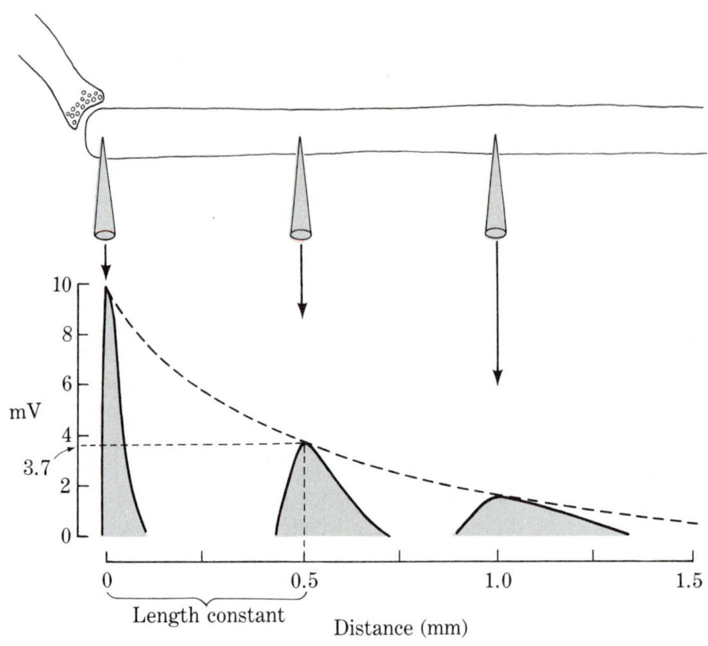

FIGURE 20. Spatial decay of postsynaptic potentials. Three recordings of the same hypothetical EPSP, at three different distances from its origin at the end of a dendrite. The height of the EPSP as a function of distance is graphed. The distance required for the EPSP (or any other passively decaying voltage) to decay to 37% (or $1/e$) of its original height is defined as one length constant of this dendrite. It can also be seen that as EPSPs, or other passively decaying voltage signals, move away from their source, they spread out in time.

Electrotonic decay also introduces problems of interpretation of recorded postsynaptic potentials. For instance, if we impale a dendrite and record there a large EPSP, we may be tempted by this large size to regard the signal as having a fairly strong effect in evoking action potentials in this cell. However, the large size may simply reflect the proximity of the synapse of origin to our recording electrode. If the trigger zone is far way, the EPSP would greatly decay before reaching it, and thus its effect on action potential production would be small. By contrast, if we record with this same electrode a small EPSP, we may be tempted to regard it as of minimal effect on the production of action potentials. Yet this EPSP may have originated near the trigger zone and, there, may have been quite large and of substantial effect. Because we would have no way of knowing the location of the synapse of origin of the recorded EPSP, we could not use the size of this signal as a guide to its effectiveness on the distant trigger zone.

In order to interpret the effectiveness of an EPSP in evoking action potentials, then, it would be best to make one's recording right at the neuron's trigger zone. However, one does not know *a priori* for a given cell where the trigger zone is located. Moreover, it can be a difficult task to find out, requiring recordings from several different locations in the same neuron. In fact, given the small diameters of most of a cell's neurites, one often is restricted to making recordings from the cell body and possibly one or two especially large neurites, not necessarily near the trigger zone. One must thus bear in mind, when interpreting intracellular recordings, these limitations imposed by electrode location.

A synapse whose individual EPSPs each is large enough at the trigger zone to evoke by itself an action potential is called a RELAY SYNAPSE. Relay synapses are very rare. More commonly, several EPSPs are required to evoke an action potential. When two or more EPSPs from *different* synapses occur simultaneously within a neuron, their voltages add together, a process called SPATIAL SUMMATION (Figure 21A). Because each EPSP lasts for many milliseconds, it is also possible for two successive EPSPs from the *same* synapse to add their voltages together, a process called TEMPORAL SUMMATION (Figure 21B). By virtue of spatial and temporal summation, EPSPs that are unable by themselves to depolarize a trigger zone to threshold can bring the cell to threshold when they act together. Both spatial and temporal summation can occur with IPSPs as well as EPSPs (Figure 21C and D). Of the hundreds or thousands of input synapses that most neurons receive, a substantial proportion are likely to be active at any given instant, their PSPs summing both spatially and temporally. Thus, a neuron's transmembrane potential in the general region of its input synapses, including its trigger zone, is constantly fluctuating, rather than holding steady at resting potential. Consequently, whether the trigger zone reaches threshold and produces an action potential depends upon the summed activity of many synapses. If the summation

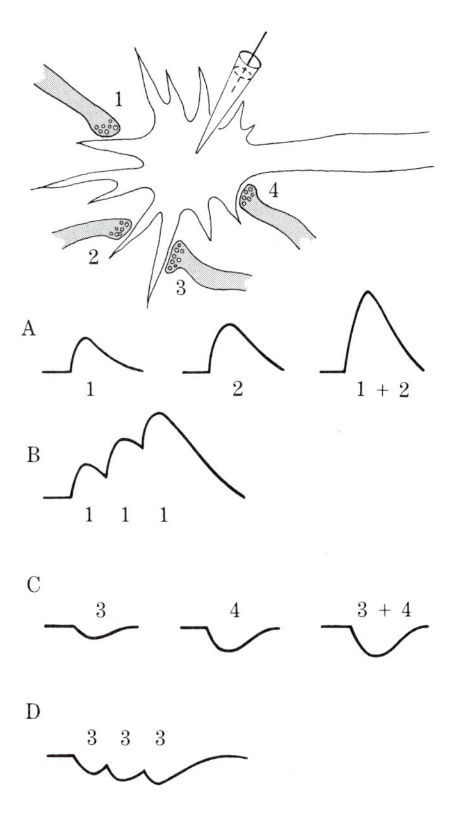

FIGURE 21.

Summation of synaptic potentials. A postsynaptic neuron is shown with synapses from four presynaptic neurons. Synapses 1 and 2 are excitatory; synapses 3 and 4 are inhibitory. A. The EPSP from synapse 1 alone, then from 2 alone, then spatial summation of 1 and 2 together. B. Temporal summation of three EPSPs from synapse 1. C. The IPSP from synapse 3 alone, then from 4 alone, then spatial summation of 3 and 4 together. D. Temporal summation of three IPSPs from synapse 3.

is great enough and prolonged enough to depolarize the trigger zone for some time to well *beyond* the threshold voltage, a vigorous barrage of action potentials will generally be evoked. It is just such a barrage, or burst, of activity that a neuron typically shows when it is engaged in evoking a behavioral act.

Summary

Though neurons vary greatly in structure, almost all are characterized by the presence of both a cell body and neurites. Microinjection of visualizable substances enables one to observe the geometry of a neuron, including its complex dendritic tree. The functional contact points between neurons are of two types: electrical synapses and chemical synapses. These differ greatly in both structure and function.

Neuronal signals are based upon changes in the permeability of the plasma membrane to specific types of ions. These changes result from the opening of specific active ion channels in the membrane. In the resting neuron, a resting potential of about -60 millivolts results primarily from the permeability of the membrane to K^+ ions. This resting potential constitutes an electric gradient across the membrane, which exerts a force on all ions: positive ions experience an inward force and negative ions an outward force. In addition, each type of ion experiences a second force, based upon its own chemical gradient across the membrane. These two forces, electrical and chemical, add together to produce the driving force, which controls the movements of a given ion that occur if and when a membrane channel opens and permits this type of ion to pass. Action potentials, chemical postsynaptic potentials, and receptor potentials all involve the opening of specific ion channels.

Excitatory postsynaptic potentials (EPSPs) sum together and, if they depolarize the trigger zone to threshold, evoke an action potential. Inhibitory postsynaptic potentials (IPSPs) counteract this threshold depolarization. The closer a synapse is to the trigger zone, the greater its effectiveness on the production of an action potential.

Questions for Thought and Discussion

1. Suppose that you are studying a synaptic connection between two large neurons, each of which you can impale with microelectrodes. Having established that an action potential in one of these neurons produces a postsynaptic potential in the other, you now wish to find the *location* of the synapse on the postsynaptic cell's dendritic tree and/or cell body.
   A. Design a set of observations or experiments using strictly *anatomical* techniques that you think might solve this problem. Remember that synapses can be seen only under the electron microscope. Indicate how you would identify the presynaptic and the postsynaptic neurons from among all the other neurons in the nervous system, when viewed under the electron mi-

croscope. How would you determine which part of each neuron you were viewing under this microscope?

B. Design a set of experiments using *physiological* techniques that you think might solve this problem. Assume that several different parts of the postsynaptic neuron are large enough to impale with microelectrodes.

2. In the diagram below, S represents a pair of wire hooks used to electrically stimulate the two axons drawn. One axon is many times greater in diameter than the other. A stimulus pulse having a duration of about 0.2 milliseconds evokes an action potential in each axon, which is then conducted along the length of each axon. Four pairs of recording electrodes (three extracellular and one intracellular) are distributed along the length of the axons. Each electrode pair is connected, through an amplifier, to a different trace of a four-trace oscilloscope. Make a single drawing of the four traces to show accurately the appearance on the oscilloscope screen of the action potentials that would be recorded by each of the four pairs of electrodes. Label the vertical and horizontal axes. Include everything that you think would appear on the oscilloscope trace.

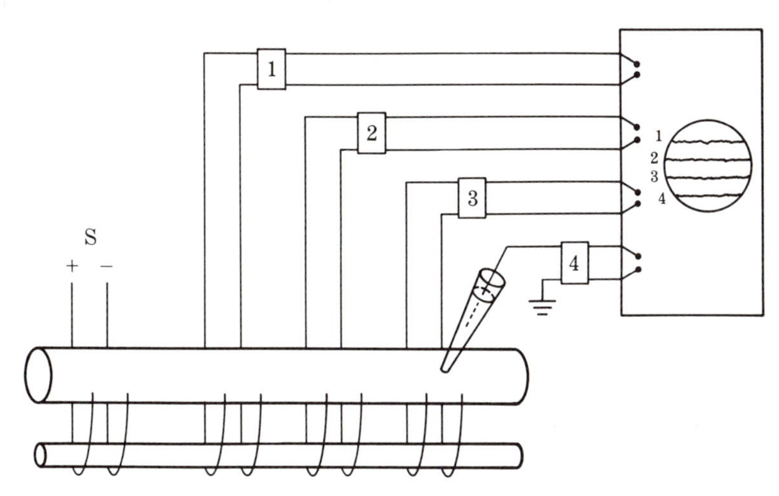

3. List as many factors as possible that you think would affect the amplitude of an action potential from a single axon as recorded with extracellular hook electrodes wrapped around a peripheral nerve. Explain briefly why each of these factors affects the amplitude.

4. The diagram below shows an intracellular recording from the cell body of an invertebrate motor neuron. The plasma membrane of this cell body is not electrogenic, so that any action potentials recorded in the cell body must have decayed electrotonically as they spread from the nearest patch of electrogenic membrane.

A. What do you suppose the four sharp peaks are? What experiment could you do to verify your supposition?

B. Is synaptic inhibition demonstrated? If so, where on the trace?

C. Is synaptic summation demonstrated? If so, where on the trace?

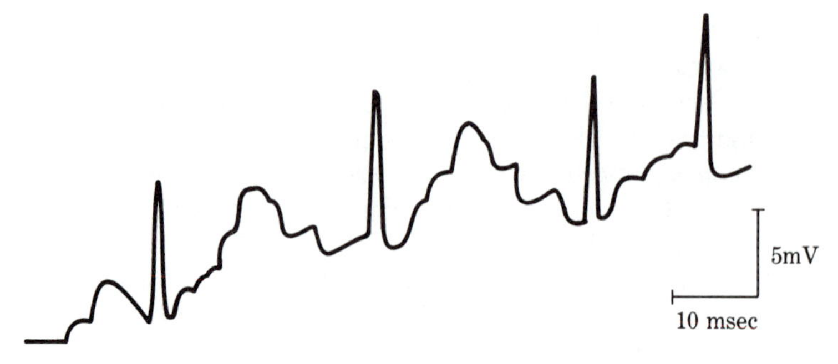

D. Is synaptic facilitation demonstrated? If so, where on the trace?
E. Is synaptic defacilitation demonstrated? If so, where on the trace?
F. What is the threshold depolarization of the cell?

5. You have impaled each of two neurons with recording electrodes. You stimulate one and the other gives the response shown in the figure below. Stimulation at high frequency gives rise to summation and to a full, overshooting action potential (not shown in the diagram). Devise a hypothesis to explain the shape of the response shown, and devise a set of experiments that would convincingly support or refute your hypothesis.

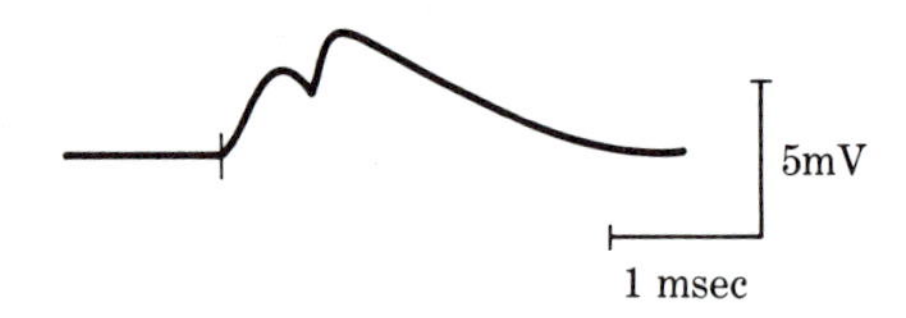

Recommended Readings

BOOKS

Heimer, L. and RoBards, M. J. (1981) *Neuroanatomical Tract-Tracing Methods*, Plenum Press, New York.
This book surveys many current neuroanatomical methods of interest to neuroethologists.

Kandel, E. R. (1976) *Cellular Basis of Behavior*. W. H. Freeman and Company, San Francisco.
This excellent book covers most of the major principles of cellular integration. Most of its examples are drawn from work on a single animal, the mollusk *Aplysia*.

Kandel, E. R. and Schwartz, J. H. (eds.). (1981) *Principles of Neural Science*. Elsevier/North-Holland, New York.
A general text which leans slightly toward the medical aspects of neurobiology.

Kuffler, S. W. and Nicholls, J. G. (1976) *From Neuron to Brain*. Sinauer Associates, Sunderland, Massachusetts.
This excellent text stresses the functioning of individual nerve cells, exploring in depth the experimental basis for our understanding of neural functioning.

Schmidt, F. O. and Worden, F. G. (eds.). (1979) *The Neurosciences: Fourth Study Program.* M.I.T. Press, Cambridge, Massachusetts.
A huge volume containing introductory treatments by leaders in a variety of neurobiological subdisciplines.
Scientific American. (1979) *The Brain. Sci. Am.* 241(3).
A collection of excellent introductory articles on a variety of neurobiological subjects.

Appendix: Tests on Synaptic Connections

Suppose that one records an action potential in one neuron and a resulting postsynaptic potential in another neuron. Often an experimenter's next step would be to characterize the synaptic connection between these two cells. Three specific questions generally arise: (1) Is the connection MONOSYNAPTIC (that is, is the first neuron synapsing *directly* on the second) or POLYSYNAPTIC (are one or more neurons interposed between these two cells)? (2) Is the connection a chemical synapse or an electrical synapse? (3) Is the synaptic potential excitatory or inhibitory? Answering these three questions is important if one wishes to define clearly the nature of the interaction between these two neurons, a necessary step in understanding the way that these cells might function to regulate behavior. We will examine here some of the more important experimental techniques employed to solve these three problems.

MONOSYNAPTIC VERSUS POLYSYNAPTIC CONNECTION

In practice, it is surprisingly difficult to distinguish between a monosynaptic and a polysynaptic connection. The experimental problem is to discern between the situations shown in the left versus the right panel in Figure 22A, that is, to determine whether there exists a neuron (or more than one) interposed between the cells (1 and 2) from which one is recording. If one or more neurons are interposed, this indicates that there are as yet unknown cells in the circuit, which thus is not yet fully understood. Therefore, an ultimate goal is to be able to specify each neuron and each monosynaptic connection of the neural circuit.

If one could actually see the neurons and their neurites, the problem might be somewhat simplified; one would just look to see whether or not cell 1 contacts cell 2. But, in general, one cannot see the neurons recorded from because they are often small and usually transparent and are embedded among many other cells. Moreover, the synapses themselves are too small to see without using an electron microscope. In fact, even if one were to find anatomical evidence of a monosynaptic connection, this would not prove that the particular structures found actually *function* to produce the observed postsynaptic potential. This PSP could instead result from some other, perhaps polysynaptic, connection between the two cells.

There are five physiological tests that are generally employed to help distinguish between monosynaptic and polysynaptic connections. As we shall see, there are limitations associated with each of these tests (especially the first two presented below). Taken as a group, however, these tests can offer substantial evidence pointing toward or away from a monosynaptic connection.

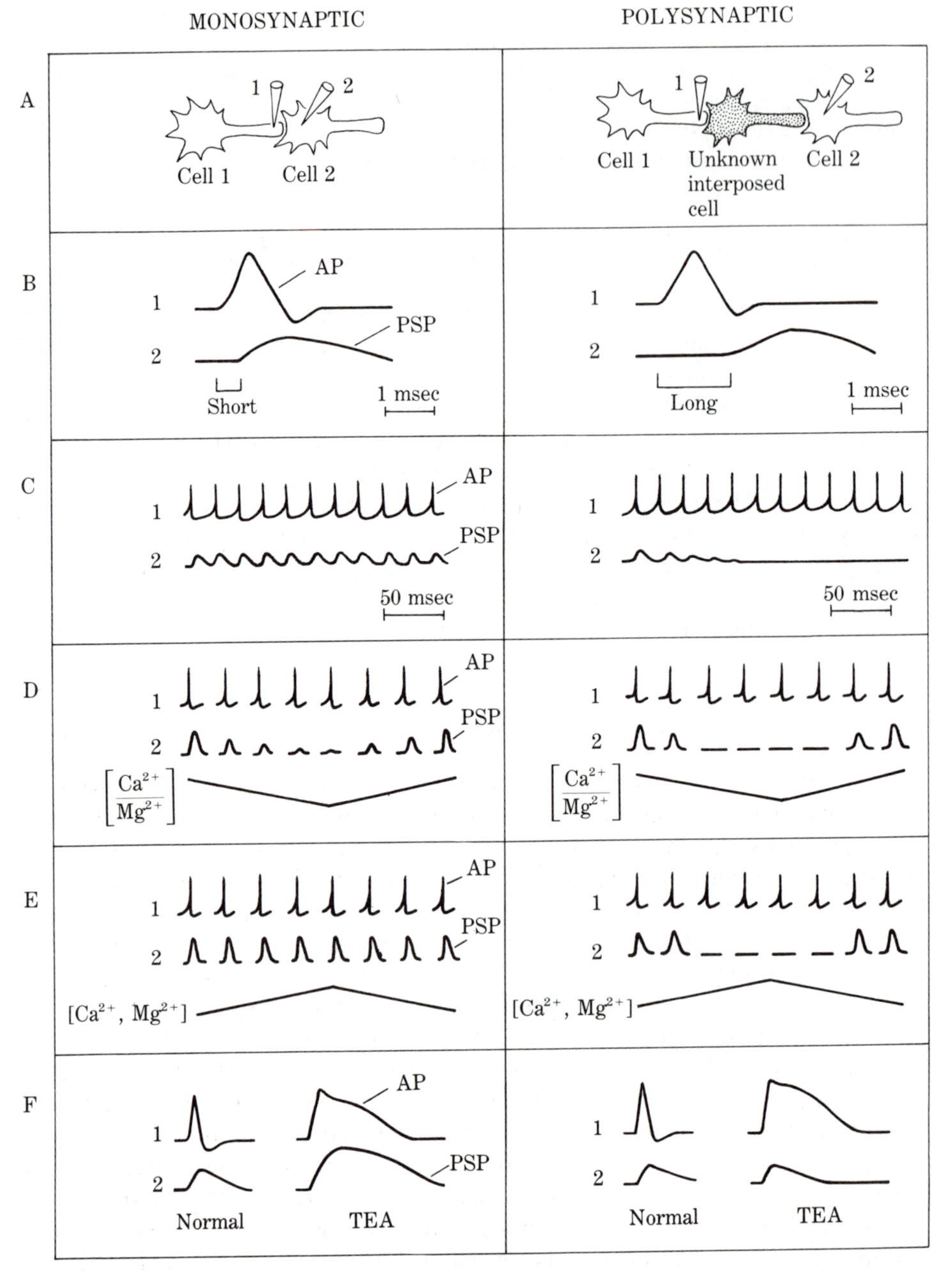

FIGURE 22. Five tests for distinguishing between a monosynaptic and a polysynaptic connection. (See text for explanation.)

The first test is to measure the synaptic delay—the interval from the arrival of an action potential at the axon terminal of cell 1 to the start of the postsynaptic potential in cell 2 (Figure 22B). This measurement could be made directly from the recordings by electrodes 1 and 2 in Figure 22A.[13] If the delay is very short, a monosynaptic connection is suggested because there might be insufficient time for the action of an interposed neuron. But how short is short enough to be certain? Recall that synaptic transmission at electrical synapses is briefer than at chemical synapses. Typical measured synaptic delays are from 0 to 0.2 milliseconds for an electrical synapse and 0.3 to 0.7 milliseconds for a chemical synapse. Suppose, then, that the measured synaptic delay is 0.5 milliseconds. This could be a monosynaptic chemical connection. But it could also be a DISYNAPTIC connection (one with a single interposed cell, as in Figure 22A, right) in which both synapses are electrical. This is possible especially if the interposed neuron is one that is short and does not produce action potentials, so that this time-consuming step is eliminated. Thus, although the recorded synaptic delay might be suggestive of a monosynaptic connection, it is by no means conclusive.

The second test is to measure the degree of faithfulness of the response in cell 2 to a prolonged, high-frequency barrage of action potentials in cell 1 (Figure 22C). The underlying principle is that the more intervening physiological processes lie between electrodes 1 and 2 (especially chemical synapses, which often exhibit defacilitation in response to high-frequency presynaptic action potentials), the greater is the likelihood that the postsynaptic potential at electrode 2 will fail. In practice, the ability of these postsynaptic potentials to follow one-for-one action potentials in cell 1 at frequencies above perhaps 50 action potentials per second can be taken as supporting, but not conclusive, evidence for a monosynaptic connection.

A third test involves the replacement of the Ca^{2+} ions in the bathing medium with magnesium ions (Mg^{2+}).[14] Recall that chemical synapses require the presynaptic entry of Ca^{2+} for the release of the neurotransmitter. Mg^{2+} ions can enter through the Ca^{2+} channels but do not induce transmitter release. Thus, by replacing the Ca^{2+} of the bathing fluid with Mg^{2+}, one blocks chemical synaptic transmission.[15] In the present experiment, the Ca^{2+} is replaced by Mg^{2+}, not suddenly, but rather gradually. The expected result, if the connection is monosynaptic, is a *gradual* decrease in the size of the postsynaptic potential, reaching zero only after the extracellular Ca^{2+} concentration has been reduced to near zero. At this concentration, no neurotransmitter molecules are being released in response to the presynaptic action potentials. Gradually restoring the normal Ca^{2+}:Mg^{2+} ratio would gradually return the postsynaptic potential to its normal size (Figure 22D, left panel).

[13]In practice, it is usually impossible to impale an axon terminal with a microelectrode because these terminals are usually tiny. In general, then, one must record from the cell body or some other fairly large region of the neuron and calculate the synaptic delay, having measured the speed of conduction of the action potential from the recording site to the axon terminal.

[14]The reason that the removed Ca^{2+} is replaced by another type of ion (Mg^{2+}) is that if the overall concentration of the bathing fluid were to be reduced, an osmotic disturbance to the cells would result.

[15]An alternative way to achieve this blockage is to leave the Ca^{2+} and Mg^{2+} concentrations unchanged but to add cobalt ions to the bathing fluid. Cobalt binds to, and blocks, the Ca^{2+} channels and thus achieves the same goal—blockage of chemical synaptic transmission.

If the connection were polysynaptic, one would expect in this test at first a gradual drop and then a *sudden* drop to zero in the size of the postsynaptic potential (Figure 22D, right). To understand this, let us assume that the synapse from cell 1 to the interposed neuron (Figure 22A, right) is a chemical synapse, as are most synapses, and that the interposed cell, like most neurons, uses action potentials. When one begins gradually to lower the Ca^{2+}:Mg^{2+} ratio, the EPSP in the interposed cell (not recorded) would gradually become smaller. At some instant, these EPSPs would become too small to evoke action potentials in the interposed cell. Thus, at this moment, the postsynaptic potential in cell 3 would suddenly fail to appear. Upon gradually restoring the Ca^{2+}:Mg^{2+} ratio, the postsynaptic potential in cell 3 would suddenly reappear when the EPSP in the interposed cell again became large enough to evoke an action potential.

Thus, the criterion for a monosynaptic connection, using this test, is a purely gradual and not a precipitous drop in the size of the recorded postsynaptic potential. And for a polysynaptic connection, it is a gradual *and then a sudden* drop to zero of the PSP. A limitation of this test is that it requires that the synapses be chemical, not electrical, and that the interposed neuron use action potentials. Because the interposed neuron is an unknown entity (in fact, whether or not it exists is just what the experiment is supposed to find out), one cannot verify whether such assumptions about this cell are valid. Though this is a serious limitation, the fact that most synapses are chemical and most neurons use action potentials permits one to use this test with a modicum of confidence.

A fourth test involves *increasing* both the Ca^{2+} and the Mg^{2+} concentrations in the bathing solution, while keeping the *ratio* of Ca^{2+} to Mg^{2+} concentrations fixed. With this fixed relationship, there is no effect on chemical synaptic transmission because the ratio of activation to blockage of synaptic transmission is held constant. What is affected is the *threshold voltage* of nerve cells; virtually all neurons tested show an elevation of threshold when placed in high Ca^{2+} concentrations, irrespective of the Mg^{2+} concentration. Thus, a larger depolarization is now required to evoke an action potential. Consequently, all the cells—numbers 1 and 2 and the interposed cell if it exists—will experience an elevated threshold. (For this reason, throughout this experiment one may need to give especially strong depolarizing stimuli to cell 1 in order for this cell's transmembrane potential to reach its elevated threshold and thus produce the action potentials needed to carry out the test.)

Here are the expected results of this test. If the connection is monosynaptic, no change is expected in the postsynaptic potential of cell 2 (Figure 22E, left). This is because, with normal action potentials being evoked by strong electrical stimuli to cell 1 and with the synapse from cell 1 to cell 2 unaffected, all physiological events leading up to the postsynaptic potential have remained unchanged. But if the connection is polysynaptic, one would expect a sudden drop to zero in the size of the postsynaptic potential in cell 2 (Figure 22E, right). The reason is that, with the threshold of the interposed cell gradually increasing, at some moment the EPSPs in this cell would suddenly fail to reach this elevated threshold and thus would fail to excite an action potential. Cell 2 would then suddenly fail to receive any input. Restoring the normal Ca^{2+} and Mg^{2+} concentrations should lead to a sudden restoration of the postsynaptic potential in cell 2. In this test, then, one assumes that the interposed cell is one that uses action potentials, but one does not need to assume that the synapses are chemical.

A final test involves the injection into cell 1 of the drug tetraethylammonium (TEA). This drug binds with, and blocks, the active K^+ channels that open during the action potential. As a result of this drug, then, K^+ leaves the neuron much more slowly, so that the duration of the action potential is very greatly lengthened. Because the transmembrane potential of cell 1 thus remains depolarized longer for each of its action potentials, more neurotransmitter is released from its chemical synapses, producing a larger and more prolonged postsynaptic potential.

Here, then, are the expected results of this test. If the connection is monosynaptic, the postsynaptic potential in cell 2 should become larger and more prolonged after the TEA injection into cell 1 (Figure 22F, left). If the connection is polysynaptic, no change in the postsynaptic potential of cell 2 is expected (Figure 22F, right). This is because the enlarged EPSP in the interposed cell, once it reaches threshold, would evoke an action potential in this cell, which would be a *normal* action potential, because the interposed cell has not been injected with TEA. This normal action potential would then evoke a normal EPSP in cell 2.

For this experiment, one must assume that the synapse from cell 1 is chemical because TEA is known to diffuse across electrical synapses. Its diffusion into an interposed cell would lengthen this cell's action potentials and thus invalidate the result. Moreover, one must assume that the interposed cell, if it exists, is a long neuron that uses action potentials. Otherwise, the enlarged EPSP in the interposed cell would directly evoke an enlarged postsynaptic potential in cell 2.

CHEMICAL VERSUS ELECTRICAL SYNAPSES

Suppose that we have identified a particular connection as monosynaptic and that we now wish to determine whether the synapse is chemical or electrical. As we have seen, chemical and electrical EPSPs are sufficiently similar in appearance (Figures 10 and 11) that special tests would be required to make this determination.

Some hint on this matter may be obtained, however, by measuring the synaptic delay. As we have seen, synaptic delays of less than about 0.2 milliseconds imply an electrical synapse and those over about 0.3 milliseconds imply a chemical synapse. However, actual measured synaptic delays usually are longer than these values because the measurements include some time for conduction of the presynaptic potential to the axon terminal. This introduces sufficient uncertainty into the measurement that one simply cannot be confident on the basis of synaptic delay alone whether a chemical or an electrical synapse is involved.

A further hint can be obtained by observing whether the postsynaptic potential follows faithfully presynaptic action potentials that repeat at more than perhaps 50 per second. At these fairly high frequencies, many chemical synapses fail to transmit with complete faithfulness, whereas most electrical synapses continue to function normally. However, this observation also is at best suggestive.

A more useful test is to replace the Ca^{2+} in the bathing medium with Mg^{2+} and see whether this blocks synaptic transmission (see Footnotes 14 and 15 above). Because chemical, but not electrical, transmission requires a presynaptic influx of Ca^{2+}, synaptic blockage resulting from removal of Ca^{2+} implies a chemical synapse. An absence of blockage, however, might result from a failure of the calcium-free solution to reach the particular synapse within the time of the experiment. This would need to be checked by showing that other, neighboring chemical synapses were in fact blocked.

The most compelling evidence, however, is derived from a test designed to determine whether the postsynaptic potential exhibits a phenomenon called a REVERSAL POTENTIAL, or INVERSION POTENTIAL. Reversal potentials occur at chemical, not electrical, synapses. To understand this concept, imagine that we have two electrodes in the postsynaptic neuron, one for recording an EPSP and the other for passing current to depolarize the cell. First, while *not* passing any current to depolarize the postsynaptic cell, we stimulate the presynaptic cell to give an action potential and we record postsynaptically an EPSP of, say, 15 mV (Figure 23, bottom trace). Now we repeat the presynaptic stimulus, but just before doing so, we pass a current to depolarize the postsynaptic cell, perhaps to −40 mV. We now find that the EPSP is reduced in size to perhaps 10 mV (Figure 23, second-to-bottom trace). The reason is that our depolarization of the postsynaptic cell to −40 mV has altered this cell's electrical gradient and in so doing has altered the driving forces on Na^+ and K^+ ions. The inward driving force on Na^+ is reduced and the outward driving force on K^+ is increased. Therefore, when the synaptic Na^+/K^+ channels open, fewer Na^+ ions enter and more K^+ ions leave than if we had not depolarized the neuron. This leads to a reduced EPSP.

Now we evoke a third presynaptic action potential, but prior to this one we depolarize the postsynaptic cell even more.[16] The more we depolarize, the lower the inward sodium driving force and the greater the outward potassium driving force, and, therefore, the smaller the EPSP. Ultimately, we reach a transmembrane potential at which the driving forces on Na^+ and K^+ are equal and opposite. At this potential, when the synaptic Na^+/K^+ channels open, an equal number of Na^+ ions enter as K^+ ions leave the cell. Thus, there is no voltage change, that is, the EPSP has been reduced to zero (Figure 23, third trace from top). If we depolarize the postsynaptic cell even beyond this voltage, the outward driving force on K^+ becomes greater than the inward driving force on Na^+. Therefore, when the Na^+/K^+ channels open, more K^+ ions leave than Na^+ ions enter. Thus, the synaptic potential is now in the hyperpolarizing direction, the reverse of a normal EPSP. The particular value of transmembrane potential at which the EPSP reverses its direction is called the reversal potential. It equals about −10 mV (or about 50 mV depolarized from rest) for most chemical EPSPs.[17]

[16]If one depolarizes a neuron gradually, by gradually increasing the depolarizing current, the neuron's threshold increases, a process called ACCOMMODATION. The more one depolarizes, the higher the threshold becomes. As a result, the membrane potential may never actually reach threshold. In this manner, very large depolarizations can be achieved without producing any action potentials. This is why in Figure 23, even when the membrane is depolarized all the way to +20 mV, the neuron is not firing action potentials.

[17]Chemical inhibitory as well as chemical excitatory synapses have a reversal potential. For an inhibitory synapse, the reversal potential is always at a voltage more hyperpolarized than the cell's threshold voltage.

FIGURE 23.

A demonstration of the reversal potential of a chemical EPSP. (See text for explanation.)

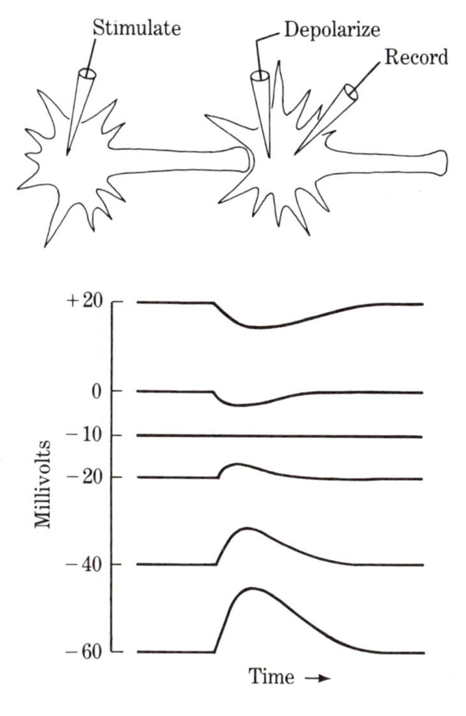

Electrical EPSPs do not show a reversal of direction in an experiment like that of Figure 23. That is, they have no reversal potential. This is because the electrical EPSP, rather than resulting from an opening of active channels for Na^+ and K^+, results from a direct flow of current from the presynaptic to the postsynaptic neuron. No matter what the depolarization of the postsynaptic neuron, there is no way that a presynaptic action potential could induce a flow of positive ions from the postsynaptic to the presynaptic neuron. Therefore, there is no way that an electrical EPSP could become a hyperpolarization in response to depolarizing the postsynaptic cell. To summarize, then, if the synapse in question is chemical, one should find a reversal potential, and if it is electrical, one should not.

EXCITATORY VERSUS INHIBITORY SYNAPSES

Because EPSPs and depolarizing IPSPs look alike (Figures 11 and 13A), one must carry out a test to determine whether a given depolarizing postsynaptic potential is excitatory (EPSP) or inhibitory (IPSP). As a reminder, by excitatory we mean that the PSP contributes to the production of an action potential; and by inhibitory, we mean that the PSP contributes to counteracting the production of an action potential. (If the postsynaptic cell does not use action potentials, then the criterion becomes the production or counteraction of action potentials in a neuron onto which this cell synapses.) As Figure 13A and B shows, if a depolarizing IPSP occurs during a large depolarization from some other source (in this case, from a large EPSP), the IPSP reverses its direction. This can be understood as an outcome of the reversal potential of the IPSP.[17] By contrast, an EPSP occurring during another EPSP would remain in the depolarizing direction; and if the summed depolarization of these two EPSPs were to reach threshold, it would evoke an action potential.

A test, then, to distinguish between an EPSP and a depolarizing IPSP is to depolarize the postsynaptic cell to just below threshold and to observe whether the postsynaptic potential now reverses direction. The most controlled way to depolarize this cell is to pass depolarizing current through an intracellular microelectrode; a second microelectrode is then used to record the postsynaptic potential.

This same experimental arrangement can provide two additional, useful pieces of information. First, if a presynaptic action potential evokes no postsynaptic potential in a normal resting neuron, we have seen that there could still be strong inhibition (Figure 13C and D). Such inhibition would be revealed as a hyperpolarizing IPSP when the postsynaptic cell has been depolarized to just below threshold. Second, if the postsynaptic potential is an EPSP, one often wishes to test whether this EPSP is large enough and/or close enough to the trigger zone to actually contribute at all to evoking an action potential. By depolarizing the postsynaptic cell with a microelectrode to *just* below threshold and then immediately exciting the presynaptic cell one can observe whether or not the EPSP produced is adequate to evoke an action potential. If so, this demonstrates that at least some synaptic current from this synapse reaches the cell's trigger zone.

CHAPTER 4

A Case Study in Neuroethology: The Escape System of the Cockroach

Having examined some key concepts of animal behavior in Chapter 2 and of cellular neurobiology in Chapter 3, we now turn to a neuroethological blending of these two subjects. In this chapter, we shall view a variety of problems faced by a single animal, the cockroach *Periplaneta americana*, in escaping from predators. As in many neuroethological studies, in those described here behavioral and neurophysiological experiments complement each other. Among the issues we will touch upon are the detection of biologically significant sensory signals, the discrimination of these signals from background noise, the localization of sensory cues, and the development and modifiability of behavior and of its neural controls. In subsequent chapters, most of these themes, as well as others, will be explored more deeply, each in a different animal species.

The Escape Behavior

Cockroaches, though all too common as cohabitors of human dwellings in the world's temperate zones, actually evolved in tropical Africa (Roth and Willis, 1960). They are still highly abundant in tropical rain forests, where they are nocturnally active (Schal, 1982). Colored dark brown, they are cryptic against the ground or against the bark of trees; nevertheless, they are visible, especially when they move, to a variety of predators.

Tests have been carried out on the behavioral responses of cockroaches to the strike of a natural predator, the toad *Bufo marinus* (Camhi *et al.*, 1978). This toad is generally sedentary and waits for prey to walk into its field of view. When a cockroach enters this field, it elicits a predatory strike. The insect responds to the strike by first turning its body away from the toad and then running (Figures 1 and 2). The cockroach's turn begins while the toad is still several centimeters away and before this predator's sticky tongue has begun to protrude from its mouth. Thus, the cockroach detects some cue or cues from the approaching toad and on this basis is notified of the presence and direction of this approach.

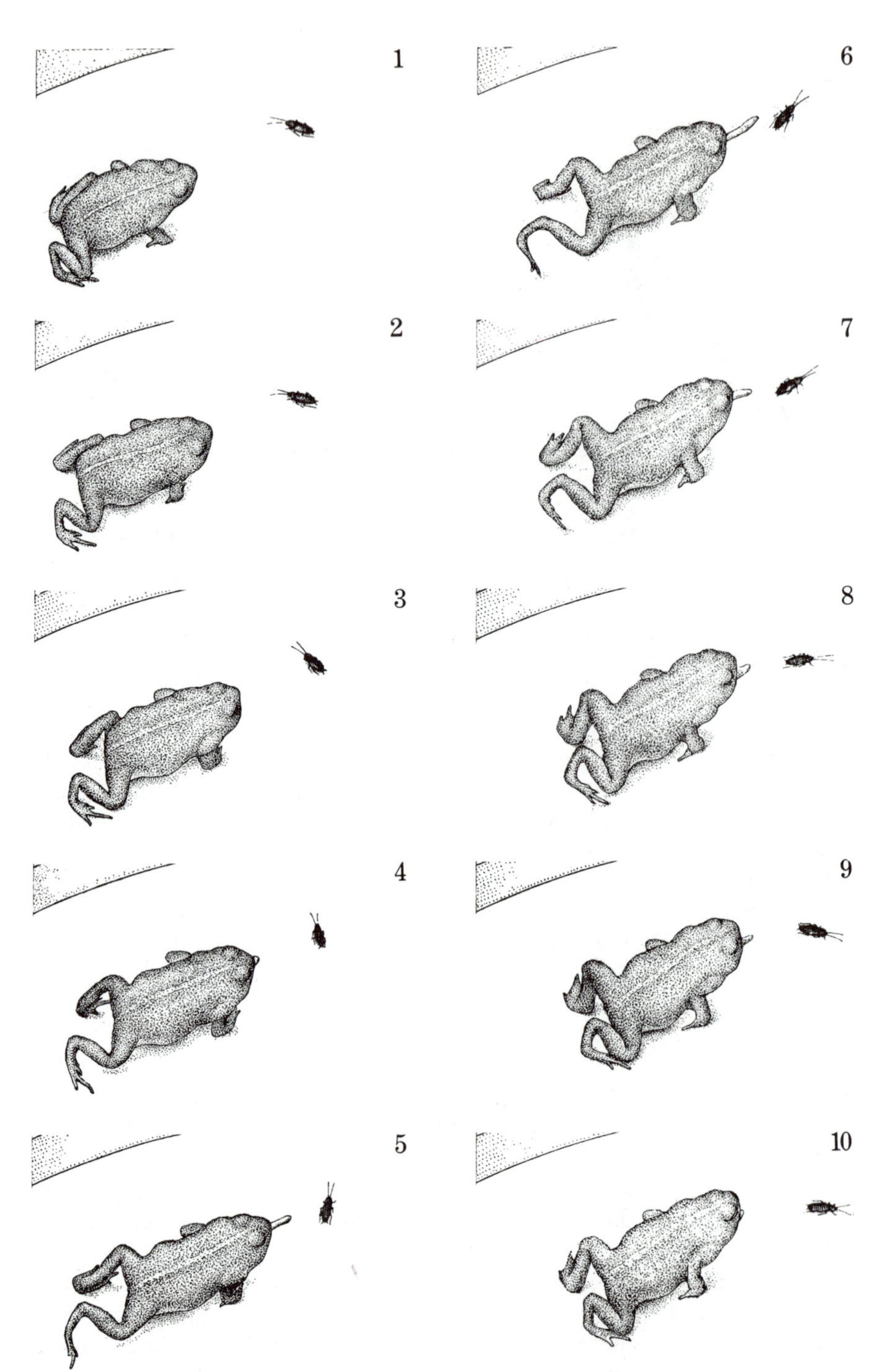

FIGURE 1. The escape of a cockroach from a toad (drawings from a motion picture sequence). The interval between frames is about 16 milliseconds. The toad has begun to lunge forward by frame 2. By frame 3, the cockroach has already begun to turn away. This turn causes the toad's tongue to miss its target. (From "The Escape System of the Cockroach" by J. Camhi.

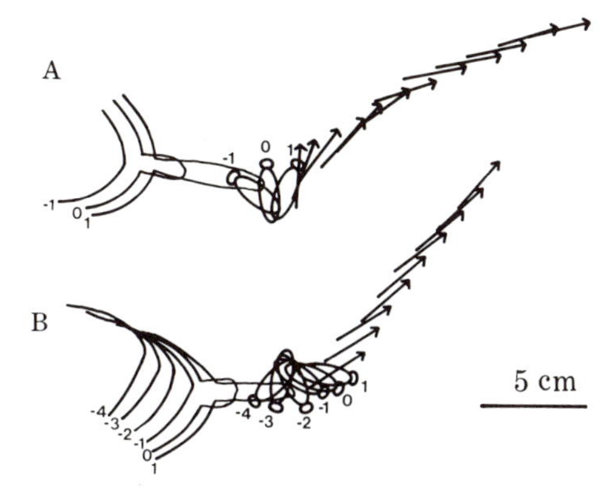

FIGURE 2. Close-up view of two strikes by toads at cockroaches. Shown are the outlines of the toad's head and tongue and of the cockroach's body. Profiles are derived from filmed frames preceding and including the frame showing maximal extension of the toad's tongue. Corresponding numbers on toad and cockroach indicate corresponding times; the frame on which the toad's tongue is first visible is labeled "0." Following this, the positions of the cockroach's body on successive frames taken during its run are shown by the series of arrows. (From Camhi *et al.*, 1978.)

BY WHAT SENSORY CUE DOES THE COCKROACH DETECT THE TOAD?

Attacks on cockroaches by toads in nature are likely to occur at night, when both predator and prey are active. Because little light is available and because the toad's strike is nearly silent, visual and auditory cues might be expected to play relatively small roles in the cockroach's detection of the approaching danger.[1] Moreover, cockroaches are not repelled by the odor of a toad, as they will literally climb upon a sedentary individual. One cue by which a cockroach might detect an approaching toad is the tiny wind gust created by the predator's movement. In fact, numerous hairlike wind-sensitive receptors are found on a pair of posterior antennalike appendages called cerci (singular: cercus) (Roeder, 1948, 1963). These are shown in Figure 3.

The possibility of wind as a cue was tested first by comparing the escape success of three groups of cockroaches: (1) normal individuals; (2) those that have had their cerci covered with glue and, thus, the cercal hairs inactivated; and (3) control animals with intact cerci but with glue placed on the ventral abdominal surface (Camhi *et al.*, 1978). The normal cockroaches escaped from 55% of the strikes and the control individuals from 47%—values that are not statistically different. By contrast, cockroaches with covered cerci escaped from only 8% of the strikes, a significantly lower percentage. Thus, something on the cerci greatly

[1]In spite of the darkness, toads hunt their prey visually at night. However, unlike the cockroach, which must react to its predator very quickly after the strike has begun, the toad has considerable time to gather visual information as the prey walks into its field of view.

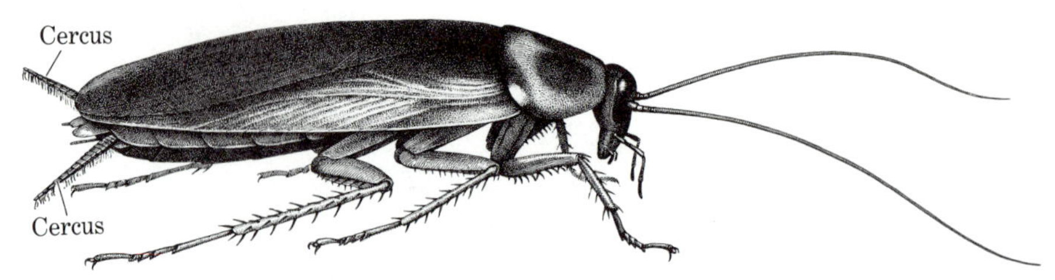

FIGURE 3. Adult male cockroach *Periplaneta americana*, showing the two cerci which bear the wind-receptive hairs on their ventral surfaces. (From "The Escape System of the Cockroach" by J. Camhi. Copyright 1980 by Scientific American, Inc. All rights reserved.)

aids a cockroach in escaping from a toad. This something might well be the wind-receptive hairs, though this remained to be proved conclusively.

Further evidence for wind as a cue to the approaching predator came when it was found that cockroaches made virtually identical movements in response to a toad as to a gust of air from a wind stimulator (Camhi and Tom, 1978; Camhi *et al.*, 1978). But the most conclusive evidence came when the wind created by the toad's strike was measured and was compared with the minimal puff size (produced by a wind stimulator) needed to evoke an evasive run (Camhi *et al.*, 1978; Camhi and Nolen, 1981; Plummer and Camhi, 1981). To measure the wind generated by a lunging toad, a cockroach was anesthetized with CO_2, tied to the end of a miniature fishing pole, and cast about on the floor of a chamber harboring the toad (toads strike only at objects that are moving or have just moved). As soon as the toad became alerted, the cockroach was positioned within a centimeter of a wind meter's measuring probe, which had previously been placed on the floor in the middle of the chamber. When the toad struck, the wind delivered to the cockroach was measured. The wind velocity was read out as the magnitude of a voltage displayed on an oscilloscope. Both the oscilloscope screen and the moving toad were viewed by a high-speed motion picture camera, so that on each frame of film, both the toad's position and the wind that it produced at the cockroach could be determined.

As the record of Figure 4 shows, the cockroach received wind throughout the entire strike of the toad, and the wind velocity increased at least until the toad's tongue emerged from its mouth (time 0). It had been learned from many previous filmed sequences of toads striking at unanesthetized, free-ranging cockroaches that the insect begins its escape movement at a mean time of 16 milliseconds before the emergence of the toad's tongue. Therefore, the wind intensity prior to this moment was a matter of crucial interest. But what precise moment was of interest, and what wind speed at this moment would indicate that the wind had evoked the cockroach's response? To determine this, let us turn for a moment to behavioral tests on cockroaches, tests in which a controlled wind stimulator was used in place of the toad.

Tests with controlled wind puffs showed that both a minimal wind velocity averaging 12 millimeters/second and a minimal wind acceleration averaging 600

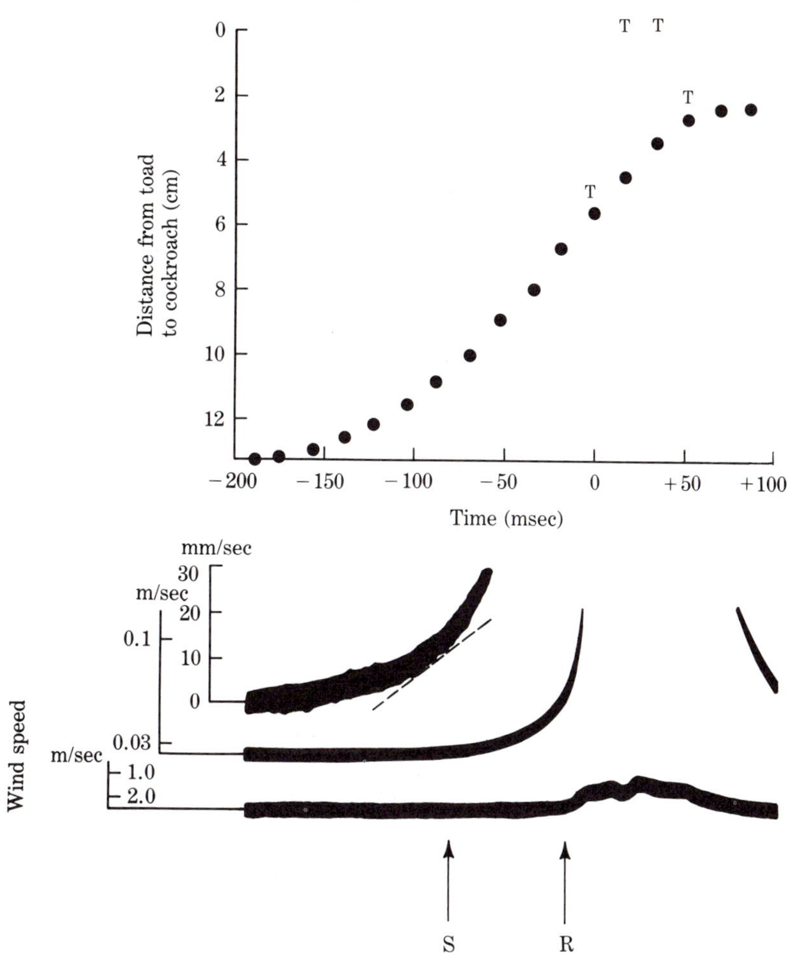

FIGURE 4. The wind delivered to a cockroach by a toad's predatory strike. The top graph shows the position of the front of the toad's head as it moves toward the cockroach. "T" represents the position of the tip of the toad's tongue, first visible at the time labeled "0" on the *X*-axis. The three traces below are three simultaneous recordings of the wind produced by the toad, made at three different sensitivities. "S" indicates the moment that the wind reached its critical magnitude for evoking running—at least 12 mm/sec velocity and at least 600 mm/sec^2 acceleration. (The slope of the dashed line represents an acceleration of 600 mm/sec^2.) "R" indicates the moment (derived by averaging data from many individual responses of cockroaches) that normal cockroaches respond to toad strikes (16 msec before the toad's tongue is first visible). Time scale is the same for wind traces as for graph. (From Camhi *et al.*, 1978.)

millimeters/second^2 were required to evoke an evasive run by the cockroach. (These wind values were obtained only if the stimulus was delivered either while the cockroach was walking slowly about or within 0.5 second after a pause in walking. If the insect was resting, greater stimuli were required; Camhi and

Nolen, 1981.) With these minimal puffs, the mean time from the onset of the wind stimulus at the cerci to the onset of the cockroach's movement response, called the BEHAVIORAL LATENCY, was 44 milliseconds (Plummer and Camhi, 1981).

Given this information, let us now reexamine the wind recording of Figure 4. If the wind from the toad's lunge were the *only* stimulus by which the cockroach detected the approaching toad, this wind should reach its critical value of 12 millimeters/second and 600 millimeters/second^2 about 44 milliseconds before the moment when cockroaches begin their movement response. The average moment of response (16 milliseconds before the tongue emerged) is labeled "R" below the wind traces of Figure 4. The critical value of wind was reached at the moment labeled "S." The interval from S to R in the particular experiment illustrated in Figure 4 is 56 milliseconds; but the *mean* interval from S to R, based on a large number of similar trials, was 41 milliseconds. This close agreement between the two values of latency (44 milliseconds for the wind stimulus *vs* 41 milliseconds for the toad) indicates that the wind generated by the toad's strike is a sufficient stimulus to account for the cockroach's response to the toad. This result was confirmed in independent tests in which toads struck at walking, unanesthetized cockroaches. Again the toad's wind was measured. But here, the latency from the arrival of the critical wind (12mm/sec; 600mm/sec^2) to the onset of behavior could be measured directly. The mean latency again was 41 milliseconds (Plummer and Camhi, 1981).

Attempts to find cues other than wind that came from a striking toad and that could evoke escape by a cockroach have met with no success (Camhi *et al.*, 1978). The cues investigated included the sight of a toad striking from behind a Plexiglas barrier, the odor of a toad, and air-borne sounds or ground-borne vibrations of even much greater intensity than those made by the striking toad. Thus, the wind produced by the lunge appears to be the necessary stimulus for evoking the cockroach's escape from an approaching toad. Although wind is not usually thought of as containing much information, one can see that wind both contains and conveys to the cockroach information about the approach and the direction of a predator.

HOW DOES THE COCKROACH DISCRIMINATE SIGNALS FROM NOISE?

We have seen that wind stimuli whose peak velocity is just 12 millimeters/second evoke running behavior. This raises a substantial problem for the cockroach: how to discriminate between such small wind signals coming from a predator and other, nonthreatening sources of wind, which constitute the background noise[2] in this sensory channel. There are two major sources of noise. First is atmospheric wind, which, as measured at night in tropical rain forests, generally ranges up to about 150 millimeters/second (Schal, 1982). Second is wind created by the cockroach itself while it walks. A typical walking speed of 100 millimeters/second causes a wind of the same speed to flow posteriorly over the body—just as when

[2]The term *noise* as used here does not mean sounds, but rather energy of the same physical type as the signal, in the face of which the signal must be detected.

FIGURE 5. Wind gusts produced by the stepping movements of a cockroach's right hindleg. The wind was recorded at the right cercus. Stepping movements (bottom trace) were recorded by having the leg interrupt a beam of light shining on a tiny photocell each time the leg moved. The dots below the trace indicate the approximate moments when the leg reached its backwardmost position. These gusts were superimposed upon a roughly 80 mm/sec wind that represents the animal's forward progress at this speed through the air. (From Plummer and Camhi, 1981.)

you run, you create a headwind. These two sources of noise together, then, can range up to about 250 millimeters/second, or roughly 20 times greater than the critical windspeed for evoking a run.[3] Moreover, the velocity of this self-generated wind fluctuates as the animal walks; each step of a hindleg produces a wind fluctuation which, as recorded at the nearer cercus, can be greater than the 12 millimeters/second critical wind speed (Figure 5).

Somehow the cockroach manages not to respond behaviorally to the large and complex background noise, and yet to respond with a run to the much smaller wind signal. A clue as to how the insect accomplishes this discrimination comes from the observation that the *acceleration* of the wind signal (the rate of change of its wind speed) is actually greater than the wind accelerations contained in the background noise. The maximal atmospheric wind acceleration, measured at night in tropical rain forests, is less than a few millimeters/second^2 (Camhi *et al.*, 1978). And the acceleration of the wind gusts produced by the stepping legs of a cockroach is below 300 millimeters/second^2 (Plummer and Camhi, 1981). By contrast, we have seen that a toad's strike delivers wind with accelerations of at least 600 millimeters/second^2 (Figure 4).

Is this difference in wind acceleration in fact the cue by which the cockroach discriminates signal from noise? To test this, wind puffs all having the same peak velocity (40 millimeters/second) but differing in acceleration (between 150 and 1700 millimeters/second^2) were delivered to slowly walking cockroaches (Plummer and Camhi, 1981). The higher the acceleration, the greater the percentage of running responses (Figure 6). The lower the acceleration, the greater the percentage of trials with no response. Intermediate accelerations often evoked a pause in the cockroach's walking. These findings were independent of wind direction. Thus, the cockroach appears to attend specifically to the acceleration of the stimulus in order to weather windy conditions without jamming its prey-detecting system.

[3]Slowly walking cockroaches respond, though not by running, to wind stimuli even smaller than 12 millimeters/second. Specifically, puffs of just 3 millimeters/second consistently evoke a pause in walking. This pause presumably aids the cryptically colored cockroach in remaining undetected. This 3 millimeters/second signal, then, is about 80 times smaller than the background noise.

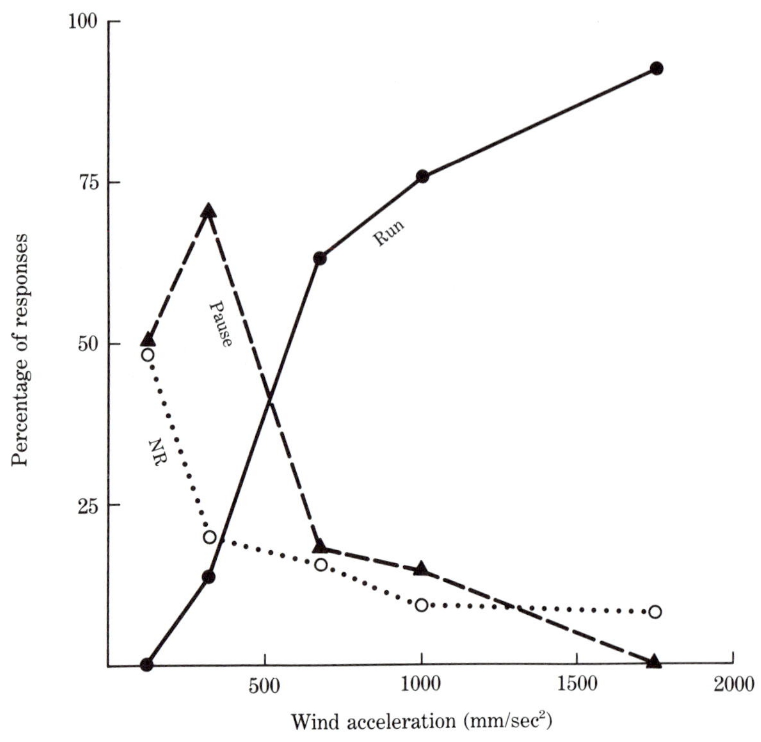

FIGURE 6. Effect of wind accelerations on the cockroach's response to wind stimuli. NR, no response. See text for explanation. (From Plummer and Camhi, 1981.)

How Does the Cockroach Nervous System Encode the Direction of an Approaching Toad?

Because cockroaches regularly turn away from an approaching toad or other source of rapidly accelerating wind, it is clear that this insect's nervous system can determine wind direction. Some of the neurons that appear to be involved in this determination are shown in Figure 7A. At the base of each of approximately 220 wind-receptive hairs on a cercus is the dendrite and cell body of a single sensory neuron. The axon of each sensory cell projects to the central nervous system. Like most higher invertebrates, the cockroach's central nervous system is subdivided into GANGLIA (singular: ganglion) and CONNECTIVES. The ganglia contain cell bodies, dendrites, and axons, whereas the connectives contain only axons. In the cockroach's terminal ganglion, the wind receptor cells excite, apparently monosynaptically, a group of GIANT INTERNEURONS (GIs), whose axons are of substantially greater diameter than any other interneurons in this nervous system (Roeder, 1948; Westin *et al.*, 1977). Figure 2 of Chapter 3 shows one of these GIs filled with cobalt sulfate.

The cockroach has seven bilateral pairs of GIs. Each GI can be identified

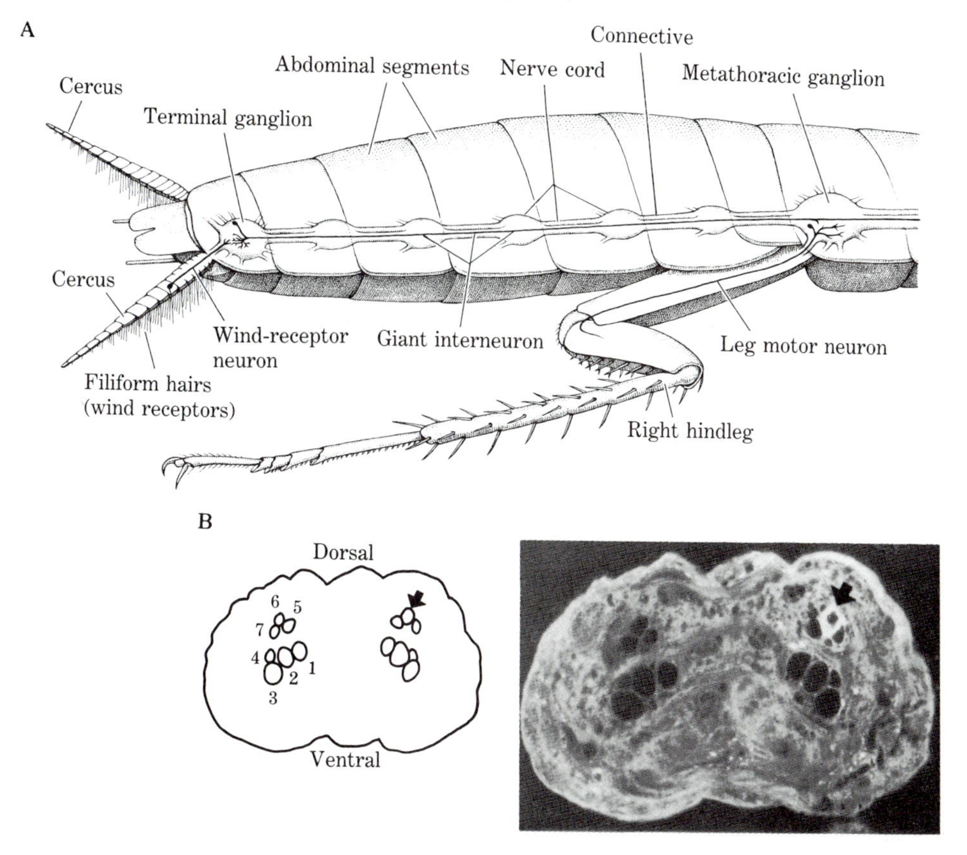

FIGURE 7. Some of the neurons of the cockroach escape system. A. Sketch showing one cercal sensory cell, one giant interneuron, and one leg motor neuron in their actual locations. B. On the right is a cross section through an abdominal ganglion. The seven large profiles on each side (drawn in isolation in the sketch on the left) are the cross sections of the seven bilateral pairs of giant interneurons (GIs). The right GI 6 (arrow) has been filled with the dye Procion yellow, following a recording from its axon. (A from "The Escape System of the Cockroach" by J. Camhi. Copyright 1980 by Scientific American, Inc. All rights reserved. B from Westin *et al.*, 1977.)

individually in all cockroaches of this species, on the basis of the shape of its dendritic tree (Daley *et al.*, 1981) or by the relative position of its axon as seen in a cross section of an abdominal ganglion (Figure 7B). The axons of the GIs ascend the nerve cord to the head (Spira *et al.*, 1969), passing through the thoracic ganglia. In the thorax, excitatory synaptic contacts (which have not been well characterized) are made onto motor neurons of those leg muscles that produce running behavior.

Where in this neural circuit does the cockroach's determination of wind direction occur? It begins with the individual cercal wind-receptive hairs. Close inspection shows that the roughly 220 hairs on an adult cercus are organized

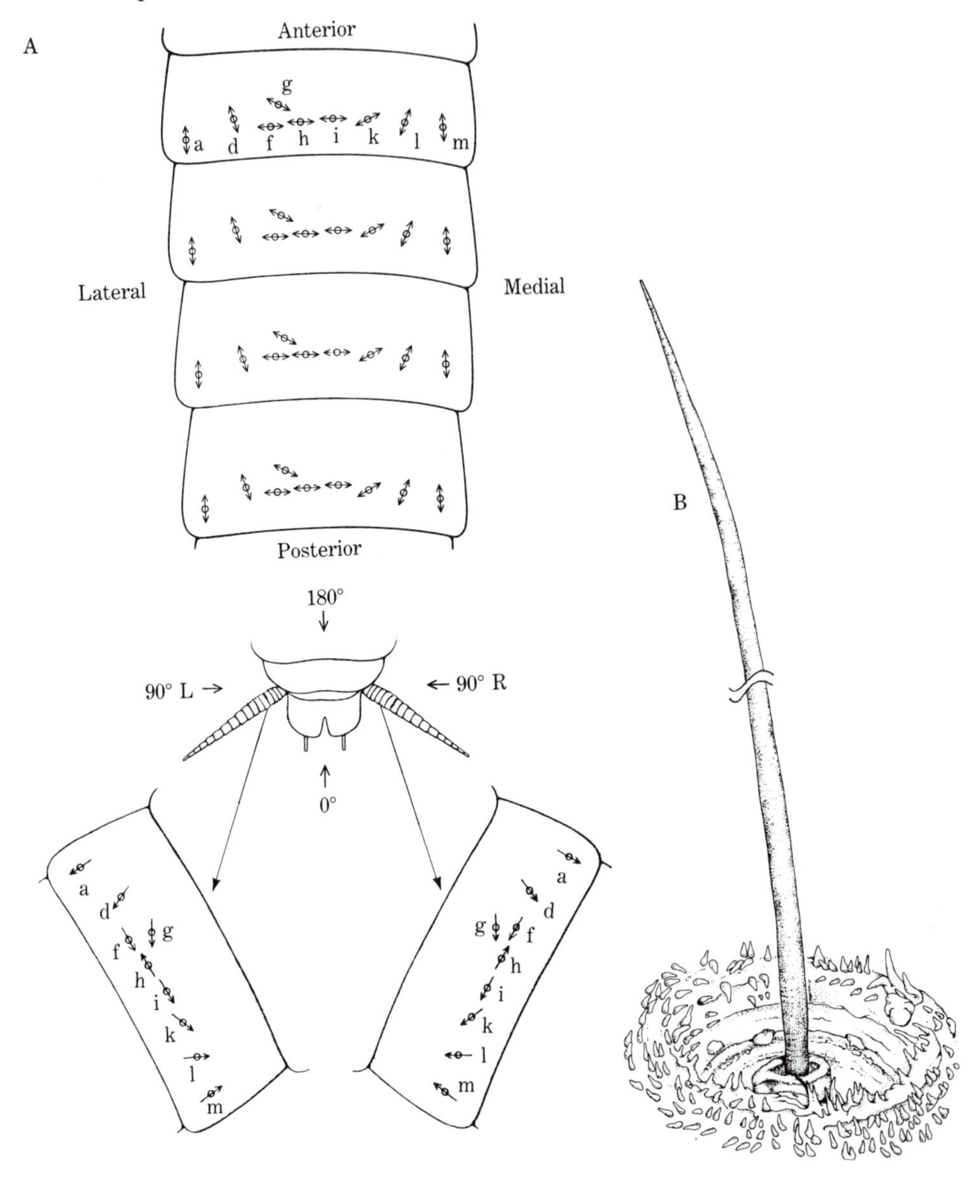

FIGURE 8. Organization of wind-receptive hairs on the cerci. A. Top: Underside of four of the 19 segments from an adult cercus. Each circle shows the position of a wind-receptive hair. Columns a, d, f, h, etc. run along most of the cercal length. Double-headed arrows show the two directions of maximal mechanical pliancy of each hair. Bottom: A single segment from each cercus. The single-headed arrows show the optimal wind direction for each hair. All hairs of a given column have the same optimal wind direction. B. A single wind-receptive hair, drawn to scale. At the hatch mark, three-fourths of the actual length of the hair has been omitted. (A from "The Escape System of the Cockroach" by J. Camhi. Copyright 1980 by Scientific American, Inc. All rights reserved.)

spatially into columns and rows (Figure 8). Each of the 19 cercal segments of the adult (except those near the tip and base) has a row of nine long hairs, which range from 0.5 to 1 millimeter in length. (Other, shorter hairs that are apparently not wind-responsive, as well as other types of sensory structures, are also found on the cerci.) The similarly placed hairs on each segment can be regarded as members of a column running from the cercal base to its tip. All the hairs in a given column are functionally alike in that they can be deflected easily in either of two (opposite) directions (double-headed arrows in Figure 8, top) and less easily at right angles to these directions. The directions of maximal pliancy are different from column to column (Nicklaus, 1965).

By recording with microelectrodes from individual sensory cells, it was found that pushing the hair in one of its two most pliant directions evokes a maximal number of action potentials. Pushing it in the other most pliant direction (180° away from the first) produces a maximal inhibition of spontaneously occurring action potentials (Nicklaus, 1965). The excitatory direction is thought to stretch the dendrite of the sensory cell and thus evoke a receptor potential and the resulting action potentials. The opposite (inhibitory) deflection is thought to slacken the sensory cell's dendrites (Figure 9).

The responses of a sensory cell to wind resemble these directional responses to the pushing of its hair from side to side. This was determined from extracellular recordings of the whole sensory nerve, made after all the cercal hairs except those in one column had been covered with glue (Dagan and Camhi, 1979). By giving wind puffs from many different directions within the horizontal plane, a single most excitatory wind direction for each column of hairs was identified (Figure 8, bottom, single-headed arrows). In other experiments, recordings were made intracellularly from the axons of individual sensory cells, again in response to wind puffs from different horizontal directions (Westin, 1979). Again, each cell showed a single wind direction that was most excitatory and an opposite direction that inhibited action potentials. However, the overall range of directions to which a given sensory neuron responded was found to be quite broad (Figure 10). Given

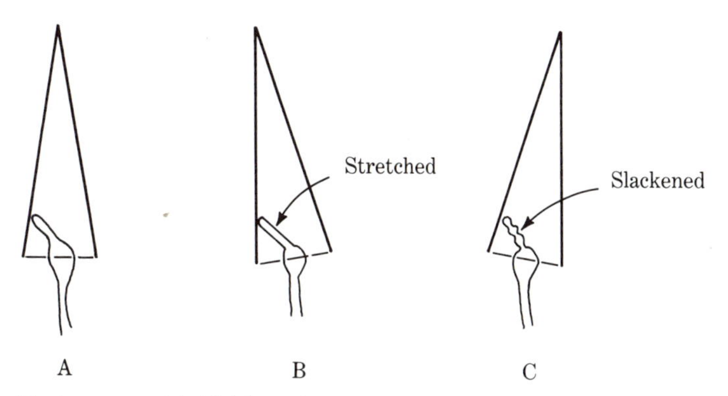

FIGURE 9. Excitation and inhibition of a wind-receptor cell by opposite deflections of its hair. A. Undeflected hair. B. A deflection that stretches the dendrite and excites the cell. C. A deflection that slackens the dendrite and inhibits the cell.

the many different best excitatory wind directions represented on each cercus (Figure 8, bottom) and the broad directional response of each sensory cell (Figure 10), a wind puff from any direction must excite sensory cells from several different columns on each cercus. A slight shift of wind direction, then, would result in different relative amounts of excitation of these various columns of sensory cells. Thus, each direction of wind inscribes on the sensory nerve its own signature in the form of the relative numbers of action potentials evoked in the different sensory columns.

Do the GIs, excited by the cercal neurons, preserve the directional information encoded by these sense cells, or is this directional information transformed or even lost in crossing the synapse to the GIs? These questions were answered by making intracellular recordings from each GI and by delivering identical wind stimuli from different horizontal directions (Westin *et al.*, 1977). Just after com-

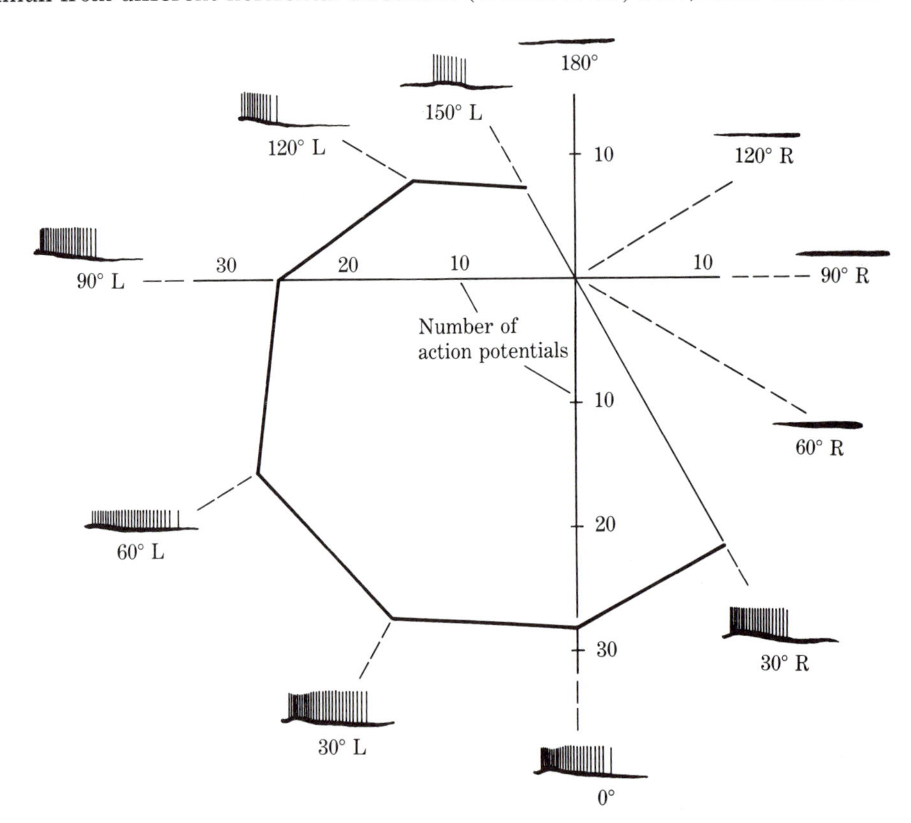

FIGURE 10. Directional response of a single wind-receptor cell. Constant wind puffs were delivered to the cockroach from different angles within the horizontal plane. The mean number of action potentials evoked from each angle is plotted. Representative recordings for each angle tested are also shown. This cell responded best to wind from the cockroach's left rear quadrant. Wind from the opposite quadrant (right front) evoked no action potentials. Though not shown here, the latter direction also inhibited any spontaneously occurring action potentials in the cell. (After Westin, 1979.)

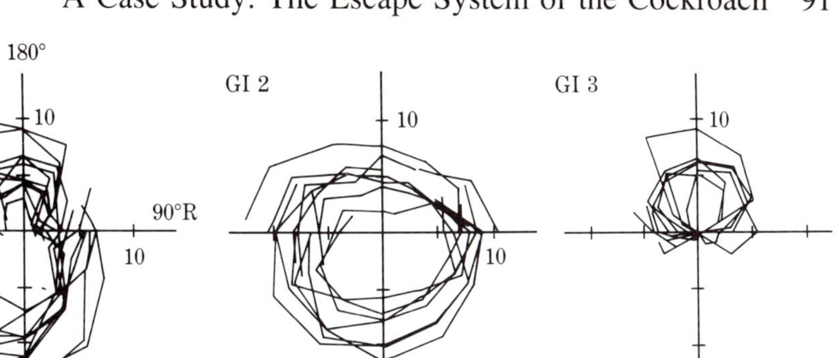

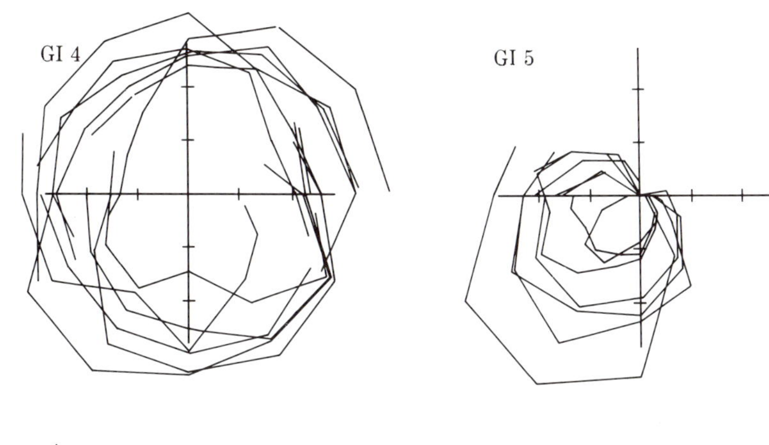

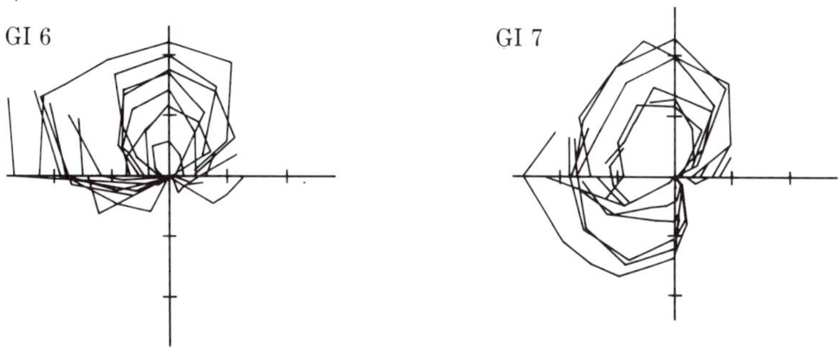

FIGURE 11. Directional responses of the seven left GIs to wind stimuli. (A left GI is one whose axon runs on the left side of the nerve cord, though its cell body is on the right. The seven right GIs give mirror image responses.) Each line on any of the seven sets of axes represents the responses of a single cell. Thus, the same cell type from several animals is plotted on each set of axes; numbers on axes represent number of action potentials. The method of plotting is the same as that in Figure 10. For technical reasons, smaller wind puffs were used for stimuli from the front than from the rear, producing discontinuities in many of the graphs. (From Westin *et al.*, 1977.)

pleting a set of such recordings from a given cell, Procion yellow, Lucifer yellow, or cobalt was injected intracellularly from the micropipette electrode to enable subsequent identification of the particular GI recorded.

Of the seven GIs on each side of the nerve cord, two were found to respond with roughly the same number of action potentials to wind from any direction (GIs 2 and 4; Figure 11). Two others (GIs 1 and 7) respond most strongly to wind coming from the same side as the position of the GI's axon. The three remaining GIs on each side (numbers 3, 5, and 6) give responses that are directionally more restricted. Considering the GIs as a group, then, one can see that directional information is retained but in a form somewhat different from that of the sensory neurons.

The nature of this transformation has been studied by recording intracellularly from the cell bodies of giant interneurons after plucking out or covering up all the cercal hairs except a limited number from a single column (Daley and Camhi, 1980; Daley, 1982). In most of these experiments, six hairs of a column were left uncovered, and this number was then reduced one by one, by covering additional hairs. The experimenter in this way determined the number of hairs required to produce an action potential in a given GI, in response to a standard wind puff from a fixed direction. In different experiments, different combinations

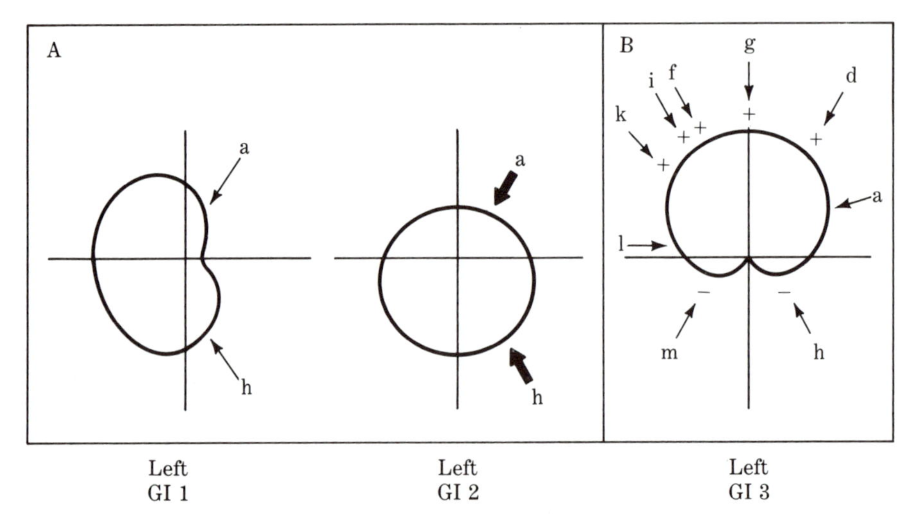

FIGURE 12. Mechanisms underlying the shapes of the directional response curves for three GIs. The directional curves for the left GIs 1, 2, and 3 are as shown in Figure 11. Lower case letters indicate particular columns of sensory hairs on the left cercus. The arrows indicate the optimal wind direction for each column, taken from Figure 8. A. The sensory input to left GI 1 differs from that to left GI 2 primarily in the functional strength of activation by two sensory columns, a and h. Their input is stronger to GI 2 (thicker arrows). B. Left GI 3 receives excitatory input (+) from sensory columns on the left cercus that encode wind from the front, and inhibitory input (−) from those columns that encode wind from the rear. Those that encode wind from the sides (a and l) evoke no detectable response. (Courtesy of D. Daley.)

of column and GI were tested. It was found, for instance, that GIs 1 and 2 receive excitatory input from all nine columns of hairs on the ipsilateral cercus (that is, the cercus on the same side as the GI's axon). The differently shaped directional responses of these two GIs (Figure 11) appear to result from the different synaptic strengths that the two GIs receive from certain columns of sensory hairs.[4] In particular, columns a and h (Figure 8) from the left cercus excite the left GI 1 weakly but the left GI 2 strongly. This appears to account for the difference between the directional responses of these two GIs (Figure 12A). By contrast, GI 3 receives excitatory input only from those columns of sensory cells whose best wind direction is from somewhere near the front (columns d, f, g, i, and k). This GI receives inhibitory input from sensory cells of columns h and m, whose best excitatory directions are near the rear (Figure 12B). Columns a and l provide no detectable input to GI 3.

In summary, then, the directional coding of the sensory neurons results from the mechanical properties of their hairs and of the dendritic connections to these hairs. The transformation of these directional properties to the GIs involves the use of excitatory synaptic connections of different functional strengths, as well as inhibitory connections.

From Giant Interneurons to Behavior: How is the Directional Information Decoded?

Although directional information is available to the cockroach through its GIs, this does not prove that the insect actually uses these GIs to elicit the escape response. Indeed, there could be other, unidentified neurons that play this role. As we shall now see, however, there is good evidence that the GIs are important mediators of the escape behavior and that the directional information that they convey is crucial for a properly oriented turning response to wind. We will also see something of how this encoded directional information is decoded to execute a proper turn.

On theoretical grounds, one might suspect that the GIs are involved in the escape behavior because this behavior can occur very quickly and may thus require the rapid conduction of action potentials afforded by axons of especially large diameter (Chapter 3). Although the behavioral latency in response to the onset of a toad's wind was over 40 milliseconds (Figure 4), if one uses an even more rapidly accelerating wind puff as the stimulus, behavioral latencies as short as 11 milliseconds can be recorded (Camhi and Nolen, 1981). This interval, together with measured escape behaviors of a few other animals (Chapter 8), are among the shortest behavioral latencies known.

More direct evidence for a role of the GIs in escape behavior comes from experiments in which one or two GIs are stimulated electrically through intracellular electrodes. Each stimulus consists of a rapidly repeating train of current

[4]As we saw in Chapter 3, the functional strength of a synapse can be evaluated only by its effect on the trigger zone of the postsynaptic cell. In the present experiments, it was just such an effect that was determined, because the parameter measured was the production of an action potential.

pulses. The stimulus evokes a rapidly repeating train of action potentials, a pattern that resembles the barrage with which a GI responds to natural wind puffs. In order to prevent movements of the insect, which could dislodge the microelectrode from the stimulated GI, the cockroach's body was pinned to the substrate. Because the pinned insect could not move, any locomotory consequences that the GI stimulation might evoke were monitored by recording extracellularly from the axons of motor neurons to muscles of a hindleg. It was found that several of the GIs, when stimulated alone or together with one another, evoked bursts of action potentials in these axons (Ritzmann and Camhi, 1978; Ritzmann, 1981; Ritzmann and Pollack, 1981). The motor neurons excited are the very ones used to produce running behavior.

Perhaps the strongest evidence for a role of the GIs in escape comes from experiments in which selected GIs were killed and the cockroach's escape behavior tested the next day. The selective killing of a particular GI was accomplished by injecting this cell, through an intracellular microelectrode, with the proteolytic enzyme Pronase. Pronase digests the cell—that is, destroys it. Moreover, because this enzyme, like all enzymes, is itself a protein, its proteolytic actions soon cause its own destruction, so the neuron-killing ability of Pronase ceases before it has had an opportunity to destroy other cells. Figure 13 shows examples of three nerve cords in which Pronase was used to kill, respectively, the right GI #1, the left GI #2, and the left GIs 1 and 2. So far, this Pronase technique has been applied only to the three largest GIs, numbers 1, 2, and 3 (Figure 7B).

Before describing the behavioral results of the Pronase-lesioning of specific GIs, let us consider theoretically what results we might expect if the cockroach uses its GIs to determine the wind direction. First, we might expect that killing some or all of the GIs 1, 2, and 3, but leaving intact the smaller GIs 4–7, should interfere with proper turning. The reason concerns differences in the times of arrival at the thoracic ganglia of those action potentials carried by GIs 1, 2, and 3 *vs* those carried by the smaller GIs. Giant interneurons 1, 2, and 3, having nearly twice the diameter of the others, conduct action potentials from the terminal ganglion to the thorax in about 1 millisecond less than do the remaining GIs. Moreover, GIs 1, 2, and 3 are activated by wind about 5 milliseconds before the smaller GIs (Westin *et al.*, 1977). This total difference of about 6 milliseconds all but rules out GIs other than numbers 1, 2, and 3 as initiators of the escape response, at least on those occasions when the behavioral latency is at its minimum of 11 milliseconds (Camhi and Nolen, 1981). And since the initial part of the escape response is a turn, it would seem that GIs 1, 2, and 3 alone can initiate a proper direction of turning.

Suppose that this idea of left and right GIs 1, 2, and 3 as the turn initiators is true. Then these six cells must contain sufficient information to distinguish among at least four different wind directions—left front, left rear, right front, and right rear—each of which results in a different initial turning response. (For instance, wind from the left front produces a right turn of high angular velocity; and wind from the left rear produces a right turn of low angular velocity; Camhi and Tom, 1978). In fact, left and right GIs 1, 2, and 3 do encode sufficient information to discriminate among at least these four quadrants of horizontal space. As is shown in Figure 14A, when the wind comes from anywhere on the left side,

A

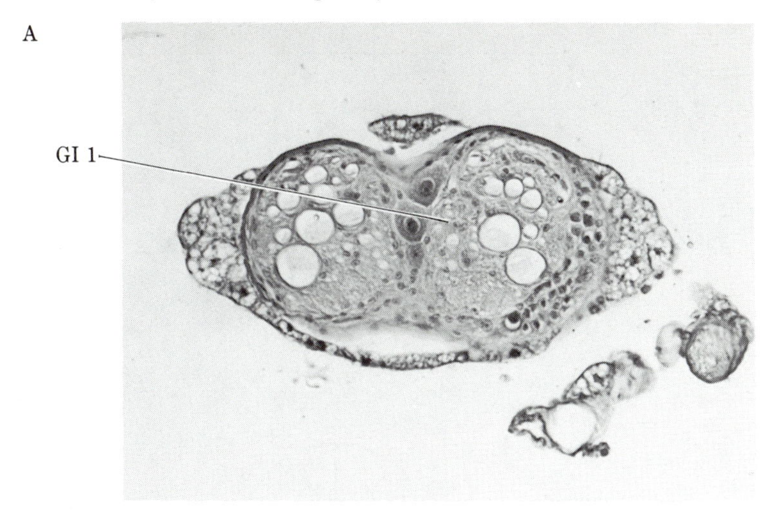

B

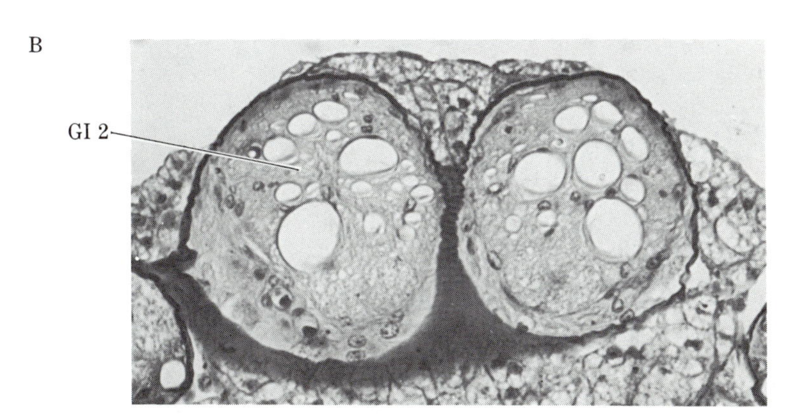

C

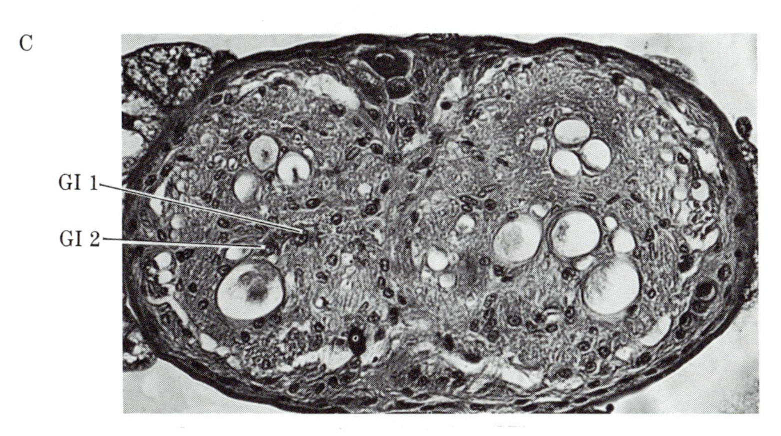

FIGURE 13. Cross sections of nerve cords in the abdominal regions of three different cockroaches. A. Right GI 1 has been injected with Pronase and is missing. B. Left GI 2 is missing after Pronase injection. C. Left GIs 1 and 2 are both missing after Pronase injection. Arrows point to the expected locations of the missing axons of these GIs.

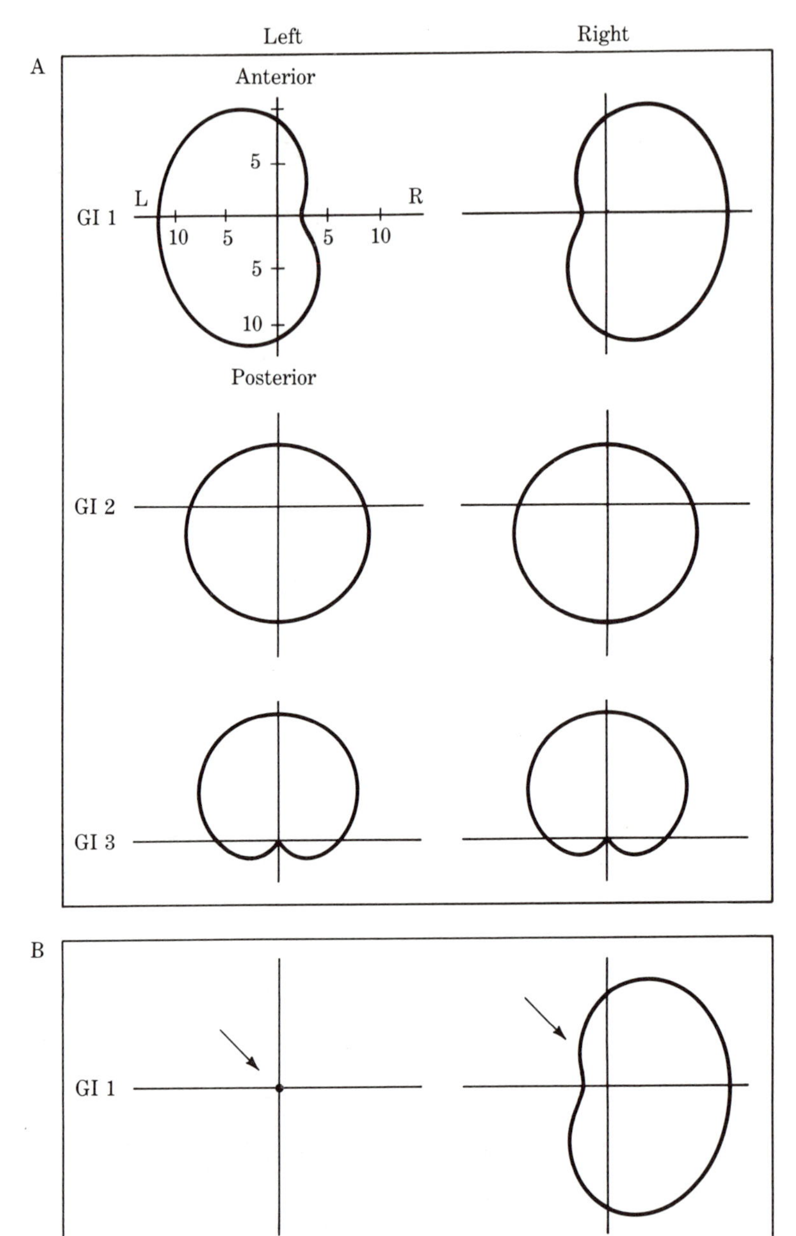

FIGURE 14. Hypothesis for directional specification by the left and right GIs 1, 2, and 3. The directional curves for these GIs are as shown in Figure 11. A. As explained in the text, these GIs contain sufficient information to specify at least the four quadrants of horizontal space around the cockroach. B. If the left GI 1 is destroyed, wind from the left (arrows) will excite right GI 1 but not left GI 1. Numbers on axes represent number of action potentials.

the left GI 1 gives more action potentials than the right GI 1.[5] And when the wind comes specifically from the left *front*, both GIs 3 are activated; whereas if it is from the left *rear*, the GIs 3 are silent. Because the responses of the GIs 2 are not directionally selective, one might guess that these cells play no role in specifying wind direction.

Thus, a possible mechanism for decoding directional information from the GIs would involve some form of comparison by the nervous system between the responses of the left and right GIs 1, as well as a determination of whether or not the GIs 3 are active. If this were the actual mechanism, one would predict that after killing just the left GI 1 with Pronase, the animal would turn to the left not only when the wind comes from the right, but even when it comes from the left. The basis of this prediction is that a wind puff from the left would evoke a few action potentials in the right GI 1 but none, of course, in the killed left GI 1 (Figure 14B, arrows). Thus, a comparison mechanism would mistakenly calculate that a puff from the left had actually come from the right and would thus evoke a turn to the left.

In fact, cockroaches with their left GI 1 killed do turn left significantly more often than normal animals in response to wind from the left. But this occurred on only 25% of the trials, not on the great majority as had been predicted (Figure 15). (The remaining 75% of the trials produced right turns.) This surprisingly low incidence of left turns suggests that additional cells besides the two GIs 1 must be involved in determining whether a wind puff was from the left or the right. Because the directional curves of GIs 2 and 3 are bilaterally symmetrical (Figure 14A), they would not appear capable of a left–right discrimination. One would therefore predict that a cockroach with just one of these GIs killed would show unaltered behavioral discrimination of left *vs* right winds. As predicted, killing either of these cells did not significantly effect the turning direction (Figure 15). Surprisingly, however, when both left GI 1 and left GI 2 were killed, there was a much higher incidence of mistaken left turns than when just left GI 1 had been killed (Figure 15). In these experiments it was necessary to control for possible damage to neurons other than the one or two GIs one intended to kill. Such unintended damage could be caused by Pronase leaking from the electrode or from the injected GI, and could conceivably contribute to producing the altered turning directions. Injecting, not into a GI, but rather extracellularly in the vicinity of GIs 1, 2, and 3, did not significantly effect the turning direction (Figure 15, control).

It is not yet clear how killing a cell that has a symmetrical directional response (GI 2) adds to the disruption of the behavior's directionality. Whether also killing the left GI 3 would further increase the percentage of left turns remains to be

[5]Although these graphs show number of action potentials as a function of wind direction, similarly shaped graphs result from plots of other parameters, such as the average frequency of action potentials within the evoked bursts, the duration of the bursts, and 1/latency to the first action potential of the burst (Westin *et al.*, 1979). The latency parameter is of particular interest because, by comparing the latencies among the different GIs, the cockroach could obtain directional information as soon as the GI bursts have begun, rather than waiting for these bursts to terminate. In fact, it is known that the cockroach can begin its escape movements in response to just the first one or two action potentials in each of the GIs 1, 2, and 3 (Camhi and Nolen, 1981).

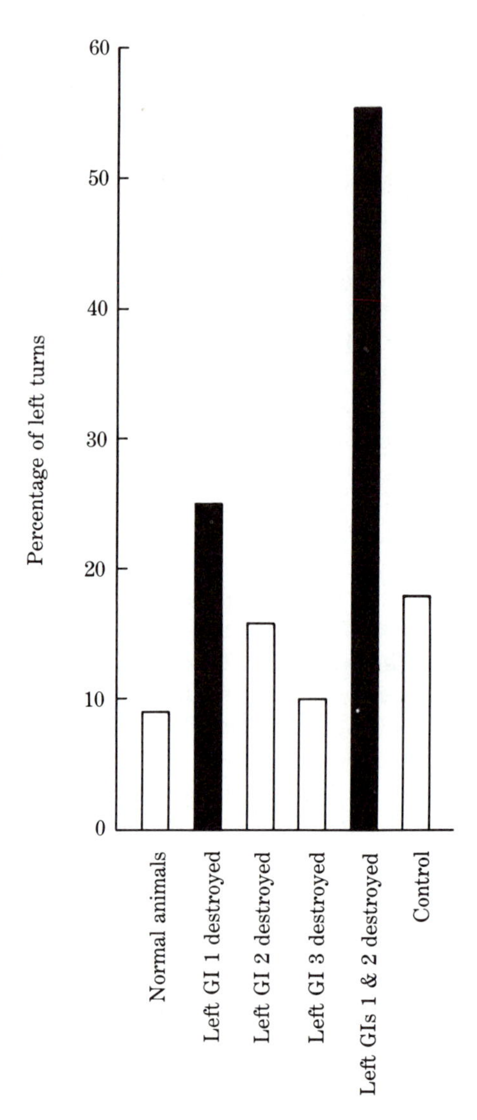

FIGURE 15.

Effects of destroying left GIs 1 and/or 2 on the direction of turning by cockroaches in response to wind from the left. All wind puffs were from within a range of 45° left of front. Only the solid bars represent percentages significantly different from those of normal animals. All responses that were not turns to the left were turns to the right. Wind puffs from the right front (not represented on this graph) usually evoked turns to the left, with no significant differences after destruction of any of the left GIs 1, 2, or 3. (Courtesy of C. Comer.)

seen. Nevertheless, one can already see that GIs 1 and 2 play an important role in reading the wind direction and in prescribing the initial turning direction. An understanding of the specific mechanisms by which the comparisons among the GIs are carried out must await a study of the synaptic interactions from these GIs onto their postsynaptic neurons.

IS THE COCKROACH BORN WITH A FUNCTIONAL ESCAPE SYSTEM?

When a cockroach hatches from an egg case, each of its cerci bears, not 220 wind-receptive hairs as in the adult, but just two (Figure 16). One can record the wind-evoked responses of the single sensory cell under each of the animal's four hairs. Its directional response (Figure 17) is similar to that of a sensory hair in the adult (Figure 10). Each of the four overlapping graphs for the hatchling shows

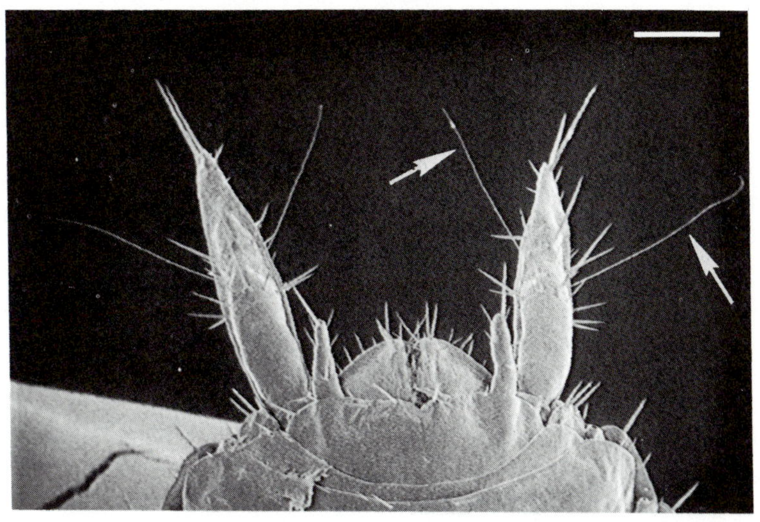

FIGURE 16. Scanning electron micrograph of the cerci of a newly hatched cockroach *Periplaneta americana*. Only two wind-receptive hairs like those found in the adult are present on each cercus (arrows). Calibration bar = 200 mm. (From Dagan and Volman, 1982.)

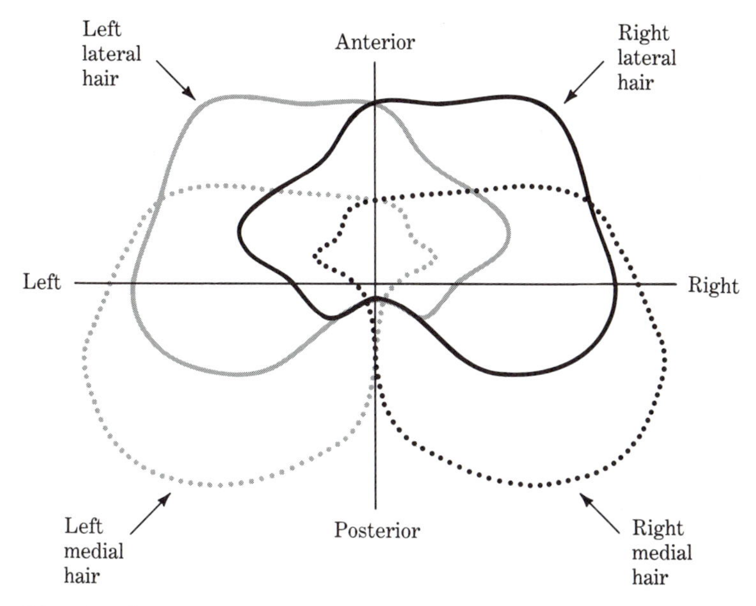

FIGURE 17. Directional responses of each of four wind-receptive hairs of a newly hatched cockroach. Each hair appears responsive to a different quadrant of horizontal space. (From Dagan and Volman, 1982.)

a different best excitatory wind direction, a result indicating that different wind angles will produce different relative amounts of activity in these four sensory cells (Dagan and Volman, 1982). Thus, these cells encode sufficient information to enable the hatchlings to turn away from a wind source. (All seven bilateral pairs of GIs are also present, though their responses to wind have not yet been recorded.)

In fact, behavioral tests like those performed on adults show that the hatchlings are at least as accurate as adults in turning away from a wind source (Dagan and Volman, 1982). Even individuals tested just after hatching, having received no prior wind stimuli, turn away from the first puff they experience. Thus, the cockroach is born with a functional escape behavior.

When the four cercal hairs are plucked out, a hatchling cockroach does not respond at all to gentle wind puffs. Moreover, if just the two hairs on the left side are plucked, the animal responds by making turns to the left, not only when the wind comes from the right, but also on 75% of the trials when the wind comes from the left. These observations indicate that the four cercal wind receptors, and not some other, unknown receptors, provide the sensory information that evokes the behavior. Interestingly, if three hairs are plucked, leaving, for instance, just the right lateral hair, wind stimuli still evoke the escape behavior. Now, 90% of the responses to wind from the left are turns to the left. This, then, is an example of neuroethological simplicity in the extreme, a single sensory neuron evoking an entire complex turning and running behavior.

Plasticity in the Escape System

Because both the newly hatched and the adult cockroach have similar, functional escape behaviors, it might seem that no modification of this behavior occurs during the animal's life. However, under certain emergency conditions, these insects may need to make alterations in their escape behavior. Doing so would require making alterations in the underlying neural circuitry. For instance, one often finds cockroaches with part or all of a cercus missing. (Because aggressive encounters among cockroaches often entail biting each other, these lost cerci may result from intraspecific aggression.) One can predict that removing most or all of a cercus, just like ablating the only two hairs from one cercus in a newborn cockroach, would lead to serious errors in the direction of wind-evoked turns. In such a case, it would be to the cockroach's advantage to modify its neural circuitry in such a way as to correct its turning behavior. Such restorative changes in a behavior and its neural controls, occurring in response to injury or other forms of disabling experience, are called behavioral and neuronal PLASTICITY.

In a group of just subadult cockroaches, the turning responses to wind were tested; then the left cercus was ablated, and the behavior was retested on the next day (Vardi and Camhi, 1982a). Of particular interest were the responses to wind puffs from the left, the side of the missing cercus. As compared with the normal animals tested before the ablation (Figure 18A), one day after the ablation wind from the left evoked many turns to the left, that is, toward the wind instead

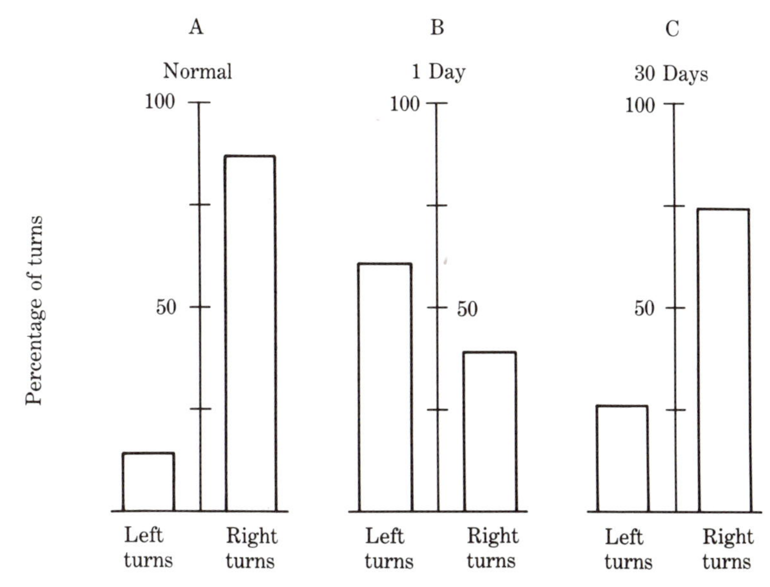

FIGURE 18. Plasticity in cockroach escape behavior. A. In intact cockroaches, wind from the left front usually evokes a turn to the right. B. One day after ablating the left cercus, wind from the left front usually evokes a turn to the left. C. Thirty days later, wind from the left front again usually evokes a turn to the right. Wind from the right (not represented in this graph) usually evokes a turn to the left, with no significant differences among the normal, 1-day, and 30-day animals. (After Vardi and Camhi, 1982a.)

of away (Figure 18B). (The direction of turns in response to wind from the right was not significantly affected by a left cercal ablation.)

These mistaken turns occurred in spite of the availability of correct directional information in the sensory cells of the intact right cercus (Figure 8); this information, though available, apparently is not used to produce correct turns one day after the ablation of the opposite cercus. By 30 days after the left cercal ablation, however, the percentage of left turns in response to wind from the left had significantly dropped, being replaced by correct, right turns (Figure 18C). This improvement developed gradually over the 30-day period. (Again, the turns in response to wind from the right showed no significant change.) The improvement did not require the cockroach to practice responding to wind puffs during the 30 days.

During the 30-day correction period, the cockroach had not developed any new sensory hairs.[6] Thus, a cockroach tested 1 day after ablation of a cercus, or tested 30 days later, possessed the identical sensory equipment; but something

[6]Insects develop new sensory hairs only when they molt. Most of the cockroaches in this experiment did not molt during the 30-day period of behavioral correction. Some did molt, however, and developed a small cercal bud on the left side, containing a limited number of sensory hairs. However, this bud and its hairs were ablated within 1 day of the molt. Thus, at the time of testing, there were no cercal hairs on the left side.

had changed in the way the nervous system was reading out the sensory information. What is the neuronal basis of this change?

Recordings from the GIs show that most left GIs are largely unresponsive to wind puffs 1 day after a left cercal ablation. This is shown in Figure 19A for left GIs 1, 2, and 3, which, as we have seen, appear largely responsible for setting the initial turning direction (Westin *et al.*, 1977; Vardi and Camhi, 1982b). As a result, on day 1, there is a strong left–right imbalance in the GI response; wind from any direction gives more action potentials in the right than the left GIs, presumably leading to the preponderance of left turns. Figure 19B shows that by day 30 the responses of the left GIs have increased substantially, reaching approximately 25% of the number of action potentials of the GIs from normal animals. By contrast, there was no significant increase in the responses of the right GIs. Thus, the normal left–right GI balance was partially restored, a restoration that presumably contributed to the partial correction of the turning behavior.

In addition to a change in the number of action potentials, there was also a substantial restoration of the relative latencies of the left *vs* the right GIs. In normal animals, in response to a wind puff from the left front, the first GI to be excited is a left GI on 92% of the trials. One day after ablating the left cercus, this drops to only 15% of the trials; but by 30 days, it returns to 63%. This partial restoration of the normal left-then-right sequence of the initial GI action potentials may also contribute to the partial behavioral correction.

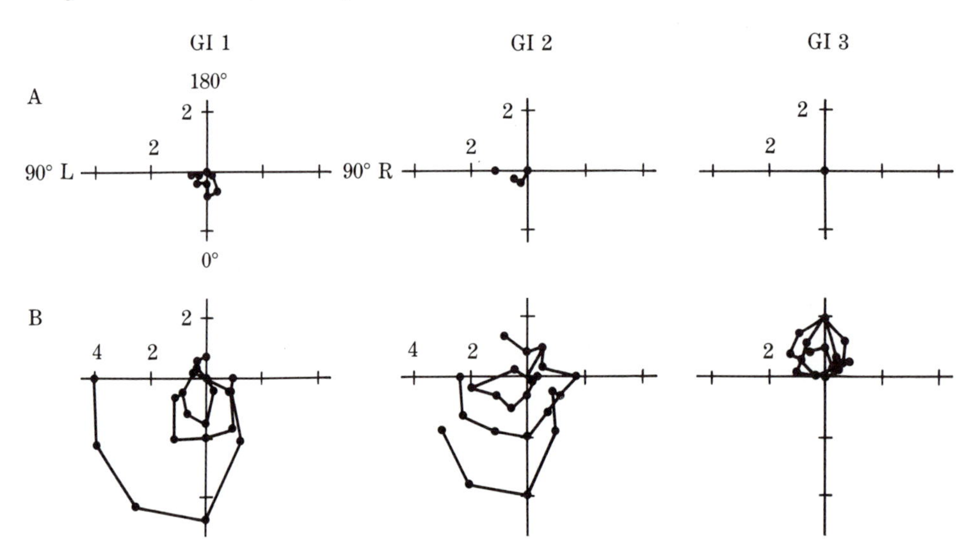

FIGURE 19. Plasticity in the responses to wind of GIs 1, 2, and 3. The method of plotting these graphs is the same as in Figure 11. A. One day after ablating the left cercus, these and other left GIs are almost unresponsive to wind. B. Thirty days later, these GI responses have made a significant recovery. Each graph shows the results from three different neurons, tested at 12 different angles; number of action potentials is indicated on the axes. Thus, for instance, the single point on the graph in A for GI 3 indicates that this GI gave very close to zero action potentials for all wind directions. (From Vardi and Camhi, 1982a.)

All the sensory input to the left GIs on day 30 appears to come from the right cercus because covering this cercus silenced all the GI responses to wind. One can conclude, then, that ablating the left cercus causes some type of strengthening of the functional connections between the right cercal wind receptors and the left GIs. So far, the cellular mechanism of this strengthening is not known.

What is the specific cue that informs the neural circuit that, with the left cercus missing, the functional connections from the right sensory cells to the left GIs should now be strengthened? One possibility is that degeneration of the left cercal nerve, which is known to follow cercal ablation, produces chemical breakdown products that are sensed by neurons in the terminal ganglion and initiate the functional changes. However, degeneration does not appear to be necessary for these changes to occur. If one covers with glue (rather than ablates) the left cercus, the cockroach's behavior and physiology show the same plasticity in response to wind from the left: on day 1, mistaken turns to the left and nearly silent left GI's; by day 30, significant improvement of both the turning and the left GI responses. The glue on the left cercus does not kill the wind receptor neurons in the left cercal nerve (and thus does not lead to their degeneration), as determined both by axonal counts carried out with an electron microscope and by physiological recordings from the sensory axons. Thus, merely the lack of wind-evoked sensory activity on the left side is a sufficient condition to bring about a functional strengthening of the connections from right cercus to left GIs.

During the 30-day correction period, the intact, right cercal nerve must be normally active in order for the behavioral and neural correction to occur (Volman *et al.*, 1980). For instance, if the left cercus is ablated and the right cercus is covered with a plastic tube or with glue for 30 days, when this covering is then removed, no behavioral recovery is seen.[7] Moreover, recordings from left GIs 1, 2, and 3 now show that, though the numbers of action potentials have slightly increased, this increase is smaller than if the right cercus had been exposed to the air during the 30-day correction period (Volman and Camhi, 1982). It seems, then, that the cue for the functional strengthening of the connections from right cercus to left GIs is a long-term comparison of the activity occurring in the left *vs* that in the right cercal nerves. Precisely where, and by what cellular mechanisms, this long-term comparison is carried out should prove to be an interesting study in neuroethology.

Summary

A parallel set of behavioral and neurobiological experiments has uncovered several principles of neuroethological organization of the cockroach's escape be-

[7]To remove the glue, one waits for the animal to molt. (Only animals that molted between 30 and 35 days after the left cercal ablation and right cercal covering were used in this experiment.) Upon molting, the glue is shed and a fresh set of sensory hairs appears on the right cercus. Most of these are innervated by the same sensory cells as existed before the molt, but a small population of new hairs is innervated by newly developed sensory cells. On the left side, a small cercal bud develops where the cercus had been ablated. This usually contains a small number of sensory hairs. However, the bud with its hairs was ablated 1 day before carrying out the behavioral or physiological tests.

havior. Cockroaches respond to the approach of a natural predator by turning and then running away. The wind made by an approaching toad provides the cockroach's cue to this predator's approach. Hundreds of wind-receptive hairs on the cerci detect the wind and, as a group, encode its direction. Background wind is filtered out of this sensory system on the basis of its having a lower rate of acceleration than the wind coming from the toad. Giant interneurons (GIs) activated by the wind receptors retain a modified version of the receptors' code for wind direction. Bilateral comparisons between the wind-evoked activity in particular pairs of GIs determine, at least in part, whether the turn will be to the left or to the right.

Newly hatched cockroaches have just two wind-receptive hairs on each cercus. These are sufficient for evoking turns away from a wind source, turns that are as directionally accurate as those of adults. However, even though hatchlings and adults show the same behavior, there can be changes in this behavior during the animal's life. One such change occurs in response to the loss of a cercus. Removing or covering one cercus leads to mistaken turning directions, which appear to result from a left–right imbalance in the responses of the GIs. Within 30 days, both the GI responses and the behavior are substantially corrected. At least one signal that leads the nervous system to carry out this gradual correction is a long-term comparison of the impulse activity originating in the two cercal nerves.

Questions for Thought and Discussion

1. List as many possible mechanisms as you can by which the functional connections from the right cercal sensory cells to the left GIs could become strengthened, as occurs after ablating a cockroach's left cercus. Design a set of experiments that would help you determine which of your suggested mechanisms these cells actually use.
2. In response to the wind of an approaching toad, certain species of cricket first turn away, just as a cockroach does, but then crickets jump repetitively instead of running. Design a set of experiments to determine which of these insects has the more successful escape strategy. What assumptions do you need to make? How can you justify these assumptions?
3. Suppose that someone suggests to you that the repetitive jumping of the cricket described in question 2 is disadvantageous to the jumping individual's own survival, because a single jump would be enough to escape from the predator and subsequent jumps would only draw the predator's continuing attention. How could you test this question? If this suggestion is correct, what hypothesis can you offer for a possible adaptive function of the repetitive jumps? How could you test your hypothesis?
4. If a prey species is very successful in escaping from a given species of predator, this predator may abstain from striking at this prey. Two advantages of *not* striking would be that the predator could save energy, and could avoid detection by its own enemies. If the predator does in fact abstain from striking, its rapidly escaping prey would derive a double advantage from its escape behavior; discouragement of attack by predators and effective evasion in

the event such an attack does occur. Design an experiment to determine whether cockroaches are doubly protected in this way. Include all necessary controls.

5. Design a set of experiments to determine where in the neural circuit for the cockroach's escape behavior this insect's discrimination of low *vs* high wind acceleration takes place.
6. Draw a set of synaptic connections from specific GIs onto their postsynaptic neurons that you think could give rise to the cockroach's directional turning responses to wind puffs. Design a set of experiments that could determine what the actual connections are. Specify what types of electrodes you would use, exactly where you would place them, and precisely what you would look for on the oscilloscope trace.

Recommended Readings

BOOKS

Roeder, K. D. (1963) *Nerve Cells and Insect Behavior.* Harvard University Press, Cambridge, Massachusetts.
The publication of this beautifully written little book provided a considerable stimulus to the then infant field of neuroethology. Though now somewhat outdated, it still makes excellent scientific reading. It contains one chapter on the cockroach escape system, which was first studied in depth by Roeder.

Ewert, J. P., Capranica, R. R. and Ingle, D. S. (eds.). (1983) *Advances in Vertebrate Neuroethology.* Plenum Press, New York.
This recent symposium volume, though devoted exclusively to vertebrate studies, offers a broad view of current research topics in neuroethology.

ARTICLES

Camhi, J. M., Tom, W. and Volman, S. (1978) The escape behavior of the cockroach *Periplaneta americana.* II. Detection of natural predators by air displacement. *J. Comp. Physiol.* 128: 203–212.
This article establishes that wind from an approaching predator is the cue used by a cockroach to detect the predator. Modifications of this argument are presented in the article (cited below) by Plummer and Camhi.

Plummer, M. R. and Camhi, J. M. (1981) Discrimination of sensory signals from noise in the escape system of the cockroach: The role of wind acceleration. *J. Comp. Physiol.* 142: 347–357.
A description of the cockroach's highly sensitive wind-receptive system and how it avoids being jammed by background and self-generated winds.

Ritzmann, R. E., Tobias, M. L. and Fourtner, C. R. (1980) Flight activity initiated via giant interneurons of the cockroach: Evidence for bifunctional trigger interneurons. *Science* 210: 443–445.
This article demonstrates that under certain conditions, action potentials in certain of a cockroach's giant interneurons lead not to running, but to flying.

Westin, J., Langberg, J. J. and Camhi, J. M. (1977) Responses of giant interneurons of the cockroach *Periplaneta americana* to wind puffs of different directions and velocities. *J. Comp. Physiol.* 121: 307–324.
Demonstrates how the direction and other parameters of a wind puff are encoded by the cockroach's giant interneurons.

PART II SENSORY WORLDS

Each animal has its own Merkwelt *(perceptual world) and this world differs from its environment as we perceive it, that is to say, from our own* Merkwelt.
(N. Tinbergen, 1951)

Animals are exposed to environmental energy in a rich variety of forms. A partial list of these forms would include mechanical energy such as sound, ground-borne vibrations, wind, and water currents; chemical energy such as that contained in air-borne odorant molecules and in foods; electromagnetic energy such as visible light; the forces of gravity and of the earth's magnetic field; and electric currents detected by specialized electroreceptors of some fishes. Each of these several forms of energy can occur with a wide variety of physical parameters. For instance, as explained in detail in this section of the book, light and sound energy are wave phenomena, and each can have a wide range of wavelengths. No animal is capable of responding to all wavelengths of light or of sound or to all the parameters of any form of energy. Rather, sensory systems are selective, evolution having adapted a given species of animal to respond only to certain parameters that are available in its natural habitat and that contain information needed to carry out its necessary life functions. This perceptual world, called the UMWELT or *Merkwelt* of a species can be so different from our own that it can be a source of wonder to discover how some animals make their way effortlessly in worlds within which we ourselves are helplessly blind. We shall encounter some notable examples in these chapters.

This section of the book is about the energy in the environment and how animals make use of it to carry out behavior. In order to treat this monumental subject on a manageable scale, we will consider just two forms of energy: light and sound. Chapter 5, on vision, describes (1) the nature of light energy in an animal's environment, (2) how animals utilize this energy to guide their behavior, and (3) how light energy is encoded by the receptors of the eye. Chapter 6 deals with a parallel set of questions for sound energy and the ear. Chapter 7 describes how visual cues, and to a lesser extent auditory cues, encoded by sensory receptors are analyzed by the brain in order to create for the animal a meaningful perceptual world.

Although only vision and hearing are treated here, studies on other sensory systems have contributed importantly to sensory neuroethology. Interested readers should consult the following sources: for chemical stimuli, Schneider (1974), Adler (1976), Muller-Schwarze and Mozelle (1977), Papi *et al.* (1978), Carterette and Friedman (1978); for electric detection in fish, Hopkins (1974, 1980), Heiligenberg (1977), Bodznick and Northcutt (1981); for magnetic detection, Keeton (1971), Blakemore (1975), Phillips (1977), Kalmijn (1978), Gould et al. (1978), Walcott *et al.* (1979); for various senses, Kreithen (1978). Much valuable material can be found in the *Handbook of Sensory Physiology* and in issues of the *Journal of Comparative Physiology*, among other journals.

CHAPTER 5

Visual Worlds

Most animals use light energy to obtain information about their surroundings. This information may include the presence, identity, and location of important items such as food, suitable shelter, potential mates, approaching predators, and landmarks for orientation and navigation. Animals have evolved different strategies for employing light in these tasks, depending upon the physical parameters of available light energy and the nature of the task to be performed.

This chapter is about the use of light energy in animal behavior and the encoding of this energy by photoreceptors. We will begin with a discussion of the physical nature of light, which provides a basis for understanding both visually elicited behavior and the structure and function of photoreceptors. We will then consider in detail the visual behavior and photoreception of one especially well studied animal, the honeybee. This treatment will touch upon several key principles of visual neuroethology. For comparison, there follows a discussion of visually guided behavior and photoreception in vertebrate animals.

Physical Properties of Light

WAVE PROPERTIES

Imagine that you could hold two electrostatically charged particles, one charged positively and the other negatively, that is, a DIPOLE. Suppose further that you could oscillate these two particles, moving them together and apart repeatedly at high frequency. This movement would generate a wave of energy that would propagate outward in all directions perpendicular to the line connecting the two charges. One such propagation path is shown in Figure 1. The wave would consist of an electric force and a magnetic force. The electric force would oscillate in a direction parallel to the oscillation of the two charged particles. The vector (i.e., direction and magnitude) of this electric component is called the E-VECTOR. Oscillating at right angles to the E-vector would be the magnetic force, the direction and magnitude of which is expressed by the H-VECTOR. Both the E-vector and the H-vector are oriented perpendicular to the direction that the wave propagates. For both vectors the FREQUENCY (the number of peaks or

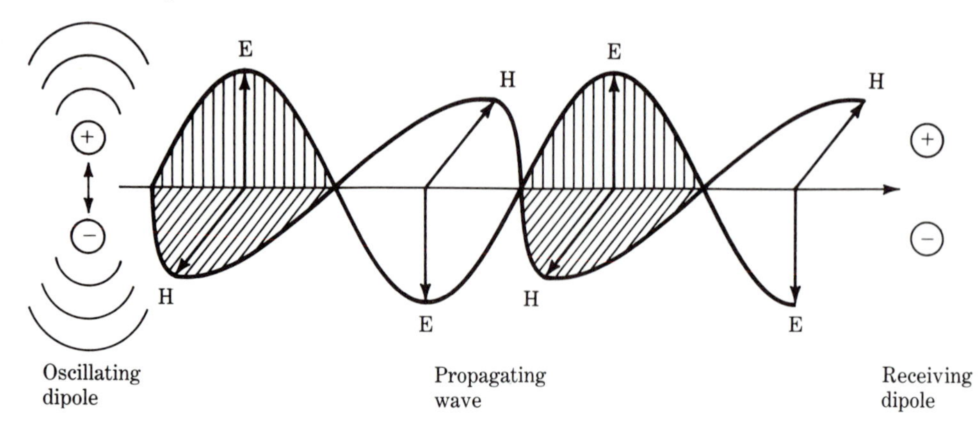

FIGURE 1. Propagation of electromagnetic radiation from an oscillating dipole. The distribution of E- and H-vectors at one instant in time is shown. Energy would also radiate from the oscillating dipole toward the left, and both in and out of the page—in fact, in all directions within the plane perpendicular to the line between the two oscillating charges. A receiving dipole is also shown; it is oriented properly for absorbing energy from the wave shown.

troughs per second, measured at a fixed point) would equal the frequency with which you had oscillated the charged particles. The whole wave would propagate away from the source at the speed of light.

Now you place a second dipole in the path of this propagating wave. This second, "receiving," dipole can be set into oscillation by the wave, but only if it is parallel to the dipole that generated the wave and thus parallel to the E-vector of the wave. By oscillating in response to the wave, the receiving dipole absorbs some of the wave's energy.

What we have just described is the generation, propagation, and absorption of ELECTROMAGNETIC, or EM, RADIATION. The phenomenon we call visible light is just such an EM radiation for which the frequency lies between about 4.3×10^{14} and 7.5×10^{14} cycles/sec. EM radiations, however, are most commonly designated not by their frequencies but by WAVELENGTH (the physical distance between neighboring peaks of the E-vector). There is a fixed relationship between the wavelength *lambda* (λ), the frequency *nu* (ν), and the speed of propagation[1] of light s, namely,

$$\lambda \text{ (meters)} = \frac{s \text{ (meters/second)}}{\nu \text{ (second}^{-1}\text{)}} \tag{1}$$

In terms of wavelength, the spectrum of human vision extends from about 400×10^{-9} to 700×10^{-9} meters [that is, 400–700 nanometers, (nm)]. Within this

[1]The speed of light s depends upon the medium through which it is propagating. This speed is maximal in a vacuum, namely, 3×10^{8} m/sec, a value usually designated as c.

range, we perceive different wavelengths as different colors of light. Wavelengths somewhat shorter than 400 nm are called ULTRAVIOLET (UV); those somewhat longer than 700 nm are called INFRARED (IR). Even though the human eye is blind to these radiations, wavelengths as short as about 300 nm and as long as 1100 nm do actually arrive at the earth's surface from the sun. (However, even this wider range of EM wavelengths on earth is but a tiny fraction of the overall spectrum of EM radiation that exists in the universe; Figure 2.) As we shall see, there are numerous animals that respond to EM radiations in the UV range and a few that respond to radiations in the IR range.

The description of oscillating charges applies well to the earth's major source of light, the sun. Inside the sun, numerous electrons are stripped from their nuclei by the intense thermal energy. These nuclei thus carry net positive charges. Through thermal agitation, the positive nuclei and the negative electrons move relative to each other, generating EM radiation.

Because the movements of these charged particles in the sun occur in random directions, EM radiation is emitted outward in all directions. Moreover, all possible E-vector orientations occur. This latter point can be appreciated in terms of Figure 1 by imagining a pair of charged particles that happens to be oscillating not upward and downward, but rather, perpendicular to the plane of the page. This would produce E-vectors that are exactly like the H-vectors in the diagram, and vice versa. Intermediate E-vector orientations would be produced by intermediate dipole orientations, oblique to the page. Such a range of E-vector orientations is shown in Figure 3A for one propagation path from the sun to the earth. Given this range of E-vectors, any receiving dipole on the earth's surface that is oriented perpendicular to the path of the sunlight will be exposed to some light energy whose E-vector orientation matches the receiving dipole's orientation. Thus, absorption of light can occur. Such absorption, occurring within the photoreceptor cells of an animal's eye, initiates the process of vision.

Some sunlight arrives at the earth's surface only indirectly, being scattered by particles in the earth's atmosphere. The scattering particles can be thought of as tiny reflectors, although physically this is not strictly accurate. The scattered light does not contain a random assortment of all possible E-vector orientations. Rather it has an abundance of certain E-vectors relative to others, a condition referred to as PLANE POLARIZATION. This is illustrated in Figure 3B, where a ray of sunlight is seen to be scattered by an atmospheric particle downward to the earth. Scattering in other directions also occurs but is not shown. The light path labeled A, from the sun to the atmospheric particle, contains all the E-vector orientations perpendicular to this path. (Only two of these E-vectors are shown in the figure: E_1 within the plane of the page and E_2 perpendicular to the plane of the page.) For path B, namely, from the atmospheric particle to the earth's surface, the direction of propagation is parallel to vector E_1. Because, as we have seen, all E-vectors must be perpendicular to the path of propagation, light of vector E_1 cannot end up in path B. Path B is left, then, with only vector E_2, which is perpendicular to both propagation paths A and B. (Vector E_1 would be scattered by the particle in the directions into and out of the page, but this light would never reach the earth and therefore is not of biological interest.) The E-vector orientation of scattered light in the daytime sky is important in vision

ELECTROMAGNETIC
WAVELENGTHS IN
THE UNIVERSE

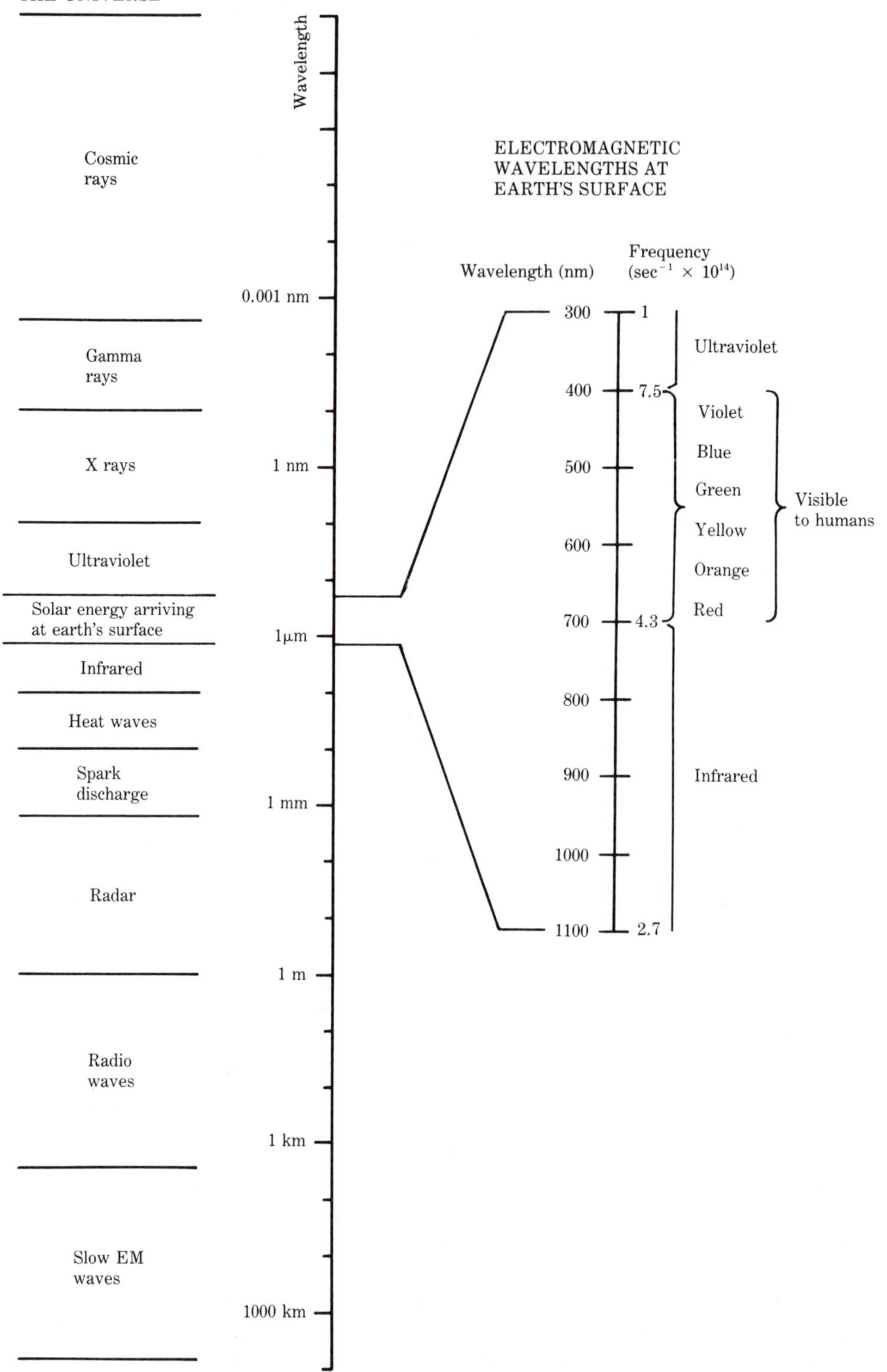
Wavelength
Cosmic
rays
0.001 nm
Gamma
rays
X rays
1 nm
Ultraviolet
Solar energy arriving
at earth's surface
1μm
Infrared
Heat waves
Spark
discharge
1 mm
Radar
1 m
Radio
waves
1 km
Slow EM
waves
1000 km
ELECTROMAGNETIC
WAVELENGTHS AT
EARTH'S SURFACE
Wavelength (nm)
Frequency
(sec⁻¹ × 10¹⁴)
300
1
Ultraviolet
400
7.5
Violet
Blue
500
Green
Visible
to humans
Yellow
600
Orange
Red
700
4.3
800
900
Infrared
1000
1100
2.7

◀ FIGURE 2. Scale of electromagnetic (EM) radiation. Only a tiny fraction of the wavelengths of EM radiation in the universe arrives at the earth, most wavelengths being blocked by the earth's atmosphere. Two additional symbols for wavelength that are sometimes used are the millimicron (mμ), which equals 1 nm, and the angstrom (Å), which equals 0.1 nm.

because, as we shall see, numerous animals are able to detect the orientation of this E-vector and use it to orient their behavior. It is the light propagating directly from the sun (Figure 3A) and that from the sky (Figure 3B) that illuminates our daytime world. We see objects in our environment by virtue of their ability to reflect sunlight and skylight into our eyes.

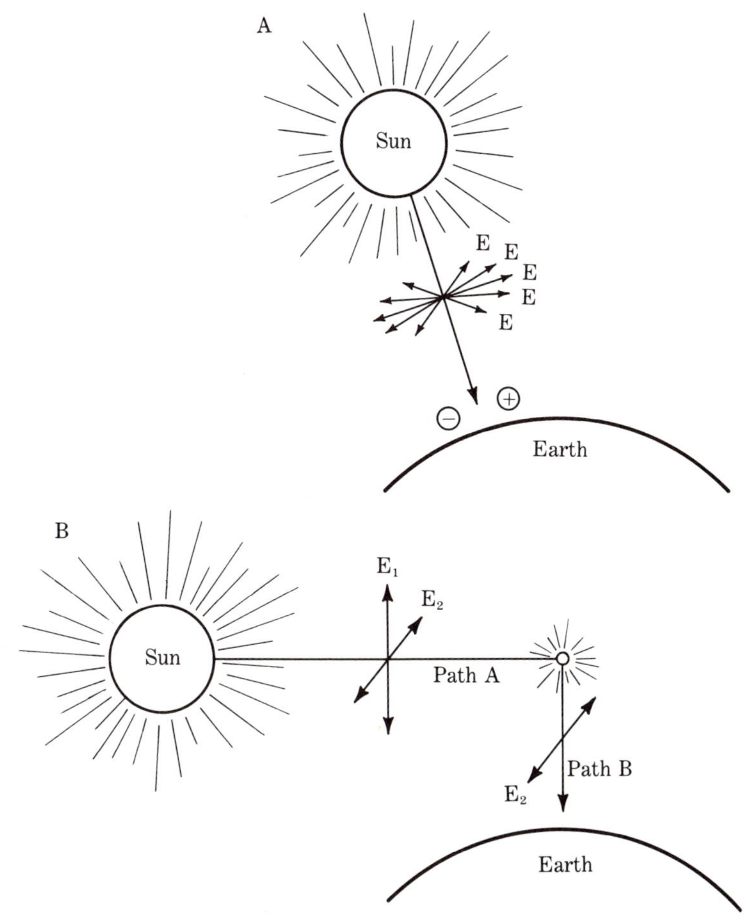

FIGURE 3. E-vector orientations of sunlight. (H-vectors are not shown.) A. All E-vectors perpendicular to the propagation path are contained in direct sunlight. Therefore, any dipole on the earth oriented perpendicularly to the path of light from the sun can absorb light. One such dipole on earth is shown. B. Light scattered to the earth by an atmospheric particle is plane polarized (explained in the text).

The receiving dipoles of vision are molecules called VISUAL PIGMENTS. In all animals so far tested, all visual pigments are derivatives of vitamin A linked to a protein called opsin. For instance, in vertebrate photoreceptors used for perception of dim light, the vitamin A in the isomeric form called 11-*cis*-retinal is linked to opsin, forming the visual pigment called RHODOPSIN (Figure 4). The light energy is actually absorbed by the vitamin A component of the visual pigment. Such a light-absorbing component is referred to as the CHROMOPHORE of the molecule. The chromophore's receiving dipole consists of a chain of alternating single and double carbon–carbon bonds. Within this chain, certain electrons are free to travel in molecular orbitals, circling not just one, but rather all the carbon nuclei. Because these molecular orbitals conform to the elongated shape of the molecular chain, a given electron has, at any instant, a higher probability of moving parallel to the long axis of the chain than perpendicular to it. Thus, light whose E-vector is oriented parallel to the chromophore's chain (arrow heads in Figure 4) is most likely to be absorbed.

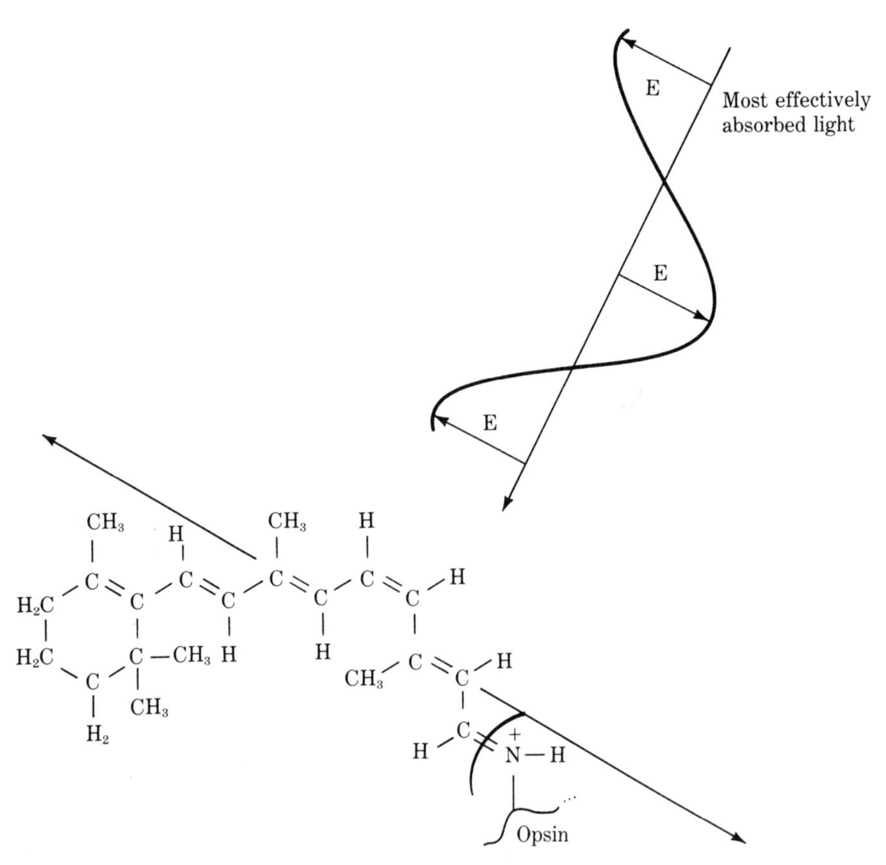

FIGURE 4. Chemical structure of 11-*cis*-retinal, the chromophore of rhodopsin, showing its dipole orientation (double-headed arrow) and the E-vector orientation of light most effectively absorbed (E-vector parallel to the double-headed arrow). The protein opsin is normally linked to the aldehyde at the end of the carbon chain (bottom, right).

QUANTAL PROPERTIES OF LIGHT

Whether or not a given molecular dipole such as that of rhodopsin will absorb energy from EM radiation depends not only on its proper dipole orientation but also on the amount of energy contained in the radiation. In order for absorption to occur, the amount of energy in the EM radiation must be matched to certain discrete possible energy states of the molecule. These energy states can be envisioned largely in terms of the positions of the molecular orbitals of electrons in the dipole portion of the molecule. The absorption of EM energy results in a slight shift of the positions of the molecular orbitals. However, only certain discrete orbital positions are possible, and correspondingly only certain discrete amounts of energy can be absorbed. (In addition, the positions of atomic nuclei in the dipole portion, as well as the positions of other charged parts of the molecule contribute slightly to determine the allowed energy states.)

The amount of energy contained in EM radiation depends upon the wavelength λ of the radiation [or the frequency ν, which is inversely proportional to λ; Equation (1)]. This is best understood by considering EM radiation not as a wave phenomenon but as a stream of particles called QUANTA or, in the special case of EM radiation in the visible range, PHOTONS. Both the wave and the particle descriptions are in accord with physical observations on EM radiation. For instance, a higher intensity of light can be thought of either as a greater amplitude of E-vectors and H-vectors (Figure 1) or as a greater number of quanta per unit time. However, certain phenomena are best understood in terms of one or the other of these alternative descriptions.

The energy E of a quantum is related to the frequency ν of its radiation by

$$E = h\nu \tag{2}$$

(where h is a constant called Planck's constant). Thus, another way of saying that a particular visual pigment molecule would require a particular amount of energy E in order for absorption to occur is to say that the molecule would require a certain value of $h\nu$. But because h is a constant, the molecule would require a particular value of ν; or alternatively, a particular value of wavelength λ, which is inversely proportional to ν [Equation (1)]. Expressed in this way, it is possible to understand one reason why vision is limited to such a narrow range of wavelengths. Light of wavelengths greater than about 800 nm cannot be absorbed by visual pigment molecules because the light has insufficient energy. Light of wavelengths below about 300 nm possesses sufficient energy to damage delicate receptor cells exposed to this energy for long periods. The vertebrate retina is in fact protected from exposure to wavelengths shorter than about 400 nm because the proteins that are abundant in the lens absorb almost all of this potentially damaging energy (Lythgoe, 1979).

One can represent the relationship between the wavelength λ of EM radiation and absorption by a given type of dipole such as rhodopsin by means of a graph called an ABSORPTION SPECTRUM (Figure 5). To obtain the data for such an absorption spectrum, one first exposes a solution of rhodopsin molecules to EM radiation of a single wavelength, called MONOCHROMATIC radiation. One then uses a spectrophotometer to determine the percentage of the arriving energy that is

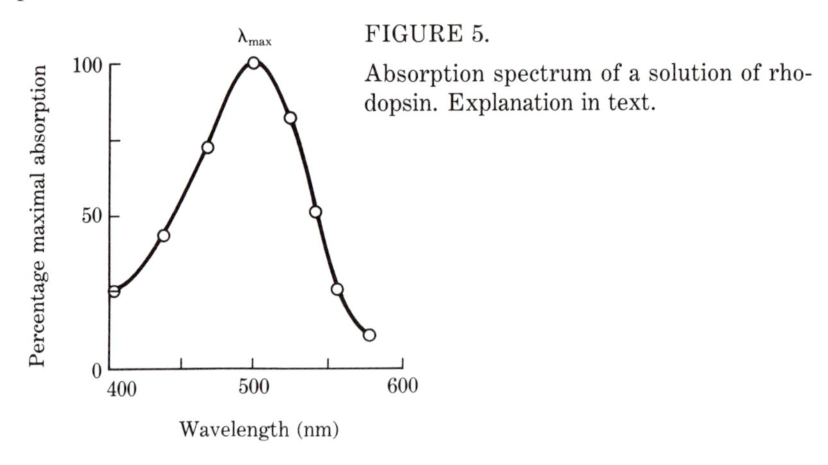

FIGURE 5.

Absorption spectrum of a solution of rhodopsin. Explanation in text.

absorbed by the rhodopsin. Then one switches to several different wavelengths, each time repeating the measurement of percentage absorption. As Figure 5 shows, the closer the wavelength is to some particular value, called λ_{max}, the higher the percentage absorption. At this wavelength, a maximal number of rhodopsin molecules are absorbing the light. For rhodopsin, λ_{max} equals about 500 nm. At much shorter and longer wavelengths, very few rhodopsin molecules absorb EM energy. This characteristic absorption by rhodopsin would determine the range of wavelengths visible to an animal that would use only this one visual pigment. As we shall see, however, many animals use more than just one type of visual pigment. When graphed, the absorption spectra of these additional pigments are very similar in shape to that of rhodopsin. But owing to slight differences in their chemical structure, each would be shifted to the left or right of the rhodopsin graphed in Figure 5, and thus each would have a different λ_{max}.

THE DIRECTION OF LIGHT PROPAGATION

Besides detecting the presence of an object in the field of view, an animal must often determine the location and the structure (i.e., the spatial arrangement of parts) of the object. The ability of eyes to do so depends critically upon the fact that light, in general, propagates in straight paths. As a consequence, light rays that enter the pupil of an eye from a range of different regions in space strike a corresponding range of different retinal positions (Figure 6). Thus, a two-dimensional map of visual space is projected onto the retinal surface. In consequence, a photoreceptor cell at a particular location in the retina is affected only by light arriving at the eye from a particular angle. As we shall see in Chapter 7, this point-to-point mapping of visual space onto the retina has important consequences for the structural and functional organization of the visual pathways within the brain.

But light doesn't always travel in straight paths, and neither do other wave energies, including sound (the subject of the next chapter). If such waves are made to pass through a small aperture, the aperture width being less than the

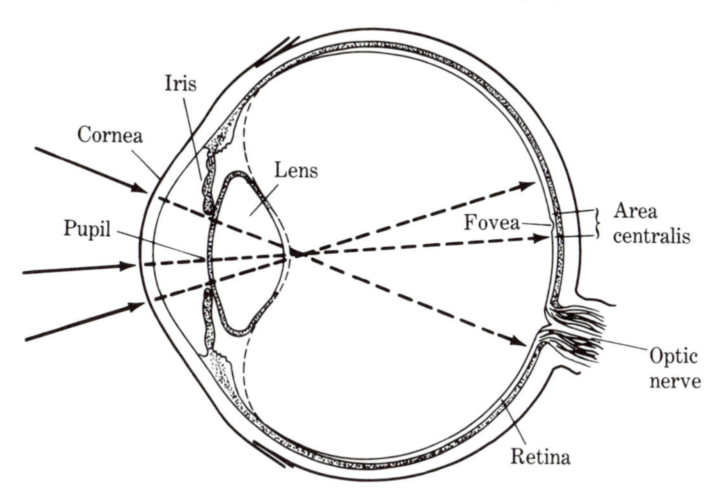

FIGURE 6. Correspondence between the angle of light entering the pupil of the human eye and the location where the light strikes the retina. (After Walls, 1942.)

size of one wavelength of the light, the waves are diverted from a straight path—a phenomenon called DIFFRACTION.[2] Diffraction can be demonstrated by examining surface ripples on a body of water. If one divides an aquarium in half by a partition that has a small aperture, ripples that are in straight alignment on one side of the partition spread out to all regions on the other side of the partition (Figure 7). However, if the aperture is much larger than the distance of one wavelength (the distance between neighboring wave peaks), this diffraction does not occur. Instead, the waves continue straight to the far wall of the aquarium, striking an area only as wide as the aperture (though in practice some slight diffraction does

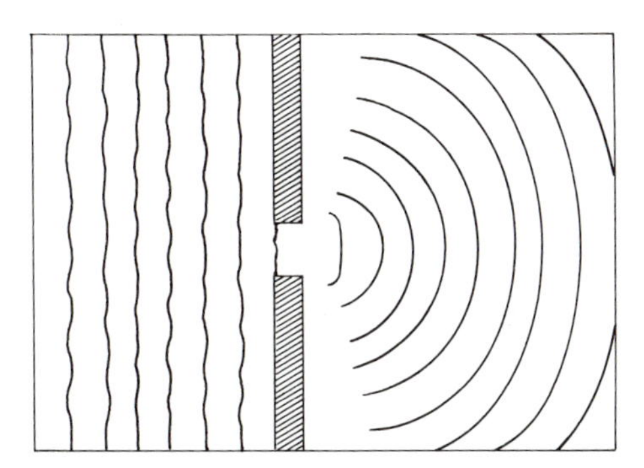

FIGURE 7. Diffraction of surface waves in an aquarium, viewed from above. Straight waves produced at the left pass through a small opening in a partition and spread out to all regions on the right side of the opening.

[2]Diffraction should not be confused with *refraction*, the slowing of light propagation as it enters a denser medium (or speeding as it enters a less dense medium) and the consequent bending of the rays of light, as in a lens.

occur at the aperture's edges). Because the aperture of the eye (namely, the pupil) is much larger than the wavelengths of visible light, aperture size does not introduce diffraction. Therefore, light from a given point in space does not spread to all parts of the retina but only to one retinal point (though in practice the edges of the pupil do introduce some slight diffraction). However, as we shall see in the next chapter, because the wavelength of sound is very great, diffraction of sound waves introduces serious problems for spatial localization of acoustic signals, which auditory systems have solved in interesting ways.

The Visual World of the Honeybee

Having established some physical characteristics of light, we will now consider in depth the acquisition and use of visual information by one animal, the honeybee *Apis mellifera.* This bee is a native of tropical Africa; however, its honey-producing habit and its consequent wide commercial use has resulted in its dispersal throughout most of the world's temperate zones. The visual world of this insect has been explored perhaps more fully than that of any other animal, except our own species, and this research offers considerable insight into the principles of visual neuroethology.

Worker bees (that is, the sterile females) regularly fly out from the colony in search of food—the nectar or pollen of flowering plants. Upon returning home from a rich source of food, a worker arouses other workers to fly out and search for the same food. The returning bee (the FORAGER) communicates to the other bees that it aroused (the RECRUITS), the location of the food source even if, to take an extreme case, this source is as far away as 12 kilometers (von Frisch, 1967; Gould, 1975). In this communication and in the recruit's subsequent finding of the food source, the bees call upon elaborate specializations of their visual capabilities. In an elegant set of experiments for which he was awarded the Nobel prize in 1973, Karl von Frisch (1967) and his colleagues, working in Germany, elucidated many of these capabilities.

In von Frisch's experiments, the bees were housed in a glass-enclosed hive having a single layer of honeycomb and were therefore readily visible to an observer (Figure 8). For different experiments, the hive could be oriented hori-

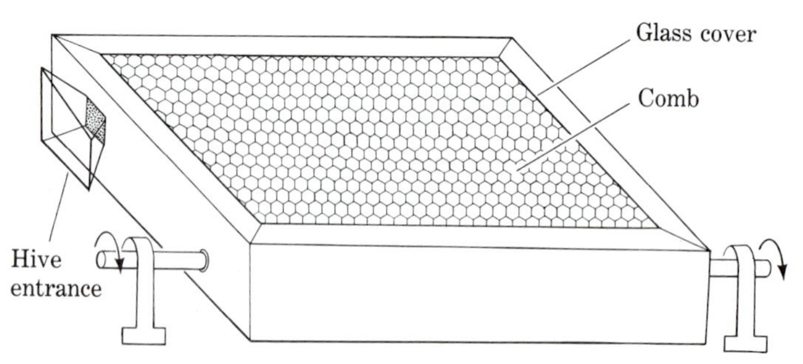

FIGURE 8. Honeybee observation hive, in a horizontal position. The hive can be rotated to oblique or vertical positions. The behavior of the bees can be observed on the comb, under the glass cover.

zontally, vertically, or obliquely; could be either exposed to the light of the sky or covered; and could be transported to different ecological settings as needed. In more recent experiments by others (Brines and Gould, 1979; Edrich, 1979), the bees in the hive have been exposed to an artificial light source of controlled intensity, wavelength, and polarization.

COMMUNICATING THE DIRECTION OF A FOOD SOURCE

In the following experiment, von Frisch demonstrated the fact that foragers communicate to recruits in some manner the location of the food source (von Frisch, 1967, page 129). First he placed in front of the hive a table with a dish containing a weak sucrose solution (0.5 molar). Many bees readily found this "feeding station," drank there, then returned to the hive, and subsequently made repeated trips back and forth to the feeding station. Gradually the station was moved to ever greater distances from the hive, and the bees continued to fly out and back to it. During this period, each bee that arrived at the feeding station was individually labeled with a dab of paint. When fed with these low concentrations of sucrose, the same small group of individual bees, identified by their labels, kept coming back to the feeding station. These foragers apparently were not yet arousing any recruits to fly out for food. When the station had been moved about 300 meters from the hive (position a on Figure 9A), von Frisch changed the food to a more concentrated sucrose solution (2 molar). Also, to simulate a real flower, he added to the food a floral scent, oil of lavender. Meanwhile, four other tables had been set out in other locations (positions b through e in Figure 9A). These "nonfeeding" tables had similar dishes, also scented with oil of lavender, but no sucrose.

Shortly after the change to 2 molar sucrose, numerous recruits, recognizable by being unlabeled, were seen at the feeding station. There and at each of the nonfeeding tables, observers counted the number of recruits that arrived in the next hour. Those recruits that arrived at the feeding station were captured to prevent their returning to the hive and, in turn, arousing yet other recruits. The foragers, however, were permitted to make repeated trips back and forth. By far the greatest number of recruits arrived at the feeding station (a) and at the nonfeeding table nearest to the feeding station (b). The next day the experiment was repeated, but this time the feeding station was located in nearly the opposite direction (at position f in Figure 9A). Again, the recruits selected the direction where the foragers had fed. Figure 9B plots the results of these two experiments.

How the foragers actually communicate to recruits the location of a distant food source became clear when von Frisch observed the behavior of foragers returning to the hive after they had fed on the concentrated sucrose at a distant feeding station (von Frisch, 1967, page 57). Upon entering the hive, a forager performs a stereotyped locomotory behavior called the WAGGLE DANCE—a circling to one side followed by a forward run, then a circling to the other side and again the forward run (Figure 10). The entire sequence was repeated many times. The forward running component was accompanied by a sound and by a repeated sideways waggling of the body. Several of the worker bees were seen to gather around and follow the dancing bee. (When returning from a rich source of food *close* to the hive, a forager did a different dance, the ROUND DANCE.)

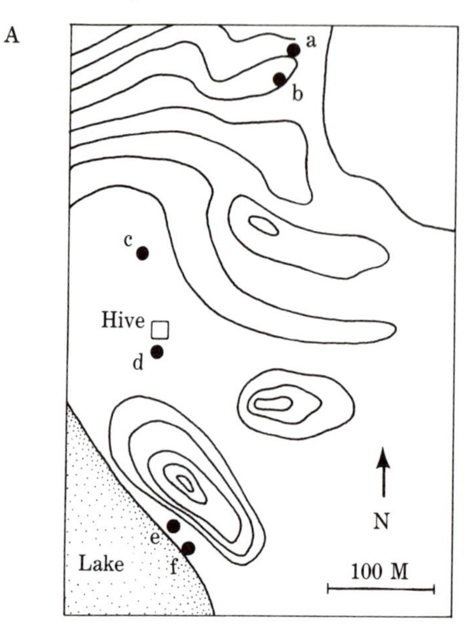

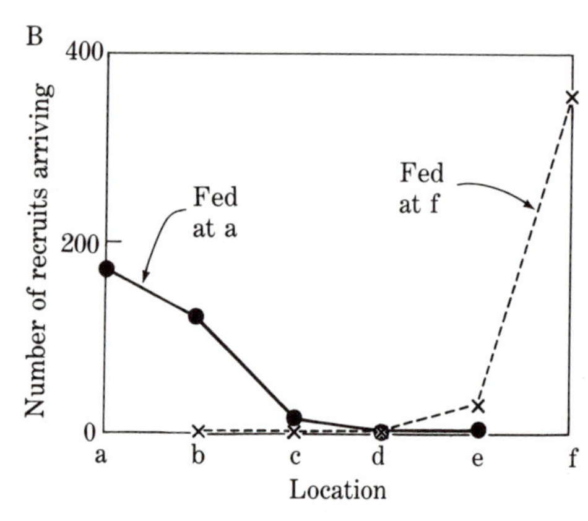

FIGURE 9.

Communication of the location of the food source by foragers to recruits. A. Observation hive, feeding stations a and f, and four nonfeeding tables b–e in a hilly terrain. In the first experiment, bees were fed at a; in the second experiment at f. Wavy and circular topographic lines indicate points of equal elevation. B. When foragers were fed at a, most recruits arrived at a and b. When foragers were fed at f, most recruits arrived at f. (After von Frisch, 1967.)

Von Frisch noticed that if the hive were oriented horizontally and, importantly, if the dancing bee could see the cloudless sky, the dance was oriented such that the forward waggle segment pointed directly toward the feeding station, even if this was very distant and out of sight. (The combs of natural hives are in fact usually oriented vertically, not horizontally, and are shielded from the sky by the hive's outer covering. Nevertheless, much experimental work has been carried out using horizontally oriented hives because of the bee's clear use in such situations of visual cues to direct their dances. Dancing on a horizontal surface does sometimes occur in nature and sometimes in clear view of the sky; Lindauer, 1961.) Bees dancing on a horizontal hive under an open sky were quite persistent in pointing their waggle runs toward the food source, even if the experimenter rotated the hive and thus turned the bee's dance floor under her. This persistent pointing shows clearly that information about the location of food is contained in the waggle dance. However, this does not prove that this information is com-

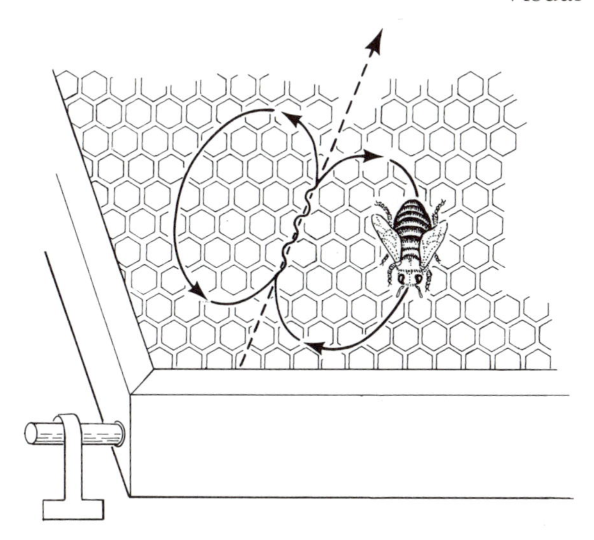

FIGURE 10. The waggle dance of the honeybee, performed on the comb inside a horizontal observation hive. The long arrow indicates the dance component specifying the location of the food source.

municated to and is actually used by the recruits to find the food. By the terminology of Chapter 2 (Figure 3), the forager is transmitting by its dance a coded signal. But are the recruits detecting and decoding this particular signal? This was shown to be the case by experiments in which it was possible to trick foragers into indicating by their waggle dances the wrong direction to the food. Recruits responded by flying out in this wrong direction (Gould, 1975).

CELESTIAL CUES: SUN COMPASS AND POLARIZED LIGHT

In what way is the forager's vision involved in her communicating to recruits by means of the waggle dance? If one blocks a forager's view of the sky while she is dancing on a horizontal hive, although the dance may persist, the waggle portion ceases to point in any consistent direction (von Frisch, 1967, page 131). Thus, some visual cue in the sky seems to be important for properly oriented dancing. One such important cue is the sun, because, if one shields the bee's view of the entire sky except for the sun, normally oriented dancing continues. Shielding the entire sky but providing a single artificial lamp in the position of the sun also permits proper orientation. Now if one moves this lamp, for instance, by 90° around the bee, the dance orientation shifts by this same 90°. Thus, the forager dancing on the horizontal hive appears to point to a food source by positioning its body at a particular angle relative to the sun; in fact, she points at the same angle relative to the sun that she would be pointing if flying straight from the hive to the feeding station. Such orientation by bees and other animals at a selected angle to the sun, because it is similar to the way we might orient at some selected angle relative to the needle of a magnetic compass, is called SUN COMPASS ORIENTATION.

Surprisingly, if one shields the sun but not the rest of the clear blue sky from bees on a horizontal hive, their dances are still properly oriented to the food source. Even just a few small patches of blue in an otherwise shielded or overcast sky are adequate for normally oriented dances (Rossel *et al.*, 1978). But if these patches of sky also become clouded over, the dances become irregularly oriented. Thus, some feature of the blue sky provides directional information that the bees can use. The feature that they use is the E-vector orientation of light scattered from the sky, as was shown by changing the E-vector orientation of the light striking the hive; when the overhead light was passed through a rotated polarizing filter, such as that used to make Polaroid sunglasses, the bees made corresponding rotations in the orientation of their dances (von Frisch, 1967, page 384).

The clear sky contains a map in the form of a spatial pattern of the E-vectors of light scattered to the earth from atmospheric particles in different locations. This is because the E-vector orientation of light scattered from each point in the sky to an observer on earth depends upon that point's position relative to the sun; the E-vector orientation is perpendicular to the line from the sun to that point in the sky (Figure 3B).

One can think of the daytime sky as a celestial hemisphere containing the sun and all visible points in the atmosphere (Figure 11). Consider first the light scattered to an observer on the earth from points along the SOLAR MERIDIAN—the great circle[3] through the sun and the zenith of the sky. The light from all these points will have the same E-vector, namely, perpendicular to the solar meridian and tangential to the surface of the celestial hemisphere. Likewise, points

[3]A great circle drawn on this celestial hemisphere really means a semicircle with the earth at its center.

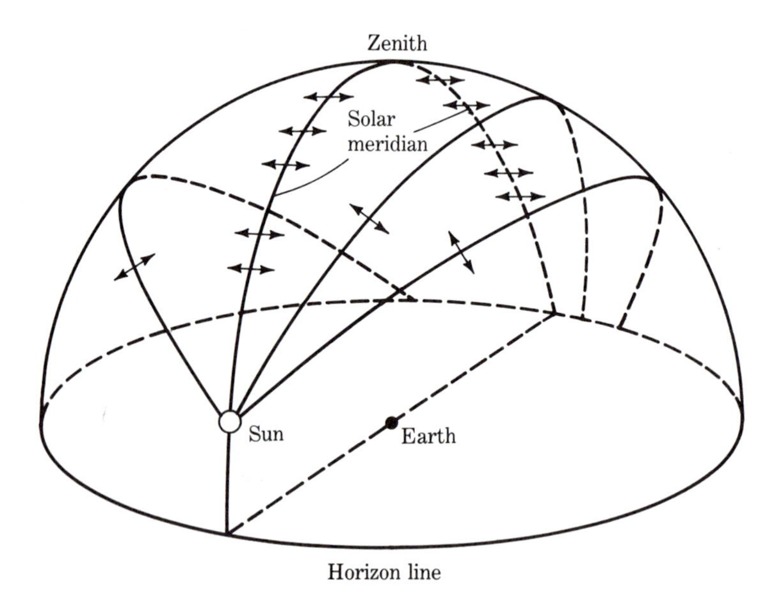

FIGURE 11. Partial map of E-vector orientation of the daytime sky, drawn as a hemisphere viewed from a distant point. The solar meridian and parts of three other great circles passing through the sun are shown in dark lines. Explanation in text. (After Wehner, 1976.)

in the sky along any great circle have E-vectors perpendicular to that circle and tangential to the hemisphere. (Parts of three additional great circles are shown in Figure 11.) As the hours pass and the sun moves through the sky relative to a point on earth, the sky's E-vector map shifts.

It is not known precisely how bees read the sky's map of polarized light. One possibility would be for them to note, while on a foraging flight, the E-vector orientation of every point in the sky and the relationship of each of these to the direction of the food. Then, when they view from the hive any part of the blue sky, they could derive the proper direction to the food source. An alternative possibility (Kirschfeld *et al.*, 1975) would be for them to use the sky's map of E-vector orientation to determine the location of the sun even when it is hidden behind clouds. Then they could use the sun's position to direct their dances in the normal manner. In theory, the sun's position could be calculated by the bees if they determine the E-vector orientation at just two points in the sky. The bee would then need to calculate the great circle line perpendicular to the E-vector orientation at each of these two points. The intersection of these two great circle lines above the horizon line is the position of the sun (Figure 11). Other strategies for finding the sun on the basis of the sky's E-vector information are also theoretically available to the bees (Wehner, 1976). There is, in fact, evidence that under some conditions bees can use the sky's polarization pattern to determine the position of the sun when it is obscured from their view (von Frisch, 1967, page 392; Brines and Gould, 1979).

DANCING ON A VERTICAL HIVE

From the inside of a natural bee hive there is usually no view of sun or sky. Thus, one might predict that any waggle dances in natural hives would be disoriented and, hence, useless. Moreover, the combs of natural nests are oriented vertically, not horizontally. For a bee dancing on a vertical surface, it would be possible to aim the waggle run at the horizon in only two directions, not all compass directions. In fact, though, when von Frisch positioned his observation hive vertically (von Frisch, 1967, page 137), foraging bees did perform waggle dances whose orientations were consistently related to their feeding location. Again, the position of the sun was important, but in a more complex way than with the horizontal hive. Consider first in Figure 12 feeding station 1, located along the sun's AZIMUTH line (an imaginary line along the ground from a given observation point on earth to the intersection of the horizon with the solar meridian). Bees returning from this feeding station pointed the waggle portion of their dance vertically upward. If the feeder lay on a line, for instance, 80° to the left of the sun's azimuth (feeding station 2), the waggle run pointed 80° left of vertical. And if the feeder lay in the direction exactly opposite the sun (feeding station 3), the waggle run pointed directly downward. Thus, bees use the direction of the sun as a reference line for food location. But with the view of the sun unavailable inside the hive, they transpose instead to a vertical reference line. Presumably they can determine the vertical direction by measuring the direction of the earth's gravitational force. Recruits responding to the waggle dance on a vertical hive are just about as accurate in finding the food source as are those responding to the waggle dance on a horizontal hive.

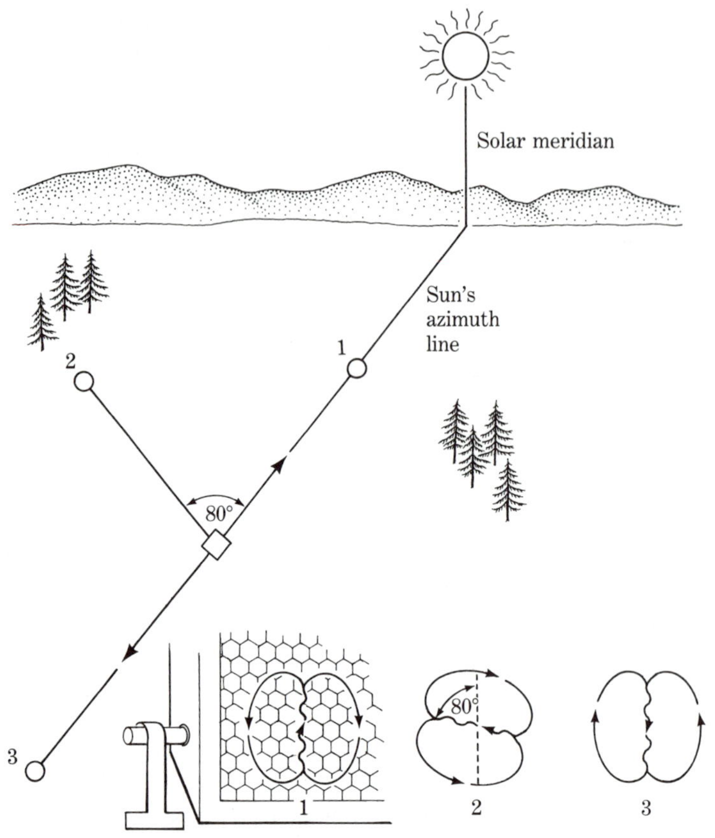

FIGURE 12. Waggle dance in a vertical observation hive. Points 1, 2, and 3 are the locations of three different feeding stations. For each, the orientation of a forager's dances is shown. (After von Frisch, 1967.)

COMPENSATION FOR THE SUN'S MOVEMENT

There is a serious complication in using the sun as a compass. Unlike the magnetic north pole, which provides a fixed reference point for a magnetic compass, the sun's position with respect to an observer on earth is anything but fixed. Thus, foraging bees must constantly update their information on the sun's location in order to use it as a reference point in the waggle dance. That bees actually do this has been shown by several approaches, one of which is observations of so-called marathon dancers (von Frisch, 1967, page 350). These are bees, observed only occasionally, that dance in hives for especially long periods after their last foraging trip of the day. For instance, one of the many marathon dancers studied had been feeding late one afternoon, then returned to the hive, danced for a while, and then stopped dancing overnight. This bee resumed dancing the next morning even though the exit of the hive had been blocked by the experimeter all night long and remained so in the morning. The bee danced from 7:05 to 10:46 AM, during which time the sun's movement was such that its azimuth line shifted by 54.5° toward the west. The bee could not have observed this shift because the hive was inside a tightly closed room and the bee could not leave the hive. Nevertheless, dancing on a vertical hive, the bee gradually shifted the orientation of its waggle dances by 53.5° toward the *east*. Thus, by adjusting its dances it took account of the sun's movement, a phenomenon referred to as TIME COMPENSATION of the sun compass. The result was that the bee continued to signal the same direction to the food source over more than 3 hours. This indicates that the bee possessed accurate information about the sun's movement, even though it couldn't see the sun or the sky. Though such prolonged marathon dances are unusual, the need for time compensation actually arises very commonly. At some times of the day and year and in some locations, the sun's azimuth line can shift by a sufficient angle in just 5 to 10 minutes to require a change of direction of the waggle dance. This is less time than is often occupied by an outbound flight from the hive plus the return flight (Mautz, 1971).

Bees need complex information concerning the sun's course through the sky because this course varies with latitude, season, and time of day (Brines and Gould, 1979). It is difficult (though not impossible) to imagine that all the complex information that would be required concerning the rate of change of the sun's azimuth for all latitudes, seasons, and times of day could be permanently stored as fixed information in the bee's brain. Apparently it is not. Rather, foraging bees appear to measure visually, a few times per hour, the rate of change of the sun's azimuth and use this updated information for time compensation of their sun compass (Gould, 1980). Given this finding, it seems likely that the rate at which the marathon dancer mentioned earlier shifted its dance orientation was determined by its final observation of the moving sun at dusk on the previous afternoon.

WAVELENGTH-SPECIFIC BEHAVIORS

Any given type of visual pigment molecule absorbs only a rather narrow range of wavelengths (Figure 5) compared to the broad range of wavelengths that reaches the earth from the sun and the blue sky (Figure 2). Therefore, one can ask which wavelengths from among this broad range the bee uses to detect the sun, or the polarized light scattered from the sky, or other features of its visual world. This question has been studied using a horizontally oriented hive that was covered by an opaque screen so that the bees could not see the sky at all but instead were presented with artificial, controlled light sources (Edrich *et al.*, 1979; Brines and Gould, 1979). The bees were fed at a distant feeding station on a sunny day. While the dances of returning foragers were occurring in the hive, they were presented with a small spot of artificial light, simulating the sun. If this light was unpolarized (as is the case for direct sunlight; Figure 3A) and contained wavelengths between 400 and 600 nm, the bees set their dance orientations to point properly toward the food source. Thus, the beesbees appear to interpret this spot as the sun.

As mentioned earlier, when bees received no light in the horizontal hive, dances were misoriented. A very low intensity spot of light gave some proportion

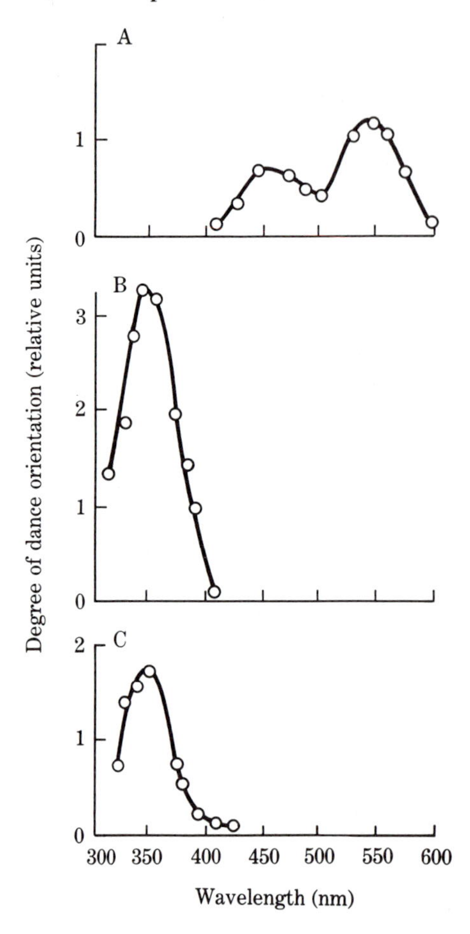

FIGURE 13.

Spectral sensitivity curves of oriented dancing on a horizontal hive. A. Responses to unpolarized light of between 400 and 600 nm. B. Responses to unpolarized light of between 300 and 400 nm; dances were oriented in a direction opposite the food source. C. Responses to a large spot of polarized light of between 300 and 400 nm.

of normally oriented dances; the rest were misoriented. Therefore, by using a dim spot of monochromatic light whose wavelength could be varied, it was possible to determine to which wavelengths, between 400 and 600 nm, the bee was most sensitive in producing properly oriented dances. These results can be represented by a SPECTRAL SENSITIVITY FUNCTION—a graph that plots the strength of some response as a function of the wavelength of monochromatic light evoking the response. (When the response whose strength is plotted is a behavior, as in this case, the graph is sometimes called an ACTION SPECTRUM.) The spectral sensitivity function shows that the bees gave their highest percentage of properly oriented dances at two different wavelengths: 450 and 550 nm (Figure 13A). Peaks of sensitivity at these same two wavelengths have also been found in other behavioral tests (Edrich, 1979). Because all known visual pigments have just one absorption peak in this range of wavelengths, these findings suggest that bees have at least two visual pigments: one with a λ_{max} at 450 nm and the other at 550 nm.

When unpolarized light of wavelengths between 300 and 400 nm was presented to the bees in a horizontal hive, surprisingly they danced in a direction exactly opposite that of the food source. They thus appear to regard such light not as the sun but as skylight opposite the sun. There is, in fact, a unique sky location, called Arago's point, which is directly opposite the sun's position in the sky and which, like the sun's light, is unpolarized. Thus, the bees appear to interpret this light as coming from Arago's point (Edrich *et al.*, 1979). Using a dim spot of unpolarized monochromatic light between 300 and 400 nm to plot the percentage of dances oriented opposite the food source, a maximal percentage was found for wavelengths near 350 nm (Figure 13B). Thus, there appears to be a third visual pigment, with a λ_{max} at this wavelength.

If bees in the horizontal hive were offered a large spot of polarized light, the dances were properly oriented only for wavelengths below 400 nm (von Helversen and Edrich, 1974). These dances pointed directly to the food source. Upon dimming the light, there was a maximal percentage of these properly oriented dances at 350 nm (Figure 13C).

Thus, bees appear to use ultraviolet light to detect the light scattered from the sky (Figure 13B and C) and higher wavelengths to detect the sun itself (Figure 13A). This wavelength specificity apparently is an adaptation to the physical parameters of these two categories of light. Only about 8% of the quanta of direct sunlight are in the UV range, whereas 20–30% of the quanta of scattered skylight are in this range. Moreover, E-vector orientation is less disturbed by certain forms of atmospheric irregularity for UV light than for light of longer wavelengths (von Frisch, 1967).

In summary, specific wavelengths are used by the bee to carry out specific tasks. The spectral sensitivity functions of Figure 13 suggest that there may be three separate groups of photoreceptor neurons, each with a different visual pigment having λ_{max} values of 350, 450, and 550 nm, respectively. Moreover, these behavioral experiments suggest that there may be two types of ultraviolet (350 nm) receptors: one that is sensitive to the plane of polarized light (Figure 13C) and the other that is not (Figure 13B). As we shall see a bit later in this chapter, there is physiological evidence supporting all of these predictions.

A DIGRESSION ON HUMAN COLOR VISION

As an introduction to color vision in bees, it will be useful to establish some general principles of color vision. For this purpose, we will digress briefly to consider vision in vertebrate animals, especially humans. If all of an animal's photoreceptors had just one type of visual pigment molecule—a condition called MONOCHROMACY—this animal should be unable to distinguish unambiguously among different wavelengths of light. This can be seen from the absorption spectrum of rhodopsin in Figure 5. For instance, because the percentage absorption by rhodopsin is the same for approximately 450 and 550 nm, all photoreceptor neurons containing just this pigment molecule would be equally excited by these two wavelengths. Thus, the animal could not distinguish which of these two wavelengths it was exposed to. Moreover, for any wavelength of monochromatic light, different light intensities would lead to different amounts of absorption, so that any given percentage absorption could occur at any wavelength. Thus, the percentage absorption of any one kind of visual pigment molecule (and consequently the physiological responses of a photoreceptor cell that contains just this type of molecule) cannot by itself be used to identify wavelength.

Actually, most vertebrates, including humans, use just one type of visual pigments (rhodopsin) contained in one type of photoreceptor cells (called RODS) to see in dim light. This is referred to as SCOTOPIC vision. Using our scotopic vision we are color blind, all visual features appearing in dim light as different shades of gray. This differs from vision in bright light (called PHOTOPIC vision), for which photoreceptors in the vertebrates are called CONES. In photopic vision we use more than one type of cones each with a different type of visual pigment molecule. This permits wavelength discrimination, or color vision, as will be explained presently.

Some animals, such as the ground squirrel, have just two classes of cones, each with a different type of visual pigment molecule, a condition called DICHROMACY. In such animals, some degree of wavelength discrimination is possible. This is because the *relative* amount of absorption of the two pigments would be different for different wavelengths. For instance, considering the two absorption spectra on the right in Figure 14, light of about 500 nm is absorbed more by one pigment and light of about 600 nm more by the other pigment. This difference in *relative* absorption will persist regardless of the intensity of the monochromatic light. A given *relative* absorption of the two types of pigments will lead to a corresponding relative physiological activation of the two types of cones, which the brain could then interpret as a particular wavelength. Subtle differences in the relative absorption, for nearby wavelengths, could permit finer wavelength discrimination.

The presence of a third population of cones with yet a different visual pigment—called TRICHROMACY—would extend the range of wavelengths that could be detected and discriminated from one another (Figure 14).[4] Human photopic

[4]Trichromacy also solves another problem inherent in dichromacy, namely, distinguishing between white light (that is, light containing all visible wavelengths) and monochromatic light of that wavelength where the spectral curves of the two receptor types cross. If an animal used the two receptor types indicated by the right and middle curves of Figure 14, light of about 550 nm would excite equally both receptor types and would thus be indistinguishable from white light. But if it also had the left-most receptor type, this would be excited by white light and not by light of 550 nm; such an arrangement would enable this discrimination to be made.

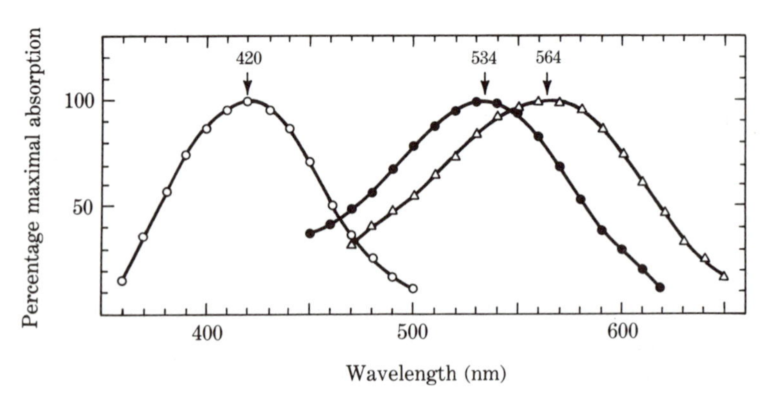

FIGURE 14. Absorption spectra of the three visual pigments of human cones, studied *in situ* by microspectrophotometry. The three spectra have been normalized to the same value of 100% absorption.

vision, and apparently that of many other species, is trichromatic. This idea was originally suggested by Thomas Young early in the nineteenth century, was reiterated by L. von Helmholtz later in that century, and was confirmed only fairly recently by studies on single human cones *in situ* using a microspectrophotometer (Brown and Wald, 1964; Bowmaker and Dartnall, 1980). The absorption spectra of Figure 14 are in fact those of the three classes of human cones.

In spite of what has just been said, it is possible under special conditions for an animal to use just one visual pigment and still discriminate among different wavelengths. This is thought to occur in pigeons, for instance, where cones that all appear to use the same visual pigment also contain colored oil droplets that filter the light before it arrives at the visual pigment molecules. In one large region of the retina, each cone has one of three colors of oil droplets. Thus, in different cones the visual pigment molecules receive one of three different spectra of light, providing a form of trichromatic vision (King-Smith, 1969; Martin and Muntz, 1978).

When a human eye receives monochromatic light of about 640 nm, primarily just one of the eye's three groups of cones is excited—those having the right-most absorption spectrum in Figure 14. This light stimulus results in the sensation of seeing red. If one looks at monochromatic light of about 420 nm, primarily the cones of the left-most curve of Figure 14 are excited, and one sees blue or violet. Monochromatic light of about 530 nm excites primarily cones of the middle curve of Figure 14 (though cones of the right curve are also excited fairly strongly). This stimulus gives the sensation of green. Other wavelengths of monochromatic light give other color sensations; for instance, 500 nm gives blue-green and 580 nm gives yellow (Figure 15A).

Blue, green, and red are often referred to as the PRIMARY COLORS of human vision because, by superimposing on a viewing screen monochromatic light beams of 420, 530, and 640 nm, one can create the sensation of any color in the spectrum.[5] For instance, by superimposing particular intensities of light beams of 640 nm (red) and 530 nm (green), one produces the sensation of yellow that is indistinguishable from the yellow sensation of monochromatic light at 580 nm (Figure 15A). By varying the relative intensities of these two beams, one can produce the whole spectral range of color sensations from green to yellow to orange to red. In like manner, by superimposing beams of 530 nm (green) and 440 nm (blue), one produces the sensation of blue-green indistinguishable from that produced by monochromatic light at 500 nm. Again, different relative intensities produce different spectral color sensations. What surprised early investigators more was that by superimposing wavelengths from near the two ends of the visible spectrum—640 nm (red) and 440 nm (blue)—one produces purple, a color sensation that cannot be evoked by any single wavelength. Moreover, if one superimposes particular intensities of all three monochromatic primaries, this mixture is seen as white. These relationships are represented graphically in the color circle of Figure 15A (Wright, 1972; von Frisch, 1967, page 476).

[5]In fact, there are other sets of three wavelengths that can also interact in this way, as long as these wavelengths are sufficiently far apart on the spectrum. However, the three wavelengths mentioned here are the most commonly used in color-mixing experiments and the simplest to understand, and therefore they are usually designated as the primary colors (Wright, 1972; Clayton, 1977).

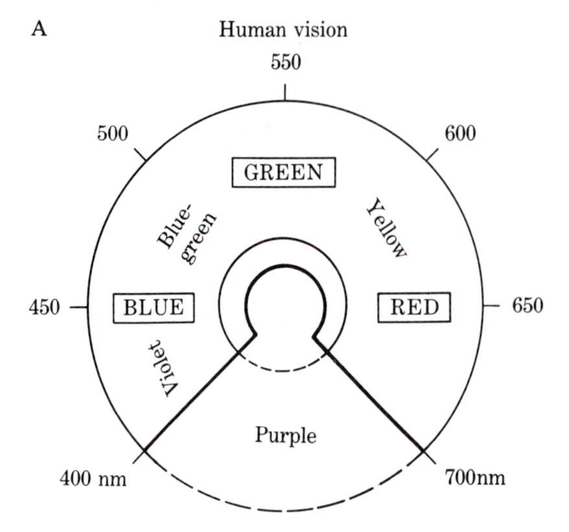

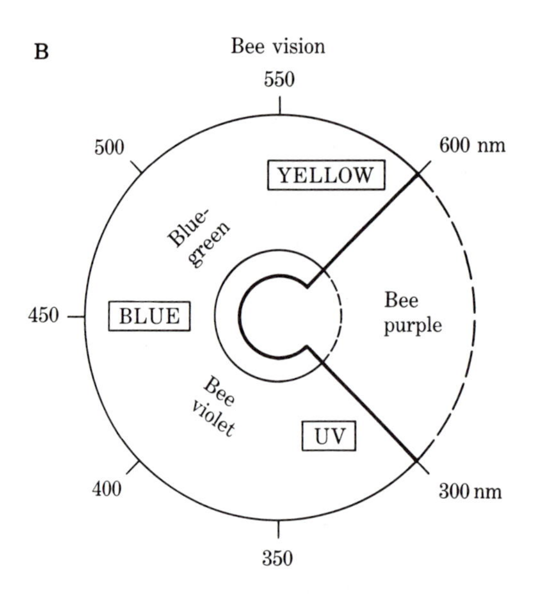

FIGURE 15.
Color circles for human vision and bee vision. Explanation in text.

These and other color-mixing experiments permit a reasonable understanding of the input–output relationship between activity of the three classes of cone receptors and human perception of simple patches of color. However, studies on the perception of intricate color patterns demonstrate the existence of highly complex processing of color information by the brain, much of which is not clearly understood (Land, 1977). Yet even the indirect view of the brain's processing that is afforded by color-mixing experiments reveals a notable aspect of the brain's handling of information coded by the cones. The fact that when two or three groups of cones are simultaneously excited we see not two or three colors, but only one, means that the brain makes a unitary experience out of these two or three separate sensory inputs. A different situation obtains in the human auditory system, for instance, where two different wavelengths of sound, which stimulate two different groups of receptors, result in our hearing simultaneously two separate sounds, as in a musical chord.

COLOR VISION IN BEES

How does color vision in bees compare with human color vision? We have already seen behavioral evidence suggesting that bees, like ourselves, possess three separate spectral classes of receptors (Figure 13). However, in bees the three spectral response peaks occur at shorter wavelengths than in human vision (compare Figures 13 and 14), making bees sensitive to UV light but insensitive to red. Daumer (1956) showed in an elegant set of experiments enploying the classical conditioning techniques (Chapter 2) that bees have good color vision. He trained bees to feed at flower-shaped feeders, each of which consisted in part of a sheet of quartz illuminated from below by light of controlled intensity and wavelength, between 300 and 600 nm. (Quartz, unlike glass, is transparent to UV light.) Once trained to one wavelength, the bees preferentially returned, in a choice situation, to the feeder of the same wavelength. They had come to associate that wavelength with food. The trained bees were resolute in returning to the same wavelength, even if the light at this feeder was made either more or less intense than the light at the other feeders. This invariance of the bee's selection regardless of light intensity is a crucial point. If, for instance, the bee had been trained at a wavelength to which its eyes are especially sensitive, its continued behavioral selection of this wavelength might have resulted not from any discrimination of wavelength *per se* but rather from the fact that this wavelength would appear brightest to the bee's eye. However, by showing that the bee selects the learned wavelength over other wavelengths at either greater or lesser intensities, it was demonstrated that the bee was in fact discriminating wavelength, that is, it has color vision. The bees were shown to be able to discriminate in this way among many different wavelengths of monochromatic light.

Daumer also showed that color vision in bees can be described by a color circle similar to that for human vision (Figure 15B). To show this, he mixed two or three wavelengths at a flower-shaped feeder and tested the ability of bees to discriminate these from particular monochromatic lights. Thus, for instance, a mixture of certain intensities of 550 nm and 450 nm was indistinguishable to the bees from monochromatic 490 nm; that is, having been trained to feed at 490 nm, bees chose equally often to return to the 490 nm feeder as to the feeder illuminated with the 550 nm + 450 nm mixture. Likewise, 450 nm + 350 nm (UV) was indistinguishable from 400 nm, which Daumer called "bee violet." A mixture of 350 and 550 nm, that is, wavelengths near the two ends of the bee's visible spectrum, could be distinguished from any monochromatic wavelength and from all shades of gray. This mixture, by analogy to human vision, was called "bee purple." A mixture of 350, 450, and 550 nm was indistinguishable from a roughly equal energy distribution of all wavelengths between 300 and 600 nm, which Daumer called "bee white."

The bee's sensitivity to ultraviolet light is probably important in its visual responses to flowers because many species of flower strongly reflect light in the UV range. Though invisible to the naked human eye, this UV reflectance can be observed using a standard camera or a television camera outfitted with a quartz lens and a filter that passes only UV light (Eisner *et al.*, 1969). Many flowers that appear to our eyes evenly tinted in fact have a striking pattern when viewed in the UV. For instance, the flower shown in Figure 16 absorbs UV light in the center and reflects it in the periphery. Such circular markings, called nectar guides, have been shown to orient bees to the center of a flower (Daumer, 1958). Ultraviolet reflectance can also cause different species of flowers to appear dramatically different to a bee, even though they may look very similar to a human viewer. This is probably important since different species of flowers are known to produce their nectar at different times of the day. Clearly identifying a rich nectar source from a distance can save a bee time and energy on foraging flights.

Most other invertebrates studied are also sensitive to UV light and insensitive to red (Menzel, 1975; Silberglied, 1979). Many butterflies, however, can detect the entire range of wavelengths from red through UV (Crane, 1955; Bernard, 1979). Thus, like bees, they can be attracted to UV patterns on flowers; and unlike bees, they can be attracted to pure red flowers (Swihart, 1971). They can also see both the red and the UV reflectance patterns on the wings of their potential mates (Crane, 1955). UV wing patterns are known to be used by some butterfly species in courtship signaling (Obara, 1970; Rutowski, 1977; Silberglied, 1979). Red wing patterns, employed as aposematic signals by many butterflies (Chapter 2) might also be useful in intraspecific communication.

FIGURE 16. Composite photograph of the flower of the marsh marigold *(Caltha palustris)*, taken with a normal lens (A) and a quartz lens (B) that is transparent to UV light. (Courtesy of Thomas Eisner.)

STRUCTURE AND PHYSIOLOGY OF THE BEE'S EYE

As with most insects and crustacea (Bullock and Horridge, 1965), bees have compound eyes, that is, each eye is composed of numerous subunits called OMMATIDIA (Figure 17A). Each ommatidium contains a DIOPTRIC apparatus (a lens and crystalline cone for gathering and focusing light) and a small number of receptors called RETINULA CELLS. Bees have about 5500 ommatidia, each with nine retinula cells (Menzel, 1979; Wehner, 1976). These cells form a ring around a central, highly specialized structure called a RHABDOM. The rhabdom consists of numerous microvilli, projecting inward from each of the retinula cells (Figure 17C). All the neighboring microvilli from one cell are oriented in the same direction. However, of all nine retinula cells, there are only four microvillar directions, so that some cells share the same direction (Figure 17B). The membranes of each retinula cell's microvilli are believed to be the location of its visual photopigment molecules (Fuortes and O'Bryan, 1972; Snyder, 1975; Harris *et al.*, 1977). Rhabdoms are found in varying forms in the eyes of most animals belonging to the evolutionary line that includes the flatworms, mollusks, annelids, and arthropods; but rhabdoms are not present in the other major evolutionary line that includes echinoderms and vertebrates (Eakin, 1968; Salvini-Plawen and Mayr, 1977). As we shall see later in this chapter, within the eyes of vertebrates as well as invertebrates, visual pigment molecules are almost always associated with some form of folded extension of the plasma membrane, causing the light to pass through a great many membrane-bound pigment molecules.

Several of the bee's nine retinula cells are individually recognizable histologically in different ommatidia. For instance, in most parts of the eye, one cell, called cell 9, has a short cell body restricted to the proximal region of the ommatidium (Figure 17B). The other eight cells are identified in part by the characteristic positions of their nuclei. Two of these cells have their nuclei far distal and are called class I cells. Four have their nuclei midway along their cell bodies and are called class II cells, and the two remaining cells have proximal nuclei and are called class III cells (Gribakin, 1969). (Another classification scheme takes account of the differences in projection paths of the axons of different retinula cells; Ribi, 1975.) By filling a retinula cell with Procion yellow or other dyes during an intracellular recording experiment, one can often identify by subsequent histological examination the cell from which the recording was made (Menzel and Blakers, 1976).

All of the bee's retinula cells respond to flashes of light with a slow, graded depolarizing receptor potential (Naka and Eguchi, 1962). With sufficient light intensity, this gives rise to a single action potential (Figure 18A). By recording responses to monochromatic light of different wavelengths, Menzel and Blakers (1976) found three categories of cells: those having their maximal sensitivity at about 350 nm, those at 450 nm, and those at 550 nm (Figure 18B). These maxima match remarkably well the wavelengths predicted from experiments on orientation behavior and wavelength discrimination (Figures 13). By filling several of the recorded cells with Procion yellow, it was found that the cells of class I and probably cell 9 gave their peak responses to 350 nm, cells of class III to 550 nm, and among class II cells, some gave their peak response at 450 nm and others at 550 nm (Figure 17B) (Menzel and Blakers, 1976).

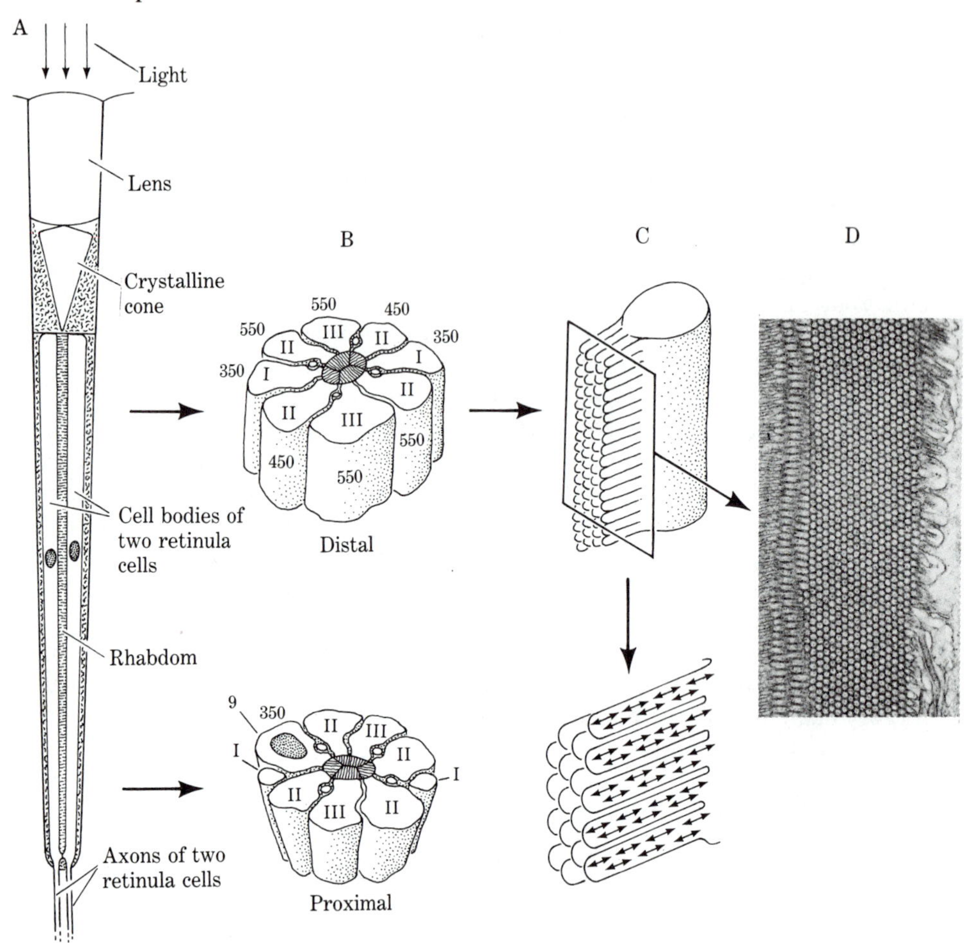

FIGURE 17. The structure of the bee's eye. A. One of the bee's approximately 5500 ommatidia, in longitudinal section. The lens and crystalline cone are parts of the dioptric apparatus for gathering and focusing light. Proximal to these, the elongate cell bodies of two of the nine retinula cells are shown. B. A more distal *(above)* and a more proximal *(below)* portion of the rhabdom showing the spatial arrangement of the cell bodies of retinula cells and their microvilli. Also indicated are the anatomical designations of each cell as a class I, II, III cell or the 9th cell, and the wavelength of maximal physiological sensitivity for each. C. A portion of one cell body *(above)* showing the microvilli of that portion, and an enlargement of some of these microvilli *(below)* with arrows to indicate the presumed orientation of the chromophores of their visual pigments. D. Electron micrograph of a section through a rhabdom. The cell had been sectioned in the orientation shown by the slice in C. The many hexagonal elements are profiles of microvilli cut in cross section. To the left of this region of hexagonal profiles are microvilli from other retinula cells of the same ommatidium. Their microvilli were cut in longitudinal section because they are oriented perpendicular to the microvilli from the cell in C. (A, B, and C after Wehner, 1976; D from Ribi, 1979.)

Which of the bee's retinula cells encode the plane of polarized light used in its orientation to skylight? The sensitivity of different retinula cells to the plane of polarization was tested by comparing the size of a cell's receptor potential to test flashes that had the same intensity and wavelength but different E-vector

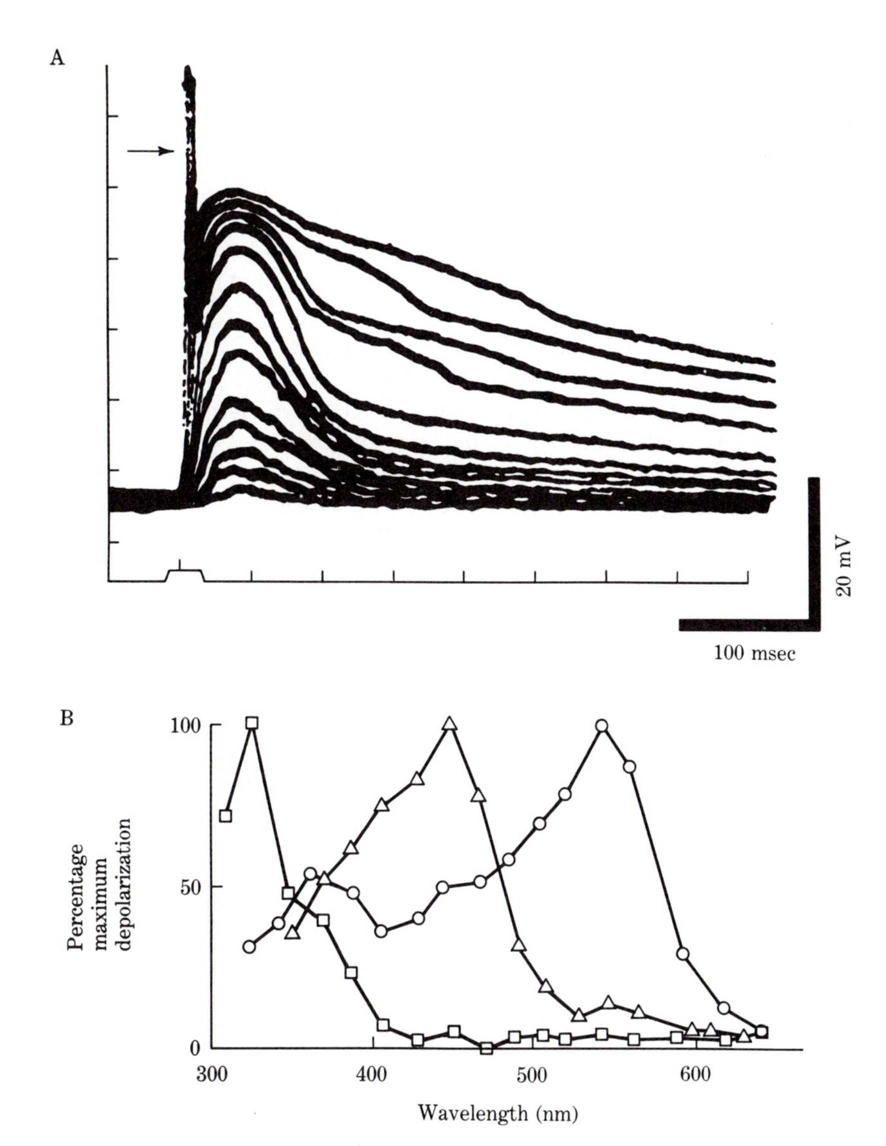

FIGURE 18. Physiology of honeybee photoreceptors. A. Intracellularly recorded responses of a retinula cell to bright flashes of light of different intensities. The duration of the flash is shown on the bottom trace. Flashes of sufficient intensity evoke overshooting action potentials. Arrow indicates 0 mV membrane potential. B. Action spectra of three different retinular cells, recorded intracellularly. (A after Baumann, 1968; B after Menzel, 1979.)

orientations (Menzel and Snyder, 1974). The ratio of the greatest sensitivity (for one E-vector orientation) to the lowest sensitivity (for another E-vector orientation, generally at right angles to the first) is called the POLARIZATION SENSITIVITY of the cell. Of the many recordings made from retinula cells, most gave low values of polarization sensitivity, between 1 and 2. Of all the cells recorded, however, three gave values of polarization sensitivity several times greater than all the rest. Although these three cells were not successfully filled with dye, there is indirect evidence suggesting each may have been a cell 9—the short, proximal cell of the ommatidium (Menzel and Snyder, 1974). This is consistent with the situation in fly ommatidia, where the short, proximal retinula cells have been shown clearly to have much higher values of polarization sensitivities than the long cells (Hardie, 1979). In view of the importance of polarized light in the behavior of honeybees, it is of great interest to identify positively the polarization-sensitive receptors, which will require that additional intracellular recordings be made.

How might a retinula cell, and in particular a cell 9, achieve a high polarization sensitivity? To do so requires that the cell display DICHROISM—a greater absorption of light of some E-vectors than of other E-vectors. Dichroism implies that there is an abundance of visual pigment molecules whose chromophores have one particular orientation, namely, parallel to the E-vector absorbed (Goldsmith, 1973; Laughlin *et al.*, 1975; Menzel, 1975; Waterman, 1975a,b; Ribi, 1979). In invertebrates generally, the maximal light absorption occurs when the E-vector is parallel to the microvilli of the retinula cell recorded (Waterman, 1975b). Therefore, the visual pigment chromophores are thought to be oriented parallel to this long microvillar axis (Figure 17C). There is no clear explanation, however, as to how some retinula cells (putatively cell 9) achieve a particularly high value of polarization sensitivity while others have only low values. Though a number of explanations have been proposed, this is a matter of current controversy (Wehner, 1976; Laughlin *et al.*, 1975; Ribi, 1979).

HONEYBEE VISION: SUMMARY AND PERSPECTIVES

The vision of honeybees has been studied in both the field and the laboratory, by investigators using both behavioral and physiological techniques. The findings from these two different research settings and strategies show a remarkable correspondence, particularly with regard to wavelength sensitivity and the utilization of polarized light. In terms of its effective bridging of field and laboratory approaches, this work on honeybee vision represents a model of neuroethological research.

Through behavioral studies it was found that in communicating the location of a food source to its neighbors a bee relies upon either the position of the sun (aided by time compensation) or the sky's map of E-vector orientation. This deployment of *alternative* orientation strategies reflects the bee's flexibility in dealing with the vagaries of the natural environment. Moreover, bees demonstrate remarkable sophistication in their use of complex visual information such as that contained in the sky's E-vector map. Wavelength-specific behaviors suggested the presence of three different visual pigments, a finding corroborated by intracellular recordings from the photoreceptors. Bees also display trichromatic color vision and, by virtue of the presence of a humanlike color circle, appear to integrate color information in a manner similar to our own. However, bees can see wavelengths shorter than those we can see, being sensitive to ultraviolet light, but cannot see wave lengths as long as those we can see, being insensitive to red. The rhabdomeric organization of the bee's photoreceptors and the apparent alignment of its visual pigment molecules parallel to the microvilli confer upon these receptors the property of dichroism and, thus, polarization sensitivity. It is this polarization sensitivity that enables foraging bees to read the sky's E-vector map and navigate by reading its complexly coded directional information.

Like most research areas, there are limits to what can most profitably be studied in bee vision. In particular, the central nervous system of the bee does not afford the advantages of large, uniquely identifiable neurons, easy to impale with microelectrodes, as are found in some other invertebrate central nervous systems, described in later chapters of this book. Therefore, until new techniques are developed, it seems likely that the major additional progress on bee vision will come through continued studies of this insect's behavior and photoreceptors, and not of its brain.

Vertebrate Vision: A Comparison

Having established some principles of visually guided behavior and photoreceptor organization in the honeybee, we will now compare these principles to those found in vertebrate vision. Many points of similarity, as well as some variations on the theme of the honeybee, will be found.

STRUCTURE AND ORGANIZATION OF VERTEBRATE PHOTORECEPTORS

Vertebrate rods and cones share with invertebrate retinula cells an elongate structure and multiple-folded, pigment-containing membranes, both adaptations that increase the capture of light (Eakin, 1968; Cohen, 1972). However, vertebrate receptors are not clustered together around a rhabdom but rather lie separate from one another (Figure 19). The membranes containing the visual pigments are not out-pocketed microvilli, but rather are internal disks, or "lamellae," contained in the cell's most distal region, the OUTER SEGMENT (Figure 20). The lamellae, in fact, are thought to develop by an infolding of the receptor's plasma membrane. In rods, these infolded membranes become severed during development from the cell's outer surface, leaving the lamellae freely floating within the outer segment. In cones, however, at least some of the lamellae remain permanently attached to the plasma membrane.

The outer segment of a rod or a cone is connected to the receptor's more proximal region, the INNER SEGMENT, by a thin stalk. This stalk contains microtubules that connect to a centriole located in the inner segment. In cross section, the microtubules are seen to be arranged as nine peripheral doublets (Figure 20). This microtubular arrangement is very much like that of cilia and flagella (although,

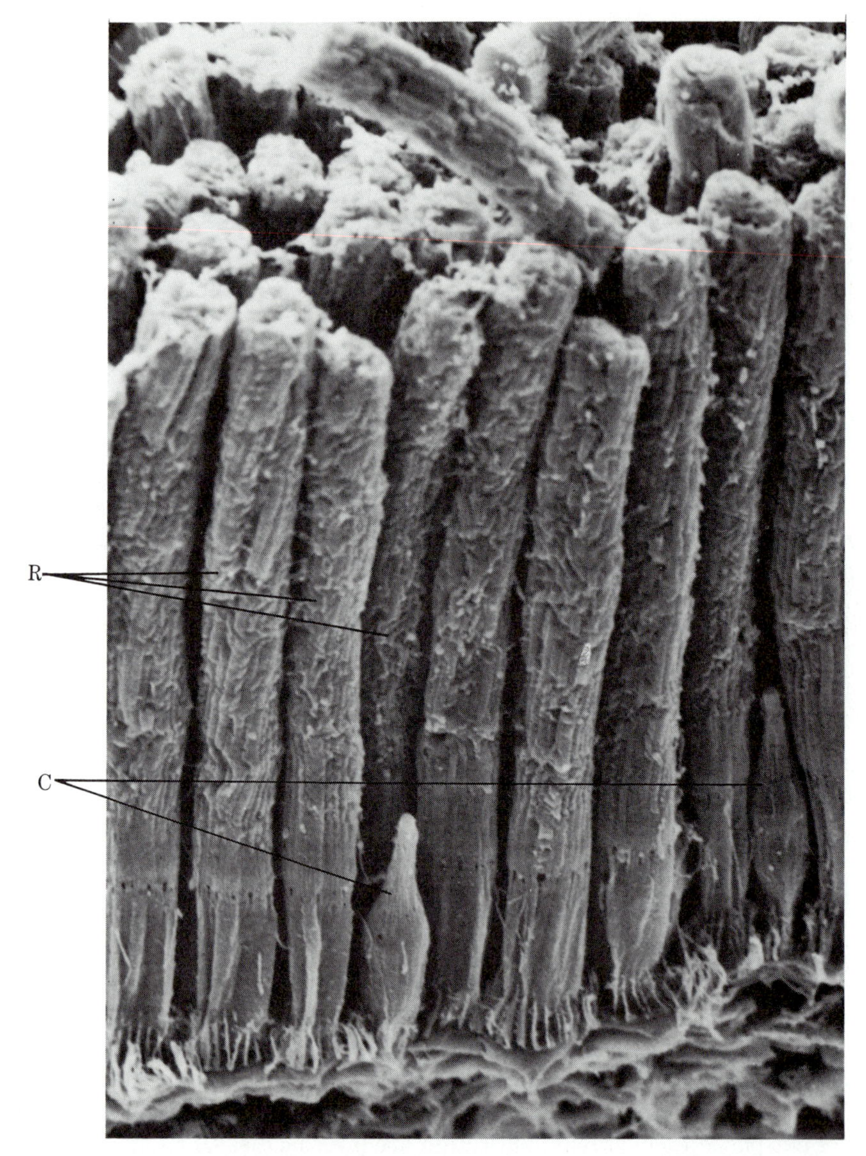

FIGURE 19. Scanning electron micrographs of a toad retina. Longitudinal section showing rods (R) and cones (C). 1200×. (Courtesy of Dr. W. Miller.)

unlike these organelles, the receptors lack both a central pair of microtubules and crossbridges extending among neighboring doublets). Based upon this resemblance, vertebrate photoreceptors, in common with numerous mechanoreceptors as well as some other receptors, are thought to have evolved from cells bearing motile cilia (Prosser, 1973). At the base of the inner segment is the nucleus and beyond this the synaptic terminals for connecting with other retinal cells.

Surprisingly, in vertebrate retinae, the tips of the outer segments point not toward the source of light but rather directly away from it. Before reaching the

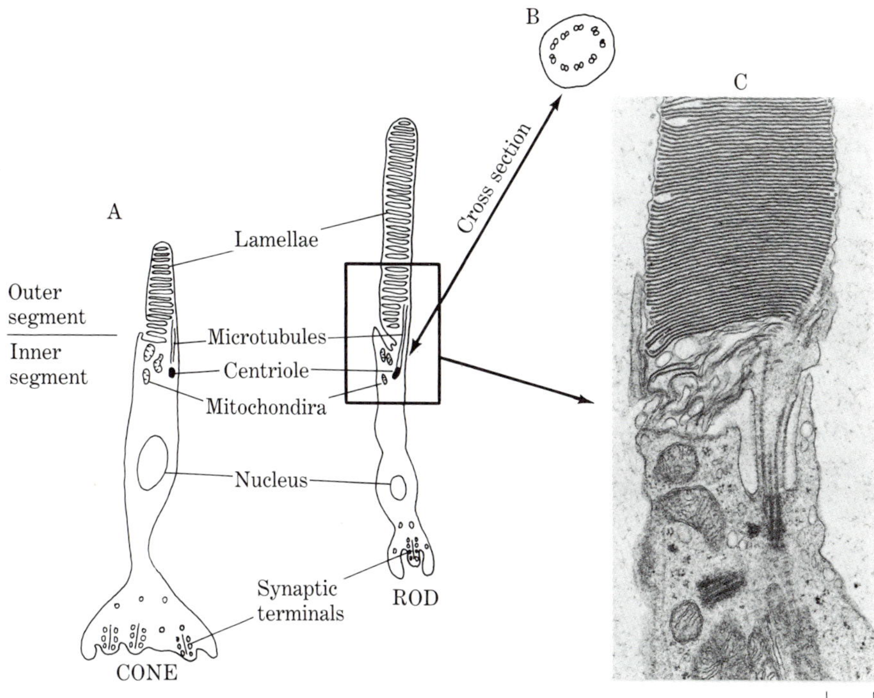

FIGURE 20. A. Schematic drawing of an amphibian cone and rod. B. Cross section through the midportion of the rod, showing the arrangement of microtubular doublets. C. Electron micrograph of a longitudinal section through the midportion of a rod from a toad's eye. This micrograph shows part of the outer segment, part of the inner segment, and the connecting stalk, at 28,000×. (Drawings after Cohen, 1972; electron micrograph courtesy of T. Kuwabara.)

receptor membranes, then, the light must pass through at least two layers of retinal cells as well as through the inner segments of the receptor cells (Figure 21). However, because in most animals all these cells are transparent, this apparently presents little problem.

As mentioned earlier in this chapter, vertebrate animals generally use rods for scotopic vision and cones for photopic vision. This principle is usually referred to as the DUPLICITY THEORY (Walls, 1942; Davson, 1962). The receptors synapse onto two types of cells: BIPOLAR CELLS, which carry information centrally (that is, toward the optic nerve) and HORIZONTAL CELLS, which carry information not centrally but rather laterally within the retina (Figure 21). In general, the rods and the cones synapse on two separate populations of bipolar cells. The bipolar cells in turn synapse on AMACRINE CELLS and GANGLION CELLS (Figure 21). In general, bipolar cells that are driven by rods and those driven by cones synapse on two separate populations of ganglion cells. Therefore, one population of ganglion cells is excited when rods are stimulated and thus is used in scotopic vision,

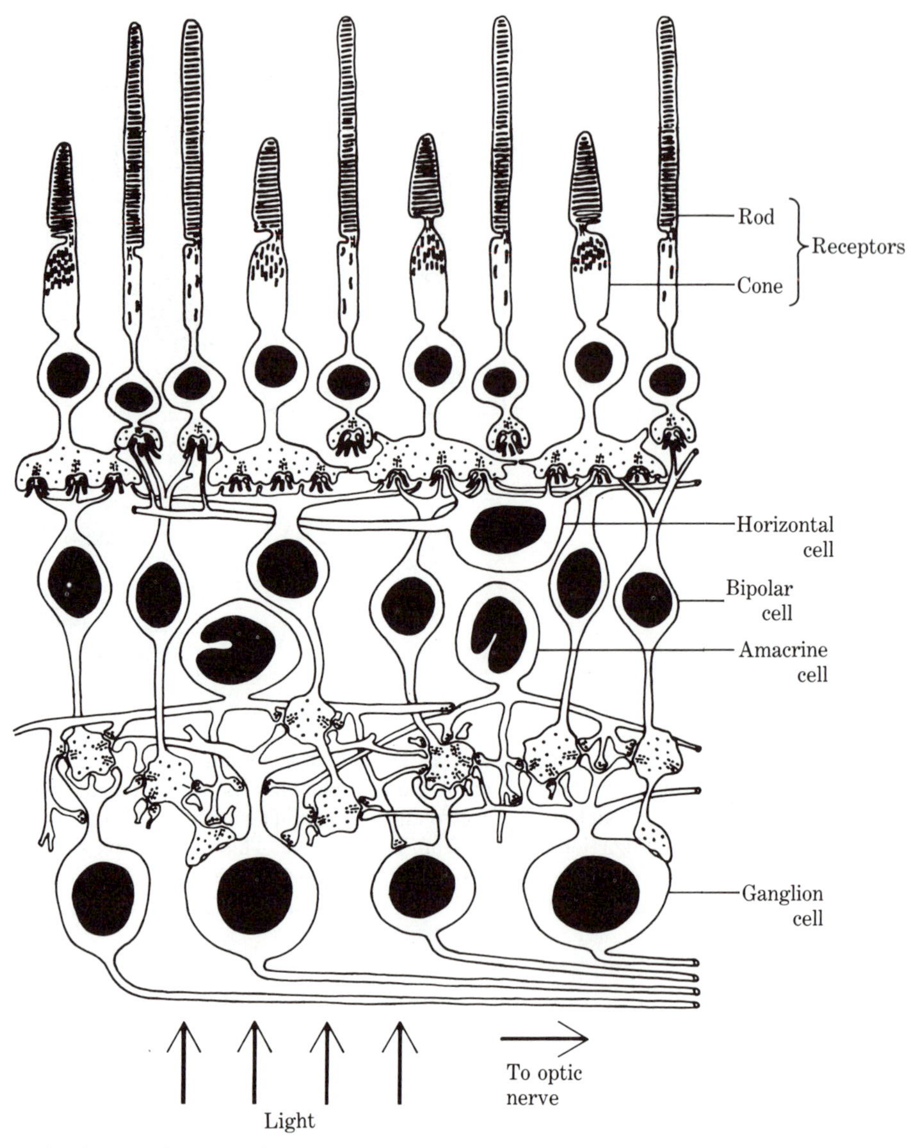

FIGURE 21. Structural organization of the primate retina. For simplicity, not all the synaptic connections are shown. A generally similar organization is found in most vertebrates. Explanation in text. (After Kuffler and Nicholls, 1976.)

whereas a second population is excited when cones are stimulated and is used in photopic vision. The axons of the ganglion cells constitute the optic nerve, which carries all the retinal information that goes to the brain.

The advantage of using rods in dim light apparently is that, having longer outer segments (Figures 19, 20, and 21), they are able to capture a higher proportion than are cones of the light quanta that impinge upon them. Nocturnal animals, with their special requirement for capturing a high proportion of incident light, have a high ratio of rods to cones in their eyes. In fact, some nocturnal animals (for instance, bats and some pelagic fish) have only rods. By contrast, diurnal animals have a high proportion of cones to rods, and some of these (for instance, some snakes, lizards, and birds) have only cones (Walls, 1942).

In bright illumination, a rapid stream of photons arrives at an animal's eye from every point in an observed scene, making high visual acuity possible. However, in very dim light, the eye receives from each location only occasional photons, thus creating only a rough-grained retinal image. Cones specialize in vision of high acuity. They do this partly by virtue of the connections they make with their postsynaptic retinal neurons. In general, the number of cones converging onto any one bipolar cell is smaller than the number of rods converging on a single bipolar cell. Also, bipolars that are driven by cones converge onto fewer ganglion cells than do bipolars driven by rods. The result is that a ganglion cell that is activated when cones are illuminated registers the illumination from a smaller region of the visual world than does a rod-driven ganglion cell. Thus, the cones provide the brain with a finer grained analysis of the visual environment than do the rods.

A special adaptation for high visual acuity (found in the eyes of some diurnal vertebrates, including ourselves) is a thickened region of the retina called an AREA CENTRALIS. The eyes of some birds have two such areas, one for medial vision and one for lateral vision (Walls, 1942). In its most extreme form, an area centralis contains a depression or pit, called a FOVEA (Figure 6). In most cases an area centralis contains only cones, which are especially densely packed. There is also an abundance of underlying bipolar cells, each of which is driven by an unusually small number of cones (sometimes just one) aiding in high visual acuity. Both the dense packing of cones and the large number of bipolar cells needed to service these through private, or nearly private, neural channels gives rise to the thickening of the area centralis. The function of the foveal pit is unknown, though several speculations have been offered (Hughes, 1977).

PHYSIOLOGY OF VERTEBRATE PHOTORECEPTORS

Intracellular recordings from vertebrate photoreceptors have been made primarily from fish, amphibia, and reptiles, in which these cells are relatively large. Surprisingly, the response to light of all vertebrate photoreceptors studies is not a depolarization but a *hyperpolarizaton* (Werblin and Dowling, 1969; Tomita, 1970; Baylor *et al.*, 1979). The size of this hyperpolarization varies with the intensity of the light stimulus (Figure 22). These receptors, then, do not respond with action potentials. In fact, action potentials also do not occur in the bipolar and horizontal cells but do occur in the amacrine and ganglion cells (Werblin and Dowling, 1969). In general, action potentials are unnecessary in cells with short axons, where the spatial decrement of voltage signals from one end of the cell to the other is small (Chapter 3). But action potentials are essential in the ganglion cells, whose long axons constitute the optic nerve, which carries retinal information to the brain.

The hyperpolarizing response to light of the photoreceptors results from a

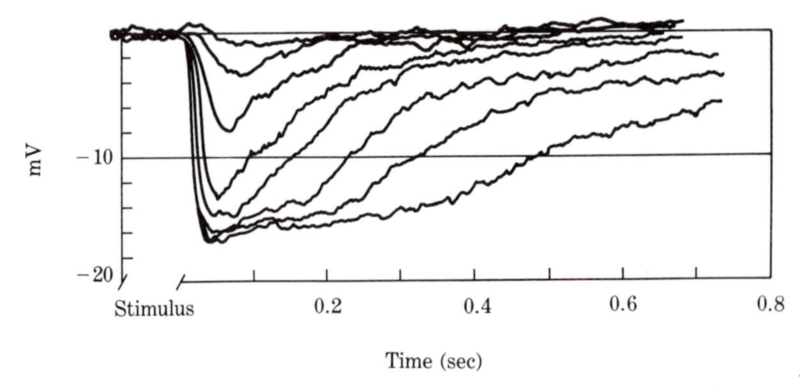

FIGURE 22. Graded, hyperpolarizing responses recorded intracellularly from a turtle cone. Stimuli were 0.1-second flashes of light of different intensities. (After Baylor and Fuortes, 1970.)

decrease in the permeability of these cells to sodium (Na^+) ions. This is contrary to the situation in most invertebrate photorceptors (Fuortes and O'Bryan, 1972), including those of honeybees (Baumann, 1968), and in most vertebrate or invertebrate receptors for modalities other than vision. In these various receptors, the Na^+ (and K^+) permeability *increases* in response to the effective sensory stimulus (Chapter 3). A vertebrate photoreceptor kept in the dark shows an unusually high resting Na^+ permeability, which serves to maintain the darkened cell somewhat depolarized. This depolarization results in a steady release of neurotransmitter molecules from the receptor's synaptic terminals. When stimulation by light decreases the Na^+ permeability, the cell is left with a high permeability only to potassium (K^+) ions. (The K^+ permeability is not altered by light.) Therefore, the membrane potential shifts toward the K^+ equilibrium potential. That is, the cell hyperpolarizes. This hyperpolarization results in a *reduction* of release of neurotransmitter by the receptor cell. It is this reduction of release (an unusual form of synaptic message) that signals to the bipolar and horizontal cells the presence of a light stimulus (Werblin and Dowling, 1969; Dacheux and Miller, 1976). Subsequent stages of retinal and central processing of visual information in vertebrates are discussed in Chapter 7.

WAVELENGTHS AND VISION

Most vertebrate eyes, like our own, are sensitive to red light but insensitive to UV (Autrum and Thomas, 1973). However, hummingbirds, pigeons, and some amphibians are able to detect UV light (Goldsmith, 1980; Kreithen and Eisner, 1978; Dietz, 1972). It is not yet clear in what way these vertebrates use their UV vision, but in the nectar-feeding hummingbirds, responsiveness to floral patterns of UV reflectance could confer the same advantages of floral recognition as in honeybees.

The eyes of most animals are insensitive to infrared radiation because the energy of an IR quantum is insufficient to alter electron orbitals and thus cannot be absorbed by known visual pigment molecules. However, IR radiation can be detected by nonphotochemical means. As any object is mildly warmed, its molecules undergo increased translational and rotational motions. Because some of these molecules are charged, their motions give rise to EM radiation. If the object is only mildly warm, there may be no radiation in the range of human vision, though there may be considerable radiation in the IR range. For instance, an animal whose body temperature is 38°C radiates from its body IR energy of wavelengths 1000 nm and longer, with maximal IR energy occurring at 10,000 nm (Bullock and Diecke, 1956).

When IR radiation strikes some surface, it sets the molecules of this surface into motion. We sense these molecular motions as heat when a warm object is held near our skin. Our skin's thermal receptors are not especially sensitive. However, IR receptors that are very much more sensitive are found in the facial pit organs of the crotaline group of snakes including copperheads, moccasins, and rattlesnakes, and in the labial pit organs of some snakes of the boid group, including boa constrictors (Barrett *et al.*, 1970; Newman and Hartline, 1982). The sensitivity of these receptor organs is adequate to detect a rat from a distance of 1 meter (Noble and Schmidt, 1937). In fact, many of these snakes appear to localize such prey animals prior to their predatory strike exclusively by the use of these pit organs, detecting the prey's body heat in the absence of any light within the visible range (Noble and Schmidt, 1937; Hartline, 1974).

POLARIZED LIGHT AND VERTEBRATE VISION

As we saw earlier, the chromophores of a bee's photoreceptors appear to be oriented preferentially parallel to the long axis of the microvilli. By contrast, the chromophores of vertebrate rods and cones, lying nearly flat within the plane of their lamellar membranes, are free to rotate within this plane (Liebman, 1962; Knowles and Dartnall, 1977b). Therefore, unlike retinula cells, the chromophores of rod and cone visual pigments have no preferred orientation within the lamellar membranes (Figure 23A). For this reason, vertebrate photoreceptors show little or no dichroism (a greater absorption of light of some E-vectors than of others) and thus little or no polarization sensitivity (a greater physiological response to light of some E-vectors than of others) when exposed to light propagating along its normal path, parallel to the long axis of the photoreceptor cell.

It is surprising, therefore, that behavioral experiments have shown several vertebrate animals to be capable of discriminating among different E-vector orientations. These animals include fish (Waterman, 1975a,b), pigeons (Kreithen and Keeton, 1974), salamanders (Adler and Taylor, 1973; Taylor and Adler, 1973, 1978), and, to a very limited degree, humans (e.g., Waterman, 1975a,b). The mechanism of polarization sensitivity is still a matter of speculation (Waterman, 1975a,b).

Perhaps the most surprising case of vertebrate polarization sensitivity is that of salamanders because it has been demonstrated that in these animals the eyes are not the polarization-sensitive organs. Rather, salamanders use a photoreceptor organ that is contained inside the skull (Adler and Taylor, 1973). This intracranial polarization-sensitive element apparently is the well-known PINEAL ORGAN, a dorsal evagination of the roof of the brain (Taylor and Adler, 1978).

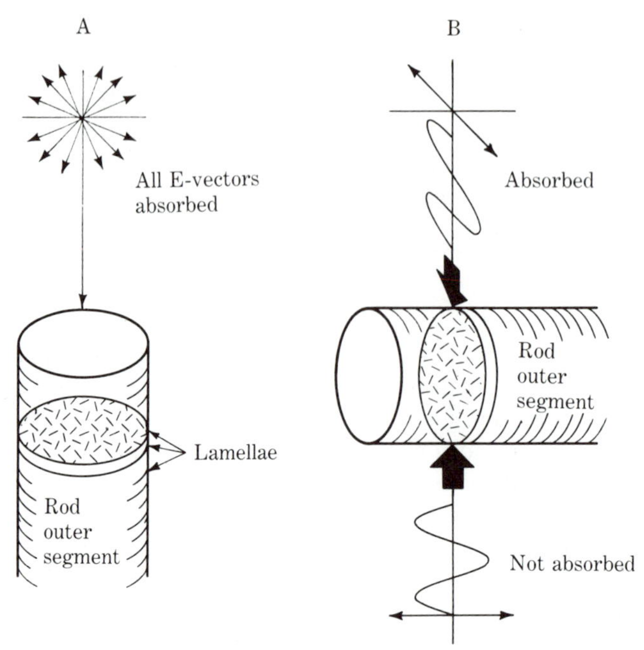

FIGURE 23. Polarized light and chromophore orientation in vertebrate photoreceptors. Chromophores of visual pigments are indicated by the short lines drawn on the outermost lamella of each receptor. A. The receptor is not dichroic for longitudinally propagating light; that is, all these E-vectors are absorbed. B. The receptor *is* dichroic for light arriving from the side. That is, if the E-vector of light is parallel to the lamellae (light ray shown on top), it is parallel to some of the chromophores and thus can be absorbed by them. If the E-vector of light is perpendicular to the lamellae (light ray shown below), it is parallel to none of the chromophores and therefore will not be absorbed. (After Davson, 1962.)

Sufficient light is known to penetrate the skull to stimulate its photoreceptors.

To demonstrate the nonocular location of their polarization-sensitive photoreceptors, salamanders were trained under polarized light to orient toward an artificial shore line in a large aquarium. Following the training period, the direction that they swam from a central release point in a circular arena shifted in correspondence with an imposed rotation of the E-vector orientation of the overhead light. This shifted orientation persisted after the eyes were removed and when a small transparent plastic sheet was inserted under the skin on top of the head. But the orientation became random when the transparent sheet was replaced with an opaque sheet (Adler and Taylor, 1973). In experiments carried out under less defined optical conditions out-of-doors, removal of the pineal organ had a similar randomizing effect on the learned orientation (Taylor and Adler, 1978).

Anatomical studies of the salamander's pineal organ suggest a possible mechanism for this animal's polarization sensitivity. Within this organ, all of the roughly 175 photoreceptor cells are nearly identical in structure to vertebrate rods. The cells have rodlike outer segments with quite normal lamellar organization. However, unlike the rods of the eye, most of these pineal photoreceptors are so oriented that their long axes lie nearly perpendicular to the direction of the light arriving from overhead (Korf, 1976). This means that light strikes these receptor cells on their sides, not their ends (Figure 23B). Given this situation, these receptor cells should be strongly dichroic (assuming that the chromophores lie flat within the lamellar membranes as in ocular photoreceptors). That is, light whose E-vector is oriented parallel to a lamellar membrane will be aligned with some of the chromophore dipoles, whereas light whose E-vector is oriented perpendicular to this lamellar membrane will be aligned with none. A group of such photoreceptors, all perpendicular to the direction of light propagation but each having different lamellar orientations (as occurs in the pineal organ), could encode the E-vector orientation of light because different receptor cells would be excited by different E-vectors (Figure 24). This arrangement has a superficial, though striking, resemblance to an invertebrate rhabdom (Figure 17B).

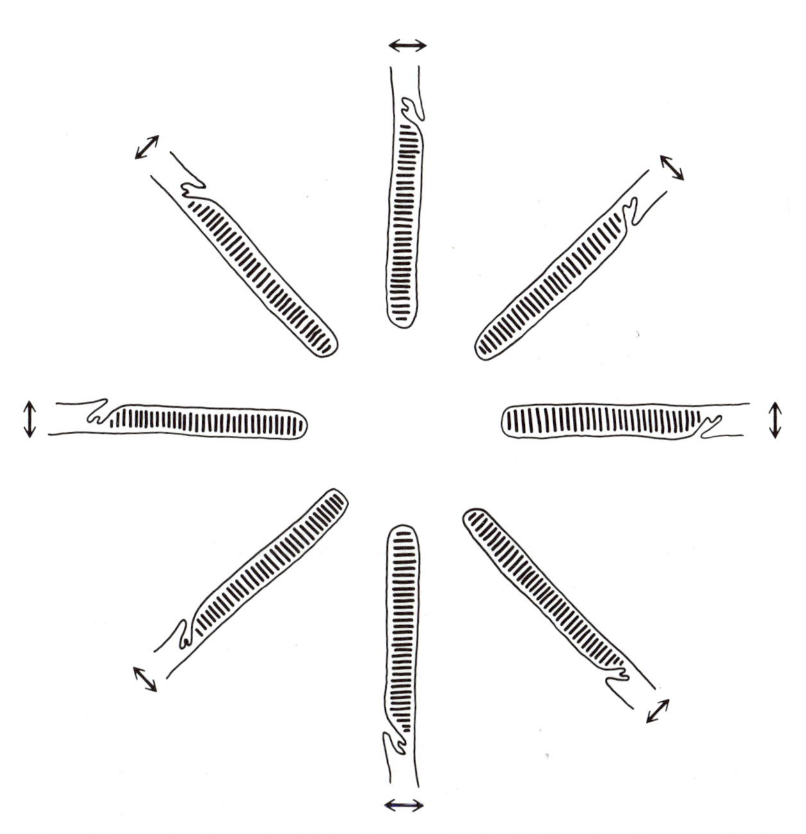

FIGURE 24. Possible mechanism of polarization sensitivity in the pineal organ of salamanders. Schematic view, from above the head, of the arrangement of photoreceptors. The double arrow adjacent to each receptor indicates the E-vector of light from above the head that should maximally excite that cell.

CELESTIAL CUES AND BIOLOGICAL CLOCKS IN VERTEBRATE ANIMALS

As we have seen, honeybees use a time-compensated sun compass to orient their waggle dances. The bees appear to determine how much they should time-compensate at any given moment (that is, by what amount they should increment their angle relative to the sun, to account for the sun's movement) by making repeated visual measurements of the sun's changing azimuth (Gould, 1980). Numerous other animals also employ sun compass orientation. However, some of these employ a different method for time compensation of their sun compass. This method consists of first determining the time of day by reference to the animal's internal time sense, or BIOLOGICAL CLOCK. With this information, the animal can then determine the compass direction of the sun. For instance, within the northern hemisphere the animal could know, as we do, that the sun's azimuth lies near the east at dawn, to the south at noon and near the west at dusk. The animal could then adjust its sun compass orientation according to its knowledge of the sun's compass direction. For instance, if it is noon and the animal intends to head eastward, it must simply orient its body so that the sun is on its right side.

The biological clocks of most animals have an important feature that has facilitated studies of their role in time compensation—the clocks can be reset (Bunning, 1973). To reset an animal's biological clock, one first places the subject in an environmental chamber in which the natural 24-hour periodicity of light and dark (or other cues with a 24-hour periodicity) is recreated by lamps and a timer. A mouse or rat, for example, kept under these conditions continues to become active during each dark period, just as it does in nature (Figure 25A, Days 1 to 6). It can be shown that the timing of this locomotory activity is produced by the animal's internal biological clock, rather than by the light–dark cycle of the lamps and timer. The strongest evidence for this is that if one leaves the lights permanently either on or off the animal continues to become active on a cyclic schedule. One might argue that this continued cycling could be caused by various daily environmental periodicities other than that of light and darkness that might somehow affect the animal even inside its environmental chamber. However, this apparently is not the case because the timing of the animal's daily locomotory period in constant light or darkness is almost never *exactly* 24 hours but rather is consistently slightly longer. (In some species it is slightly shorter than 24 hours.) This means that over the course of several days, the time of the animal's locomotory activity drifts relative to local time, the peak activity ultimately coming during the day rather than the night (Figure 25B). Because of this drifting property, such biological clocks are usually referred to as CIRCADIAN CLOCKS (circadian meaning "about a day"). It is because there are no known environmental cues that drift continually forward or backward in this way with respect to local time that the drifting under constant illumination implicates the animal's internal circadian rhythm, rather than some periodic sensory cue, in timing the animal's daily rhythm of activity.

Exposing the animal to a normal 24-hour light–dark cycle, either in nature or in an environmental chamber, synchronizes its circadian clock to this 24-hour periodicity. It is this ability of animals to synchronize their circadian clocks with external cues that enables an experimenter to reset an animal's clock. Such re-

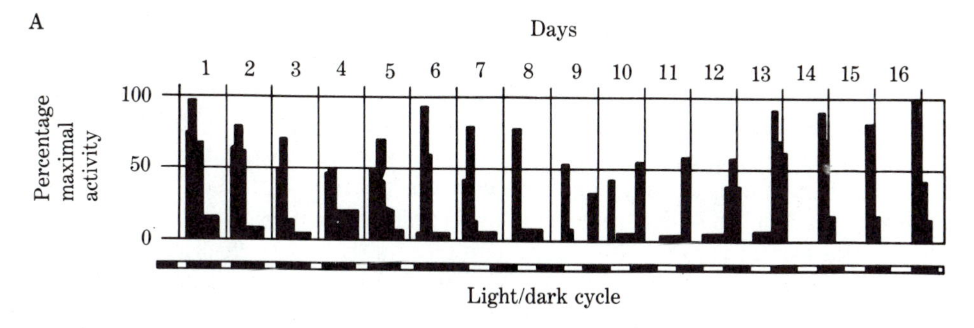

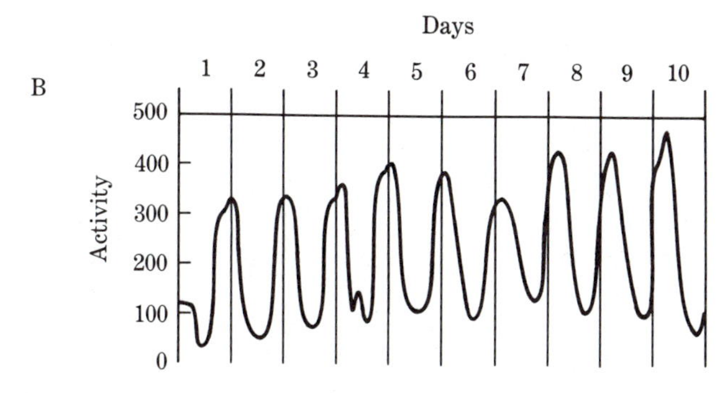

FIGURE 25. Activity rhythms timed by biological clocks. A. A mouse under controlled light–dark cycle. On the time line below, the white sections indicate the periods of illumination; black sections indicate the periods of darkness. B. A rat in constant illumination. (A after Bunning, 1973; B after Aschoff, 1960.)

setting can be achieved by shifting the timing of the light–dark cycle in an environmental chamber relative to local time; for instance, by creating within the environmental chamber an unnaturally long dark period followed by a 24-hour periodicity that is normal, except that it begins and ends each day at a time later than local time (see Day 7 of Figure 25A). For a while the animal continues to become active at the regular daily time dictated by its circadian clock (Days 7 to 10); but ultimately the clock is reset (by about Day 11) and activity occurs only during the dark period. Such resetting of the circadian clock is called CLOCK SHIFTING. We will now consider studies employing this clock shifting technique to elucidate the time compensation of celestial orientation in two much-studied birds, the homing pigeon and the indigo bunting.

CELESTIAL ORIENTATION IN THE HOMING PIGEON

Homing pigeons are legendary for their ability to fly directly to their home lofts through unfamiliar terrain over distances of hundreds of miles. How they accomplish this has been the subject of numerous studies over the past 30 years or more (Keeton, 1979a). In an early set of studies on the use of the sun as a compass, pigeons (Schmidt-Koenig, 1960) or other birds (Kramer, 1957; Hoffmann,

1960) were housed individually in outdoor circular cages, each with six or more feeding cups evenly spaced about its perimeter. Each bird was trained to feed at a cup in a particular compass direction. Once trained, such a bird continued to feed in the same compass direction, even when the entire cage and its attached feeding cups were rotated. To perform this orientation, however, the bird needed to see the sun. If a stationary light bulb was substituted for the sun, the bird fed at a regular sequence of different cups, progressing counterclockwise around the cage at the same average rate as the sun's azimuth moves clockwise—15° per hour. This finding, just like the observation of angular shifts in the marathon dances of bees, demonstrates a time-compensated sun compass.

The pigeon uses its circadian clock rather than periodic sightings of the sun's movement as in bees to achieve the time compensation of its sun compass while feeding in the circular cages. This was demonstrated by first observing the feeding orientation of trained pigeons, then clock shifting them (that is, changing the setting of the circadian clocks as in Figure 25A) and observing their orientation again. (Because these birds remember their training instructions for months, there was ample time for the clock shifting to take hold.) After being clock shifted, the pigeons oriented their feeding not in the compass direction to which they had been trained (in this case east) but in a different direction. For instance, those birds that were clock shifted 6 hours ahead (with their artificial day–night cycle beginning and ending 6 hours earlier than local time) oriented their feeding 90° counterclockwise to the training direction (Figure 26A).

One can understand this 90° counterclockwise orientation by considering that in the early morning, when the sun was in the east, the pigeons' circadian clocks misinformed them that it was already about noon and that the sun was therefore in the south. Thus, in an effort to orient eastward, they turned 90° counterclockwise to the sun. But this direction was actually north, not east. By continued time compensation throughout the day, they continued to orient to the north.

Those pigeons that were clock shifted 6 hours behind (with their artificial day–night cycle beginning and ending 6 hours later than local time) oriented 90° clockwise of their training direction (Figure 26B); and those that were clock shifted by 12 hours oriented in the direction opposite their training (Figure 26C). Because shifting the circadian clock caused misorientations in the expected directions and by the expected amounts, the pigeons were clearly using their circadian clocks for time compensation.

It remains to ask whether pigeons use their circadian clocks for time compensation of the sun compass not only when trained to a feeding direction in a cage but, more importantly, when attempting to fly home from a distant release site. To test this question, normal and clock-shifted pigeons were released individually several miles from home, and the experimenter noted each bird's VANISHING BEARING—its position when last visible through binoculars, a distance of about a mile. Control pigeons (not clock shifted) gave vanishing bearings in the direction of home. But the vanishing bearings of the birds that were clock shifted, for instance 6 hours ahead of local time, were deflected counterclockwise for all release locations (Figure 27). In fact, even when released within 1 mile of the home loft (an area with many landmarks familiar to these birds), the clock-shifted

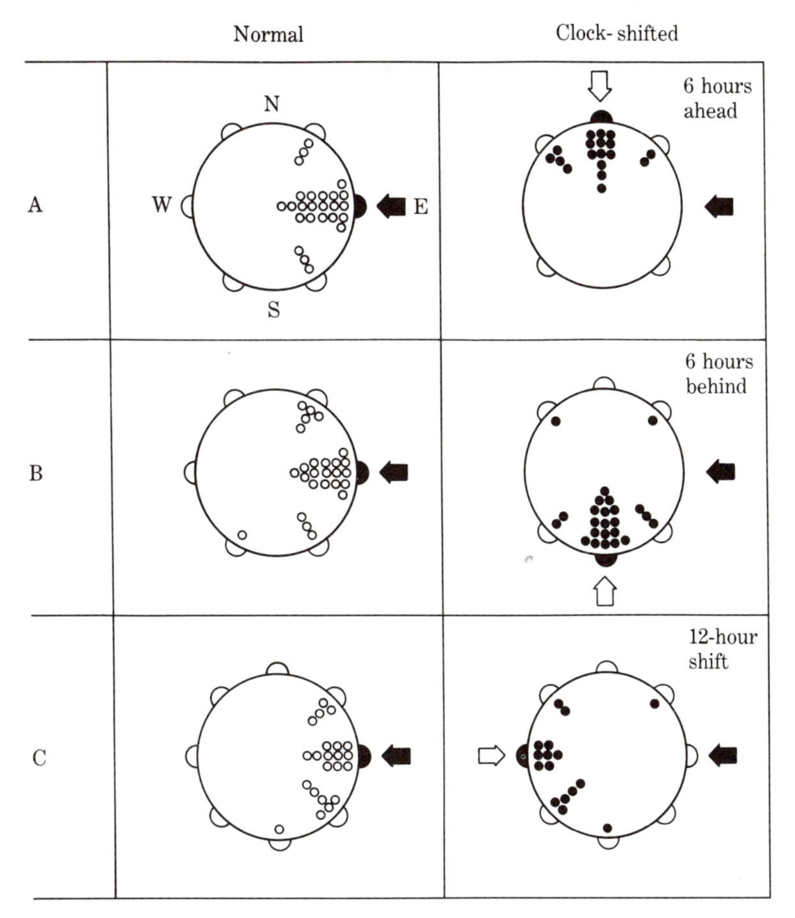

FIGURE 26. Effect of clock shifting on the feeding orientation of pigeons in circular cages. All the birds had been trained to feed in the east. Clock shifts of (A) 6 hours ahead of local time, (B) 6 hours behind local time, or (C) 12 hours from local time. Each clock shift produced the expected deviations in feeding orientation. Each circle represents one oriented feeding movement. (After Schmidt-Koenig, 1960.)

individuals flew off in the wrong direction, attesting to the overriding importance of the sun as an orientation cue (Keeton, 1979b).

It is not yet fully clear whether pigeons, like bees, can use polarized light cues from the sky's E-vector map (Figure 11) to orient home when the sun is obscured. Early studies suggest that the birds cannot use polarized light for orientation (Hoffmann, 1960), though more recently pigeons have been found capable at least of detecting the plane of polarized light (Kreithen and Keeton, 1974). More work is needed to clarify this point.

To the initial surprise of investigators, experienced pigeons were found capable of flying directly homeward after being released on totally overcast days. They were not guided during these flights by familiar landmarks because many of the release sites were at locations that were new to these pigeons. Initially it

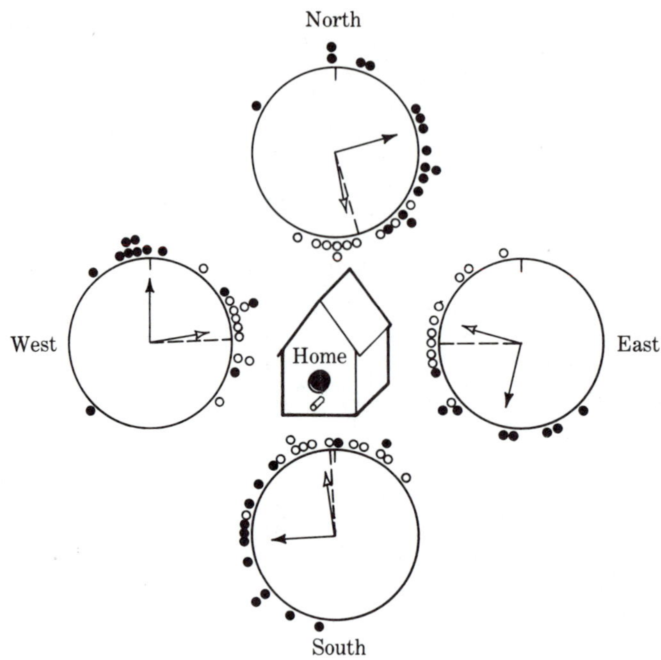

FIGURE 27. Vanishing bearings of pigeons released from four different locations about their home loft. From each of the four sites, both normal pigeons (open circles) and those clock shifted 6 hours fast (closed circles) were released. In each case the clock shifted pigeon was misoriented by about 90° counterclockwise. Arrows indicate mean directions for normal pigeons (open arrows) and those that were clock shifted (closed arrows). Dashed line indicates exact directions to the home loft. (After Keeton, 1979a.)

was thought that these birds might be better able than their investigator to see the sun through the clouds and so may have been performing their normal sun compass orientation. However, clock-shifted as well as normal birds gave homeward vanishing bearings flying in overcast skies (Figure 28). Thus, the cue or cues they were using could not have been ones requiring time compensation, as does the sun (Keeton, 1969; Keeton and Gobert, 1970). What additional environmental cue or cues the pigeons may be using is a matter of current experimental interest. Among the cues suggested are the earth's magnetic field, spatial patterns of odorants from the ground, and intense sounds of extremely low frequency that travel over hundreds of miles from stationary sound sources such as ocean waves crashing on the shore (Chapter 6). Pigeons are capable of detecting all these cues, and there is good evidence that at least the geomagnetic field is used in homing orientation (Keeton, 1979a,b).

Pigeons, then, like bees, can utilize different sensory cues as needed under different environmental conditions, affording them an obvious advantage in finding their home loft. This observation raises an important caution for experimenters. If one wants to determine whether a particular cue is used to direct a given behavior, one common approach is to remove that cue from the animal's surroundings (or wait for nature to do so, as by clouding over the sky). If the animal

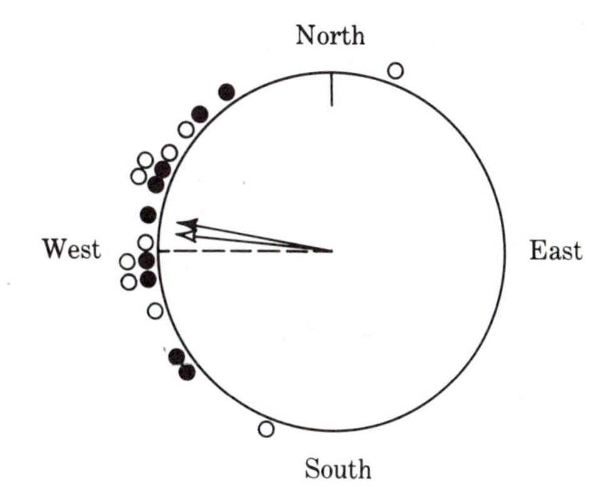

FIGURE 28.

Vanishing bearings of pigeons released under a totally overcast sky. Both normal pigeons (open circles and open arrow) and those clock shifted 6 hours fast (filled circles and filled arrow) were properly oriented homeward. The homeward direction is shown by the dashed line. (After Keeton, 1969.)

continues to perform the behavior normally, one may be tempted to conclude that the now absent cue (the sun) had not been important in directing the animal's behavior. But the ability of animals to switch to new cues as the conditions change confounds this experimental logic.

THE STAR COMPASS ORIENTATION OF THE INDIGO BUNTING

Many migratory birds carry out long-distance, biannual flights not only in daylight but also at night. Surprisingly, birds use visual cues to orient these nocturnal flights, even on moonless nights. It has been possible to study these nocturnal migratory orientations of birds within circular cages not only by training them to feed in particular directions but also by simply recording the direction toward which they most often oriented their restless fluttering behavior during a migratory season. Such studies have been carried out on several nocturnal migrants (Sauer and Sauer, 1960), including the indigo bunting, which is a common songbird resident in the northeastern United States. Buntings migrate at night from their North American breeding areas to Central America, most individuals crossing the Gulf of Mexico (Emlen, 1967a,b).

Caged buntings were placed out-of-doors on clear, moonless nights. In the spring, most of the caged birds showed a strong orientation tendency toward the north, and in the autumn toward the south, the normal migratory directions for these northern hemisphere birds. By contrast, other species of birds that are not nocturnal migrants showed no such nocturnal orientation in cages. When the sky became clouded over, the buntings' orientation tendency deteriorated toward randomness. Studies carried out on caged buntings in a planetarium provided further insights. If during the fall migratory season one projected onto the planetarium's dome the normal star pattern of the fall sky, the birds showed a normal southward orientation tendency within their cages. When the whole axis of the sky was shifted by 180° (the North Star now appearing in the southern part of the planetarium dome), the birds reversed to a northward orientation. By analogy to the sun compass of bees and pigeons, then, buntings can be said to employ STAR COMPASS ORIENTATION (Emlen, 1967a,b).

As with a sun compass, one would expect a star compass to require time compensation because the positions of the stars in the sky change throughout the night, relative to an observer on earth. As in studies on the sun compass, time compensation of the star compass was tested by arranging that the birds' circadian clocks be out of synchrony with the moving celestial cues, in this case the stars. However, in these studies the desynchronization was achieved not by clock shifting the birds but rather by shifting the planetarium's display of the sky, advancing or retarding the star's positions relative to those expected at the actual time of the test. (Such a shift of the sky obviously was impossible in the outdoor studies on pigeon homing!) If the buntings employed time compensation of their star compass, they should have oriented in different directions for the normal, advanced, and retarded sky patterns. Surprisingly, however, they oriented generally in the same direction for all these tests. Thus, they had to be using some form of stellar cue that does not require time compensation.

What stellar cue in the northern hemisphere might require no time compensation? Only cues from that part of the sky whose apparent position does not change at night, namely, due north (Figure 29). Any star located in exactly this position would provide an orientation cue that is fixed relative to the earth and thus would not require time compensation. The nearest bright star to this sky location is the North Star, Polaris, located about 2° of arc away from due north. Polaris, then, shows very little apparent nocturnal movement, and other stars at slightly greater deviations from due north show only slightly more apparent movement. If one removes from the planetarium's display only Polaris, the bunt-

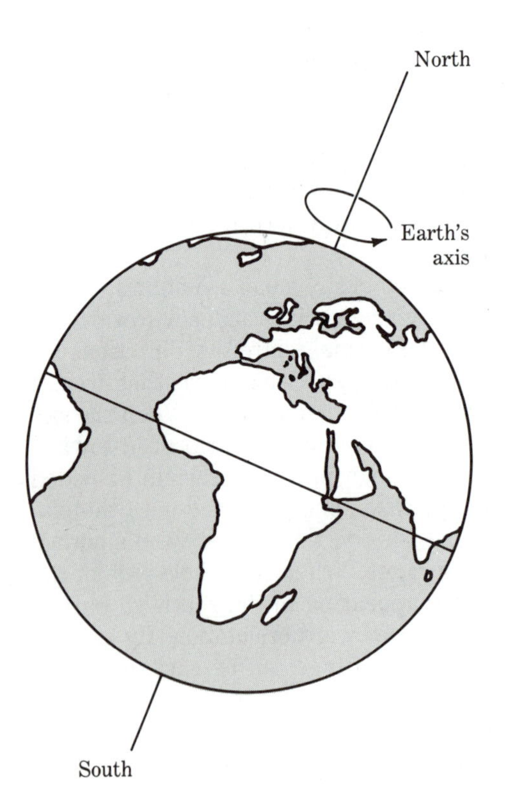

FIGURE 29.
Rotation of the earth about its axis.

ings continue to orient normally. However, if one removes from the display all the stars of the entire northern sky, covering a range within 35° of due north, the buntings' orientation greatly deteriorates. Thus, these birds appear able to identify the northern sky region by identifying its unique stellar pattern, just as we do by viewing the Little Dipper and other northern constellations.

The northern sky region contains numerous identifiable constellations. These appear, to an observer on earth, to rotate about due north (very near Polaris) during the course of the night. However, all these stars bear a fixed spatial relationship to one another, and thus rotate as a unit, as do the different points on a phonograph record spinning on a turntable. Thus, the buntings may be able to identify due north by viewing these constellations (Figure 30) and by using these as a directional reference point, even if the immediate due north area is clouded over. Whether buntings actually make this computation or whether they identify only the *general* region of the northern sky by recognizing its star patterns is unclear. Removing from the planetarium's display individual northern constellations as well as Polaris does not disturb the buntings' orientation, as they apparently then use the remaining visible parts of the northern sky. Therefore, the buntings can employ multiple star cues in their orientation. In addition, buntings, like pigeons, are known to use a variety of other directional cues in navigation, including the sight of the sunset and of major landmarks, wind direction, and geomagnetism.

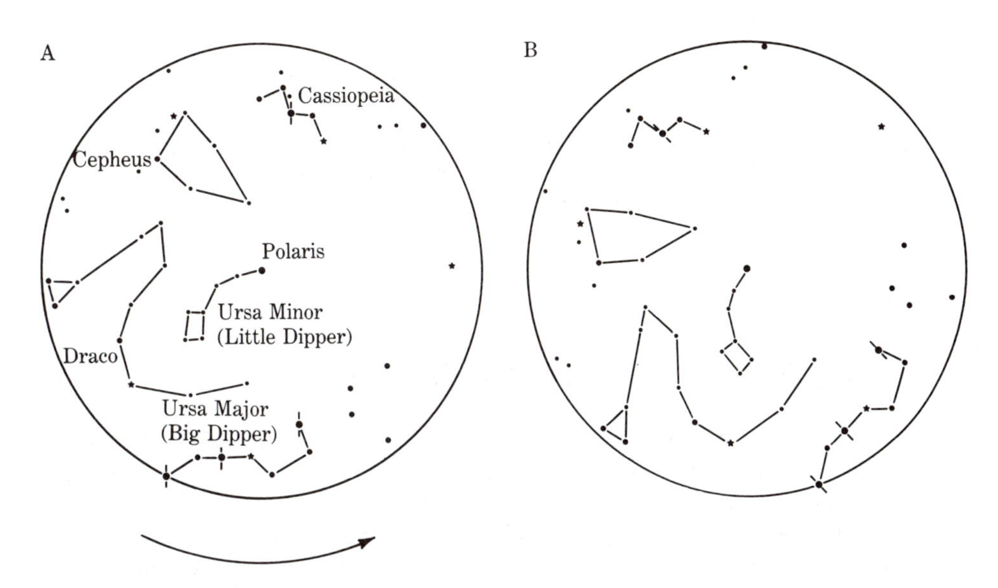

FIGURE 30. Constellations in the northern sky, within 35° of Polaris. The position of the constellations as shown in A is 3 hours earlier than in B. (After Emlen, 1967b.)

Summary: Vision in Vertebrates Compared to Vision in Honeybees

Vertebrate photoreceptors, like the retinula cells of bees, are elongate structures with multiple pigment-bearing membranes. However, these two types of photoreceptors are organized along fundamentally different structural plans, rhabdomic for the honeybee and many other invertebrates *vs* lamellar for the vertebrates. One consequence of this structural difference is that some invertebrate photoreceptors are highly dichroic, whereas vertebrate photoreceptors are not normally dichroic for light entering the pupil. Thus, special mechanisms must be utilized if a vertebrate animal is to detect the E-vector orientation of light. An example is found in the pineal organ of salamanders.

Like bees, many vertebrates have trichromatic vision. However, only photopic vision, which utilizes cones, is normally trichromatic. Scotopic vision, using rods, is monochromatic. Also, the light detectable by a vertebrate eye typically covers somewhat longer wavelengths than in bees and many other invertebrates, which are sensitive to UV and insensitive to red. Certain snakes are sensitive to IR, though they use a nonphotochemical mechanism to achieve this sensitivity. Unlike the depolarizing responses to light and the action potentials of the bee's retinula cells and of most invertebrate photoreceptors, vertebrate rods and cones have a hyperpolarizing response. This results in a decreased release of neurotransmitter, which signals the postsynaptic retinal cells at the onset of light.

Like bees, pigeons use a time-compensated sun compass to orient their homing flights on sunny days. The mechanism of time compensation, however, seems to be different from that of the bees. The pigeons use their circadian clocks as a reference for calculating the sun's movement, whereas the bees appear to measure periodically the actual movement of the sun. Indigo buntings use a star compass to orient their nocturnal migratory flights. This behavior does not require time compensation for the moving star pattern because it is the northerly region of the sky, which shows relatively little apparent movement, that serves as the buntings' directional cue. These birds identify this northerly sky area by its unique star pattern.

Questions for Thought and Discussion

1. As described in this chapter, honeybees must determine the position of the sun while they are foraging for food, in order to communicate to recruits in the hive the location of a food source. Design an experiment to determine whether they determine the sun's position while they are flying from the hive to the food source, from the food source back to the hive, or both.
2. How might the structure and physiology of a bee's ommatidium give rise to a high polarization sensitivity for the short cell (cell 9) but a low polarization sensitivity for the eight long cells? Suggest structural or physiological features of the ommatidium that could help, even if these features are unknown.

3. Design an experiment, complete with necessary controls, that could demonstrate convincingly (a) whether a given vertebrate animal can discriminate between different E-vectors of polarized light; (b) whether a given vertebrate animal has color vision. You will need to make certain assumptions about the behavior of each animal in order to decide on your experimental design.
4. Green plants absorb light by means of the pigment molecule chlorophyll. Chlorophyll molecules are located in the lamellar membranes of chloroplasts, an arrangement much like that of vertebrate photoreceptors. In response to light stimuli, some plants bend toward the light (positive phototropism). Can plants therefore see? Can honeybees see? Can your best friend see? (How do you know?) Can you see? (How do you know?) Make your answers to all four parts of this question consistent.
5. Suggest as many visual cues as you can from which animals could theoretically derive directional information for long-distance migration. Indicate any special requirements for the use of each of these cues in direction-finding.
6. Suggest as many *nonvisual* cues as you can from which animals could theoretically derive directional information for long-distance migration. Specify exactly what information you think would be available from each of these cues. Suggest how a receptor cell might be designed to detect each of these cues?

Recommended Readings

BOOKS

Clayton, R.K. (1977) *Light and Living Matter.* Vol. 1. *The Physical Part.* Vol. 2. *The Biological Part.* R. E. Krieger Publishing Company, Huntington, New York.
An excellent elementary treatment of the biophysics of vision and other photobiological processes.

von Frisch, K. (1967) *The Dance Language and Orientation of Bees.* Harvard University Press, Cambridge, Massachusetts.
An outstanding account of 50 years of experimental work carried out by this Nobel laureate and his co-workers. Although some sections are now outdated, a careful reading of this book is a must for those interested in visual neuroethology.

Lindauer, M. (1961) *Communication Among Social Bees.* Harvard University Press, Cambridge, Massachusetts.
A fine, shorter account of selected topics of honeybee ethology.

Schmidt-Koenig, K. and Keeton, W. T. (eds.). (1978) *Animal Migration, Navigation and Homing.* Springer-Verlag, New York.
This book contains the proceedings of a symposium and includes contributions by most of the leaders in the field of animal navigation and orientation.

ARTICLES

Gould, J. L. (1975) Honeybee recruitment: The dance language controversy. *Science* 189: 685–693.
An excellent experimental verification of the use of the waggle dance by the honeybee in its communication of the location of distant food sources.

Menzel, R. (1979) Spectral sensitivity and color vision in invertebrates. *Handbook of Sensory Physiology*. Vol. VII 6A. *Comparative Physiology and Evolution of Vision in Invertebrates: Invertebrate Photoreception*. H. Autrum (ed.). Springer-Verlag, Berlin.
This review gives a good, comprehensive overview of invertebrate color vision.

Menzel, R. and Snyder, A.W. (1974) Polarized light detection in the bee, *Apis mellifera*. *J. Comp. Physiol*. 88: 247–270.
This article describes the measurements of polarization sensitivity of the bee's retinula cells.

Wehner, R. (1976) Polarized light navigation by insects. *Sci. Am*. 235(1): 106–115.
This highly readable account of E-vector orientation presents one interesting model of ommatidial structure that could account for the polarization sensitivity of cell 9 and the lack of sensitivity of the other eight cells. The article by Ribi (next on the list), however, claims to disprove this model.

Ribi, W. A. (1979) Do the rhabdomeric structures in bees and flies really twist? *J. Comp. Physiol*. 134: 109–112.
An attempt to disprove the theory of polarization sensitivity by ommatidial twisting, which is presented in the article by Wehner.

Adler, K. and Taylor, D. H. (1973) Extraocular perception of polarized light by orienting salamanders. *J. Comp. Physiol*. 87: 203–212.
A description of interesting experiments showing that salamanders can use an intracranial receptor to detect the plane of polarized light and can orient their behavior by this directional cue.

Keeton, W. T. (1979). Avian orientation and navigation. *Annu. Rev. Physiol*. 41: 353–366.
This brief article by the late leader of experimental work on pigeon homing surveys current knowledge and ideas on sensory cues in homing and bird migration.

Emlen, S. (1967) Migratory orientation in the indigo bunting, *Passerina cyanea*. Part II: Mechanisms of celestial orientation. *Auk* 84: 463–489.
Describes a set of clever experiments in which the stellar display on a planetarium dome was altered to determine the visual cues by indigo buntings in their nocturnal migrations.

CHAPTER 6

Auditory Worlds

Although the sense of vision is found in animals of all phyla, only arthropods and vertebrates have a sense of hearing. And although the light for almost all vision originates from one source—the sun, the sounds that are important to animals originate from multiple sources, most notably from other animals. Examples include intraspecific vocalization signals and rustling noises that may inform a predator and a prey of each other's presence.

An animal has relatively little control over the types of sound that it makes while rustling through the underbrush. However, very specific sound parameters can be designed into its vocalization signals. As we shall see, different sound parameters offer different advantages in vocalization. For instance, sounds of low frequency travel farther than those of very high frequency. Also, sounds of some frequencies are easier than others for a listener to localize. Moreover, sounds containing a narrow range of frequencies may be easier for a listener to detect, whereas those containing a broad range of frequencies may carry much more information. A given animal, vocalizing under a given set of conditions, may attempt to match the parameters of its emitted sounds to those that are optimal for its needs at the moment. A successful listener, then, must possess an ear whose properties permit it to detect and analyze these particular sound parameters.

The major concerns of this chapter are the structure of vocalization signals and the design of the ears used to hear them. We will first consider in some detail the physical properties of sound. We shall then turn to a detailed discussion of the emission and perception of sounds by bats, which display perhaps the most sophisticated auditory capabilities of all animals. Next we shall consider a range of mechanisms used by animals to localize sound sources. And finally, we shall examine briefly a highly unusual auditory specialization found in caterpillars.

Physical Properties of Sound

DISPLACEMENT AND PRESSURE COMPONENTS OF SOUND

Suppose that a membrane, such as the cone of a phonograph speaker, is caused to vibrate back and forth. This vibration will produce two different effects

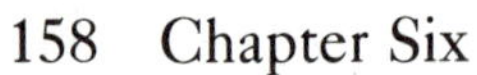

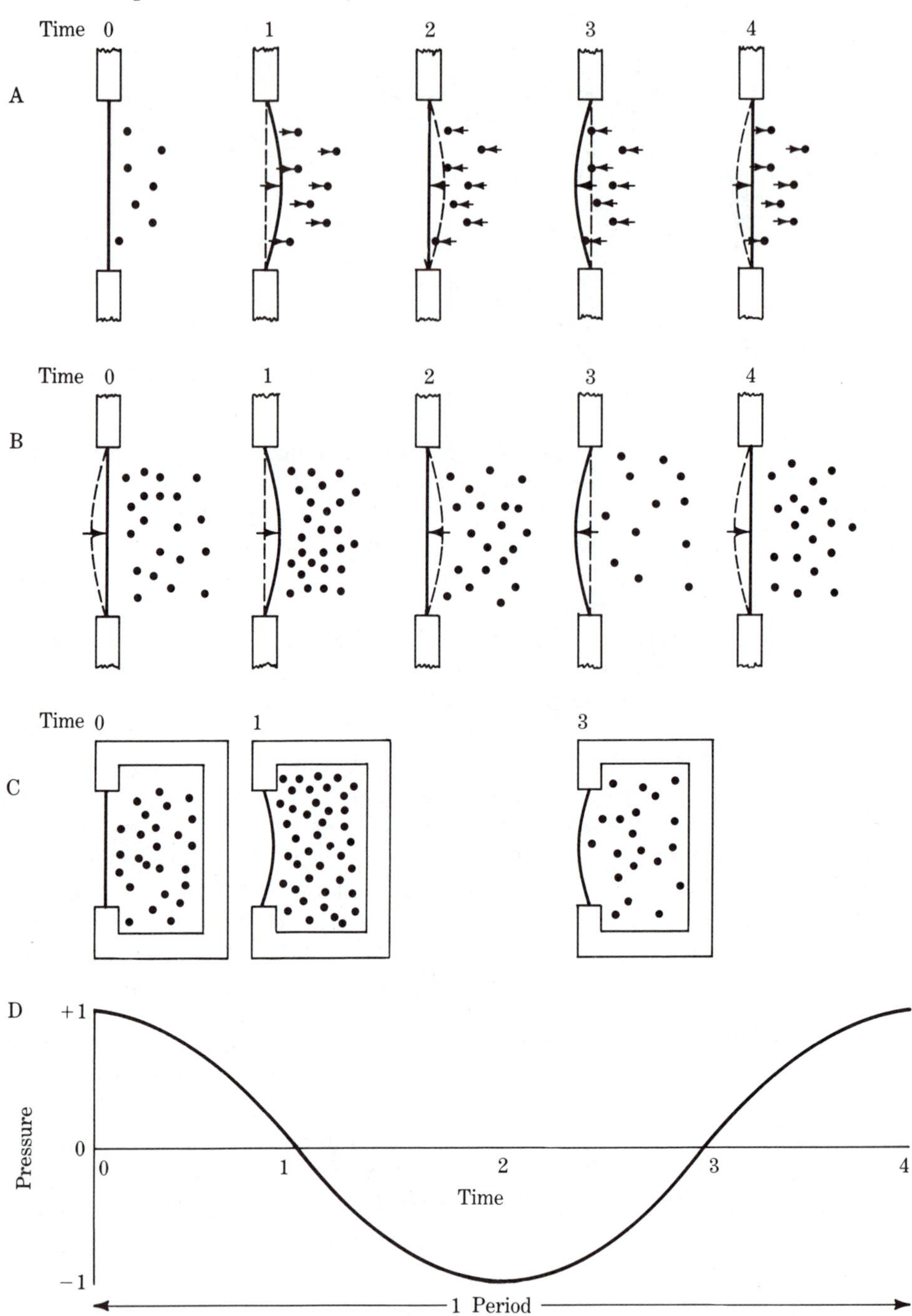

upon the air molecules just in front of the membrane. One effect is shown in Figure 1A: when the membrane moves toward the right (at time 1), it pushes the air molecules toward the right. Then, when the membrane returns to its starting position (at time 2), it draws these molecules back to their original lo-

◀ FIGURE 1. Molecular movements comprising a sound. A. The displacement component of sound. A membrane oscillates to the right and left. This causes nearby air molecules on the right side of the membrane (shown by the dots) to be displaced in the same directions as the membrane's movement. B.–D. The pressure component of sound. B. As the membrane moves to the right, the molecules on the right side are compressed; and as it moves to the left, these molecules become separated. C. The molecular compressions and rarefactions from part B are illustrated, as though the membrane were compressing or expanding the space inside a closed box. D. Pressure as a function of time. Note that the peak pressure occurs at the moment that the membrane is moving from left to right through its midposition (at times 0 and 4). This is the moment that the membrane is moving to the right at its maximal velocity. Thus, the molecular compression is also occurring with its maximal velocity. Pressure is proportional to the velocity of these compressive movements. Likewise, the minimal pressure occurs when the membrane is moving in the opposite direction at its maximal velocity (at time 2).

cations. The membrane's leftward and return movements (times 3 and 4) move the molecules to the left and then back again. This movement, or displacement, of the molecules for this cycle of events is called the DISPLACEMENT COMPONENT of sound.

The second effect of the membrane's movement is shown in Figure 1B. When the membrane moves to the right (at times 0 and 1), the molecules on the right of the membrane are pushed closer together. This molecular COMPRESSION causes a local increase in air *pressure*, just as though the membrane was pushing into a closed box (Figure 1C). When the membrane moves back toward its starting position (at time 2), the molecules return toward their original spacing and thus the air pressure drops. When the membrane moves to the left (time 3), this RAREFACTION, or spreading, of the molecules continues. The molecular spacing again returns toward its original value when the membrane returns toward its original position (time 4). This fluctuation of pressure is called the PRESSURE COMPONENT of sound. Much more is known about the pressure component than the displacement component of sound. This is partly because most ears, including our own, are designed to detect specifically pressure. Therefore, our discussion will stress the pressure component.

Let us examine the pressure fluctuations indicated in Figure 1B. We can graph these as a cycle of high, then low, then high air pressure (Figure 1D). This cycle of pressure changes will repeat itself over and over, assuming that the membrane continues to vibrate back and forth. However, as long as the membrane's vibration is of constant form, the single pressure cycle of Figure 1D is sufficient to describe the entire sequence of pressure changes. Let us assume, for simplicity, that the membrane's movement is of a very simple, smooth form, like the movement of a pendulum swinging side to side. This is called a SIMPLE HARMONIC, or SINUSOIDAL, motion. The resulting displacement of the air molecules would have the form of a sine wave, as is shown in Figure 1D. Like all sine waves, the graph of Figure 1D can be described fully by stating its AMPLITUDE, or height[1], and its PERIOD, which is the time required to go through one complete cycle. The period is inversely related to the FREQUENCY, which is the number of cycles per unit time. Thus,

$$\text{period (seconds)} = \frac{1}{\text{frequency (seconds}^{-1}\text{)}} \qquad (1)$$

Sounds are usually designated by their frequency in cycles per second, or HERTZ (Hz). A thousand hertz is called one KILOHERTZ (kHz).

AMPLITUDE AND FREQUENCY

For any given frequency, the greater the extent of the membrane's movement to the right and left, the greater the amplitude of the pressure fluctuations that it would produce. We hear this greater amplitude as a louder sound. Loudness, then, is roughly analogous to brightness in vision.[2] That is, both loudness and brightness are subjective impressions that we derive from the physical property of amplitude.

[1]The graph of Figure 1D plots the *instantaneous* pressure as a function of time. There are several ways to describe the amplitude of this graph. One way is simply by the graph itself, which shows the instantaneous amplitudes throughout an entire period. Another way is to specify the PEAK AMPLITUDE—the height of the upward peaks (arbitrarily designated +1 on Figure 1B). An alternative measurement is the PEAK-TO-PEAK AMPLITUDE—the total distance from the top of the positive peak to the bottom of the negative peak (this equals 2 in Figure 1D). Another description is called the ROOT-MEAN-SQUARE (RMS) AMPLITUDE. This method specifies an average of the instantaneous amplitude, irrespective of its direction, throughout the whole period. It would seem that the simplest way to measure this average amplitude would be to add the instantaneous air pressures at each moment throughout the entire period and then divide by the number of moments sampled. However, for a sine wave such as that of Figure 1D, this sum would always equal zero because each positive value of pressure on the graph is matched by an equal and opposite negative pressure. To avoid this computational difficulty, one begins by taking the square of each value of instantaneous pressure. This changes all negative values to positive values. One then takes the mean, or average, of all these squared values (by adding them all together and dividing by the number of values). And finally one computes the square root of this mean. Thus,

$$\text{RMS amplitude} = \sqrt{\left(\sum a_i^2\right)/n} \qquad (i = 1, \ldots, n)$$

$\Sigma\, a_1^2$ represents the sum of the squares of all instantaneous amplitudes, from the first to the nth, sampled during the period. In the case of a sine wave, the RMS amplitude equals approximately 0.707 times the peak amplitude. The RMS values are the most commonly used way to describe sound amplitudes, as they provide a reasonable estimate of the average amplitude through a whole period for a wave of any form. This is important because not all sounds consist of regular oscillations such as that of Figure 1D; some might have very large peaks with no energy at all between the peaks. Measuring peak amplitude, or peak-to-peak amplitude, for such a wave would imply more total energy through time than is actually present.

[2]Brightness can also be explained in terms of the number of quanta of light per unit time (Chapter 5). This explanation has no parallel in the case of sounds.

As with many sensory modalities, the range of sound amplitudes that animals can hear is enormous. For humans, it is a range of about 10^{10} times, from the softest audible sounds to the loudest noises that we can stand. In fact, the way that a given amplitude of sound is most commonly specified is by comparing it to the sound that is just audible, on average, to human listeners. (This human threshold value is a sound pressure of 20 micropascals RMS.) By comparing the amplitude of a particular sound to the threshold sound for human hearing, one derives a parameter called the SOUND PRESSURE LEVEL, or SPL. One measures the SPL of a sound by means of a sound pressure meter. The unit of measurement is a DECIBEL, or dB. Specifically, the number of decibels of sound pressure level for a given measured sound is derived as follows:

$$\text{\# of dBs} = 20 \log \frac{\text{sound pressure of measured sound } (\mu\text{pascals RMS})}{20\ \mu\text{pascals RMS}} \qquad (2)$$

The decibel, then, is a *relative* measurement; it tells you how much greater the sound pressure is for a given sound than for a standard reference sound, namely, the average threshold for human hearing. As Equation (2) shows, the sound pressure level that is just equal to the human hearing threshold (namely, 20 micropascals, or μPa, RMS) is 0 dB SPL, since log (20 μPa/20 μPa) = log 1 = 0. A sound pressure level ten times greater than the human threshold would be 20 dB SPL because 20 log (200 μPa/20 μPa) = 20 log 10 = 20. And a sound 100 times greater than the human threshold would be 40 dB SPL. As a guide to typical sound pressure levels of common sounds, the background noise in a quiet room is generally about 30 dB SPL; a typical human speaking voice, heard at close range, is about 60–80 dB SPL; and rock music, as commonly played in modern times, reaches 120–150 dB and higher. (Incidentally, exposure to sounds of this SPL range causes permanent hearing loss, which often sets in only years later!) The usefulness of the log scale in Equation (2) should now be apparent; it reduces to the manageable range of just over 200 dB the actual amplitude range of 10^{10} times.

The greater the frequency (that is, the number of cycles per second) of the back and forth movements of a sound-generating membrane, the greater the frequency of the sound pressure fluctuations in air. We hear this higher frequency as a higher pitch of sound. Thus, pitch is roughly analogous to color in the visual system, which depends upon the frequency, or wavelength, of light (Chapter 5). As we shall see, however, the pitch of a sound and the color of light are encoded by sensory cells in totally different ways.

Humans are able to hear frequencies between about 20 Hz and 20 kHz. Frequencies below about 20 Hz, called INFRASOUND, are detectable by a few animals. For instance, pigeons hear reasonably well sounds with frequencies as low as about 0.1 Hz (Kreithen and Quine, 1979). Sounds of frequencies above approximately 20 kHz are called ULTRASOUND. As we shall see, many animals can hear such sounds, up to a maximum of about 200 kHz. Thus, biologically useful sounds cover a roughly two million-fold range of frequencies, from about 0.1 Hz to 200 kHz. This contrasts sharply with biologically detectable light, which covers less than a threefold frequency range (Chapter 5, Figure 2).

SOUND PROPAGATION

If the effects of a vibrating membrane upon air displacement and pressure were limited to the immediate vicinity of the membrane, as shown in Figures 1A and B, no sound would be heard beyond a very limited distance from such a sound source. In fact, both the displacement and the pressure components do propagate away from the source. However, although we may hear such a sound from a considerable distance, this does not mean that individual air molecules have traveled the full span from the membrane to our ears. Rather, any one molecule moves only a very small distance. The sound propagation, then, consists of a chain reaction of collisions of each molecule with its neighbors. It is this chain reaction of collisions that propagates from sound source to listener. To illustrate this, Figure 2 depicts each air molecule between the vibrating membrane and the listener's ear as a wood block. Each block is physically connected to its neighbor. All the blocks are resting on a frictionless surface, such as a sheet of ice.

Let us first consider the molecular movements that constitute the propa-

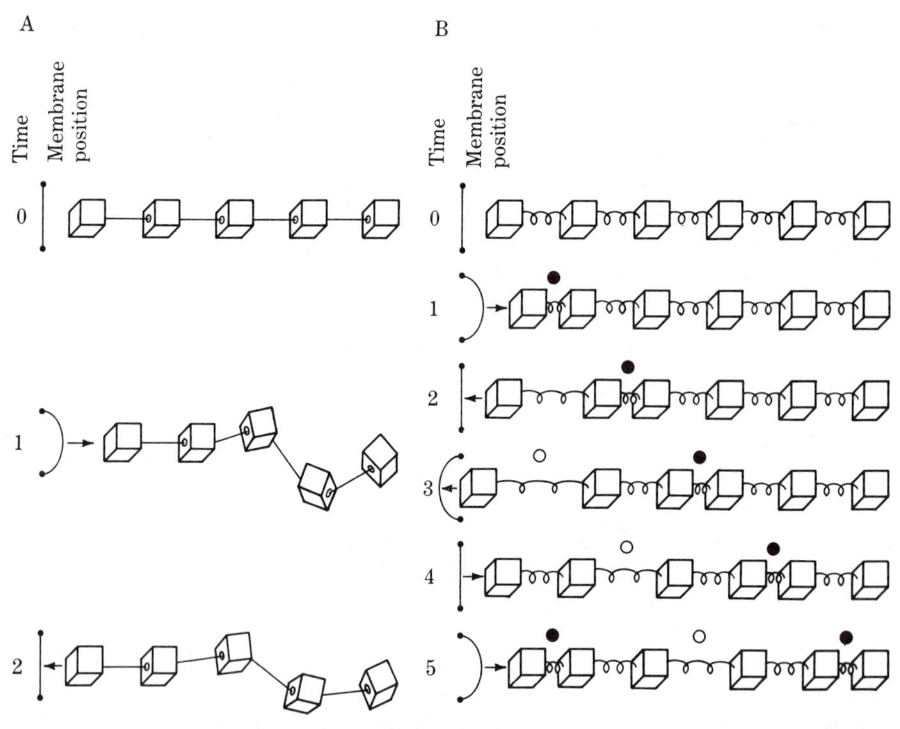

FIGURE 2. Propagation of molecular events away from a sound source, modeled as wood blocks on a frictionless surface. A. The displacement component. Only the events associated with an outward membrane movement (at time 1) and a return of the membrane (at time 2) are shown. B. The pressure component. Each intermolecular compression is indicated by a filled circle above the chain and each rarefaction by an open circle. See text for explanation.

gation through space of the *displacement* component of sound (Tautz, 1979). For this purpose, we can regard the spaces between adjacent molecules in a line as fixed. This is indicated by the bars of fixed length drawn between the blocks in Figure 2A. But each block is capable of swiveling to either side at its junction with a bar. When the membrane's movement pushes the block closest to it toward the right (at time 1), this block pushes on its neighbor, which in turn pushes on its neighbors, so that the whole chain of blocks moves to the right. This chain reaction of displacement would continue for only a limited distance, however. Beyond this distance, the regularity of the moving columns becomes disturbed, as occurs if one pushes slowly on the end ball in a row of billiard balls. The molecular displacement resulting from the subsequent leftward membrane movement is shown at time 2. Subsequent leftward and return movements of the membrane (not shown) would draw the chain of blocks to the left and back again.

Let us now consider the propagation of the pressure component of sound. To do so, we replace the rigid rods connecting the blocks with elastic springs (Figure 2B). Now, when the membrane pushes to the right, at time 1, the left-most wood block is pushed closer to the second block, compressing the spring between them. Because of the spring's elasticity, it next pushes the second block to the right and also pushes the left-most block back toward the left, at time 2. (The return of the left-most block to the left is also assisted by the return of the membrane to its starting position.) At times 1 through 5, as each block is pushed to the right, it creates a compression and leaves behind a rarefaction.

The events underlying the pressure component are shown in a more realistic (multimolecular) medium in Figure 3A. Beginning at time 1 (when the membrane

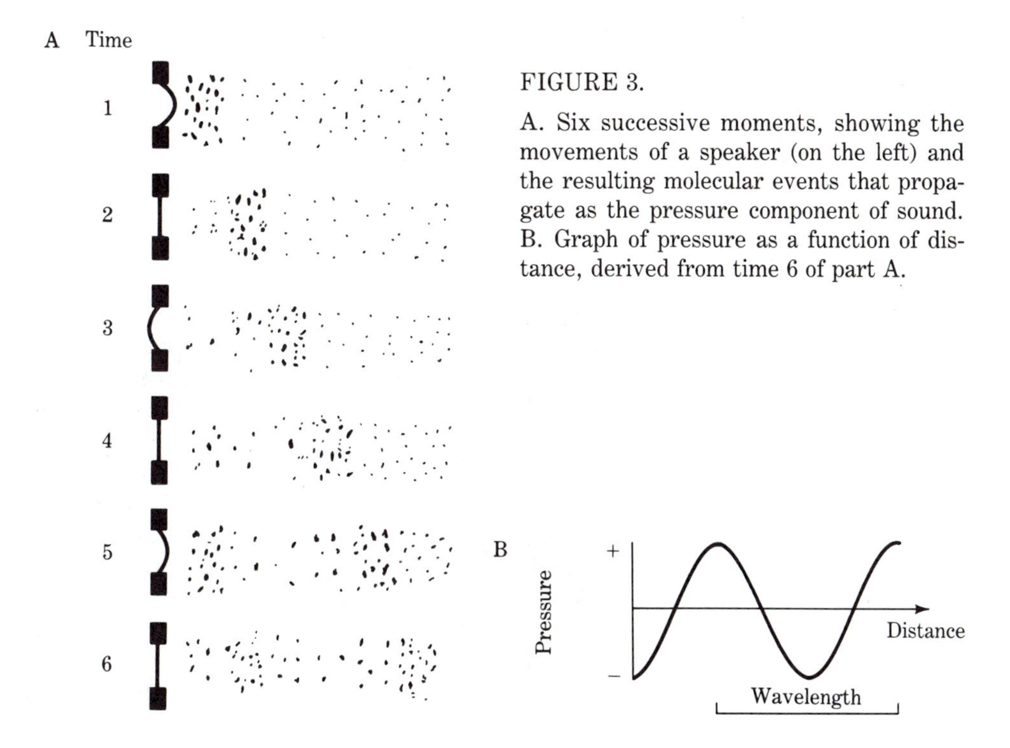

FIGURE 3.

A. Six successive moments, showing the movements of a speaker (on the left) and the resulting molecular events that propagate as the pressure component of sound. B. Graph of pressure as a function of distance, derived from time 6 of part A.

first moves) through time 6, one can see the molecular condensations and rarefactions propagating away from the sound source. One can graph these pressure variations as a function of distance, as shown in Figure 3B. Although this graph is a sine wave like that of Figure 1D, it is different from 1D in that it plots pressure as a function of distance at a given moment in time rather than pressure as a function of time at a given distance. From Figure 3B we can designate the spatial extent occupied by a single cycle of sound, namely, the WAVELENGTH. Wavelength and frequency are inversely related by the equation:

$$\lambda \text{ (meters)} = \frac{s \text{ (meters/second)}}{f \text{ (seconds}^{-1}\text{)}} \quad (3)$$

where λ = wavelength; f = frequency, and s = the propagation speed in the given medium. This equation is identical to that relating the wavelength to the frequency of light [Chapter 5, Equation (1)] except, of course, that here s represents the much slower speed of sound. In air (at 0°C) the speed of sound is 334 meters/sec.

To repeat an earlier point, the propagation of sound through distance involves only very small movements of individual molecules. It is the chain reactions of molecular displacements (for the displacement component), or the chain reaction of compressions and rarefactions (for the pressure component), that propagates through distance. Real sound sources produce both a displacement and a pressure component. Therefore, molecular movements occur as depicted both in Figure 2A and in Figure 2B. That is, molecules are moving at the same time partly as rigid chains and partly as compressible chains. However, the movements occurring as rigid chains (the displacement component) propagate only a very short distance from the sound source. In general, practically all the energy of the displacement component dissipates within the distance that it takes the pressure component to travel through only about one wavelength. Thus, for instance, for a sound frequency of 100 Hz, most of the displacement component would have dissipated by λ = (330 meters/sec)/100 sec^{-1} = 3.3 meters [Equation(3)]. For a 1000-Hz sound, the distance is only 0.33 meters; for 10 kHz, 0.033 meters; and for 100 kHz, just 0.0033 meters, or 3.3 millimeters!

This short distance of just one wavelength, within which almost all the energy of the displacement component dissipates, is called the acoustic NEAR FIELD. Some auditory receptors, as we shall see, are designed specifically to detect the displacement component of sound. We can now see that such receptors can detect a sound only within very close range of its source (Tautz and Markl, 1978). These are sometimes called near field receptors.

The space beyond roughly one wavelength's distance from a sound source is called the acoustic FAR FIELD. Because the far field is occupied almost exclusively by the pressure component, ears that are designed to detect pressure can hear a sound source from a much greater distance than those designed to detect displacement. For instance, a redwing blackbird, using a pressure sensitive ear, can hear its mate's call over a distance of 100 meters or more (Brenowitz, 1982). Because the major sound frequency of the call is about 2.5 kHz, the displacement component dissipates within just 0.13 meters of the caller [Equation(3)].

The pressure component, though it travels a long distance, does become reduced as it propagates. This reduction results primarily from two factors (Griffin, 1971; Wiley and Richards, 1978; Lawrence and Simmons, 1982). The first, called SPHERICAL SPREADING, stems from the fact that, at ever greater distances from a sound source, the given amount of sound pressure generated by the source must spread out to cover an ever increasing area of space. Spherical spreading produces a drop in sound pressure by a factor of 2 every time the distance from the sound source is doubled. This equals a drop of 6 dB SPL for each doubling of distance.[3] Thus, for instance, suppose we were to measure the sound pressure right in front of a sound source, and again at a 10 meters distance. And suppose we found that at 10 meters the sound pressure level was 20 dB less than right at the sound source. (This is called a 20-dB ATTENUATION.) At 20 meters, then, the SPL would be attenuated by 26 dB below that of the sound source, at 40 meters by 32 dB below that of the sound source, and so forth as shown on the top trace of Figure 4.

The second factor leading to a reduction through space of the pressure component is called EXCESS ATTENUATION. This is really a group of phenomena, the most important of which are absorption of the sound by the air (called ATMOSPHERIC ATTENUATION) and absorption or reflection by solid objects in the sound path. Atmospheric attenuation is greatest for sounds of high frequency—above about 10 or 20 kHz. For this reason, ultrasound travels much shorter distances than do sounds in the frequency range audible to us. For sounds in the range of 100 kHz and higher, used by many bats and several other types of animal, atmospheric attenuation is so pronounced as to restrict the usefulness of vocalizations to just a few meters' distance (Figure 4). We shall examine sound absorption and related phenomena in the next section.

REFLECTION, DIFFRACTION, REFRACTION, AND ABSORPTION OF SOUND

Like light, sound waves may be disturbed as they encounter inhomogenieties in their propagation path. For instance, when a sound propagating through air encounters a large object such as a wall, much of the sound energy is reflected, forming an echo (Figure 5A), just as light is reflected off a surface. It is by virtue of such reflected light that we see objects illuminated by a light source; and, as

[3]Consider a sound as it propagates away from its source, first through a distance x, and then through an additional distance x (so that it is now at distance $2x$ from the source). How much greater is the amount of space to be filled at distance $2x$ than at x? Consider x and $2x$ as the radii of two concentric spheres, whose center is the sound source. A sound wave that has propagated by distance x can be thought of as covering the surface area of the smaller sphere. And when this wave has propagated by distance $2x$, it now covers the surface area of this larger sphere. The ratio of the surface areas of the smaller to the larger sphere is $x^2/(2x)^2$, or 1/4. Therefore, the sound is 1/4 as "strong" at distance $2x$ as it is at x. This difference in strength represents a difference in the parameter called *sound intensity*, or sound *power density*. Thus, sound intensity at $2x$ is 1/4 that at x. Sound *pressure*, however, is proportion to $\sqrt{\text{intensity}}$. Thus, pressure varies with distance not as 1/4, but as $\sqrt{1/4}$, or as 1/2. Thus, the sound pressure level at distance x is twice that at $2x$. This is equivalent to a 6 dB difference, since $20 \log 2/1 = (20)(0.3) = 6$ dB.

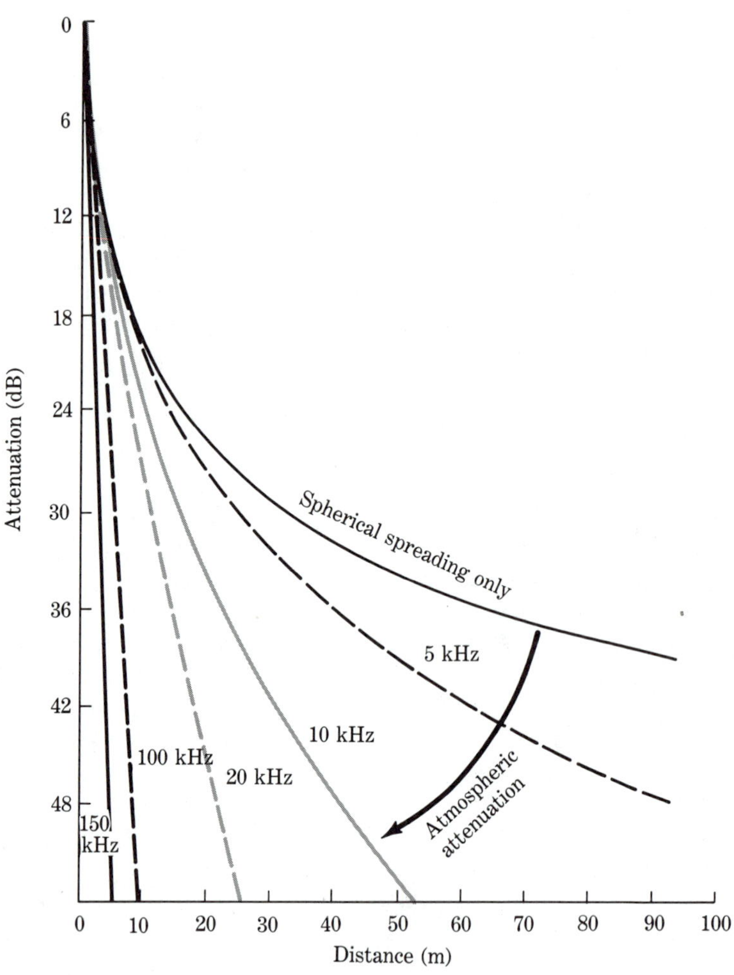

FIGURE 4. Attenuation of the pressure component of sound as a function of distance from the source. The uppermost curve shows the effect of spherical spreading alone. The other curves add to spherical spreading the effect of atmospheric attenuation, with the air containing 40% relative humidity. (Top curve from Brenowitz, 1982; other curves calculated from Griffin, 1971.)

we shall see later in this chapter, it is by means of reflected echoes that bats and a few other types of animal detect objects in their environments.

The percentage of sound energy that is reflected by an object depends in part upon whether that object is larger or smaller than the distance occupied by one wavelength of the sound. (This is true also for light waves and water waves.) For instance, the object in Figure 5A should be thought of as being many times higher and wider then one wavelength. It reflects sounds well. In Figure 5B, however, the sound wave encounters an object much smaller than one wavelength. Such objects hardly interfere at all with the sound, reflecting only a tiny fraction of its energy. (However, a sound wave of much higher frequency, whose wave-

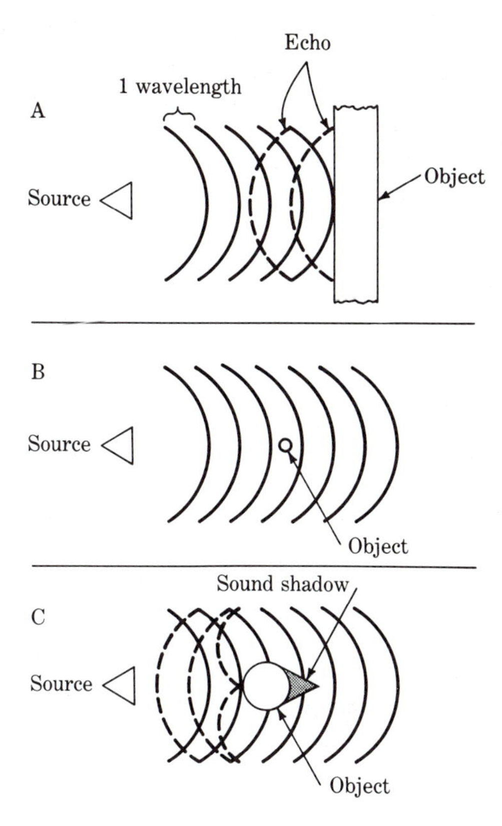

FIGURE 5.

Interaction of a sound with objects of different sizes. A. The object is much greater in height and width (perpendicular to the page) than the distance of one wavelength of the sound. Thus, the sound is strongly reflected. B. The object is much smaller than one wavelength, so essentially no sound is reflected. C. The object is slightly larger than one wavelength, so a sound shadow is produced. The sound shadow results not only from sound reflection but also from absorption and scattering by the object. (After Yost and Nielsen, 1977.)

length would therefore be much smaller, would be more effectively reflected by this object.)

Consider now a situation where sound waves encounter an object of a size equal to, or just slightly larger than, one wavelength (Figure 5C). This object is able to reflect a small amount of the sound energy. This leaves less energy to propagate past the object. In particular, at the location *immediately* beyond the object, there is a region of reduced sound energy, into which sound waves spread only slightly. This region is called a SOUND SHADOW. The waves continue uninterruptedly beyond the sound shadow. However, having lost some of their energy through reflection, the sound pressure level is lower than it would have been if the reflecting object had not been present.

The passage of sound waves around the objects in Figure 5B and C are examples of DIFFRACTION—the spreading of waves around objects, as occurs also with light and water waves (Figure 7 of Chapter 5). Diffraction around an object occurs only if the wavelength is about the size of the obstacle or larger. The sounds that we hear have much greater wavelengths than the light that we see. It is for this reason that audible sounds, but not visible light, can travel around many obstacles in the environment. Consequently, an animal in a dense forest may be able to hear, but not to see, another animal located just a few meters away. In such dense habitats, vocal communication can be much more important than visual communication.

The diffraction of sound presents a difficult problem to an animal attempting to localize a sound source. Recall that in the vertebrate eye, light waves passing through a narrow opening, the pupil, impinge upon a restricted region of the retinal surface, producing a retinal map of visual space (Figure 6 of Chapter 5). By contrast, sounds passing through a narrow opening onto an underlying sensory surface would spread out, through diffraction, to cover all regions of this surface (like the wave spreading in Figure 7 of Chapter 5). For this reason, the ear, unlike the eye, is unable to produce a sensory map of space. Rather, all the auditory receptors of both our ears are exposed to sounds originating from all directions. We shall examine later in this chapter, and in Chapter 7, how the auditory system manages to localize sound sources in spite of this absence of a sensory map of space.

In addition to diffraction, sound, like light, exhibits the property of REFRACTION—an increase in the speed of propagation whenever the waves enter a more dense (and more elastic) medium, such as water. Though sound travels about 330 meters/sec in air, it travels about 1450 meters/sec in water. In addition, just as light energy is absorbed by pigment molecules, sound energy can also be absorbed. This ABSORPTION consists of a conversion from the ordered molecular movements that constitute changes in sound pressure (Figure 2B) to random molecular movements producing heat. Objects along a sound path, as well as the medium itself, can absorb sound. Finally, like light, which can be transmitted from one transparent medium to another, sound waves also can be *transmitted* across a boundary between two media. This is what happens when sound waves pass from air to our eardrum, causing it to vibrate at the same frequency as the sound waves striking it. It is this vibration of the eardrum, transmitted to our auditory sensory neurons, that initiates the process of hearing.

SOUND AND LIGHT: A COMPARISON

Let us summarize the similarities and differences we have seen betwen light and the pressure component of sound. Both exist as wave phenomena, and thus, both have wavelength and frequency. (Light is generally described by its wavelength and sound by its frequency.) However, whereas light is an electromagnetic wave, sound is a mechanical wave, resulting from the collisions of molecules. The time required for this collision is responsible for the much lower propagation speed of sound as compared with light. Whereas the electric and magnetic forces of light are oriented perpendicular to the direction of the propagation path (Figure 1 of Chapter 5), the mechanical force of sound is oriented parallel to the propagation path (Figure 2). Finally, both light and sound exhibit the properties of reflection, diffraction, refraction, absorption, and transmission.

COMPLEX SOUNDS

Most natural sounds are more complex than those described so far in this chapter; they contain not just one frequency but several frequencies, all propagating simultaneously. This can be understood by considering the sound produced by a piano string. When the hammer strikes, the entire string is set into vibration, and therefore acts upon the surrounding air just like the vibrating membrane of Figure 1. The frequency of vibration is determined by the tautness, mass, and length of the string. In addition to this vibration of the whole string as a single unit, each half-length of the string also vibrates semi-independently of the other half and at twice the frequency at which the whole string is vibrating. In addition, each third-length vibrates at three times the frequency of the whole string, each quarter-length at four times the frequency, and so forth. All these vibrations occur simultaneously. The lowest frequency, that of the whole string, is called the FUNDAMENTAL FREQUENCY, or FIRST HARMONIC of this sound. Twice the fundamental frequency is called the SECOND HARMONIC, three times the fundamental frequency is the THIRD HARMONIC, and so forth. This combination of frequencies, each a multiple of the fundamental, is called a HARMONIC SERIES.

In such a harmonic series, each of the contributing frequencies can have a different amplitude from any of the other frequencies, the amplitude depending on the properties of the particular sound source and transmitting medium. Our ear generally identifies a musical note by its fundamental frequency, while the quality of the sound is influenced by the presence and the relative sound pressure levels of the various harmonics. Animal vocalizations are produced generally by forcing air over a membrane, causing the membrane to vibrate, which produces a fundamental and a series of harmonic frequencies. Like a piano, whose tuning pegs can be turned to tighten the strings, muscles that tighten the vocal membrane can alter its fundamental frequency and harmonics.

When a complex sound like that of a piano string or a vocalization propagates, it does so as a single complex wave, composed of all the contributing frequencies. Consider, for instance, a sound consisting of just a fundamental frequency and its second and third harmonics. These three components are graphed individually in Figure 6A; a l-kHz fundamental, a 2-kHz second harmonic having an amplitude two-thirds as great as the fundamental, and finally a 3-kHz third harmonic having an amplitude one-third that of the fundamental. A complex waveform composed of these three wave components is shown in Figure 6B. This waveform was drawn by simply adding, for each moment in time,[4] the amplitudes of all three component waves.

If the sound shown in Figure 6B were being transmitted through air, we could easily record it by means of a microphone. The microphone would convert the amplitude of the sound pressure component, at each instant in time, to a voltage that is proportional to this amplitude. We could then display this voltage on an oscilloscope. The oscilloscope screen would show exactly the waveform drawn in Figure 6B. This representation of a complex sound, showing the overall amplitude as a function of time, is termed a representation of sound in the TIME DOMAIN.

When one records a complex sound in this way, one cannot be certain, without performing some form of analysis, just what the frequencies and amplitudes of its component sine waves are. These components may constitute one harmonic

[4]These three components could be combined in an infinite number of ways, depending upon the phases of each component with respect to the others. For simplicity, the complex waveform for Figure 6B is drawn with all phase angles equal to 0°. That is, all component waves begin together at zero amplitude, as shown in part A.

series (that is, a fundamental with its harmonics) such as that of Figure 6A; or there may be additional fundamentals, each with its own harmonic series, as occurs when we play simultaneously several unrelated notes on the piano; or practically all audible frequencies may be included, as occurs in noises like that made by clapping the hands. As we shall see shortly, the ears of many animals act in such a way as to break down a complex sound wave into its separate frequency components. That is, rather than determining the shapes of the individual lumps and bumps of a complex waveform such as that in Figure 6B, such ears convert these waveforms into a list of component sound frequencies and the amplitude for each of these frequencies. This is called analyzing a sound in the FREQUENCY DOMAIN, as opposed to the time domain. To understand what the ear does, then, we must find a way to analyze the sound into its component frequencies, just as the ear does.

In the nineteenth century, Joseph Fourier devised a mathematical procedure to do just this. This procedure is now called FOURIER ANALYSIS. Using this method, any form of signal, such as a waveform of sound as complex as one likes, can be decomposed into a series of sine waves that if added together again with the proper phase relationships would exactly recreate the original waveform. Although

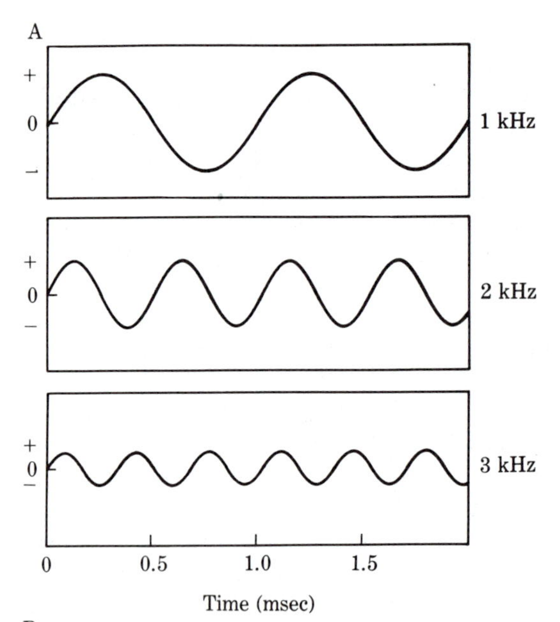

FIGURE 6.

A harmonic series. A. Three sine waves at 1, 2, and 3 kHz, respectively. The amplitude is greatest for the 1-kHz sine wave, 2/3 as great for the 2-kHz wave, and 1/3 as great for the 3-kHz wave. B. A complex sound composed of the three sine waves in A. (Courtesy of Carl Resler.)

the details of Fourier analysis are beyond the scope of this book, we can readily appreciate the importance of this type of analysis by viewing the graphical results that it produces. An example is the graph shown in Figure 7A, called an AMPLITUDE SPECTRUM. This graph plots the RMS amplitude of each component sine wave of the complex waveform of Figure 6B. An amplitude spectrum, then, is a representation of a sound in the frequency domain. It enables us to visualize readily how an animal's ears would record this sound. Suppose, for instance, that an animal is exposed to the sound in Figure 6B. We can see from Figure 7A that if one group of sensory cells were maximally sensitive to sounds of 1 kHz, they would receive maximal stimulation; if a second group of sensory cells were maximally sensitive to 2 kHz, they would receive ⅔ this amount of stimulation (because the amplitude of this peak on the amplitude spectrum is ⅔ that of the 1 kHz peak); and if a third group of sensory cells were sensitive to 3 kHz, they would receive

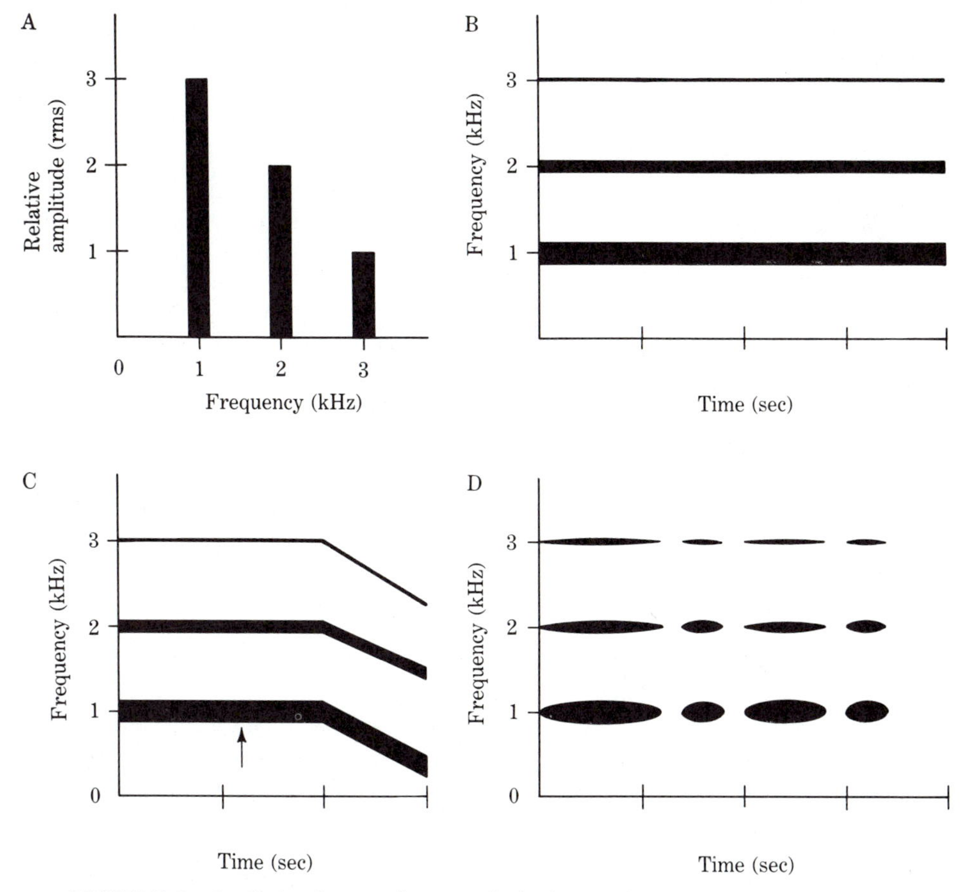

FIGURE 7. Analysis of a complex sound. A. An amplitude spectrum of the sound in Figure 6B. B. A sonogram of the same sound. C. A sonogram of the same sound, but ending with a downward FM sweep. A section made at the time shown by the arrow would produce the amplitude spectrum shown in A. D. A sonogram of an amplitude-modulated sound.

⅓ of the stimulation of the first group. Because many sounds are much more complicated than that of Figure 6B, one would have no hint at all of their frequency composition without carrying out a Fourier analysis of this sort.

Fortunately, there are instruments that perform Fourier analyses directly on sounds, eliminating the need for laborious mathematical calculations. The instrument standardly used for this purpose is the SPECTRUM ANALYZER, or SONOGRAPH, although increasingly, computers are being used instead. A sonograph produces two kinds of graphical records of sound: an amplitude spectrum like that of Figure 7A, and another graphical record called a SONOGRAM. Figure 7B shows a sonogram of the sound in Figure 6B. The sonogram plots the component frequencies of a sound through time. (It is thus a hybrid between the representation of sound in the frequency and the time domains.) The relative amplitudes of the different frequency components are indicated by the relative darkness and thickness of the different lines. Although this qualitative measurement of amplitude is much less accurate than that shown in an amplitude spectrum (Figure 7A), the sonogram has the advantage that it reveals any changes through time in the frequency composition of a sound. An example of such FREQUENCY MODULATION (FM) is shown in Figure 7C. This sound ends with a downward FM sweep. This sweep would have the sound, to us, of a decrease in pitch. A sonogram can also indicate qualitatively, by changes in the thickness of the line, fluctuations of sound pressure level, or AMPLITUDE MODULATION (AM). An example is shown in Figure 7D. We would hear this sound as a repetition of a note having a constant pitch.

Because both the frequencies and amplitudes of a sound can vary with time, one typically begins an analysis of any sounds of interest by making a sonogram. This affords an overall view of the frequencies, the relative amplitudes, and the modulations of both frequency and amplitude throughout the entire duration of the sound. If one then desires quantitative information on the sound's frequency composition at a given moment, the sonograph can produce an amplitude spectrum of the sound at that moment. For instance, the amplitude spectrum at the moment shown by the arrow in Figure 7C would be that of Figure 7A. A different amplitude spectrum would be produced at some moment during the later FM sweep. Making this conversion from a sonogram representation to an amplitude spectrum for a given moment on the sonogram record is called taking a SECTION of the sonogram.

The Auditory World of Bats

We shall now examine in some depth the remarkable vocal and auditory capabilities of bats. In the process, we shall encounter many important principles of auditory neuroethology. Bats of some species fly in the dark of night through thick forests, avoiding collision with trees and their branches, while capturing prey in the air, on the ground, or just under the surface of a pond. Many bats accomplish these remarkable feats not visually but rather by the use of elaborate auditory specializations. Though other nocturnally active vertebrates, such as owls and many small mammals, have excellent hearing, it is the bats, together with the porpoises and a few species of cave-dwelling birds, that have developed the special mechanism known as ECHOLOCATION—detecting objects in the environment by sensing the echoes of one's own emitted calls (Busnel and Fish, 1980).

BAT ECHOLOCATION

Let us briefly recount the experiments by which bat echolocation was discovered (Griffin, 1958). The brilliant Italian biologist Spallanzani (1729–1799) carried out an important early set of experiments on the flight orientation of bats. He found that individuals whose eyes had been removed would fly around a room avoiding obstacles about as well as a normal bat. Moreover, when he plugged the ears of otherwise normal bats with wax and released them to fly, they collided helplessly with obstacles in their path. To control for the possibility that the wax ear plugs were painful and that this pain had somehow been disorienting to the bats, Spallanzani used a different method to plug the ears. He first inserted into each ear a short, open-ended tube. By themselves, these tubes had no effect on the bat's ability to avoid obstacles. Then Spallanzani filled just the outer openings of the tubes with wax, an act that presumably caused the bats no pain. Now they again collided with obstacles in their flight path. Bats, then, seemed somehow to use their ears in avoiding obstacles.

This conclusion was not generally accepted for the next 130 years, largely because no one could imagine by what types of sounds (or possibly other sensory cues discerned by the unplugged ears) the bats were detecting obstacles. In fact, no sounds were audible to human listeners while the bats flew around a room. Early in this century, it was suggested that infrasounds, produced by the flapping wings of the bat, might be involved; perhaps these sounds, with a frequency equal to the wingbeat frequency, were reflected off objects and the echoes produced detected by the bat's ears. It apparently was not recognized at that time that the long wavelengths of infrasounds preclude their being well reflected from any but very large objects (Figure 5). Infrasounds are now known to play no role in bat echolocation. In 1920, the English physiologist Hartridge suggested that ultrasound might somehow be involved. But there was no evidence that bats produced, or could hear, ultrasound.

It was a Harvard undergraduate named Donald Griffin who, beginning in 1938, solved this mystery of bat echolocation. He showed that bats do indeed produce ultrasound, that they hear ultrasonic echoes from objects, and that they use these echoes as the basis for obstacle avoidance and other flight maneuvers. Griffin first persuaded the physicist G. W. Pierce, who had invented the first accurate detector of ultrasound, to direct this listening device at a bat that was sitting in a cage. Immediately, high energy pulses of ultrasound were recorded (Pierce and Griffin, 1938). The bat also produced these pulses while it flew around a room. Moreover, by taking the ultrasonic detector out-of-doors, Griffin and his colleagues later found that bats make ultrasonic pulses while flying in their natural habitats.

Ultrasound, because of its short wavelengths, should be well reflected even by small objects (Figure 5). For instance, a sound of 100 kHz, with its wavelength of just 3 millimeters [Equation (3)], should reflect well from objects as small as fine twigs. However, because of the extreme atmospheric attenuation of ultrasound

(Figure 4), it should be useful only for very limited distances. In fact, Griffin and his colleagues noted that when a flying bat approached to within only 1 or 2 meters of an obstacle, it suddenly began to emit many more ultrasonic pulses per unit time, as though it had just detected the obstacle. Rather than the 10 to 30 pulses per second that characterize flight in unobstructed air, the rate increased to between 50 and 200 per second.

Remarkably, objects as minute as a strand of wire just 1 millimeter thick, hung from the ceiling, induced this increase in pulse rate by an approaching bat. In fact, when bats flew through a grid of these wires, which were spaced 1 foot apart, they would hit a wire with their wings or body on only 24% of their passes through the grid. Griffin then plugged the ears of these bats by a variety of methods, including the use of inserted tubes that were plugged only after control flights had shown that the tubes alone had no effect. After plugging, the rate of collision with the wires increased to about 65%. This was roughly the rate to be expected by a bat-sized object traversing such a grid of wires totally at random. Unplugging the ears, or opening the plugged tubes, restored the collision rate to its normal, low value. By this quantitative measurement of behavior, then, the openness of the ears was shown to account *entirely* for the bat's ability to avoid these obstacles.

If the bat's emitted ultrasonic pulses are important for its obstacle avoidance, muting a bat should increase its rate of collision with the wires of a grid, just as plugging its ears had done. In fact, Griffin and his colleagues found that bats that had unplugged ears but that were silenced by tying shut their mouths and sealing tight their lips collided helplessly with obstacles. To avoid obstacles, then, the bats needed both to emit ultrasonic pulses and to have unimpaired hearing. Griffin concluded that what they were hearing had to be the echoes of their own pulses, reflected from the objects in front of them. He gave this process the name echolocation.

Echolocation belongs to a class of orientation behaviors, called ENERGY EMITTED ORIENTATION, in which an animal broadcasts some form of energy into its environment and detects changes in that energy imposed by the environment. Other examples include electrolocation by certain fish, which emit electric currents into the surrounding water and detect disturbances of these currents by objects in the water (Heiligenberg, 1977); the creation of surface waves by the aquatic whirligig beetle *(Gyrinus)* and the detection of wave reflections from solid objects (de Wilde, 1941); and the emission by flashlight fish of light used for vision (Morin *et al.*, 1975; McCosker, 1977).

PERCEPTUAL ABILITIES OF ECHOLOCATING BATS

How good is a bat's echolocating system at detecting small objects? Griffin tested the ability of bats to steer through grids of progressively finer and finer wires. The bats had little trouble avoiding wires of about 0.4 millimeters or more in diameter and even showed some ability to avoid direct hits with wires of less than 0.2 millimeters (Figure 8). This is remarkable because this diameter is only about 1/10 as great as the *shortest* wavelength contained in the bat's emitted cries; thus, only a tiny fraction of such sounds would be reflected by these wires.

Bats also display remarkable echolocating skills in their prey-catching behavior. When a hungry insectivorous bat flies at night, it makes numerous twists and turns along its flight path, darting after insects on the wing. Each such twist is accompanied by an abrupt increase in the rate of emission of ultrasonic pulses, just as occurs when a bat approaches an obstacle. To catch an insect, the bat need not hear the wingbeat sounds of its prey, because even totally silent insects, such as mealworms tossed gently into the air, can be successfully captured (Figure 9). Because this can occur in total darkness, but not if the bat's ears are plugged, it is clear that echolocation is employed for the capture of flying insects.

Aside from detecting and capturing insects, bats can also use their echolocation to discriminate among different forms of prey on the wing. For instance, a bat could easily be trained to chase after mealworms in total darkness while avoiding plastic disks of similar shape that were tossed upward into their flight path (Griffin *et al.*, 1965). The echoes reflected from these disks were only slightly different from those reflected by the mealworms. Thus, small differences in these reflected echoes permitted the bat to carry out this refined discrimination.

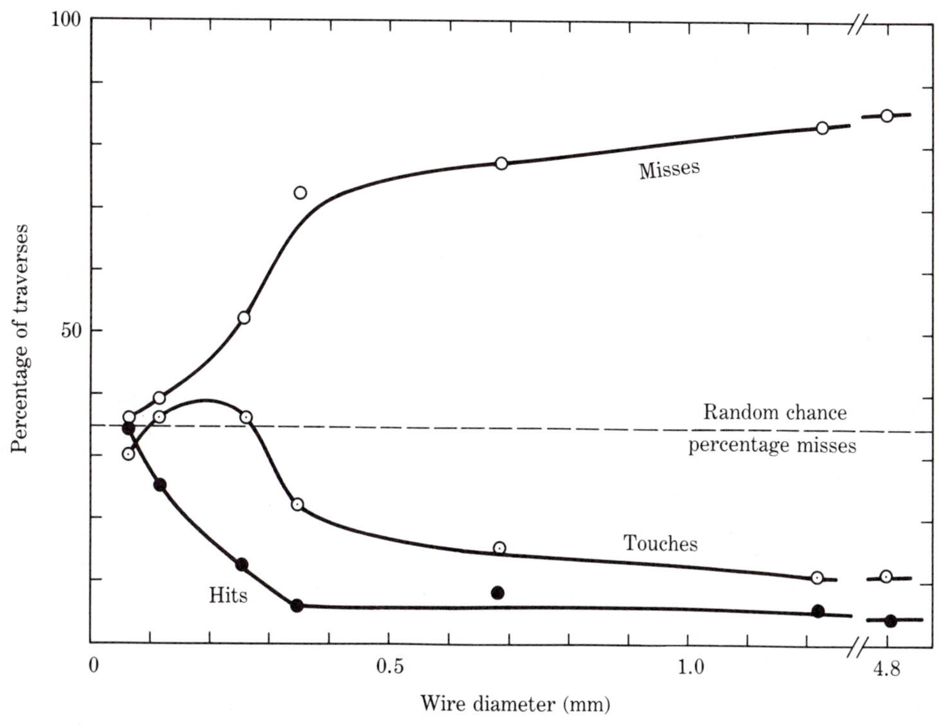

FIGURE 8. Avoidance of fine wires suspended in a grid across a bat's flight path. For greater wire diameters, the bats missed the wires more often and touched or hit them less often. A "touch" was defined as a light brushing of a wire by the tip of a wing, with the bat's flight path remaining undisturbed. A "hit" was a more direct contact, which deflected the bat from its flight path. (From Griffin, 1958.)

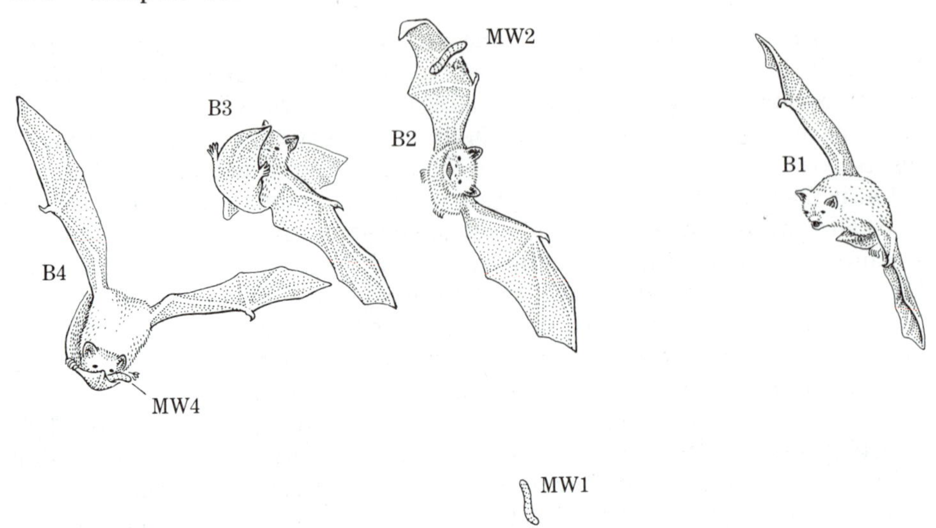

FIGURE 9. Filmed sequences of a bat *(Myotis lucifugus)* capturing a mealworm tossed into the air. Shown are the positions of the bat and the mealworm in four sequential frames of film. When the bat was at position B1, the mealworm was at position MW1 and was rising. On the second frame, the bat (B2) used its wing to deflect the mealworm (MW2) downward. It then caught the mealworm in a pouch formed by a flap of skin connecting its two legs and tail. It ducked its head into this pouch (at B3 and B4) and ate the mealworm. Bats usually catch insects either in this manner or by scooping them out of the air with their interleg skin flap. They rarely catch insects directly in the mouth. (From Webster and Griffin, 1962.)

The discriminative ability of echolocation has been explored further using a more complex training procedure (Simmons and Vernon, 1971). In this procedure, a bat (often one that had previously had both eyes surgically removed) was trained to stand on a perch, and then to fly on one of two choice platforms. Next to each platform was placed an object of a different shape, for instance, a circle at one choice platform and a square at the other. When the bat flew from its perch to that choice platform with the circle (which the experimenter had selected as the preferred platform for this particular experiment), the animal was given a food reward. During this training, sometimes the circle was placed next to the left choice platform and the square next to the right one, and sometimes the arrangement was reversed, on a random schedule. Thus, the bats were indeed learning to go to the circle and not, for instance, to the choice platform on a given side.

Using this training procedure, bats quickly learned to select one of two shapes, proving that they could discriminate between the shapes. They could also be trained to select the larger or the smaller of two objects of the same shape (as long as the two sizes differed by more than about 15%). Moreover, they could learn to discriminate between identical objects at two different distances, such as 60 *vs* 58 centimeters from the perch. In their selection of targets, the bats used their echolocation, because blinded or sighted bats could discriminate equally well, but those with their ears plugged could not discriminate.

Additional properties of echolocation were revealed only after the training procedure was modified. The modified experiments were used to explore the discrimination of target distance in bats of the genus *Eptesicus* and of various target properties in the genus *Rhinolophus*. The next two sections describe these two sets of experiments. As we shall see, *Eptesicus* carries out its distance discrimination by making a combined frequency and time analysis of the sounds, whereas *Rhinolophus* carries out certain discriminations by analyzing the sound only in the frequency domain.

EPTESICUS: ANALYSIS OF TIME AND FREQUENCY

The North American insectivorous bat *Eptesicus fuscus* makes echolocating pulses consisting of three harmonics. Each pulse lasts from about 1 to 13 milliseconds and consists primarily of a sharp downward FM sweep (Figure 10A). Because this FM sweep covers a wide range of frequencies, from about 100 to 25 kHz, it is said to have a large BANDWIDTH or, to be a BROADBAND SOUND.

These rapid FM sweeps should be especially well suited for determining target distance. This is because both the emitted pulse and any resulting echo sweep very quickly through each sound frequency. To take one sample frequency, 40 kHz, the pulse and then the echo each produce this sound frequency for just a brief instant (Figure 11). Therefore, receptors in the bat's ears that are excited primarily by sounds of 40 kHz will be stimulated precisely at these two very brief

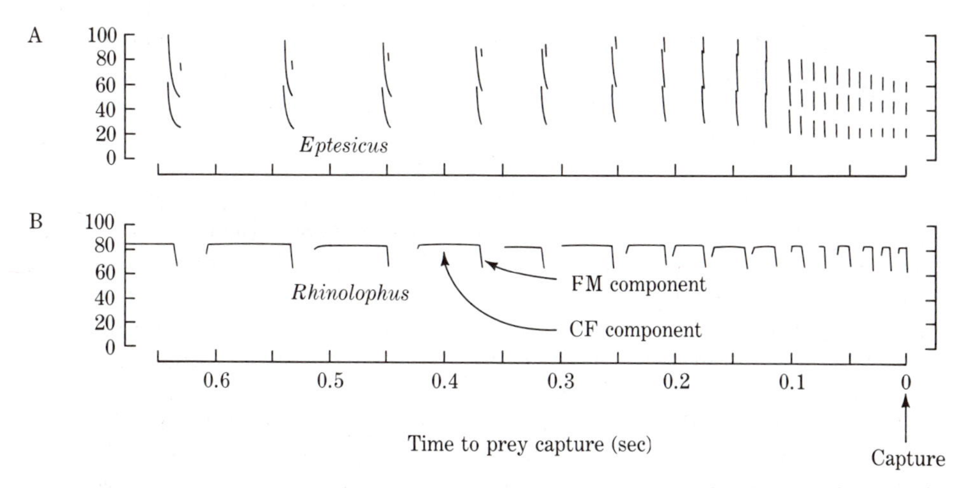

FIGURE 10. Sonograms of sequences of echolocation pulses emitted by two bats, each approaching an insect in flight. Each bat captured and ate its target insect at the end of its sequence of pulses. A. *Eptesicus fuscus*. B. *Rhinolophus ferrumequinum*. As is typical of echolocation, once an insect has been detected, the pulse rate increases, while the pulse duration decreases. (After Simmons *et al.*, 1979.)

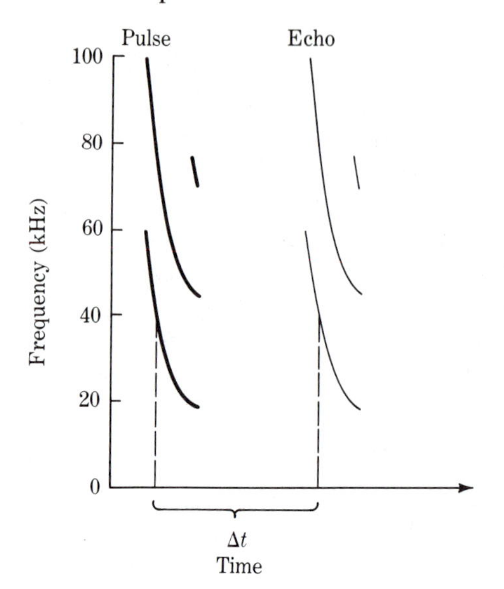

FIGURE 11.

Sonogram of an echolocation pulse and a resulting echo made by *Eptesicus fuscus*. The pulse-to-echo delay, Δt, could provide an unambiguous measurement of target distance.

moments. Because these two moments are clearly demarcated in time, the interval between them (Δt in Figure 11) should be clearly determinable by the bat's brain. The interval Δt depends strictly upon the distance from the bat to the echo-producing target; the farther away the target is, the longer it takes for the sound to travel from the bat to the target and back again. Thus, this pulse-to-echo delay could provide the bat with an unambiguous means of measuring target distance. Moreover, the measurement could be reiterated many times for a single pulse and echo pair because the pulse and echo each sweep through many different frequencies (not just 40 kHz) and because each of these frequencies excites a different group of auditory receptors. This mechanism for measuring target distance, then, would employ both a frequency analysis (the excitation of specific cells only by sounds of particular frequency) and a time analysis (the measurement of time Δt).

It should be noted, however, that other cues for target distance, besides pulse-to-echo delay, are also available to the bat and could theoretically be used. These include echo intensity (which decreases for increasing target distance) and the amplitude spectrum of the echo relative to that of the pulse; the echo's amplitude spectrum should include fewer high-frequency sounds for greater distances, owing to the greater atmospheric attenuation of high frequencies (Figure 4).

To determine whether *Eptesicus* uses pulse-to-echo delay to determine target distance, Simmons (1973) trained bats initially to jump or fly from a resting perch to the more distant of two choice platforms. Aside from their different distances from the perch (25 *vs* 30 centimeters in most experiments) these two platforms were identical. Each had a large Plexiglas triangle that reflected sounds well (Figure 12A). As in earlier training studies, the more distant platform was positioned sometimes on the left side and sometimes on the right, on a random schedule.

When a bat had learned to select consistently the more distant choice platform, the following changes were made in the experiment (Figure 12B). The Plexiglas triangles on each choice platform were removed, and these two platforms were now placed each at the same distance from the perch, namely 30 centimeters. Instead of using the Plexiglas triangles for effective echo reflection, the echoes now were provided by an electronic speaker controlled by the experimenter, permitting him to vary at will the timing of these echoes. Every time the bat emitted a pulse, the speaker emitted a delayed version of this pulse, which the bat was to regard as its echo. It was these "phantom echoes" produced by the speaker, which were reporting to the bat the presence of a "phantom target" that did not in fact exist, that the experimenter used to trick the bat into deducing that a good sound-reflecting object was present. By setting particular delays between the bat's emitted pulse and the speaker's emitted phantom echo, the experimenter tried to convince the bat that this echo was located at either a 25-centimeter or a 30-centimeter distance from its perch.

The experiment, then, was to see whether a bat that had already been trained to select a real target (a triangle) at 25 centimeters in preference to one at 30 centimeters would now select instead a phantom target that, according to its pulse-to-phantom-echo delay, was located at 25 centimeters. Would the bat prefer these phantom echoes to those signaling a distance of 30 centimeters? The experimental setup by which particular pulse-to-echo delays were delivered, signifying to the bat particular distances to the phantom target, is shown in Figure 12B. All cues were the same for the two phantom targets *except* the magnitude of the pulse-to-echo delay.

All individuals of *Eptesicus* were in fact immediately successful in selecting and flying to the phantom target at 25 centimeters. Thus, pulse-to-echo delay is an important cue in determining target distance. Next, the experimenter gradually increased the pulse-to-echo interval of the closer phantom target, making it more and more similar to the interval of the 30-centimeter target. To the bat, this was equivalent to making the distance from perch to phantom target more similar for the two phantom targets. The bat was still able, with 75% accuracy, to select the closer phantom target when it was at 29 centimeters and the farther was at 30 centimeters. The pulse-to-echo delay for this 29-centimeter phantom target was only 60 microseconds (60×10^{-6} seconds) less than that of the 30-centimeter target. Remarkably, although these 60 microseconds amount to only about 1/20 the duration of a single action potential in the bat's auditory nerve (which typically lasts about 1 millisecond), still *Eptesicus* is able to measure this minute time difference.

In fact, the bat's temporal discrimination is actually far better than 60 microseconds, as was shown in subsequent tests (Simmons, 1979). These tests were similar to those just described, but with one difference. The bat was required to discriminate not between a closer and a more distant phantom target but rather between one that was fixed at a 50-centimeter distance and another that jittered slightly, on successive echolocating pulses, about a mean distance of 50 centimeters. That is, the pulse-to-echo delay for one target was constant, whereas that for the other target varied between slightly shorter and slightly longer intervals. The bat could be trained to discriminate between the target with the fixed interval and that with the variable interval when this latter interval was

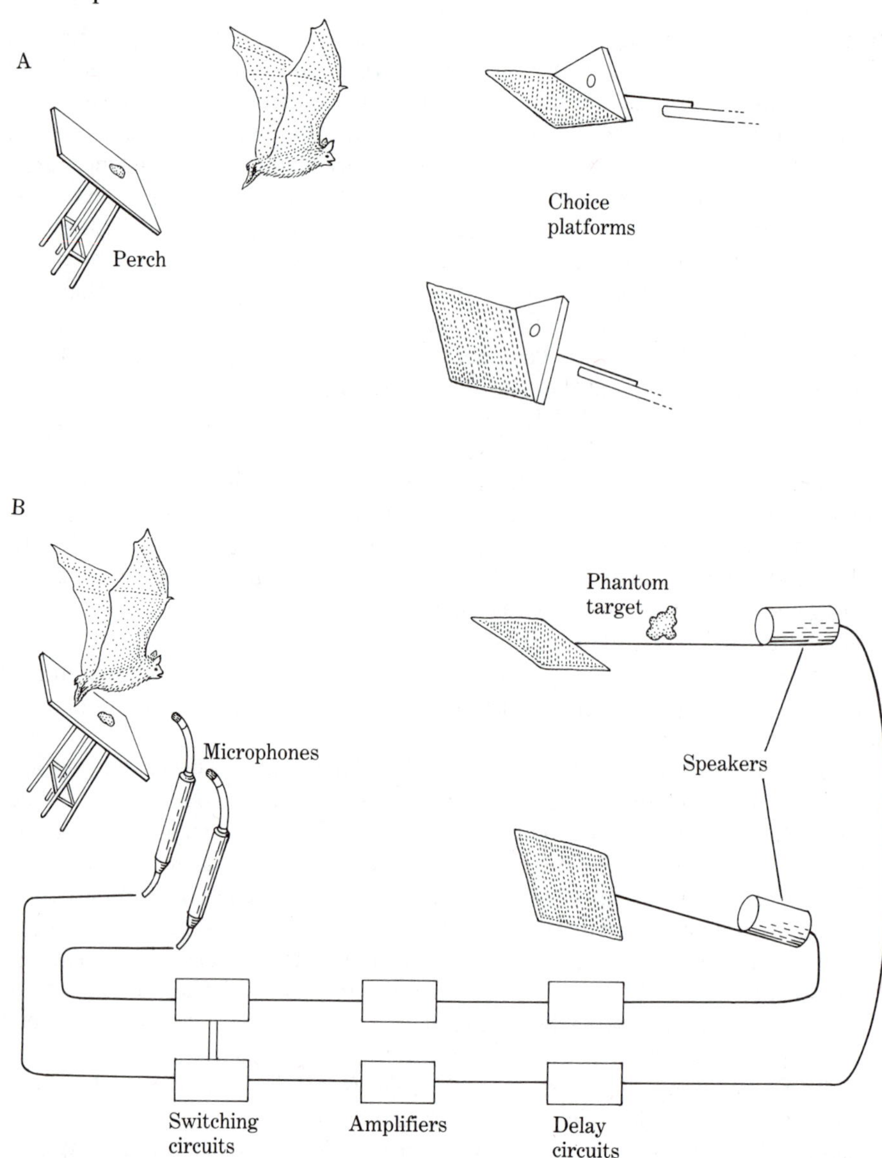

varied by only 1 microsecond (10^{-6} second)! (This was equivalent to a movement of the phantom target backward and forward by just 0.2 millimeters.) Thus, *Eptesicus* has some mechanism for measuring time differences of just 1 microsecond. This is truly astounding because 1 microsecond is only about 1/1000 the duration of a single action potential. How is it possible for the bat's nervous system to encode such brief intervals? The neural mechanisms responsible for this are entirely obscure. Individual neurons that have been found in the auditory cortex of another bat *(Pteronotus)* respond selectively to specific pulse-to-echo delays (Suga and O'Neill, 1980; O'Neill and Suga, 1982). Different cells among this group respond

◀ FIGURE 12. Bat training experiments. A. *Eptesicus* was trained to fly to the more distant of two choice platforms (the upper one in this figure). Each choice platform had an identical target, a Plexiglas triangle, for effective reflection of the bat's echolocation pulses. B. Replacement of real target by electronically produced "phantom targets." Two microphones were placed just in front of the bat's perch, each on a line with one of the choice platforms. These microphones recorded the echolocating pulses emitted by the bat while it stood on its perch and tried to decide toward which platform it should fly. In making its decision, the bat typically directed its pulses alternately toward one and then the other choice platform. When it pulsed toward a given platform, the microphone in that direction detected it as a louder sound than did the other microphone. The microphone that detected the louder sound then switched off the circuitry from the other microphone, for just a moment. Thus, only the circuitry for the "louder" microphone was operational at any given moment. This microphone activated, after a delay set by the experimenter, the speaker on its side. This speaker emitted an attenuated version of the bat's pulse that the microphone had just detected. It was by virtue of the precise delay setting (which determined the pulse-to-phantom-echo delay heard by the bat) that the experimenter attempted to inform the bat of the distance to the phantom target. Aside from these phantom echoes, the bat's emitted pulses also produced real echoes, reflected off the microphones, the platforms, the speakers, and other objects in the room. However, the trained bats appeared to ignore these echoes and to attend only to the phantom echo at 25 or 30 centimeters from the perch. It was at these distances that their training had prepared them to expect targets. (After Simmons, 1973.)

selectively to different pulse-to-echo intervals. However, no cell yet recorded is nearly accurate enough in its selection of particular pulse-to-echo intervals to account for a behavioral discrimination of 1 microsecond time differences.

In summary, then, *Eptesicus*, by performing a combined frequency and time analysis, can discriminate differences in target distances as small as 1 centimeter. It does so by detecting differences down to 60 microseconds in the pulse-to-echo interval. This bat can also discriminate a stationary from a jittery target where the jitter is as small as 0.2 millimeters, a jitter that results in a difference in the pulse-to-echo interval of just 1 microsecond. The neural mechanisms underlying this amazing temporal resolution, once they are understood, could reveal entirely new principles of neural coding.

RHINOLOPHUS: ANALYSIS IN THE FREQUENCY DOMAIN

The Eurasian insectivorous bat *Rhinolophus ferrumequinum* emits echolocating pulses that have just one harmonic (Figure 10B). The pulse has a long component with a constant frequency of 83 kHz (called the CF COMPONENT) followed by a rapid downward FM sweep (called the FM COMPONENT). *Rhinolophus*, then, can be referred to as a LONG CF–FM BAT, as compared to *Eptesicus*, an FM

BAT (Figure 10).[5] As we shall see in this section, *Rhinolophus* can carry out an analysis in the frequency domain of the CF component of its echoes. It uses this analysis to determine several important parameters of the target.

The ears of *Rhinolophus* are designed to be especially sensitive to sounds of just around 83 kHz, the same frequency as the CF component of its emitted pulses. (We shall examine in a later section the auditory specializations giving rise to this 83 kHz sensitivity.) This special auditory sensitivity to the CF frequency would seem to represent a good match of vocal and auditory design. Paradoxically, however, this match presents *Rhinolophus* with a problem. The problem arises because, when a flying bat emits a pulse of 83 kHz, the echo that returns from any object that the bat is approaching has a frequency actually *higher* than 83 kHz; and *Rhinolophus* is less sensitive to these higher frequencies. The explanation of this elevation of echo frequency lies in a phenomenon called the DOPPLER SHIFT, which applies to moving sound sources and/or moving listeners.

To understand the Doppler shift, imagine first that a *stationary* sound source is detected by a distant, *stationary* listener. Suppose that the sound has a constant frequency of 100 Hz. This means that at any given point in space, such as the point occupied by the listener's ear, there will be 100 amplitude peaks and troughs of sound per second. Now, if the listener were to walk toward the sound source, he would encounter amplitude peaks and troughs at a rate higher than 100 per second. This increase above 100 per second would result from his walking, so to speak, through the peaks and troughs of the sound. The faster he walks toward the sound source, the greater the rate at which he encounters peaks and troughs and thus, the greater the elevation above 100 Hz. Such an elevation of frequency would also occur if the listener were stationary and the sound source were moving toward him. In this situation, one can think of the whole mountain range of peaks and troughs as being pushed forward through space by the approaching sound source; thus, these peaks and troughs would arrive at the listener at a rate higher than 100 per second. In summary, the rate of arrival at the listener's ear of the peaks and troughs of sound is elevated above the emitted sound frequency whenever the sound source and the listener are in relative motion toward one another. This apparent frequency elevation is called a positive Doppler shift. (The reverse, a negative Doppler shift, occurs if the sound source and listener are moving apart.) We hear a positive Doppler shift as a rise in pitch. You may have experienced this phenomenon when traveling on a highway and hearing the rising pitch of a horn from an automobile approaching from the opposite direction. (After the automobile has arrived and then recedes, a negative Doppler shift causes a decreasing pitch.)

A bat is both a source of emitted sound and the listener to the echo of this sound. Therefore, because there is a Doppler shift in both the emitted and the reflected sound, the bat's ear detects twice the amount of Doppler shift as it

[5]There are also numerous species, referred to as SHORT CF–FM BATS, in which the CF component is only about ¼ as long as in *Rhinolophus*. Note, however, that when *Rhinolophus* or any other echolocating bat approaches a target, the overall duration of its pulses decreases. Thus, the distinction between FM, short CF–FM, and long CF–FM bats refers only to the pulses emitted while the bat is searching for targets, not while approaching close to a detected target.

would if it were merely listening to some stationary sound source that it approached at this same speed. The resulting frequency elevation can amount to several kilohertz. Thus, the CF component of the echo will have a frequency not of 83 kHz, to which the ear of *Rhinolophus* is especially sensitive, but rather of somewhere between 83 and about 87 kHz. This, then, is the paradoxical problem that arises because *Rhinolophus* matches the frequency of its emitted CF component to the frequency of its greatest auditory sensitivity; whenever this bat begins to approach an echo-reflecting object, the sound frequency of the echo tends to shift out of the range of most sensitive hearing.

Rhinolophus deals with this problem in an interesting way. Whenever this bat detects the start of a positive Doppler shift (which occurs as soon as it begins to approach an echo-reflecting object), the bat lowers the sound frequency of its emitted pulses in order to prevent the echoes from returning with too high a sound frequency. The greater the positive Doppler shift that it detects, the more *Rhinolophus* lowers its emitted frequency. In fact, it lowers this frequency by just enough to keep the echo of the CF component very close to 83 kHz (usually at 83.02 kHz). Thus, the echo frequency is continuously adjusted so as to remain in the range of most sensitive hearing. This adjustment by *Rhinolophus* of the frequency of its emitted CF component, in response to hearing a positive Doppler shift, is called DOPPLER SHIFT COMPENSATION[6] (Schnitzler, 1968, 1970; Simmons, 1974; Schuller *et al.*, 1975).

The auditory signal that evokes the bat's Doppler shift compensation response, then, is a change in the perceived frequency of the echo. Thus, because sound frequency is the stimulus for this response, the bat is performing this acoustic analysis in the frequency domain, not in the time domain. This frequency domain analysis pertains only to the bat's coding of the *CF* component of the echo. The echo's final FM sweep is probably analyzed in terms of both frequency and time and is used to determine distance and perhaps other parameters, as we have seen for the FM sweep of *Eptesicus* (Simmons *et al.*, 1975).

The use of Doppler shift compensation, then, brings the CF echo to just about 83 kHz, the frequency to which this bat's ear is most sensitive. This aids in the detection of faint echoes from fairly small or distant targets. (The long duration of the CF component also assists in this task.) An additional benefit of Doppler shift compensation is that, while the CF component of the faint echo is kept within the range of greatest auditory sensitivity, the potentially deafening emitted pulse is produced at a lower frequency, to which the bat's ear is less sensitive. One can consider the CF component of the echo as the signal and the CF component of the emitted pulse as the noise in this system, in the sense that the bat derives it crucial information from the CF echo and not from the CF pulse. Thus, by lowering the frequency of the emitted CF component, the bat increases its signal-to-noise ratio.

Probably the most remarkable property of echolocation in *Rhinolophus* con-

[6]Bats are known to Doppler shift compensate for large stationary objects such as trees, but not for insects. Thus, the echo from an insect flying among trees can return at a frequency slightly above or slightly below 83.02 kHz, depending upon whether the insect is heading toward or away from the bat. However, since most insects fly slowly, this difference from 83.02 kHz is quite small.

cerns the way that it enables the bat to detect with extraordinary sensitivity objects with jittery motion, such as the flapping wings of a flying insect. The bat detects such jitter by a mechanism completely different from the time and frequency analysis of *Eptesicus*. To take the simplest case, suppose that an insect is flying perpendicular to the flight path of the bat. Each beat of the insect's wings will sequentially shorten and then lengthen the distance from bat to insect (Figure 13A). Thus, each individual wing movement will cause a Doppler shift of the bat's echo. But unlike the bat's Doppler shift compensation for objects in its flight path, it is unable to compensate for the Doppler shifts occurring on each wingbeat of the insect; these wingbeats occur much too rapidly for the bat's vocal system to follow. [Insect wingbeats range from tens to hundreds per second, whereas the Doppler shift compensation response can only follow accurately changes as rapid as about 1 in 0.25 per second (Simmons, 1974; Schuller *et al.*, 1975).] Therefore, although the bat will be able to keep the overall echo of the insect at about 83

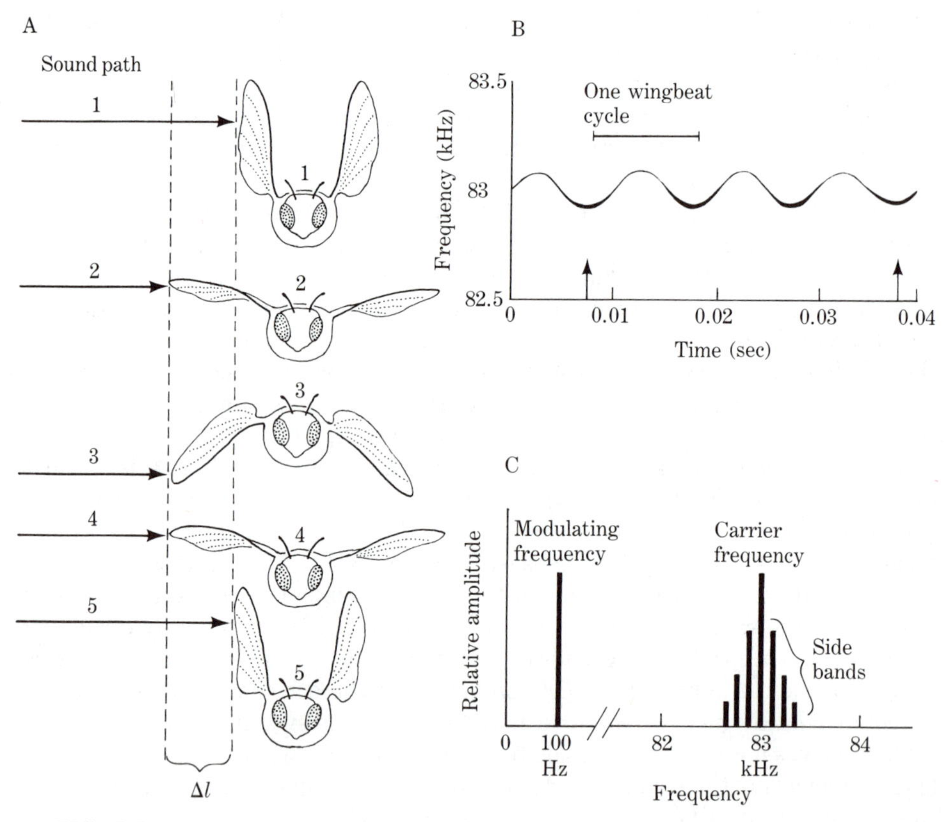

FIGURE 13. A. Change in the length of the sound path of a bat's echolocation pulse during one wingbeat of an insect. The insect is shown in five sequential positions during an entire wingbeat cycle. The pathlength difference (Δl) is shown for sounds coming from one side. B. Sonogram of the CF component of an echo returning to the bat from the flying insect shown in A. C. Amplitude spectrum of the sound shown in B. See text for explanation.

kHz, the oscillations of the insect's wings will cause the frequency of the echo's CF component to flutter up and down around this 83-kHz frequency. That is, the echo will be frequency modulated by the insect's wingbeat, as is shown by the sonogram of the CF component of an echo in Figure 13B.

Whenever an ongoing higher frequency (in this case, 83 kHz) is frequency modulated by a cyclic event of lower frequency (in this case, the insect's wingbeat), the higher frequency is called the CARRIER FREQUENCY and the lower one is called the MODULATING FREQUENCY. Such a rhythmic frequency modulation has a surprising feature, revealed by making an amplitude spectrum.[7] The amplitude spectrum shows energy not only at the carrier frequency and at the modulating frequency but also at other frequencies near the carrier frequency (Figure 13C). These additional frequencies are called SIDE BANDS. It is not possible to understand intuitively why these side bands are present. However, they are predicted by a Fourier analysis of the frequency modulated carrier wave in Figure 13B, and they are actually recorded in an amplitude spectrum of the sound. Therefore, sound energy is really present at these side band frequencies. It should be noted that on each wingbeat, at the moment when the insect's wing is positioned fully upward, a maximally large echo-reflecting surface is presented to the sound path (Figure 13A, 1 and 5). Thus, as is shown by the variation in the thickness of the line on the sonogram of Figure 13B, the bat's echo would be amplitude modulated as well as frequency modulated. Amplitude modulation also produces side bands.

Can *Rhinolophus* in fact detect the flutter of an insect's wing? These bats were trained to discriminate an oscillating target from a motionless target, by methods similar to those of Figure 12. The size of the target's oscillation was gradually reduced to determine when the bat's discriminative ability failed. For a target oscillating at a rate of 35 per second (resembling the wing movements of many insects), oscillating motions as small as 0.3 millimeters could be discriminated on 75% of the trials (Schnitzler and Henson, 1980). The Doppler shifts caused by these tiny oscillations would produce a frequency amounting to only plus or minus 30 Hz, that is, from 83,030 to 82,970 Hz. This bat can therefore discriminate sound frequencies with an accuracy of 30/83,000 or 0.036%—a truly remarkable accomplishment.

It is not yet proved how the brain of *Rhinolophus* detects frequency modulations. However, one obvious possibility is by responding to the presence of the side bands (Schuller,1979). That is, the bat's brain could know that if the echo of the CF component contains side bands in addition to the carrier frequency, the echo-reflecting target must be jittering. Because different auditory cells respond to different frequencies, the bat should have no trouble carrying out this analysis. This, of course, would be an analysis in the frequency domain. In fact, because the exact frequency of the side bands depends in part upon the modulation frequency (that is, upon the wingbeat frequency of a target insect), the bat may be able to determine the insect's wingbeat frequency. And because different types

[7]This amplitude spectrum is made, as was explained earlier in this chapter, by taking a section of the sonogram. However, one wishes to incorporate into this section a sufficient sampling duration of the sound to assure that the modulating frequency is included in the amplitude spectrum. Thus, the sampling would be carried out over a relatively long interval such as that between the two arrows in Figure 13B.

FIGURE 14. The mammalian ear. A. Diagram of the major components, roughly to scale. B. Schematic diagram of the inner ear, showing the cochlea largely uncoiled. C. Cross section through the cochlea. D. The organ of Corti, showing the outer and inner hair cells and their stereocilia. For simplicity, only one primary auditory neuron is shown connecting to each hair cell. Actually, many primary auditory neurons connect to each hair cell. However, among those primary auditory neurons that connect with the *outer* hair cells, each connects with just one. Among those that connect to the *inner* hair cells, each connects with a small number of neighboring hair cells. See text for explanation. ▶

of insects have different wingbeat frequencies, the bat may be able to identify the type of the insect, permitting it to make choices among different potential food items.

Rhinolophus, then, shows a very high degree of discrimination of sound frequency and encodes the echo of the CF component by means of a frequency domain analysis. It detects shifts of only 30 Hz away from the 83-kHz carrier frequency, amounting to a frequency discrimination with an accuracy of about 0.04%. This accuracy of frequency measurement is about 10 times greater than that performed by the human ear and 50 times greater than for most other vertebrate ears studied (Simmons, 1974). How the ear encodes sound frequencies with such precision will be described in the next two sections.

A DIGRESSION ON MAMMALIAN EARS: STRUCTURE AND FUNCTION

Before describing the adaptations of a bat's ears to its special acoustic needs, it will be useful to outline the structure and function of mammalian ears in general (Goldstein, 1968; Yost and Nielsen, 1977). Because mammals can hear sounds over distances of many wavelengths (well beyond the acoustic near field), it is clear that their ears are pressure, rather than displacement, detectors. The sound pressure waves arriving at a mammalian ear are collected by the ear flap, or PINNA, and pass down the EXTERNAL AUDITORY CANAL to the eardrum, or TYMPANUM (Figure 14A). These distal parts of the hearing organ are called the OUTER EAR. Sound pressure waves arriving at the tympanum cause it to vibrate at the same frequency or frequencies as the sound wave. The vibration of the tympanum must be transmitted to the auditory receptor cells, located in the fluid-filled COCHLEA of the INNER EAR.[8]

This transmission of vibrations from outer to inner ear is brought about through three interconnected bones, or OSSICLES—MALLEUS, INCUS, and STAPES—that lie in an air-filled space called the MIDDLE EAR. The distal end of the malleus rests on the tympanum and is caused to vibrate when the tympanum vibrates. These vibrations are transmitted to the farthest end of the stapes, which

[8]The inner ear also contains the vestibular organ, used in maintaining balance. The most prominent structures of this organ are the semicircular canals (Figure 14A). The vestibular organ is phylogenetically related to the cochlea, and the two are similar in cellular design.

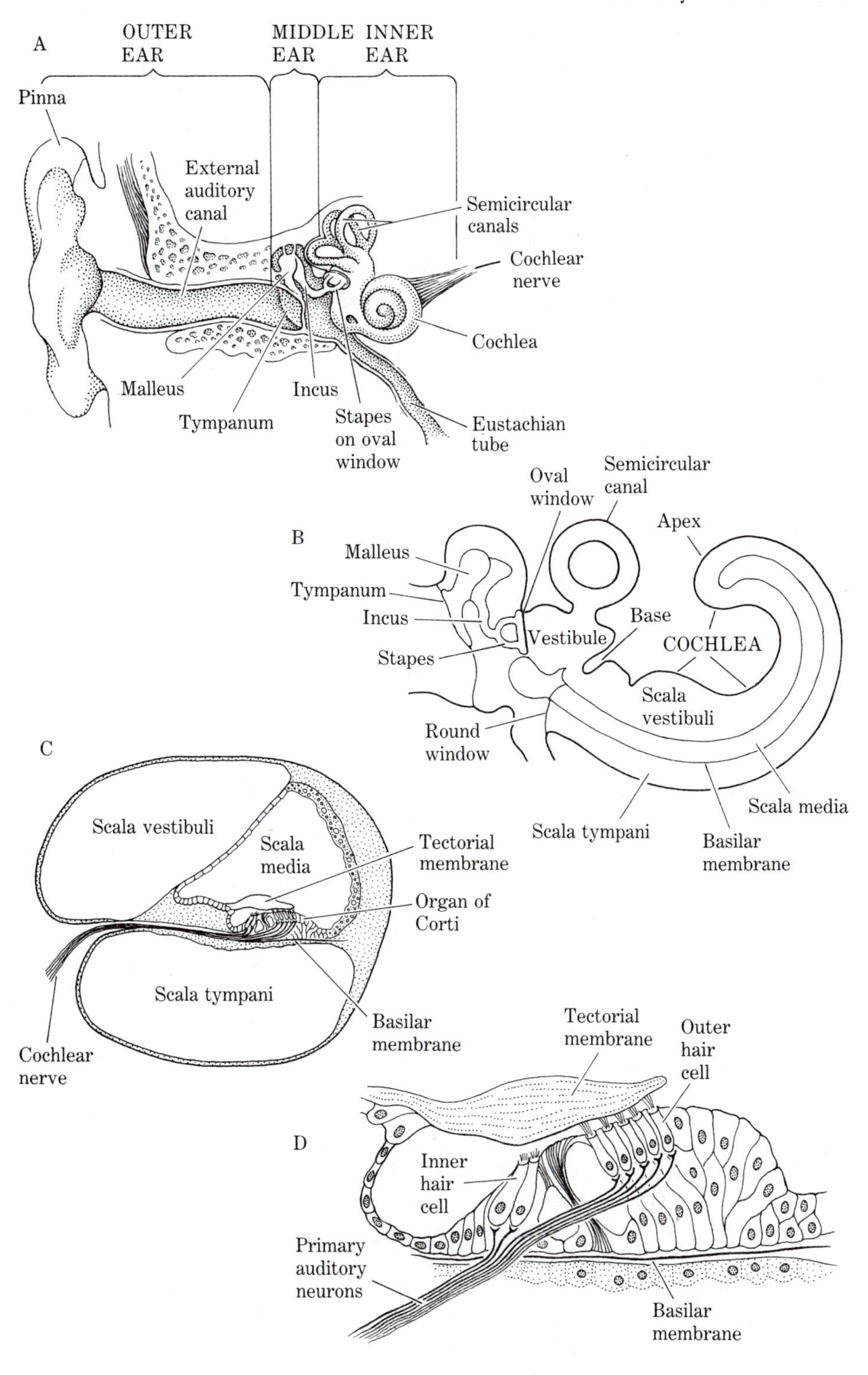
A
OUTER EAR
MIDDLE EAR
INNER EAR
Pinna
External auditory canal
Semicircular canals
Cochlear nerve
Cochlea
Malleus
Incus
Tympanum
Stapes on oval window
Eustachian tube
B
Oval window
Semicircular canal
Apex
Malleus
Tympanum
Incus
Stapes
Vestibule
Base
COCHLEA
Scala vestibuli
Round window
Scala tympani
Basilar membrane
Scala media
C
Scala vestibuli
Scala media
Tectorial membrane
Organ of Corti
Scala tympani
Basilar membrane
Cochlear nerve
D
Tectorial membrane
Outer hair cell
Inner hair cell
Primary auditory neurons
Basilar membrane

rests against a membrane called the OVAL WINDOW, the site where sound vibrations enter the cochlea. The middle ear connects with the nasal cavity via the air-filled EUSTACHIAN TUBE.

The cochlea (Figure 14B) is a helically coiled tube. In humans, this tube, if uncoiled, would be about 35 millimeters long from its BASE, near the oval window, to its APEX, at the far end. The cochlea is divided, along its length, into three chambers, each of which is filled with fluid: the SCALA VESTIBULI, which lies closest to the oval window; the SCALA MEDIA, lying in the middle; and the underlying SCALA TYMPANI. The scala media contains the auditory sense cells that, together with their supporting structures, are called the ORGAN OF CORTI (Figure 14C). This organ extends along the entire length of the scala media, from the base to the apex of the cochlea. Underlying the organ of Corti is the BASILAR MEMBRANE; and atop the organ of Corti lies the gelatinous TECTORIAL MEMBRANE. The auditory receptor cells, called INNER and OUTER HAIR CELLS, lie between these two membranes (Figure 14D). Tiny fingerlike extensions of the hair cells, called STEREOCILIA, project upward; and those of at least the outer hair cells contact the tectorial membrane. A sound causes the basilar membrane to vibrate up and down. This movement, transmitted to the overlying hair cells, causes their stereocilia to be deflected side to side. This deflection leads to a change in the membrane potential of the hair cells. The mechanism of transduction of mechanical to neuroelectrical energy resulting from deflection of the stereocilia is currently under investigation (Lim, 1980; Hudspeth, 1982).

To summarize the path of sound vibrations through the ear, sound pressure waves cause vibrations of the tympanum, which vibrates the ossicles, which vibrate the oval window, which leads to vibrations in the cochlear fluids, which causes vibrations of the basilar membrane, which results in deflections of the stereocilia of the hair cells, and thus the excitation of these receptor cells. Because the basilar membrane is the last major link in the chain of mechanical interactions prior to the stimulation of the hair cells, comprehending the vibrations of this membrane represents a key to the understanding of the sensory coding of sounds. We will therefore consider next the nature of the basilar membrane vibrations.

What is the shape of the basilar membrane's vibration, and how does this shape vary with different sound parameters? This question was answered during the 1930s and 1940s by von Békésy (1960), who won a Nobel prize for his solution to these problems. Von Békésy constructed physical models of the cochlea and studied the movements of their artificial basilar membranes. He also made mechanical measurements on the inner ears of human cadavers. More recent work has employed a variety of physical techniques to study basilar membrane movements in living animals (Lim, 1980; Zwislocki, 1980; Khanna and Leonard, 1982).

A sound of a given frequency causes the basilar membrane to vibrate up and down at this same frequency. However, the *amplitude* of the basilar membrane's vibration is different at different locations along its length, from the base to the apex of the cochlea. The location of the maximal vibratory amplitude depends upon the sound frequency; high frequencies produce maximal vibration closer to the cochlear base and low frequencies closer to the apex (Figure 15A). Actually, although a single location along the basilar membrane is maximally affected by sound of a given frequency, some vibration passes along most of the basilar mem-

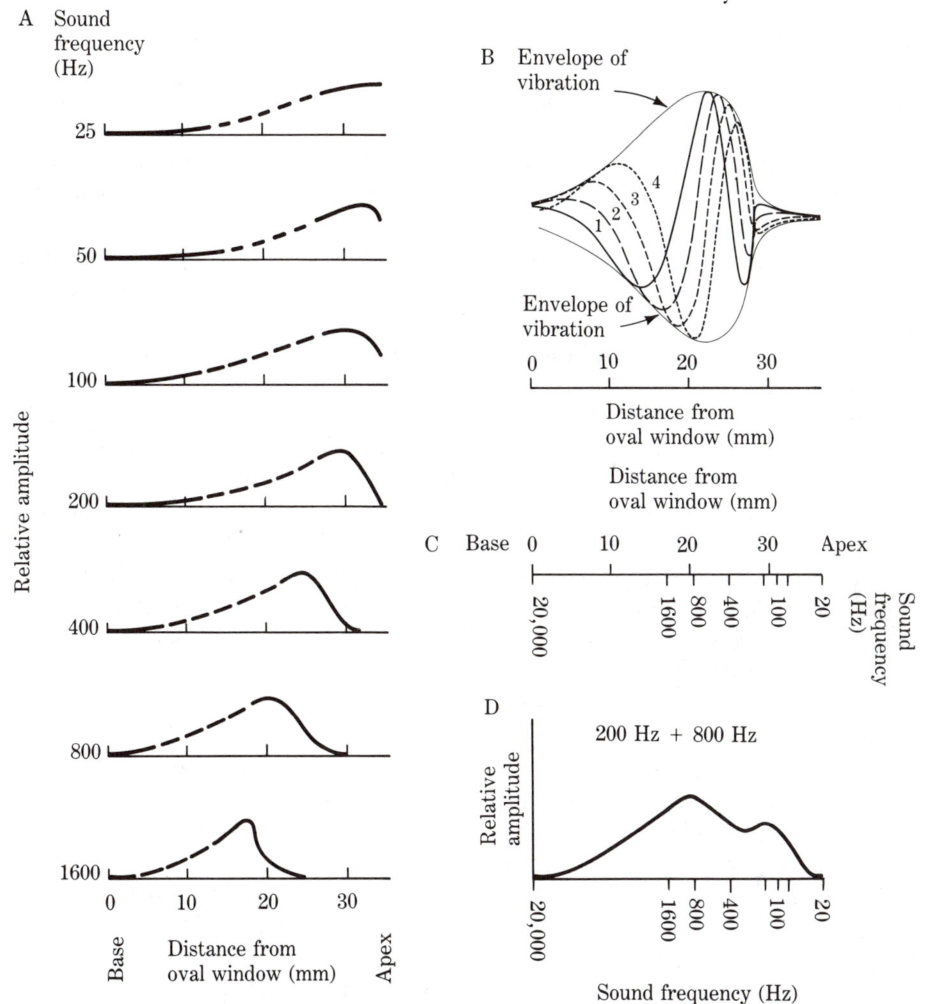

FIGURE 15. Basilar membrane vibration and sound frequency. A. Relative amplitude of a human basilar membrane's vibrations as a function of distance along the cochlea, for seven different sound frequencies. Solid lines represent von Békésy's direct measurements, and dashed lines represent his estimates. B. A traveling wave along the basilar membrane in response to sound of 400 Hz. Four successive instantaneous waveforms of the membrane's displacement are shown. The envelope (solid line) was drawn by connecting all the amplitude peaks of the instantaneous waveforms. The upper envelope line matches the 400-Hz graph of part A. C. Tonotopic map of the human cochlea. The map shows the correspondence between cochlear position (from base to apex) and the sound frequency that maximally vibrates the basilar membrane at that position. Note that about 1/2 of the map is given over to sound frequencies from just 20 to 800 Hz (which include the frequencies of human speech), whereas the other half covers from 800 to 20,000 Hz. D. Amplitude spectrum formed in the cochlea in response to a complex sound, consisting of 200-Hz and 800-Hz Fourier components. (A and B from von Békésy, 1960.)

brane's length, in the form of a TRAVELING WAVE (Figure 15B). Because different sound frequencies have their maximal effects at different points along the basilar membrane, this membrane displays along its length a map of sound frequency, or a TONOTOPIC MAP (Figure 15C). This situation contrasts with that of the retina, which contains a map not of the frequency (or wavelength) of light but rather of space (Chapter 5). The cochlear map, having the form of a line, is one-dimensional, whereas the retinal map is two-dimensional.

Let us now use the cochlea's frequency map (Figure 15C) as the x-axis on which to plot the amplitude of basilar membrane vibrations in response to a sound. For simplicity, our sound will consist of just two frequency components, 200 Hz and 800 Hz. This plot of amplitude as a function of frequency (Figure 15D) is an amplitude spectrum, like that of Figure 7A. Thus, the cochlea responds to sounds by generating an amplitude spectrum whose x-axis is the position along the basilar membrane and whose y-axis is vibratory amplitude. The basilar membrane, then, carries out a Fourier analysis of sounds. The net result is that, for any simple or complex sound, those hair cells that will be maximally stimulated are the ones that reside at the particular positions along the basilar membrane that are maximally vibrated by the frequency components contained in the sound. This idea—that a hair cell's position along the cochlea determines the sound frequency to which it will respond—is usually referred to as the PLACE THEORY of hearing.[9]

What physical properties within the cochlea determine which part of the basilar membrane will vibrate maximally for which sound frequency? This is determined largely by the physical characteristics of the basilar membrane itself. This membrane becomes gradually wider and less taut as one moves from base to apex. Thus, just like the different strings of a piano, the part of the basilar membrane used for higher frequencies has a lower mass and is more tightly strung.

Let us now consider briefly the structure and function of the cochlear hair cells. One to two rows of inner hair cells and three to five rows of outer hair cells run along the whole length of the basilar membrane (Figure 14D). These cells resemble in some respects the rods and cones of the vertebrate eye (Chapter 5). First, they have no axons. Second, they respond to their relevant physical stimulus with a receptor potential, but with no action potentials. (The response of a hair cell is a depolarization, whereas that of a rod or cone is a hyperpolarization; Chapter 5.) Third, hair cells show structural evidence of the presence of a cilium. Actually, during embryonic development each hair cell has one true cilium, called the KINOCILIUM, as well as about 40 stereocilia. Stereocilia are smaller than the kinocilium and lack the characteristic microtubular structure of a true cilium (Chapter 5,

[9]There has been a controversy as to whether the variation in the amplitude of vibration with position along the basilar membrane, for a given sound frequency, is sufficiently sharp to account for the sharp frequency selectivity that is known to occur in individual hair cells. Many investigators have postulated a SECOND FREQUENCY FILTER, located somewhere in the mechanical stages between basilar membrane movement and hair cell stimulation, which would sharpen the frequency selectivity of the hair cells (Evans and Wilson, 1975; Lim, 1980; Zwislocki, 1980). However, more refined measurements of the basilar membrane's vibrational amplitude have suggested that this parameter may be much more sharply tuned than was previously thought (Khanna and Leonard, 1982); that is, the peaks in Figure 15 would be much less spread out along the cochlear length than as shown. Thus, it may be unnecessary to postulate a second frequency filter in the cochlea.

Figure 20). During early development, the kinocilium is resorbed,[10] leaving only its centriole in place. The stereocilia remain intact and, as we have seen, function in the mature hair cells as the site of mechanoelectrical transduction. Thus, just like vertebrate rods and cones, where light is absorbed by membranous lamellae, or like invertebrate retinula cells, where light is absorbed in microvillar membranes, the sensory transduction in an auditory hair cell involves a cellular region that is rich in foldings of the plasma membrane.

Unlike the complex neural circuitry of the vertebrate retina (Chapter 5, Figure 21), however, the neuroanatomical arrangement within the cochlea is quite simple; the hair cells synapse on PRIMARY AUDITORY NEURONS, whose axons project in the COCHLEAR NERVE (the eighth cranial nerve) directly to the brain (Figure 14C and D). One group of these primary auditory neurons receives synaptic input exclusively from the outer hair cells; in fact, each receives input typically from just one outer hair cell. A second group of primary auditory neurons receives input only from the inner hair cells, each from a small neighboring group of these cells. The responses of a given primary auditory neuron, then, follow closely the responses of the one outer hair cell, or the small cluster of inner hair cells, to which it connects (Russell and Sellick, 1977; Yost and Nielsen, 1977).

A recording from the axon of a single primary auditory neuron within the cochlear nerve usually shows a fairly low, ongoing rate of action potentials, even in the absence of any sound stimuli. A sound of an appropriate frequency then increases the neuron's rate of action potentials above this ongoing level. One common method of measuring the auditory responses of such a neuron is to determine, for each of several different sound frequencies, the lowest sound pressure level that just noticeably elevates the neuron's rate of action potentials above its ongoing, resting rate. That is, one determines the threshold sound pressure level for each sound frequency. The results of two such experimental tests, each on a different primary auditory neuron, are shown in Figure 16. Each of these graphs is called a TUNING CURVE. These two tuning curves show, as expected, that a given primary auditory neuron is maximally responsive to just one sound frequency; that is, it is "tuned" to this frequency. This is called the BEST EXCITATORY FREQUENCY (BEF), or CHARACTERISTIC FREQUENCY, for that neuron. Other sound frequencies somewhat above or below the BEF also excite this neuron; but to do so these sounds must have a greater sound pressure level.

How well an animal can discriminate among different sound frequencies will depend in part upon the sharpness of the tuning curves of its different primary auditory neurons—whether the V-shapes of these curves are very narrow (that is, sharply tuned) or very broad. The degree of sharpness of tuning, as represented by a tuning curve, can be quantified by a parameter called the $Q_{10\ \mathrm{dB}}$ VALUE. This is defined as the following ratio: *best excitatory frequency* divided by the *width (in Hz) of the V-shape*, as measured at a point 10 dB above the bottom of the V. Thus, for instance, the left tuning curve of Figure 16 has a best excitatory frequency of 100 Hz, and the width at 10 dB above the bottom (dotted line) is 20 Hz. The $Q_{10\ \mathrm{dB}}$ value, then, is $^{100}/_{20}$, or 5. Note that the tuning curve on the right in Figure 16, though it has exactly the same shape, has a $Q_{10\ \mathrm{dB}}$ value of 10 because

[10]The hair cells of the vestibular and the lateral line systems of aquatic vertebrates, both developmentally related to the cochlea, retain the kinocilium in the adult stage.

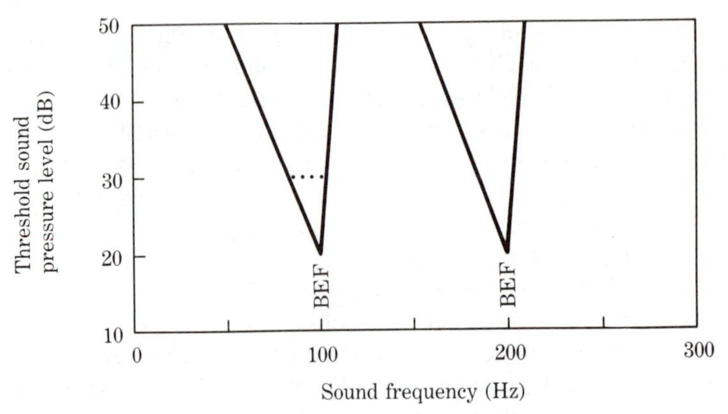

FIGURE 16. Tuning curves of two primary auditory neurons in a mammal. Each curve shows the threshold sound pressure levels for those frequencies to which its neuron is sensitive. The lowest point on each curve indicates the best excitatory frequency (BEF) for that cell. The dashed line across the left curve is used in calculating the $Q_{10\ dB}$ value. See text for explanation.

its best excitatory frequency is twice as great. The $Q_{10\ dB}$, then, indicates the *fraction* of the best excitatory frequency that is subtended by the V-shape at 10 dB above the bottom of the V.

In summary, then, sound produces a traveling wave of vibration along the basilar membrane. The maximal amplitude of this vibration occurs at a different location for different sound frequencies. This establishes a tonotopic map along the cochlea; and in the process, the cochlea carries out a Fourier analysis of the sound. A given hair cell, and a given primary auditory neuron that it excites, are maximally stimulated by the sound frequency corresponding to its location along the basilar membrane. Thus, the brain can recognize the one or more frequencies contained in a sound by taking account of which primary auditory neurons are maximally excited.

This outline of auditory design applies to mammalian ears in general. A broadly similar picture is seen in the ears of birds (Saito, 1980) and of most reptiles (Wever, 1978). However, in amphibians (Capranica, 1976) and fishes (Platt and Popper, 1981), although there is generally good discrimination of sound frequencies, there is no basilar membrane. Also, in some reptiles, such as turtles, only some of the cochlear hair cells sit atop a basilar membrane, others do not (Wever, 1978). The mechanisms of frequency discrimination in ears that lack a basilar membrane are not well understood.

THE INNER EARS OF BATS

In what ways are the ears of bats adapted for processing the highly specialized sounds used in echolocation? We will focus primarily on the inner ear of the long CF–FM bat *Rhinolophus,* whose echolocating behavior was described earlier. Recall that *Rhinolophus* analyzes the CF component of its echo in the frequency domain. This CF component, whether this bat is resting in a stationary environment or whether it is Doppler shift compensating on the wing, has a frequency very close to 83 kHz. At this frequency, as we have seen from behavioral experiments, *Rhinolophus* carries out highly refined discriminations of slight frequency changes. At the termination of the CF component, the FM component sweeps down from 83 kHz to about 65 kHz. Within this frequency range, there is no evidence for any unusual ability at frequency discrimination.

Recordings made from axons of primary auditory neurons in the cochlear nerve reveal that *Rhinolophus* has pronounced neural specializations corresponding to its refined frequency discrimination at around 83 kHz. Figure 17A shows the tuning curves of 12 such axons, each of which has a different best excitatory frequency (Suga *et al.*, 1976). The neurons whose graphs are labeled (a) through (g) have best excitatory frequencies between 10 and 60 kHz, below the frequencies contained in echolocation pulses or their echoes. These neurons are therefore not used for echolocation but rather for hearing other sounds. The neurons whose curves are labeled (h) and (i) have best excitatory frequencies in the range of the FM component of the echolocation pulse. The curves labeled (j) and (k) show BEFs right near 83 kHz, the frequency of the CF component. Curve (l) has a higher BEF. All these curves except (j) and (k) are fairly broadly tuned, with $Q_{10\ \mathrm{dB}}$ values below 20. These values are about the same as those found in the FM bat *Eptesicus* or in most other mammals so far tested. However, curves (j) and (k) are much more sharply tuned, having $Q_{10\ \mathrm{dB}}$ values of 50 and 70, respectively. Of all the primary auditory neurons that have been recorded in *Rhinolophus*, those tuned to frequencies *other than* about 83 kHz all have $Q_{10\ \mathrm{dB}}$ values below 20. But of those tuned to about 83 kHz, the average $Q_{10\ \mathrm{dB}}$ value is 140, and the highest value ever recorded is over 400 (Figure 17B). These are the highest $Q_{10\ \mathrm{dB}}$ values ever reported for primary auditory neurons in any animal (except for those of *Pteronotus*, another long CF–FM bat; Suga *et al.*, 1975). In addition to being more sharply tuned, those primary auditory neurons whose BEFs are approximately 83 kHz generally are more sensitive (that is, they have a lower threshold) than those tuned to other frequencies within the range of echolocating pulses (Suga *et al.*, 1976).

This sharp tuning of those primary auditory neurons whose best excitatory frequencies are near 83 kHz is consistent with the behaviorally demonstrated sharp frequency discrimination in this range. Because the tuning curves are very steep, even just a slight change in sound frequency will cause a change in the population of primary auditory neurons that is excited. Thus, the brain is informed of very slight frequency differences.

How does *Rhinolophus* accomplish such sharp frequency tuning of those primary auditory neurons whose best excitatory frequency is near 83 kHz? Does this sharp tuning result from some form of neuronal interaction among either the hair cells or the cochlear terminals of the primary auditory neurons? In fact, such neurally produced sharpening of tuning curves has been discovered in the brains of bats (Suga, 1965) and of other mammals (Goldstein, 1968). However, the cochleas of bats and of all other mammals examined to date appear to have no neural interactions that function as a sharpening mechanism (Pollak *et al.*, 1972; Russell and Sellick, 1977).

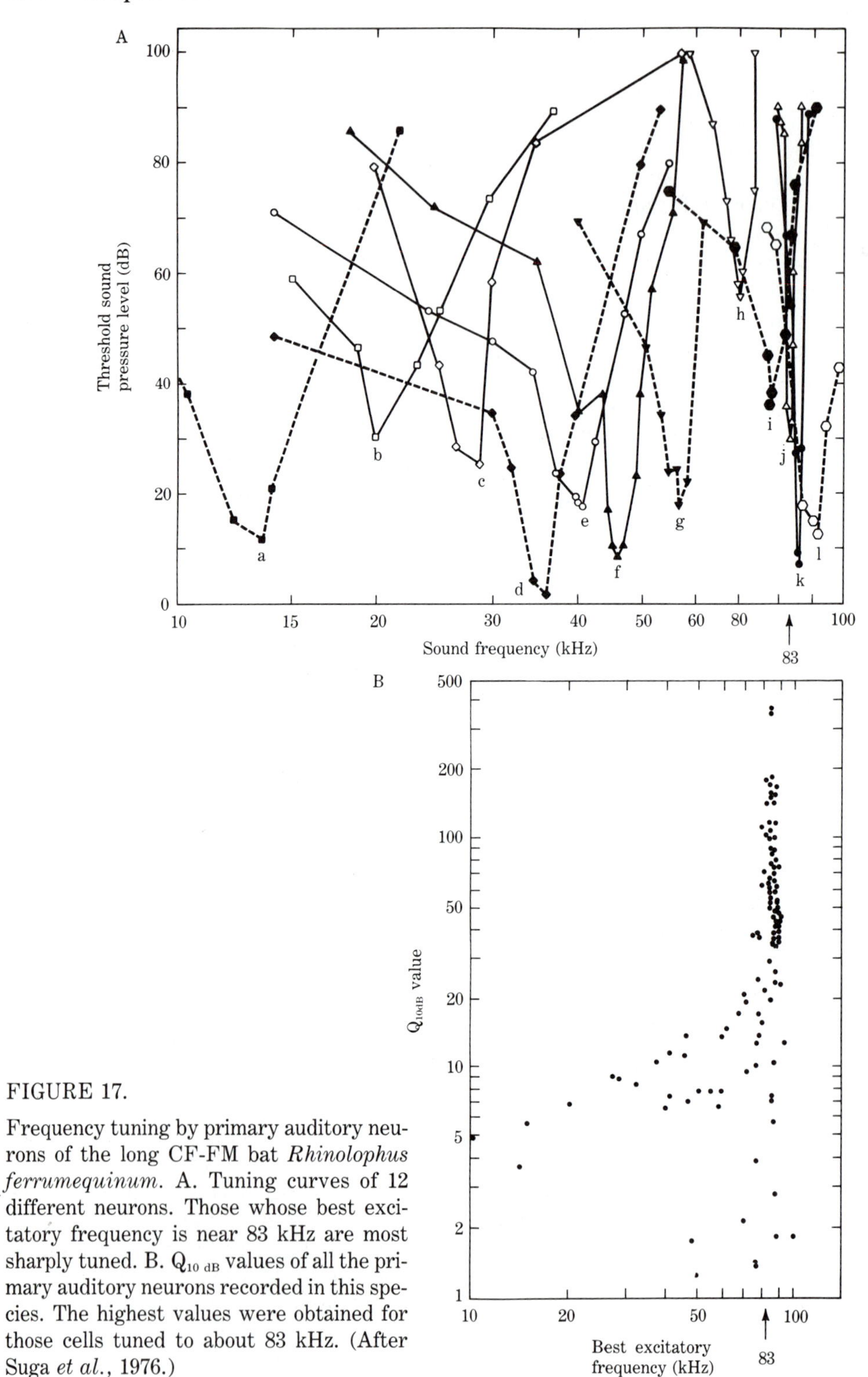

FIGURE 17.

Frequency tuning by primary auditory neurons of the long CF-FM bat *Rhinolophus ferrumequinum*. A. Tuning curves of 12 different neurons. Those whose best excitatory frequency is near 83 kHz are most sharply tuned. B. $Q_{10\ dB}$ values of all the primary auditory neurons recorded in this species. The highest values were obtained for those cells tuned to about 83 kHz. (After Suga *et al.*, 1976.)

Presumably, then, the sharp tuning at around 83 kHz has a strictly mechanical basis. But does this tuning result from specializations of the basilar membrane itself, or is there some other mechanical basis for the sharp tuning? The basilar membrane of *Rhinolophus* in fact exhibits a dramatic structural specialization that appears to account, at least in part, for the sharp tuning at 83 kHz. As one proceeds from the base to the apex of the cochlea in this bat, the thickness and the width of the basilar membrane change, not gradually as in other mammals, but abruptly. Out of a total basilar membrane length of 16 millimeters, there is a sharp discontinuity in the ratio of thickness to width at a point 4.5 millimeters from the cochlear base (Figure 18A). This thickness:width ratio of a basilar membrane at any given location is thought to be proportional to the membrane's stiffness at that location and therefore to the frequency of sound that will cause that location to vibrate maximally. Therefore, this structural discontinuity at 4.5 millimeters in *Rhinolophus* suggests something quite special about this location. Indeed, by mapping the cochlea of *Rhinolophus* with respect to sound frequency, as von Békésy had done for the human ear (Figure 15C), it was found (Bruns, 1976) that 83 kHz is coded at just about 4.5 millimeters from the base (Figure 18B). In fact, 83 kHz maps precisely onto the point where the basilar membrane's height:width ratio is *changing* most rapidly with position along the cochlea.

Thus, it is this specialized cochlear region, with its sharp discontinuity of basilar membrane properties, onto which *Rhinolophus* casts the CF component of its echoes by means of its Doppler shift compensation response. By analogy to vertebrate vision, in which an animal brings a target to bear on the specialized part of the retina called the fovea (Chapter 5), this 83 kHz region of the basilar membrane in *Rhinolophus* has been called an ACOUSTIC FOVEA.

A further specialization of the cochlea of *Rhinolophus* is an enormous expansion in length of that cochlear region devoted to sound frequencies between 83 and 86 kHz (Figure 18B). This limited frequency range occupies more than 3 millimeters, or about ⅕ of the cochlear length. It is in just this range of frequencies that a Doppler shifted echo will first return to the bat's ear from an object that has just come into echo range, before the bat has had time to carry out its Doppler shift compensation response. Thus, this frequency range just above 83 kHz is of crucial importance to *Rhinolophus*. The bat, then, deploys a disproportionately large number of sensory cells for this task of special importance. Not only the basilar membrane but also the cochlear nerve and the auditory centers of the brain devote a disproportionately large number of cells to the processing of sounds in this frequency range (Neuweiler, 1980).

Directional Localization of Sound

One important requirement of the auditory system of bats, and of many other animals, is to determine the direction of a sound source. Because the cochlea, unlike the retina, does not contain a sensory map of space, special problems are encountered in carrying out this auditory localization. We will now consider some of the ways that animals in general solve this problem. We will then consider in somewhat greater depth an intriguing mechanism of directional localization that has evolved in crickets and in a limited number of other animals.

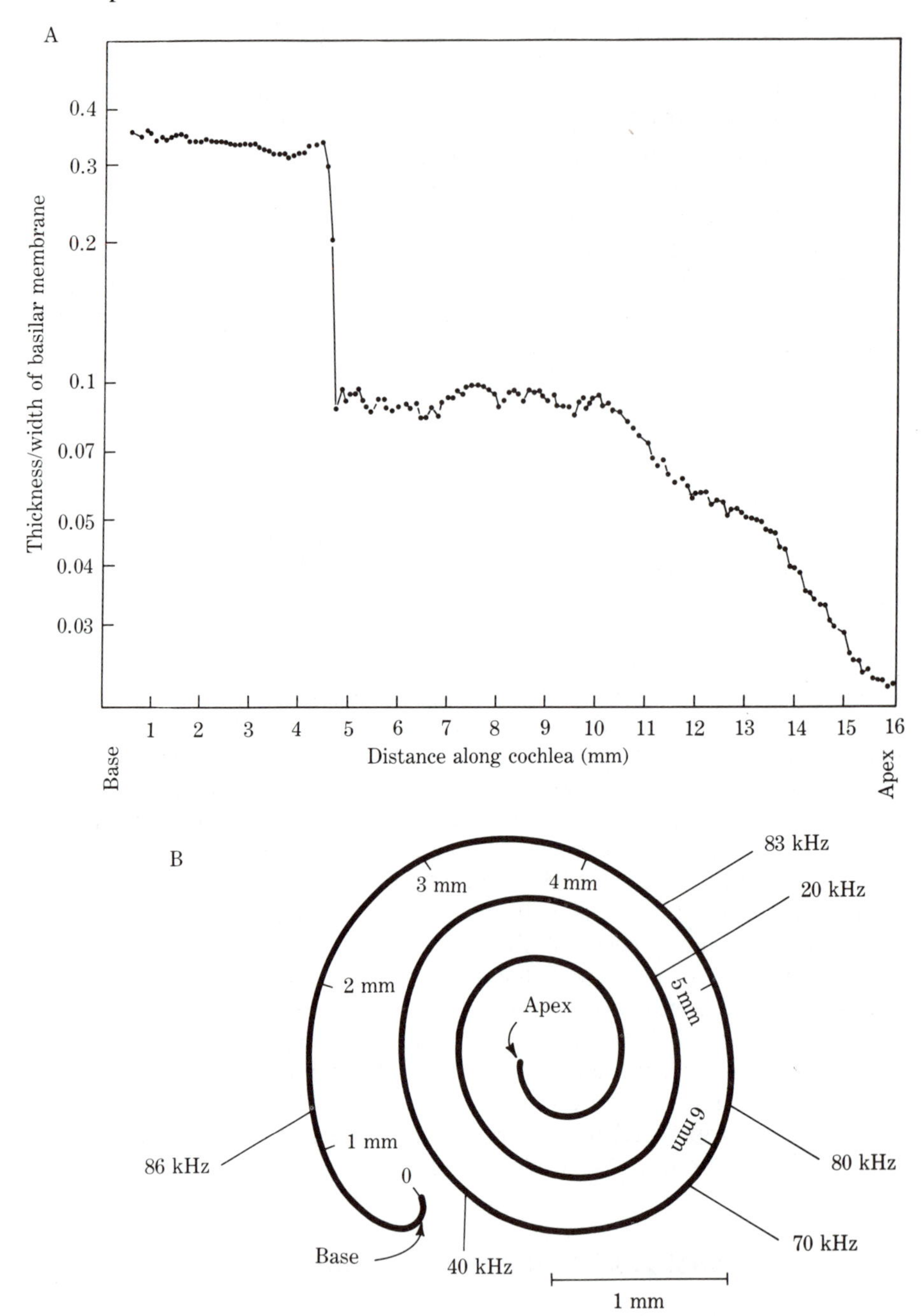

FIGURE 18. Specializations of the cochlea of *Rhinolophus*. A. The ratio of thickness to height of the basilar membrane, as a function of distance from base to apex. The sharp discontinuity occurs at approximately 4.5 millimeters from the base. B. Cochlear frequency map. A disproportionately large space is devoted to the frequencies from 83 to 86 kHz. (A after Bruns, 1976; B after Neuweiler, 1980.)

THE DIRECTIONAL INFORMATION AVAILABLE

Several acoustic parameters potentially can provide an animal with directional information (Erulkar, 1972; Konishi, 1977; Gourevitch, 1980; Knudsen, 1980). In general, however, neither ear acting *by itself* is very useful in this process. For instance, if the left ear hears a soft sound, it could be that this was produced by a weak sound source on the left side or by a strong sound source on the right. To overcome this ambiguity, both ears must be used together. A sound on the left side, then, can appear different to the two ears, and this difference can provide a clue to the location of the sound source. That is, auditory localization generally involves a BINAURAL COMPARISON of the sound.[11] The fact that an animal uses a binaural comparison for directional localization can be demonstrated by plugging or covering one ear. Many animals so treated respond, when attempting to approach a given sound source, by turning consistently toward the side of the intact ear (Feng *et al.*, 1976; Gourevitch, 1980; Knudsen, 1980).

What sound parameters might be compared by the two ears in order to localize a sound source? As we have seen, a sound produced on one side of the head will cast a sound shadow if the head is at least as large as roughly one wavelength (Figure 5). Thus, a binaural comparison of sound pressure level (usually called a BINAURAL INTENSITY COMPARISON) can help some animals in sound localization. That is, the animal's brain can deduce that the sound is coming from the side that detects the greater sound pressure level. But if the animal has a small head or if the sound frequency is low (and therefore the wavelength long), there will be no substantial sound shadow, so that binaural intensity comparisons may not be usable.

Some animals can use a binaural comparison of the TIME OF ARRIVAL of a sound at the two ears. When a sound begins on the left, for instance, the leading edge of the sound wave will reach the left ear before the right. Such a time of arrival comparison can be made most effectively if the sound stimulus begins abruptly, so that the leading edge of the sound wave has sufficient amplitude to be clearly detected by each ear. Thus, sounds such as clicks, which have sharp onsets, are readily localized by this means, whereas more gradually increasing sound stimuli are not. (Sounds that begin gradually but that contain clearly defined transients can also be located in this manner.) But if an animal's head (and thus its interaural distance) is very small, the binaural time of arrival difference will be brief. Depending upon the animal species, this brief interval may be difficult for the nervous system to measure accurately.

Some animals can also use a BINAURAL PHASE COMPARISON for sound localization. This parameter involves a time comparison, though it is not specifically the leading edge of the sound whose time of arrival is compared at the two ears. Rather, a time comparison is made by the two ears throughout the entire duration of the sound. The concept of binaural phase comparison is illustrated in Figure 19. In Figure 19A, consider first the listener with the smaller head, pictured

[11]A single ear can be used to localize sound if the head (or perhaps just the pinna) can move from side to side during the sound. The sound is then compared at two or more sequential positions of hearing. However, many animals (including ourselves) can localize a sound source without moving the head.

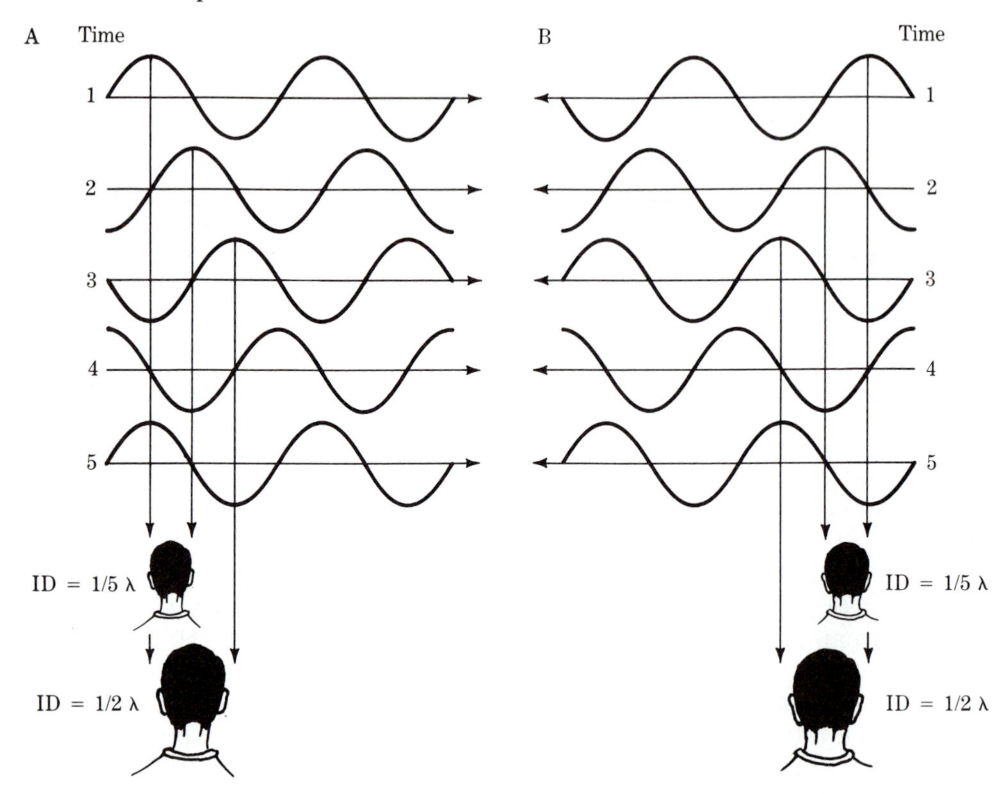

FIGURE 19. Binaural phase comparison for localizing a sound source. A. Sound from the left. B. Sound from the right. A small and a large listener, having interaural distances equal to 1/5 and 1/2 wavelength, respectively, listen to each sound. The listeners are facing into the page. See text for explanation.

below the sound waves. This listener's interaural distance equals about 1/5 of the wavelength of the sound wave drawn above. The sound wave is shown at five successive moments as it propagates from left to right. A peak of the sound wave arrives at the left ear at time 1 and at the right ear at time 2. It is not until time 5 that the next peak arrives at the left ear. This rapid left–right sequence of the arrival of peaks would indicate that sound had come from the left side. That is, the listener could localize the sound source by determining that the sound heard by the left ear was out of phase with that heard by the right ear and, specifically, that the left–right sequence was briefer than the ensuing right–left sequence. In Figure 19B, a sound from the right is detected by the smaller listener as a brief right–left sequence and a long left–right sequence.

Such a binaural phase comparison provides unambiguous directional information only if the interaural distance is short relative to the wavelength. Specifically, the interaural distance must be *less than half* the wavelength. For instance, note that in Figure 19A, the large head has an interaural distance equal to just ½ the wavelength. In this situation, the peak of the sound wave reaches the left ear at time 1 and the right ear at time 3. The next peak reaches the left ear at time 5. Thus, the left–right interval equals the right–left interval. Moreover, this same interval would result if the same sound had originated from the right side of the head (Figure 19B). This listener, then, could not localize this sound using a binaural phase comparison. One can readily verify, by means of similar drawings, that interaural distances greater than ½ wavelength also produced directional ambiguities. In summary, then, an animal can use a binaural phase comparison to localize sound only if the sound frequency is low enough (that is, if the wavelength is great enough) or if the head is narrow enough that the interaural distance equals less than ½ the wavelength of the sound.

One other directional mechanism involves the way that the pinna, or other structures that an animal's outer ear might possess, reflect sounds from different locations into the ear. For instance, in barn owls, owing to an outer ear structure called the facial ruff, the left ear collects low frequency sounds primarily from the left side and the right ear collects these sounds primarily from the right side. Thus, for these *low* frequencies, a binaural intensity comparison helps the animal to determine the side of the head from which a sound originates. But because of the different ways that the facial ruff reflects different sound frequencies, high frequencies are collected by the left ear primarily from below the head and by the right ear primarily from above the head (Knudsen, 1981). Thus, a binaural intensity comparison of *high* sound frequencies helps the owl to determine the elevation of a sound source. The high and the low frequency comparisons are carried out simultaneously but by different auditory neurons, a mechanism that permits the owl to locate a sound source in both the horizontal and the vertical axes. Such a binaural comparison of sound pressure level, carried out simultaneously but separately for different sound frequencies, is called a BINAURAL SPECTRAL COMPARISON, or an INTERAURAL SPECTRUM.

We have seen three mechanisms that can be used for determining whether a sound source is located on the left or the right of the listener; binaural comparison of intensity, of arrival time, and of phase. As we saw, the first two of these are useful for animals with large heads but are not very useful if the animal's head, and thus also its interaural distance, is very small. In fact, many small animals have problems in localizing sound sources. Some of these animals solve this problem by using ultrasound, whose short wavelength may allow the head to produce a sufficient sound shadow to permit binaural intensity comparisons. But ultrasound, because of its high excess attenuation, is useful only over relatively short distances (Figure 4). A further mechanism used for directional localization by some animals with small interaural distances involves a surprising specialization in the design of the ear. This will be considered in the next section.

THE CRICKET'S SOLUTION: PRESSURE GRADIENT EARS

In most ears that detect the pressure component of sound, fluctuations of sound pressure occurring only outside the tympanum are responsible for producing the inward and outward tympanal movements. On each sound wave, an amplitude peak pushes the tympanum inward, and then an amplitude trough draws it outward. Little or no sound energy reaches the inside of the tympanum because the skull and other tissues block sounds from reaching this location (Figure 20A).

In the design of some ears, however, one or more free paths are provided that permit considerable sound energy to reach a tympanum's inner, as well as its outer, surface. The tympanal movements, then, result from both internal and external pressure fluctuations. Ears that are so designed are called PRESSURE GRADIENT EARS. An especially well studied example is found in crickets (Figure 20B), whose ears are located on the front legs (Hill and Boyan, 1977; Kleindienst *et al.*, 1981; Larsen, 1981).

The anatomy of cricket ears reveals three possible sound paths to the inside of either tympanum; one through each of two spiracles, or respiratory pores, and a third through the opposite tympanum (Figure 20B and C). Each tympanum is backed by an air-filled tracheal tube, part of the insect's tracheal respiratory system (Hill and Boyan, 1977). The left and right tubes each extend to the midline of the body, where they meet, separated only by a very thin septum. Each of the tracheal tubes has a branch leading to one of the thoracic spiracles.

If sound were to actually reach the back of a tympanum by passing through the opposite ear or the spiracles, it could have a profound effect upon tympanal

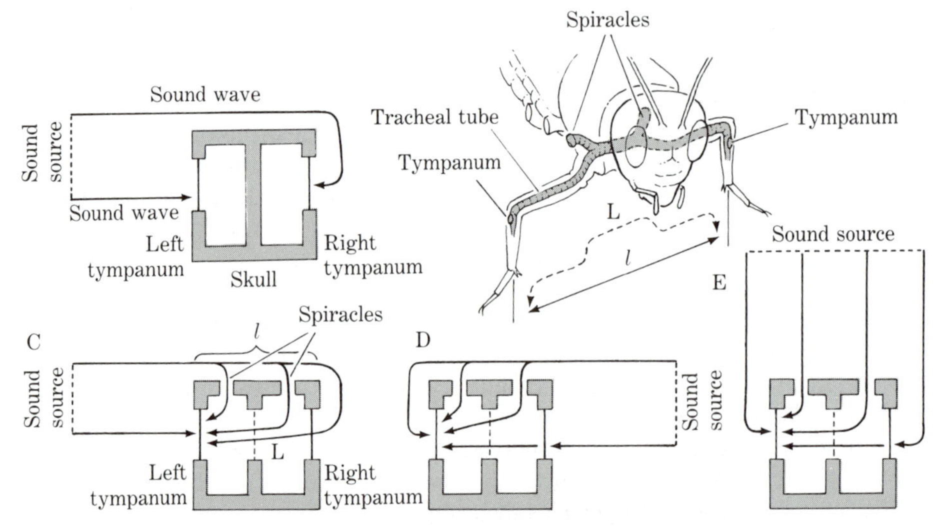

FIGURE 20. Pressure gradient sound reception. A. Arrival of sound waves at the *outside* of each tympanum in a typical vertebrate, whose head is shown here schematically. The skull and other tissues prevent any substantial sound energy from reaching the inside of either tympanum. B. Anatomy of the auditory system in the cricket *Teleogryllus commodus*. L = internal path length between the two tympana. l = external path length between the two tympana. C, D, and E. Path lengths of sound reaching the outside and inside of the left tympanum. C. Sound source on the left. To reach the *inside* of the left tympanum *via* the right ear, the sound must travel a distance $l + L$ further than sound reaching the *outside* of the left tympanum. D. Sound from the right. E. Sound from the front. (A, C, D, and E after Michelsen, 1979; B after Hill and Boyan, 1976.)

movements and, therefore, upon hearing. But does this actually occur? To answer this question, various manipulations on the ears and spiracles were carried out while recordings were made separately from the axon of each of several auditory interneurons in the nerve cord (Hill and Boyan, 1977). The axons tested were on the left side of the nerve cord. The cells to which they belong are known to be excited only by the left ear.

First, the tuning curve of a recorded interneuron was determined. For the great majority of the interneurons in the species of cricket that was studied *(Teleogryllis commodus)*, the best excitatory frequency was very close to 3.7 kHz (Figure 21, curve 1). This is the frequency of the male's calling song in this

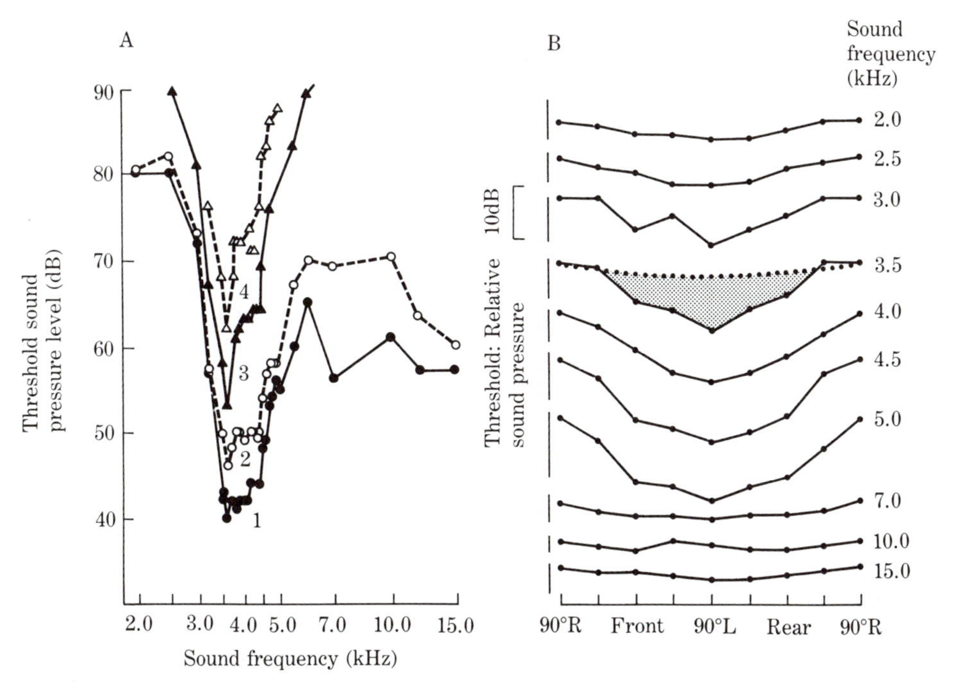

FIGURE 21. Tests for pressure gradient reception in crickets. A. Tuning curves of a left auditory interneuron, recorded under four different conditions: (1) auditory system intact; (2) after blocking access of sound to the outside of the left tympanum; (3) after covering the right tympanum and the two thoracic spiracles with petroleum jelly; (4) after cutting and filling with pertroleum jelly the left tracheal tube. B. Changes in threshold sound pressure level of a left auditory interneuron in response to sounds played from different directions. Sounds of different frequencies were used, between 2 and 15 kHz, as listed at the right. When frequencies between about 3 and 5 kHz were used, the threshold was much lower for sounds from the left side than from the right. The 3.5 kHz stimulus was repeated after the two spiracles and the right tympanum had been covered. The resulting thresholds are shown by the dotted line; the directionality was almost completely lost. Similar losses of directionality also occurred for the other sound frequencies. (After Hill and Boyan, 1977.)

species. After the tuning curve had been determined, the left ear was tightly covered by a steel cuff. This cuff prevented practically all sound from reaching the outside of the tympanum. However, because the cuff did not directly touch the tympanum but only contacted the surrounding cuticle, the tympanum remained free to vibrate if any sound actually reached it from the inside. The response to the same sound by the same interneuron was now retested. Even though practically no sound energy reached the outer tympanal surface, the interneuron was still excited. In fact, its tuning curve was almost the same as without the steel cuff, except that the threshold had been elevated by about 8 dB (Figure 21, curve 2). Thus, sound was apparently stimulating the left ear from the inside.

Covering the two spiracles and the right tympanum with petroleum jelly[12] caused a further increase in threshold (Figure 21, curve 3). Thus, sound had indeed been entering through these paths to reach the inside of the left tympanum. Because there was still some response of the interneuron, even with both spiracles and the right tympanum covered, some sound energy must have been reaching the left ear through the walls of the tracheal tubes after entering into the body and traveling through various body tissues to reach these tubes. In support of this idea, when the tracheal tube in the left leg was finally cut and filled with petroleum jelly, there was a further elevation of the interneuron's threshold (Figure 21, curve 4). Finally, removing the steel cuff that had been covering the left tympanum partially restored the tuning curve toward its original form, with a low threshold (though not as low as in the intact animal, because the internal sound path had now been disrupted). This final step showed that the elevation of threshold in the sequence 1 through 4 of Figure 21 had not resulted simply from a deterioration of the auditory system during the course of the experiment but rather had been produced by changes in the access of sounds to the tympanum.

What effect does the cricket's pressure gradient hearing have upon this insect's ability to localize a sound source? Directional localization in crickets has been studied primarily in connection with the ability of females to orient toward, and approach, the song produced by a courting male, known as the CALLING SONG (Huber, 1975; Popov and Shuvalov, 1977; Hoy, 1978; Moiseff *et al.*, 1978). This orientation behavior by the female is called positive PHONOTAXIS (Chapter 2). By comparing the lengths of the sound paths to the inside and to the outside of a tympanum, one can predict that the pressure gradient system would help a cricket to localize a male emitting its calling song (Michelsen and Nocke, 1974). This prediction can be understood with reference to Figure 20C, D, and E.

As Figure 20C shows, sound generated on the left side of the animal must travel a greater distance to reach the inside of the left tympanum than to reach the outside of this tympanum. Let us consider the sound that reaches the inside of the left tympanum by entering through the right tympanum and traveling back across the full length of the internal sound path. The additional length that this sound must travel to reach the left ear is the length l (the interaural distance)

[12]The spiracles are capped by cuticular flaps that the animal can open and close by contracting and relaxing spiracular muscles. Prior to the series of tests shown in Figure 21, these flaps had been ablated. Covering these spiracles with petroleum jelly had little effect on the animal's respiration, since it is primarily the large number of other spiracles on the body that are used for oxygen exchange.

plus the length L (the length of the internal sound path). These two lengths, l and L, are indicated accurately in Figure 20B. The summed value of $l + L$ in this species of cricket averages 3.8 centimeters. Importantly, this distance is reasonably close to ½ the wavelength of the calling song's major frequency component of 3.7 kHz; a full wavelength equals 8.9 centimeters [Equation (3)], so ½ wavelength equals about 4.5 centimeters. This means that at that moment when an amplitude peak has arrived at the outer tympanal surface, the inner tympanal surface, located about ½ wavelength away, is experiencing nearly a pressure trough. Thus, the tympanum is simultaneously being pushed inward from the outside and being pulled inward from the inside. One-half period later, when there is a pressure trough outside, there is nearly a pressure peak inside; so the tympanum is now being simultaneously pulled outward from the outside and pushed outward from the inside. Owing to this combined effect of forces on both tympanal surfaces, the amplitude of tympanal vibration in the left ear will be especially great when the calling song is produced directly on the left side. Thus, the left auditory nerve will respond strongly.

If, instead, the calling song had originated on the right side, the path lengths of the sound to the inside and outside of the left tympanum would be about equal (Figure 20D). Therefore, when there is a pressure peak outside, there would also be a pressure peak inside; and when there's a pressure trough outside, there would also be one inside. Thus, at any instant, the tympanum should experience roughly equal and opposite forces, resulting in little net movement and, thus, little neural response.

If, instead, the same sound had originated from the front or back of the animal, the path lengths to the inside and the outside of the left tympanum would differ; but this difference would be less than for a sound from the left (Figure 20E). Therefore, when there is a pressure peak outside, there would be neither a peak nor a trough inside, but an intermediate amplitude. Likewise, when there is a pressure trough outside, there would be an intermediate pressure inside. Thus, the net force that vibrates the tympanum, and the resulting neural response, would be of intermediate magnitude.[13]

It follows from these considerations that a calling song coming from a given side of the body should excite the ear on that side more than it would excite the contralateral ear. Thus, the animal could use this binaural difference in the amount of excitation to localize the sound source. Note that, although we predict a binaural

[13]This discussion has assumed that sound reaches the inside of the left tympanum by entering through the right tympanum. However, the spiracles appear to admit at least as much sound as does the right tympanum (Hill and Boyan, 1977; Larsen, 1981; Kleindienst *et al.*, 1981). A sound from the left side entering the spiracles travels a shorter distance to reach the back of the left tympanum than does the sound entering through the right tympanum (Figure 20C). Thus, when an amplitude peak arrives at the outside of the left tympanum, the sound traveling through the spiracles would produce, at the inside of this tympanum not an amplitude trough but rather some amplitude intermediate between peak and trough. The left tympanum, then, would not experience as great a net force as if all the sound reaching its inner surface had come through the right ear. Nevertheless, the difference between internal and external amplitudes at the left tympanum would still be maximal for sounds from the left and minimal for sounds from the right. Therefore, the cricket could still use its pressure gradient ears to determine the direction of the sound source.

difference in the amplitude of tympanal vibrations, this does not result from a binaural difference in sound pressure level outside the two ears. In fact, the cricket's body is too small to produce any effective sound shadow for a frequency of 3.7 kHz, so that the sound pressure levels will be essentially the same on the outside of each tympanum (Hill and Boyan, 1977). Rather, our predicted left–right differences in tympanal vibrations result strictly from the effects of sound on the inside of each tympanum.

The prediction that the internal sound path should affect the directionality of a cricket's hearing has been verified both physiologically and behaviorally. Sounds from the left side were shown to excite left interneurons most strongly, and sounds from the right excited them least strongly, as long as the sound frequency was around 3.7 kHz (Figure 21B). These tests were carried out with the right auditory nerve cut, so that any inhibitory interactions from that nerve could not be responsible for the directional response of the interneuron. As one would predict, when the right tympanum and the two spiracles were covered or when the left trachea was blocked, the left interneurons became equally responsive to sounds from all angles, that is, their directionality disappeared (Figure 21B, dotted line at 3.5 kHz). In behavioral tests on other crickets, covering one or the other of the spiracles shown in Figure 20B altered in specific ways the direction of the cricket's phonotactic behavior (Wendler *et al.*, 1980).

Pressure gradient ears are found not only in crickets but also in a few other insects (Michelsen, 1979) and in some birds, most notably the quail (Hill *et al.*, 1980; Coles *et al.*, 1980). In frogs, there is some evidence suggesting pressure gradient reception (Rheinlaender *et al.*, 1979, 1981), though some experimental results are inconsistent with this evidence.

Displacement-Sensitive Ears

By far the majority of auditory organs are designed to respond to the pressure component of sound. However, a substantial number of animals have evolved receptor organs designed to detect the displacement component (Tautz, 1979). All known displacement-sensitive receptor cells employ one or more hairlike structures that move in response to the displacement of the air or water molecules. For instance, in the LATERAL LINE ORGANS of fish and larval amphibians, a kinocilium and a group of stereocilia from each of several receptor hair cells, covered by a gelatinous cap called a CUPULA, project into the external water medium (Russell, 1976). Water displacements, occurring either as oscillations in the near field of a sound source or as individual jets of water produced by individual movements of objects in the environment, impinge on the cupula and deflect it. This causes the stereocilia and kinocilium to deflect, and this in turn produces, in a manner not yet fully understood, a receptor potential of the hair cells (Hudspeth, 1982).

One of the best-studied displacement receptors is found in the caterpillar *Barathra* (Tautz, 1979). Four sensory hairs project outward from each side of this insect's body (Figure 22A). However, unlike the kinocilium or stereocilia of a vertebrate hair cell, the caterpillar's hairs are acellular structures; they are

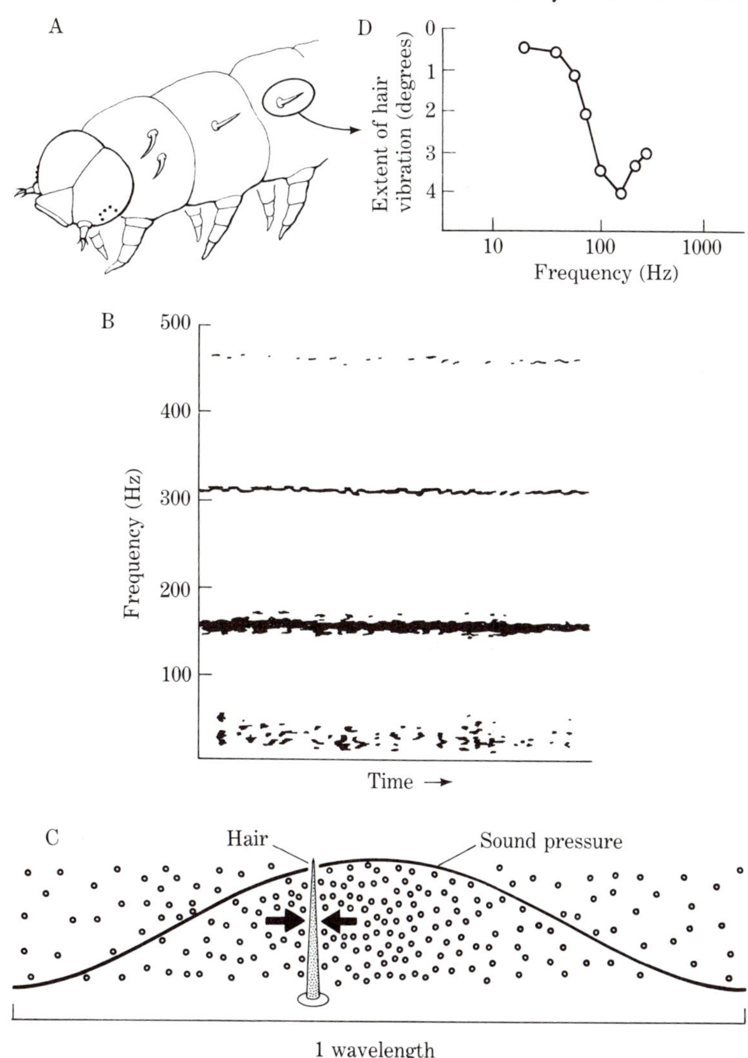

FIGURE 22. Auditory displacement receptors in the caterpillar *Barathra*. A. The anterior end of the caterpillar, showing its four sensory hairs. B. Sonogram of the wingbeat sounds of the wasp *Dolichorespula*, which preys upon *Barathra*. C. Theoretical basis for assuming that *Barathra's* sensory hairs are not deflected by the pressure component of sound. One wavelength of sound is indicated by the molecular compression and rarefactions, and by the sine wave. The thickness of the hair is very small compared to one wavelength of 150-Hz sound. (This size relationship is not drawn to scale; the hair's thickness is actually less than 1/200,000 the size of one wavelength.) Therefore, at any instant, nearly identical sound pressures will occur on either side of the hair, thus producing no net force on the hair. D. The mechanical response of one hair to sounds of constant amplitude but different frequencies. The hair was maximally deflected by sounds of about 150 Hz. By labeling the y-axis in descending order, the graph takes on the shape of a neural tuning curve. (A and D after Tautz, 1979; B from Tautz and Markl, 1978.)

cuticular hairs, nearly identical in form to those on the cerci of the cockroach (Chapter 4, Figure 8). Like the cockroach hairs, the shaft is stiff and the base is articulated in a specialized socket within which the shaft can deflect back and forth. Under the base of the hair is a single sensory cell that is excited by this deflection.

One important stimulus that activates the hair receptors of *Barathra* is the sound produced by the wingbeat of *Dolichorespula*, a tiny wasp that lays its eggs in the body of this caterpillar. When this wasp flies within about 1 meter of the caterpillar, *Barathra*'s hairs are deflected, evoking one of two behavioral responses; upon detecting low-amplitude, wingbeat sounds, *Barathra* freezes in place, whereas a higher sound amplitude causes the caterpillar to drop from the branch on which it had been standing (Tautz and Markl, 1978).

Even on theoretical grounds, one can strongly suggest that the tiny hairs of *Barathra* would be deflected only by the displacement component, and not by the pressure component, of sound. The wingbeat frequency of the wasp *Dolichorespula* is about 150 Hz, and the wings produce a sound most of whose energy is at this frequency (Figure 22B). A sound of 150 Hz in air has a wavelength of about 2 meters [Equation (3)]. Given this long wavelength and the unobstructed sound path to both the front and the back sides of a hair, at any moment there would be practically the same amount of sound pressure in front of and behind the hair (Figure 22C). Thus, the sound pressure in front and behind would exert equal and opposite forces, so these forces would not contribute to moving the hair. The displacement component, however, should easily deflect the long and slender hair because it has a low mass and, thus, a low inertia. Just like the cercal hairs of a cockroach, which are deflected by each gust of wind, *Barathra*'s hairs should be moved back and forth by the displacement component on each individual cycle of sound (Figure 1A).

The prediction that *Barathra*'s sound-receptive hairs are displacement receptors rather than pressure receptors was tested by placing individual caterpillars in a STANDING WAVE of sound. This is a wave produced by a specially constructed speaker system, such that the amplitude peaks and troughs do not propagate away from the sound source but rather occupy fixed positions in space. A special feature of a standing wave is that at certain locations the sound pressure is maximal, but the displacement is zero; at other locations the displacement is maximal, but the sound pressure is zero. Caterpillars placed at those locations having maximal sound pressure, but no displacement, gave no response to the sound. However, when placed at those locations where the displacement was maximal, these insects gave their highest percentage of responses (Markl and Tautz, 1975).

A hair of *Barathra* is more easily deflected by some frequencies of sound than by others. That is, just like a swinging pendulum, the hair has a best frequency, or RESONANCE FREQUENCY. One can determine the hair's resonance frequency by presenting different sound frequencies, all of the same amplitude, and determining under a microscope which frequency causes the hair to oscillate by the greatest amount. A graph representing the results of such an experiment has the same appearance as a neural tuning curve; this graph is, in fact, a mechanical tuning curve. The graph (Figure 22D) shows that the resonance frequency of a hair in *Barathra* is approximately 150 Hz, the same frequency as the major sound component in the wingbeat sound of the wasp *Dolichorespula* (Tautz, 1979). Thus, *Barathra*'s hairs are well designed for detecting the sounds of its predator.

Summary

Sound consists of a displacement component that propagates only very short distances and a pressure component that propagates through much longer distances. Most ears are designed to detect the pressure component. Sound pressure is a mechanical wave phenomenon and shares several properties with electromagnetic waves. These include reflection, which provides the physical basis for bat echolocation; defraction, which permits sound to travel around objects, including the head of a listener; refraction, which results in a higher propagation speed of sound in water than in air; absorption, which contributes to limiting the propagation distance of sound; and transmission, which permits the tympanum and other ear structures to vibrate in response to airborne sounds. Complex sounds, having several Fourier components, can be described in either the time domain or the frequency domain.

Echolocating bats show a remarkable ability to avoid obstacles, detect prey items, and discriminate among similar objects. The FM bat *Eptesicus* determines the distance to a target by measuring the pulse-to-echo delay. It can make this measurement with an accuracy of 1 microsecond. The long CF–FM bat *Rhinolophus* determines the velocity, relative to itself, of a moving target by a frequency domain analysis of the echo's CF component. This process involves the use of the bat's Doppler shift compensation response, which is accurate to about 0.04%. This response assists *Rhinolophus* in its signal-to-noise discrimination and in the detection of faint echoes. This bat also appears to use a frequency domain analysis to detect the jittery motion of an insect's wings.

The mammalian cochlea carries out a Fourier analysis, in the form of a tonotopic map along the basilar membrane. That is, a hair cell located at a given position along this map responds maximally to the particular sound frequency associated with this position. This confers on each hair cell and on the primary auditory neurons that it drives a best excitatory frequency and a V-shaped tuning curve.

The basilar membrane of *Rhinolophus* shows a structural specialization at that location along the cochlea that encodes sounds of 83 kHz—the frequency of the echo's CF component. This specialization consists of an abrupt change in the thickness:width ratio of the membrane. This appears to confer upon the cells whose best excitatory frequency is about 83 kHz tuning curves of extraordinary sharpness. And this presumably aids the bat in its sharp frequency discrimination at around 83 kHz.

Sound direction can be discriminated by several different mechanisms, though the usefulness of each mechanism depends upon the relationship between the sound's wavelength and the animal's interaural distance or head size. The mechanisms include binaural comparisons of intensity, time of arrival, phase, and spectral composition. In crickets, some birds, and perhaps some other animals, pressure gradient hearing can assist in directional localization.

A limited number of auditory organs detect the displacement component rather than the pressure component of sound. These include lateral line organs of fish and amphibians and the sensory hairs of the caterpillar *Barathra*.

Questions for Thought and Discussion

1. Suppose that an animal emits the courtship call whose sonogram is shown below.

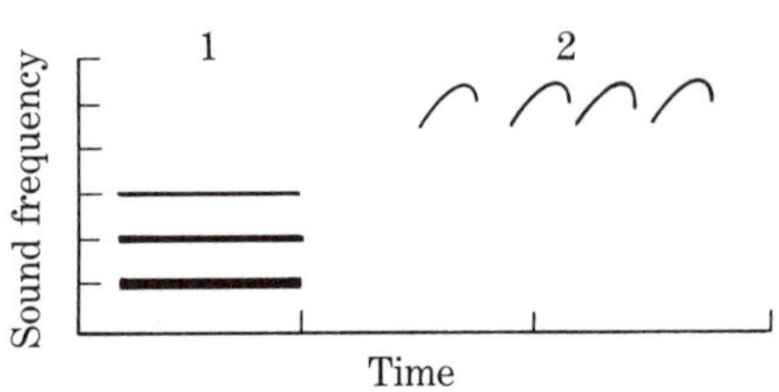

The call is divided into a first component, which has three harmonics, and a second component (subdivided into four parts), which has only a fundamental frequency. The caller stands in the middle of a thick forest. A receptive listener standing some distance away within the forest, upon hearing the call, approaches the caller. Is the first or the second component of the call likely to propagate further through the forest? Why? What quantitative experiments could you do to test which component propagates further? How could you determine experimentally over what distance within the forest the listener can hear the caller? How could you test whether the first, the second, or both components are required to elicit an approach by the listener to the caller? What further experiments can you suggest?

2. Sounds propagate through water roughly four times as fast as through air. What effect will this have on the wavelengths of sounds in water? [See Equation (3).] How might this effect the ability of aquatic animals to localize sounds? How might such animals deal with this situation?
3. Design an experiment to determine whether bats can use their echolocation to recognize individuals of their own species. (For instance, does echolocation allow a bat to discriminate between its mate and another bat of the same species or between its own offspring and those offspring of other bats?) Specify all needed controls. What components of the emitted pulses and echoes would bats be likely to use in making a discrimination of this sort? Why?
4. What are the advantages to the use of ultrasound in a bat's echolocating pulses? What are the disadvantages? What might be the advantages and disadvantages of the particular sound frequencies used for human speech (about 250 Hz to 2 kHz)?
5. Suppose you record from the axon of a primary auditory neuron in the eighth cranial nerve of some mammal and you determine this neuron's tuning curve. Now you inject some drug into the animal's bloodstream. You note the following effects on the neuron recorded: the neuron has the same best excitatory frequency; at this frequency, the neuron has the same threshold; but

the $Q_{10\ dB}$ value has substantially increased. What are some possible mechanisms by which the drug might produce this effect? How would you discriminate experimentally among these different mechanisms?

6. The ability of an animal to localize a sound source depends in part on the sound frequency. Consider an adult human subject listening to, and trying to localize, different sounds. For roughly which sound frequencies should this listener be best able to use each of the following forms of binaural comparison to localize the sound source: binaural intensity, time of arrival, or phase comparison? Based upon this, try to draw a graph of the accuracy with which a person should be able to localize a sound source, as a function of the sound frequency. How might you try experimentally to verify the shape of this graph?

Recommended Readings

BOOKS

Yost, W.A. and Nielsen, D.W. (1977) *Fundamentals of Hearing: An Introduction.* Holt, Rinehart & Winston, New York.
An extremely clear, elementary account of sound and the auditory system, primarily of humans.

von Békésy, G. (1960) *Experiments in Hearing.* McGraw-Hill, New York.
For historical interest, this book reprints the pioneering papers of this Nobel laureate, who discovered many of the mechanical properties of the basilar membrane.

Griffin, D. (1958) *Listening in the Dark.* Yale University Press, New Haven, Connecticut.
Though now dated, this prize-winning book offers an exciting account of earlier research on bat echolocation, by the researcher who discovered this process.

Busnel, R.G. and Fish, J.F. (1980) *Animal Sonar Systems.* Plenum Press, New York.
Research papers from a recent symposium devoted entirely to echolocation. This collection offers extensive reviews relating to bats and other echolocating animals. Chapters directly relevant to the material in this chapter are those by Schnitzler and Henson, Pye, Fenton, Neuweiler, and Simmons.

Popper, A.N. and Fay, R.R. (1980) Comparative Studies of Hearing in Vertebrates. Springer-Verlag, New York.
Review articles related to hearing in a variety of vertebrate species.

ARTICLES

Simmons, J.A., Fenton, B.M. and O'Farrell, M.J. (1979) Echolocation and pursuit of prey by bats. *Science* 203: 16–21.
This interesting article attempts to draw a correlation between a bat's ecological surroundings and the sound parameters it uses in its echolocation pulses.

Khanna, S.M. and Leonard, D.G.B. (1982) Basilar membrane tuning in the cat cochlea. *Science* 215: 305–306.
This brief report suggests that the frequency tuning of the basilar membrane's vibrations may be sufficient to account for the tuning of primary auditory neurons.

Michelsen, A. (1979) Insect ears as mechanical systems. *Am. Sci.* 67: 696–706.
An excellent article on the acoustic properties of a number of highly interesting auditory organs.

Hill, K.G. and Boyan, G.S. (1977) Sensitivity to frequency and direction of sound in the auditory system of crickets (Gryllidae). *J. Comp. Physiol.* 121: 79–97.
This excellent research article contains a number of experiments pointing to the importance of pressure gradient hearing in crickets.

Tautz, J. and Markl, H. (1978) Caterpillars detect flying wasps by hairs sensitive to airborne vibration. *Behav. Ecol. Sociobiol.* 4: 101–110.
A highly interesting study on the detection of the displacement component of sound and its use in natural behavior.

CHAPTER 7

Recognition by the Brain of Objects and Their Locations

In the last two chapters, we have seen that a given animal's visual or auditory receptors record only a small part of the overall range of photic or acoustic energy available from the environment. Moreover, the responsiveness of *individual* sensory cells is even more restricted, as we saw in the case of a honeybee's retinula cell, which may be activated selectively by UV light, or a bat's auditory neuron, which may be narrowly tuned to some ultrasonic frequency. An animal, then, must somehow construct a meaningful image of a world that it views through a number of narrow sensory windows. In the present chapter, we will focus upon how the information coded in sensory cells is organized within the brain so as to bring about this meaningful image.

The major specific question addressed in this chapter is, "How does an animal's brain recognize biologically significant objects detected by its sensory receptors?" For instance, how does the animal correctly identify its mates, predators, and prey? (The words *recognize* and *identify* are not intended to imply any necessary conscious awareness by the animal; rather they denote some process whereby a given stimulus object, such as prey, reliably evokes a specific, appropriate behavior, such as attack.) A related question that will emerge is, "How does the brain encode the *location* of a stimulus object?" Location is important because most behavioral responses are carried out in a particular direction relative to the position of the stimulus that evokes them.

The brain needs to recognize and localize not only sights and sounds but also numerous other sensory stimuli. The mechanisms by which this is accomplished are, to some extent, independent of the particular sensory modality involved. Thus, we can illustrate many of the principles of recognition and localization by discussing just one or two sensory systems. In this chapter, we will focus primarily on the visual system in vertebrates, with special emphasis on toad vision. For comparison, certain key aspects of central auditory processing also will be discussed briefly.

Theories of Object Recognition

The question of object recognition, especially within the human visual system, has attracted scientific interest for centuries. Useful modern contributions to the problem have come from many branches of science, including perceptual psychology (Haber and Hershenson, 1980), communications and information theory (Corcoran, 1971), computer science (Uttal, 1978), animal learning (Carterette and Friedman, 1973), classical ethology (Tinbergen, 1951), sensory neurophysiology (Hubel, 1979), and neuroethology (Ewert, 1980). Although this chapter treats the subject primarily from a neuroethological viewpoint, other approaches will aid in our discussion. For instance, the problem that the brain faces in correctly identifying an object in the environment can be illustrated by an everyday problem in human perception: How do we recognize a chair as a chair and a table as a table? Although seemingly straightforward, this is actually a very complex problem.

The image of a chair would activate a particular constellation of receptors (rods and cones) in the eye of a human viewer (Figure 1A). The image of a table would activate a different, though partly overlapping, constellation of receptors (Figure 1B). One obvious possibility for a mechanism of recognition, then, would be for all the chair-activated receptors to excite, via the retinal neurons postsynaptic to them, some particular cell or small group of cells in the brain; and for all the table-activated receptors to excite a different cell or group in the brain. Such a hypothetical chair-activated or table-activated cell or group of cells is often called a TEMPLATE. By this hypothesis, the neural circuitry would be such that only activity in *all* the receptor cells excited by the chair in Figure 1A, and not activity in any other receptors, would excite the chair template; and only activity in *all* the receptors activated in Figure 1B, and not in any others, would excite the table template. The consequent activation of the chair template would be interpreted by the brain as "chair," and activation of the table template as "table."

It is easy to see, however, that this simple template scheme does not solve the problem of recognition. For instance, the two images of chairs in Figure 1C, differing in angle of rotation, size, and position from the chair in Figure 1A, would excite different constellations of receptors and so, by our scheme, could not excite the chair template in the brain hypothesized for Figure 1A. And yet these drawings are still recognizable as chairs. Our recognition process, then, shows the property of INVARIANCE with rotation[1] and with changes of size and position; as these parameters vary, we still see the chair as a chair. This invariance requires us to seek a more comprehensive theory of object recognition.

Two relevant neurophysiological findings on the visual systems of cats and monkeys will help to direct our search for such a theory (Hubel, 1979). First, as is explained in more detail later in this chapter, most neurons in the visual system respond to light stimuli within only a restricted region of the visual world. This region of the environment within which light influences a given cell is called that cell's RECEPTIVE FIELD. For those neurons in a monkey's visual cortex (that part

[1]Not all visual objects show invariance with rotation. As one example, a map of the continent of Africa turned upside down is unrecognizable to most viewers (Rock, 1973).

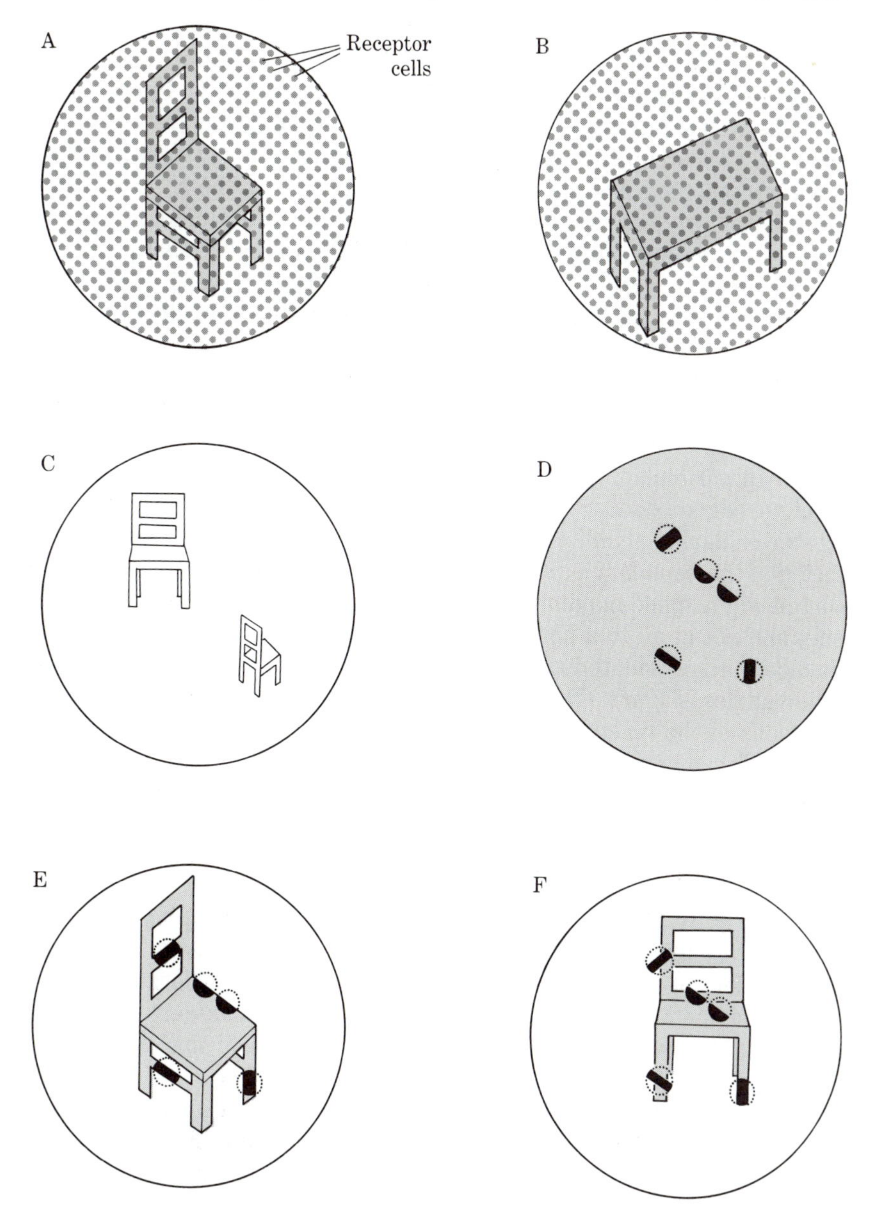

FIGURE 1. Recognition of chairs and tables by human viewers. A. A large central region of the field of view of an eye that is looking at a chair. The retinal receptors that would detect this chair, those upon which the image is projected, are indicated by the dots that fall within the chair's image. B. The same as A but for a table. C. Two other images recognizable as chairs, though exciting largely different groups of receptors. D. Receptive fields of five neurons of a type found in the visual cortex of a cat or monkey. E. All five of these same neurons would be excited by the image of the chair in A, since their receptive fields align with parts of this image. F. A chair oriented differently from that in E would strongly excite only one of these five neurons (the one on the bottom right).

of the cerebral cortex subserving vision) whose receptive fields are located in the center of gaze of the eyes, receptive field size is tiny—about 1 inch in diameter for a viewing distance of 10 feet (Hubel and Wiesel, 1974). Such receptive fields, then, as well as those of most other neurons in the visual system, are far smaller than the area occupied by a chair or table viewed at fairly close range. Therefore, none of these neurons can act as a template for a whole chair or table. Rather, the neuronal representations of these objects are multicellular mosaics, with different neurons encoding different parts of the overall visual pattern. Although no information is available on the receptive fields of cells in the human visual system, it seems likely that the general principles established in cats, monkeys, and other mammals would apply, in broad outline, to our own visual processing as well.

The second relevant physiological finding is that most cells respond selectively to specific spatial arrangements of light and dark within their receptive fields. In particular, most neurons in the visual cortex respond best to either a straight-edged boundary between light and dark or a straight stripe that is either lighter or darker than its background.[2] For all such neurons, the angular orientation of the boundary or stripe is crucial. For instance, a given cell in the visual cortex will respond maximally to a vertically oriented stripe, less so to an oblique one, and not at all to a horizontal one. Another cell will respond maximally to a boundary edge oriented at a particular oblique angle, and less so or not at all to other angles. Figure 1D indicates the positions and optimal edge or stripe orientations of the receptive fields for five hypothetical cells in a mammal's visual cortex. These neurons, then, respond selectively to rather specific features of a sensory stimulus. In this regard they are unlike rods and cones, which merely respond to the presence of light, irrespective of its spatial pattern. Such cells, displaying considerably more stimulus specificity than the receptor cells upstream from them, are called FEATURE DETECTORS.[3]

In Figure 1E, the image of a chair is shown overlapping with the receptive field representations of the same five cortical neurons as in Figure 1D. The receptive field of each of these five cells corresponds, in location and in optimal edge or stripe orientation, to the image of a particular part of the chair. Thus, each of these five cells would be excited by the view of this chair. Of course, not just these five cells, but also a great many others that are not shown, would be activated by the image of this chair, each cell detecting a small part of its structure. If now the chair were rotated to the positions shown in Figure 1F, although one of these same five neurons would still be excited (the one with its receptive field in the lower right), the other four would not. Instead, a different constellation

[2]An alternative way of describing the response properties of neurons in the visual cortex concerns the way that these cells respond not to sharp edges or stripes but to gratings in which the light intensity varies gradually across space, increasing and decreasing at regular spatial intervals (Campbell, 1974; DeValois and DeValois, 1980). These two different ways of describing the response properties of visual neurons are mathematically interrelated (Stork and Levinson, 1982).

[3]Some researchers reserve the term *feature detector* for neurons that may be selectively responsive to even much more detailed and specific stimuli, such as the sight of an animal's mate, its prey, or its predators. However, we will instead use a different term, as described shortly, for such neurons.

of neurons would now be excited. Thus, each size, shape, location,or angle of rotation of a chair activates a different array of feature-detecting neurons in the visual cortex.

We must now inquire how the brain might use these arrays of cortical feature detectors to recognize the category of object we call a chair. Of the numerous theories that have been suggested (Haber and Hershenson, 1980), three stand out as reasonable possibilities. The first of these is that all the sets of feature-detecting neurons activated by all chairs constitute, in and of themselves, the information that the perceptual process of the brain uses *directly* to recognize a chair. That is, the information used by the brain's perceptual process would consist of listings of those feature-detecting neurons that are activated. If an appropriate list is activated, the perceptual process would recognize the object as a chair. In Figure 2A, for instance, chair number 1 at the top left activates one list, a part of which is shown, consisting of the cells numbered 1, 4 and 5. The brain would somehow "know" that this list means "chair." Chair number 2 activates a different list, which includes cells 1, 7, and 9. The brain would also "know" that this list means "chair," and so forth for chair 3 and its list of cells. Any one of thousands of such lists, then, corresponding to thousands of possible sizes, angles, and positions of chairs, would be recognized by the brain's perceptual process as "chair." Also, any one of thousands of yet other lists, such as those for the three tables in Figure 2A (right side) would be recognized as "table."

It is this enormous number of lists that the brain's perceptual process would need to recognize that throws some doubt upon this first theory of object recognition. Rather, it seems more likely that the information contained in these lists is reduced in some way so that a chair of *any* size, angle, and position would cause some single, unique, physiological event within the brain. The perceptual process would use this unique event to recognize a chair.[4] The two remaining theories of object recognition that we will discuss involve such a reduction of information but differ in the way that this reduction is accomplished.

The first of these two remaining theories involves hypothetical brain cells, each referred to as a RECOGNITION NEURON[5] for a particular class of objects, that is, a cell that is activated if and only if its particular class of objects (such as a chair) is viewed, regardless of its size, angle, and position. The theory of recognition neurons does not specify that the brain contains just one neuron for *chair*, one other for *table*, and so forth. Rather there could be several recognition neurons for *chair;* but all of these would be activated together and would thus function as a unit. The implication is that each and every complete list of cortical feature-detecting neurons activated by the sight of a chair (Figure 2A, left) would activate the hypothetical chair-recognition neuron or neurons (Figure 2B, left); and each complete list activated by tables (Figure 2A, right) would activate the

[4]Although the sight of all chairs would evoke the same unique physiological event needed for chair recognition, separate information about *each* chair would also be available from the various lists of feature detectors. This information could be used to distinguish such properties as the size, shape, and color of an individual chair.

[5]A variety of alternative terms have appeared in the literature, including *object detector neuron, pontifical neuron, feature detector* (see footnote 3), and, incomprehensibly, *grandmother detector.*

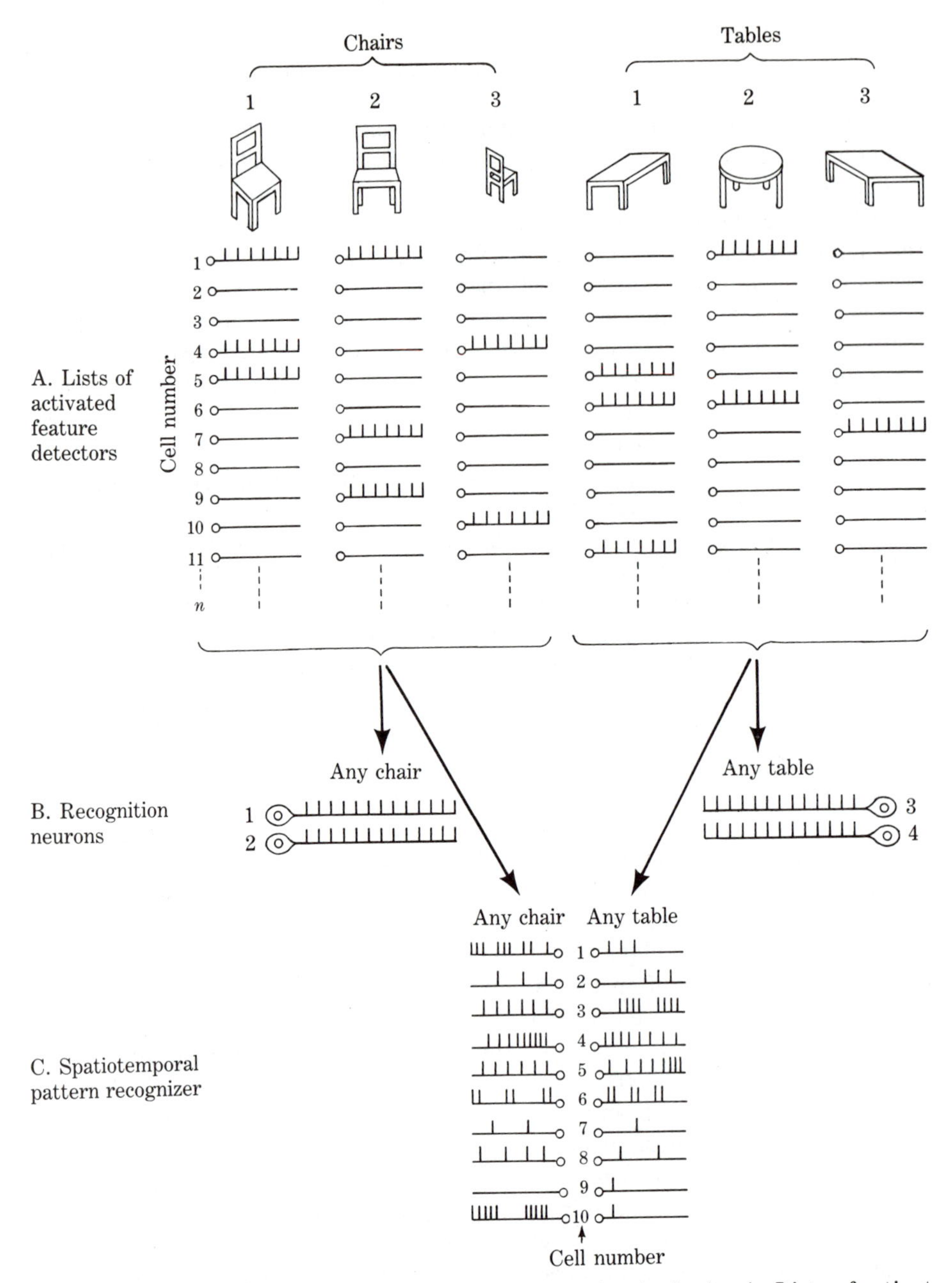

FIGURE 2. Three theories of object recognition by the brain. A. Lists of activated feature detector neurons. The 11 neurons shown constitute only a tiny fraction of the feature detector cells in the visual cortex of a mammal. Among these 11 cells, and the other feature detectors not shown, the sight of a given chair activates a small subpopulation. This subpopulation constitutes the brain's list of feature detectors specific for this particular chair stimulus. B. Hypothetical recognition neurons activated by the sight of *any* chair. C. Hypothetical spatiotemporal pattern recognizer. The ten cells shown would be activated in one temporal pattern by all chairs; and the *same* ten neurons would be activated in another temporal pattern by all tables.

table-recognition neuron or neurons (Figure 2B, right). Some mechanism would be needed to assure that only a complete list of activated feature-detecting neurons, and not combinations of parts of different lists, would cause a recognition neuron to be activated. A major part of this chapter will be devoted to evaluating evidence that some animals may use recognition neurons to identify some biologically significant objects.

The theory of recognition neurons, though representing a considerable compacting of information from lists of feature-detecting cells, still might involve some fairly cumbersome cellular machinery because at least one separate recognition neuron would be required for each of the numerous objects that are separately identifiable by the animal (perhaps many millions for human beings). Our third and final theory, then, offers a still more compact alternative. By this theory, the brain would have no recognition neurons, but instead would have a SPATIOTEMPORAL PATTERN RECOGNIZER—a limited group of neurons *all* of which would be activated by all perceived objects, but in a different temporal pattern by each class of objects. For instance, all chairs of any size, angle, and position would excite these neurons with some characteristic temporal response pattern, such as that shown in Figure 2C, left. The perceptual process of the brain would somehow determine that this temporal pattern among these cells means "chair." All tables would excite these *very same* neurons with a different temporal response pattern, such as that shown in Figure 2C, right. The brain would somehow determine that this temporal pattern means "table." By this theory, then, in contrast to the recognition neuron theory, it is not *which* cells are activated, but the *time pattern* of activity of the various cells of this group, that would provide the brain's perceptual process with its needed information. The number of different objects representable in a spatiotemporal pattern recognizer consisting of relatively few cells would be very great, just as, by rearranging the order of 26 letters of the English alphabet, one can represent nearly an infinite number of words and ideas.

The extreme compactness of a spatiotemporal pattern recognizer renders this theory very attractive. However, there are difficulties with this idea. It is not at all clear by what mechanism all lists of feature-detecting cells activated by all chairs (Figure 2A, left) could be made to produce in the cells of a spatiotemporal pattern recognizer the same temporal pattern of activity. Nor is it clear how a given temporal pattern of activity in the cells would evoke one perceptual outcome (chair recognition) and another temporal pattern a different outcome (table recognition). The problem, then, in the language of computer science, is one of INTERFACING, or connecting up the pattern recognizer with its inputs and outputs in a way that would permit it to function appropriately. These difficulties notwithstanding, spatiotemporal pattern recognizers continue to be of great theoretical interest because of their possibility for encoding enormous amounts of information using a limited number of neurons.

In summary, then, having eliminated the idea of a sensory template in the brain as too simplistic to account for object recognition, we have discussed three of the most reasonable alternative ideas: (1) the theory of lists of activated feature-detecting neurons; (2) the theory of recognition neurons; and (3) the theory of a spatiotemporal pattern recognizer. Of these three theories, the easiest to study experimentally is that of recognition neurons because only here is the neural representation of a particular recognized class of objects encoded entirely by an individual neuron. (This encoding may be reiterated in a second, or in multiple, recognition neurons for the same class of objects, as in the two hypothetical chair-recognition neurons of Figure 2B, left. Nevertheless, each of the two neurons in the figure would respond selectively to the sight of any chair.) This representation in single neurons, if it actually occurs, would mean that it ought to be possible, with a single recording electrode, to register the unique physiological event that the brain uses to recognize a given class of objects. We turn now to an evaluation of the theory of recognition neurons, as studied in the visual system of toads, primarily by J.P. Ewert and his colleagues, working at Kassel, Germany.

Toad Vision: Recognition and Localization of Prey and Enemies

BEHAVIORAL RESPONSES TO MOVING STIMULI

Toads generally sit motionless for long periods in the evening, until some moving object enters their field of view. If this object is a walking insect, a crawling worm, or some other suitable prey item, it may elicit the toad's prey-catching behavior. This consists of a fairly stereotyped sequence of acts (Figure 3A): (1) a turn of the toad's head and/or body to bring its center of vision in line with the prey; (2) sometimes walking slowly toward the moving prey; (3) holding its gaze fixed on the prey; (4) lunging forward while flicking out its sticky tongue; (5) swallowing (Ewert, 1980; Dean, 1980). Only moving objects or those that have just moved evoke this response. Although the toad turns *toward* a moving prey item, if instead it sees a large object moving overhead, such as a bird, it turns its body *away* (Figure 3B). Thus, just as a human observer can recognize chairs and distinguish these from tables, a toad recognizes suitable prey items and distinguishes these from potential enemies. Unlike a human subject, however, who can report his observations to us verbally, toads and other nonhuman subjects can inform us only by their behavioral reactions as to what objects they can recognize and distinguish from other objects.

By what visual parameters does a toad recognize suitable prey items or enemies? This question was studied using a classical ethological approach (Chapter 2) in which cardboard models simulating a prey or enemy were moved within the toad's field of view and his responses tabulated (Ewert, 1969, 1980). Specifically, the toad *Bufo bufo*, enclosed within a glass cylinder, was presented with one of three model configurations, which were moved outside the cylinder wall (Figure 4A). The configurations, shown in Figure 4B, were (1) a rectangle moving in the direction of its long axis, simulating a crawling worm or a walking insect; (2) a rectangle moving in the direction of its short axis, a stimulus configuration not often found in nature; and (3) a moving square. The rectangles used were of various lengths (but of constant width), and the squares were also of various dimensions. In the initial tests, the cardboard models were black and were presented against a white background. Toads responded to the moving models with a sequence of small, clearly definable, turning movements. The number of such separate turning

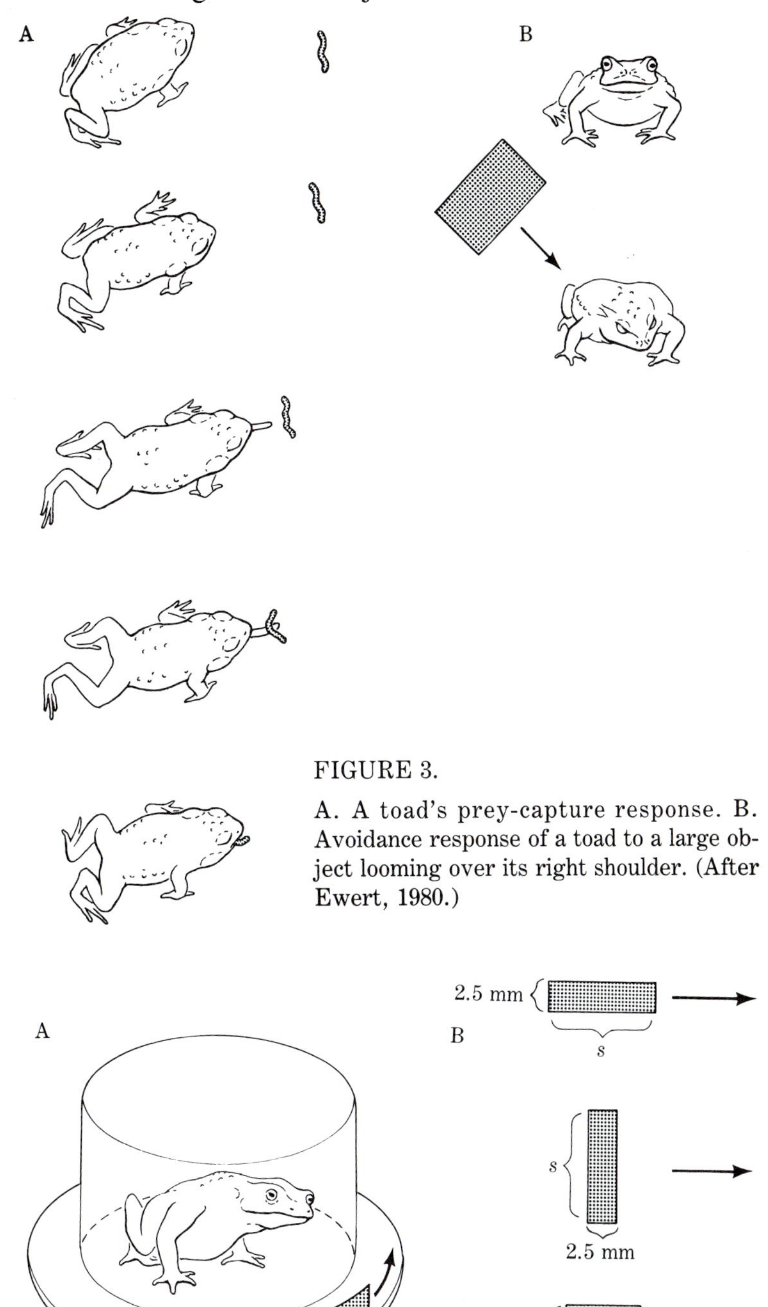

FIGURE 3.

A. A toad's prey-capture response. B. Avoidance response of a toad to a large object looming over its right shoulder. (After Ewert, 1980.)

FIGURE 4. A. Behavioral test for prey recognition in toads. B. The three classes of stimulus used. Arrows show direction of stimulus movement. For each class of stimulus, the dimension *s* was varied for different trials.

movements that the toad made toward a given moving model within a fixed time period, taken as a measure of the releasing value (Chapter 2) of that stimulus, indicated the relative effectiveness of the different models as prey stimuli.

The greatest releasing value was obtained for wormlike stimuli, that is, a rectangle moving in the direction of its long axis. The greater the length of the rectangle up to some limit, the greater the response (Figure 5A). The lowest releasing value generally was obtained for "antiworm" stimuli, that is, a rectangle moving in the direction of its short axis (Figure 5B). Small squares had releasing values similar to those of wormlike stimuli. For somewhat larger squares, releasing value decreased to zero. And still larger squares evoked turns not toward but *away* from the stimulus, as though the toad were now responding to an approaching enemy (Figure 5C). In other experiments, the direction of stimulus movement was changed from horizontal around the toad's glass cylinder to vertical or oblique. In all these experiments the result was the same—wormlike stimuli were most effective and antiworms least effective in evoking turns toward the stimulus (Figure 6).

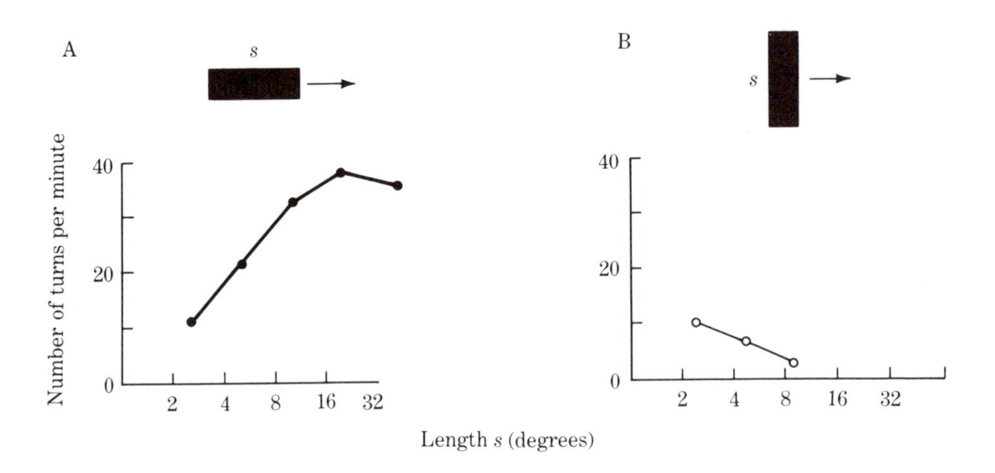

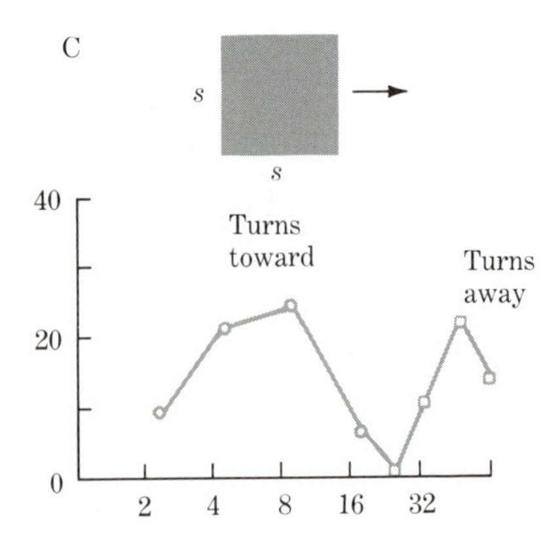

FIGURE 5.

Prey-orienting responses of toads to the three classes of moving stimulus. Large squares (graph C, right) evoked turns not toward, but away from, the stimulus. See text for explanation. (From Ewert, 1980.)

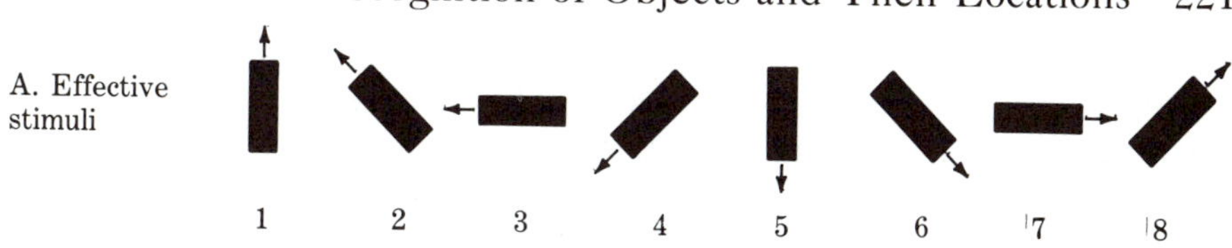

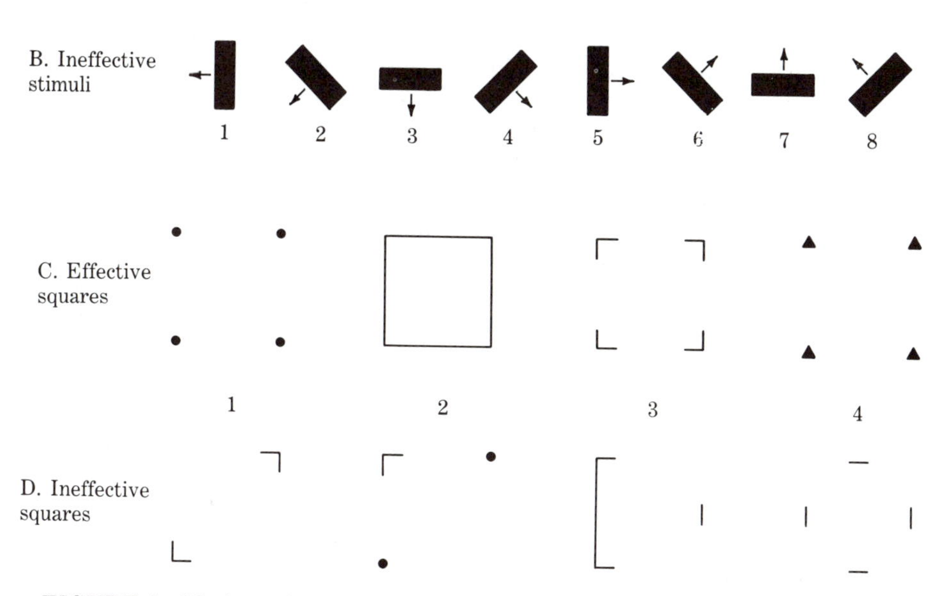

FIGURE 6. Moving stimuli that are (A) effective and (B) ineffective in evoking a toad's prey-orienting behavior. C. Four configurations that are seen by most human observers to possess the property of squareness. The only property in common among the four configurations is the spatial relationship of the four corners. D. Four configurations that are not usually perceived by human observers to possess squareness. (Configuration D4 is usually perceived as a diamond—a square tipped by 45°.) Figures D1 and D2 contain, respectively, two and three corners. A fourth corner is crucial for perception of squareness.

Further behavioral studies indicate that the toads continue to orient toward wormlike stimuli under a wide variety of stimulus conditions; either a black stimulus against a white background or a white stimulus against a black background; black or white stimuli against a checkered background; various rates of stimulus movement (tested from 2.5 to 60 degrees per second); various distances from the toad to the stimulus (tested from 4 to 16 centimeters); and different forms of stimulus movement, including continuous and jerky displacement. Because the response is strongest to worms when tested under all these different conditions, the toad's recognition of worms shows the property of invariance.

Figure 6 shows that prey items are recognized by toads not by means of stimulus shape alone because stimulus A1, which is an effective stimulus, has the same shape as B1, which is ineffective. Also, prey are not recognized by the direction of stimulus movement alone because the effective stimulus A1 has the same movement direction as the ineffective stimulus B7. Rather, it is the *relationship* between shape and movement direction—specifically, movement in the direction of the long axis—that is the effective stimulus property. Because shape alone and direction alone are each ineffective, the toad's recognition shows something more than just stimulus summation (Chapter 2), in which each of two or more separately effective stimulus properties would simply add their partial effects when combined with one another. Rather, for prey recognition in toads, there are no partial effects; it is only by adding together these separate stimulus properties that these properties become effective. The whole, then, is greater than the sum of its parts. A stimulus of this type is called a GESTALT (a German term meaning "configuration"). The Gestalt school of perceptual psychology (Koffka, 1935; Chaplin and Kraweic, 1974), centered in Germany prior to World War II, demonstrated many instances of such relational properties in human perception, one of which is shown in Figure 6C and D. Here, only by having all four corners present does the object take on the subjective appearance of a square. One, two, or three corners, or various line segments, contribute little on their own to the sense of squareness.

One helpful guideline that emerged from gestalt psychology is that it is often important to present one's experimental subject with test stimuli in which all of the major stimulus components are presented in their normal spatial and temporal relationships. As we shall see later in this chapter, however, an alternative approach used by many workers (which has produced important insights of a somewhat different sort) has been to employ only very simple, nongestalt stimuli such as tiny, stationary spots of light. However, for a toad prey recognition, because one knows from the behavioral studies that prey recognition involves a particular gestalt stimulus, it makes most sense to present this same stimulus while searching physiologically for the brain's recognition mechanism. We will now take up this physiological search.

RESPONSES OF CELLS IN A TOAD'S RETINA

How is the image of a moving wormlike stimulus encoded by a toad's visual system? Are there recognition neurons for prey stimuli and others for enemies? And how is the location of each of these stimuli encoded? We will begin our discussion of these issues by considering physiological recordings made from a toad's optic nerve, which contains the axons of the retinal ganglion cells (Chapter 5). Although one might expect not to find recognition neurons at such an early stage of visual processing as the retinal ganglion cells, two factors make this a reasonable location at least to begin our search. First, the amphibian retina is considerably more complicated both anatomically and physiologically than is the mammalian retina (Dowling, 1976). Second, unlike mammals, amphibians do not have a visual cortex. Thus, some of the more complicated visual processing, which in mammals is carried out largely at higher, cortical stages of visual circuitry, in amphibians takes place directly within the retina. In fact, a frog retina has long been known to contain some neurons with very complex response properties (Lettvin *et al.*, 1959).

The recordings of single retinal ganglion cells in toads were made from the terminals of their axons within the brain. To make these recordings, a toad is first injected with an immobilizing drug.[6] In addition, a local anesthetic is applied to the region of the head overlying the intended recording site, which is then surgically exposed. A tungsten microelectrode is then probed into the brain area containing the terminals of the optic nerve axons, until action potentials from a single, extracellularly recorded axon are registered on the oscilloscope (Figure 7). A variety of physiological criteria can be used to verify that these action potentials are from the axon terminals of a retinal ganglion cell rather than from a brain neuron (Lettvin *et al.*, 1959).

While recordings of action potentials were made from a ganglion cell, the toad was presented with the full series of moving wormlike, antiworm, and square stimuli (Figure 4B). The key question was whether the response strengths of these cells varied with stimulus type in the same manner as did the toad's turning behavior (Figure 5). This would mean that these cells recognize prey items in the same manner as the whole animal does. Such cells would then become good candidates for the animal's prey-recognizing mechanism.

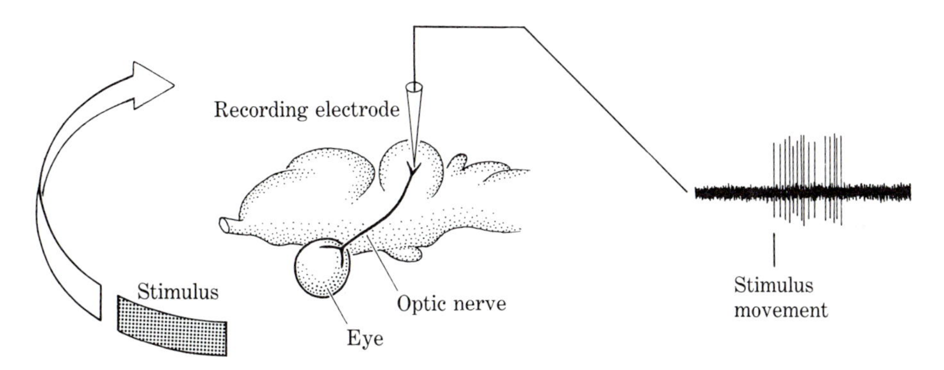

FIGURE 7. Experimental arrangement for recording from the terminals of optic nerve axons in the brain, in response to moving stimuli.

[6]The drugs usually used (succinyl choline, gallamine, or D-tubocurarine) are antagonists of the neurotransmitter acetylcholine (ACh) used by the neuromuscular junctions of skeletal muscles (Chapter 3). This prevents the muscle's normal response to the ACh released by the terminals of the motor axons and, thus, prevents the animal from moving. These drugs have the advantage that, at most, a small fraction of their molecules enter the retina or brain from the blood circulation, owing to the presence of a barrier against passage of many chemical agents, called the blood–brain barrier. Consequently, retina and brain are left pharmacologically nearly intact. This is important because the particular form of ACh receptor to which these blocking agents bind, that found at the neuromuscular junction (nicotinic, as opposed to muscarinic receptors), are abundant in the retina (Vazulla and Schmidt, 1976). Thus, were it not for the blood–brain barrier, vision might be greatly distorted by the immobilizing drugs.

Of all the retinal ganglion cells studied, none has ever been found that matched the toad's behavioral responses to different moving stimuli (Ewert, 1980). The ganglion cells, in fact, fell into three categories, called R2, R3, and R4 cells, each showing different physiological responses to the series of test stimuli (Figure 8). In the responses of all three cell types, unlike the behavioral responses, the effectiveness of wormlike stimuli did not increase with stimulus length. In fact, particularly for R3 and R4 cells, wormlike stimuli generally gave responses weaker than did antiworms and squares. Moreover, although the strength of the toad's behavioral response was different for antiworms and squares, the responses of ganglion cells to these two stimuli were nearly identical (Figure 8). Thus, recognition neurons for prey could not be found among the retinal ganglion cells; one must search for such cells deeper in the brain. However, before we progress in this central direction, it is of some interest to look backward within the retina in an attempt to understand the responses of the retinal ganglion cells that are graphed in Figure 8. In so doing, we will uncover integrative principles useful in our subsequent discussion of visual processing within the brain.

RECEPTIVE FIELDS OF VERTEBRATE RETINAL GANGLION CELLS

Parallel to the neuroethological investigations of toad vision, in which Ewert and his co-workers have sought to find prey-recognition properties in retinal ganglion cells, other investigators have carried out neurophysiological studies on these cells guided instead by an interest in the cellular interactions and synaptic integration within the retina. Much of the important work has been carried out on a different amphibian, the mudpuppy *Necturus* (Werblin and Dowling, 1969; Dowling, 1976; Miller and Dacheux, 1976), as well as on frogs and toads (Grüsser and Grüsser-Cornehls, 1976; Gold, 1979) and on other lower vertebrates (Baylor and Fettiplace, 1977; Waloga and Pak, 1978). These studies, together with some on mammals (Dowling and Boycott, 1966), permit a reasonable understanding of visual integration within the retinae of vertebrates in general. These studies also help to explain the responses of the toad's retinal ganglion cells to particular configurational stimuli.

As in the search for recognition neurons, this cellular approach also begins with recordings from the retinal ganglion cells. Now, however, one uses not configurational, or gestalt stimuli, but rather tiny spots of light that are flashed on and off in different regions within the cell's receptive field and that are usually held stationary. The goal of these studies on ganglion cells is to determine, with the use of these tiny light spots, the detailed shape of a ganglion cell's receptive field. (As a reminder, a visual cell's receptive field is that area of the visual world within which illumination can affect the cell's activity—either by exciting or by inhibiting the cell.)

In all vertebrates so far studied, most retinal ganglion cells have small, circular receptive fields (Rodieck, 1973). In fact, in most such cells, the receptive fields are composed of two concentric regions having different functional properties. For instance, in the receptive field for one retinal ganglion cell, shown in Figure 9, a tiny spot of light projected anywhere within a small central region of visual space excites the cell (Figure 9A). This region of visual space is therefore

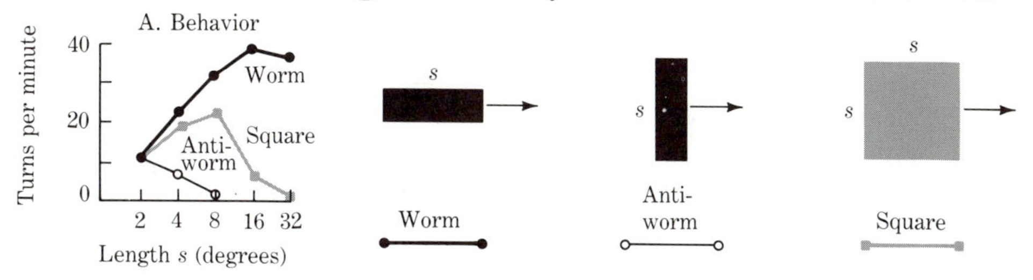

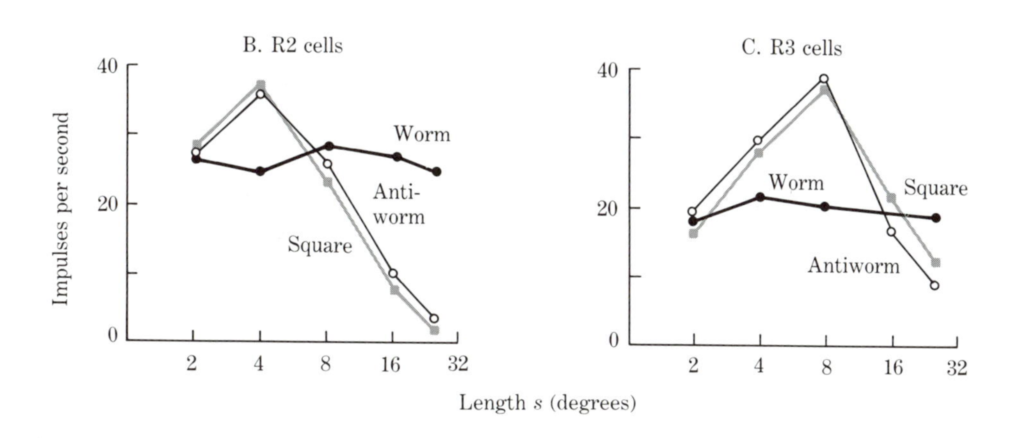

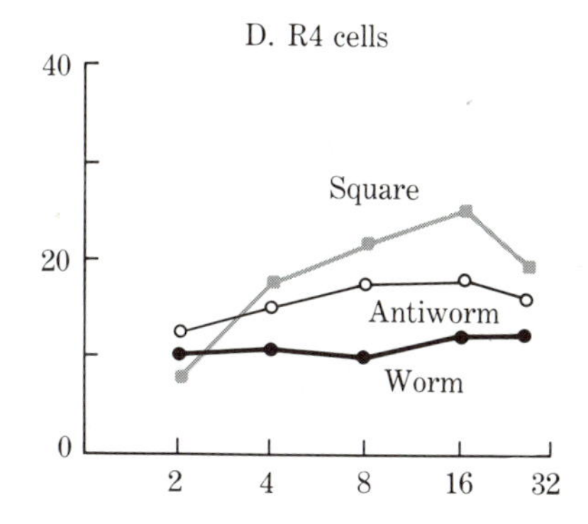

FIGURE 8.

Responses of retinal ganglion cells to moving configurational stimuli. A. Summary of behavioral response (from Figure 5). B, C, and D. Responses of the three categories of ganglion cells. None has a response profile that matches that of the behavior. (After Ewert, 1980.)

called the ON-CENTER of this neuron's receptive field. If this same spot of light were turned on together with a second spot projected anywhere within a region surrounding the center, the response would be less than that to a central spot alone (Figure 9B). The surround, therefore, exerts an inhibitory affect and is called an OFF-SURROUND. (In mammals, but not in amphibians, turning *off* a light that had been illuminating the off-surround excites the cell, as does turning *on* a light within the on-center.) The entire receptive field shown in Figure 9A and B is called an ON-CENTER OFF-SURROUND receptive field, and such a cell is often called an on-center off-surround cell. An inhibitory influence of a sensory stimulus in one region, counteracting a stimulus in a neighboring region, as is shown in

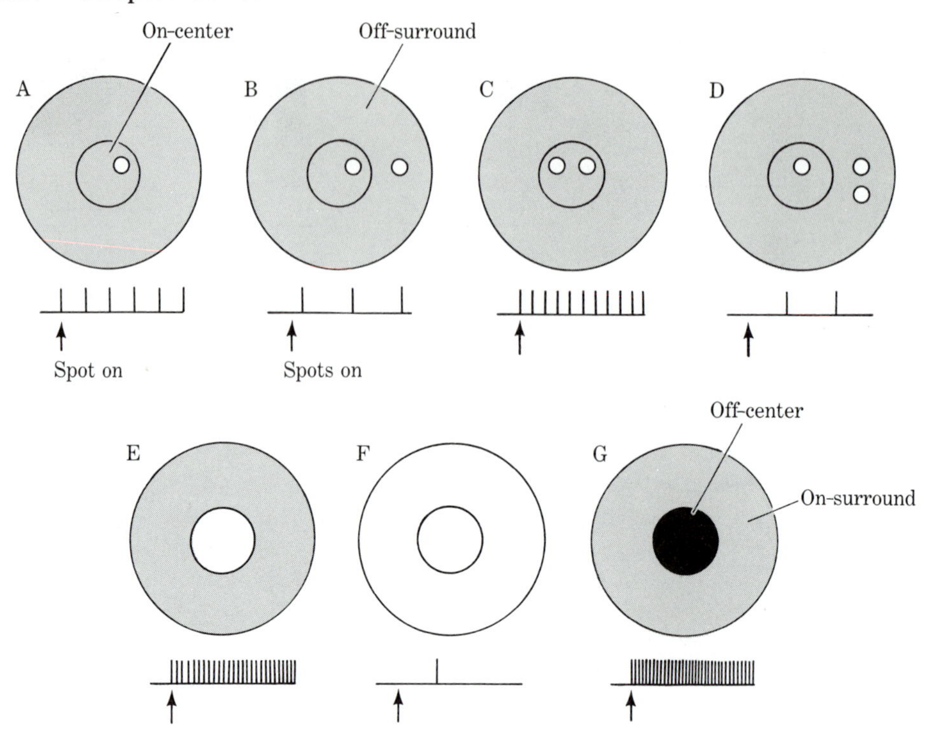

FIGURE 9. Center–surround receptive fields of retinal ganglion cells in a vertebrate animal and their physiological responses to light stimuli. See text for explanation.

Figure 9B, is called LATERAL INHIBITION (Hartline *et al.*, 1956). Lateral inhibition in a retinal ganglion cell assures that the cell responds most strongly to stimuli that are clearly inside the border of the on-center region. A light stimulus extending across this border into the off-surround would be much less excitatory to the cell. Thus, lateral inhibition has the effect of profoundly demarcating, in terms of the cell's physiological response, this border between on and off regions of the cell's receptive field. Lateral inhibition plays a similar demarcating role within receptive fields in many cells of the visual system, as well as in several other sensory systems.

If one projects two or more spots of light within the on-center region of a ganglion cell's receptive field (Figure 9C), the cell is more excited than by one spot alone (Figure 9A); and two or more spots within the off-surround (Figure 9D) give more inhibition than does one spot alone (Figure 9B). The *maximal* excitation, then, would be produced by a larger spot of light that entirely fills the on-center region but does not extend at all into the off-surround (Figure 9E). A light stimulus of much greater extent, covering the entire receptive field or even the entire retina, produces very little response in such a cell because inhibition from the off-surround almost fully counteracts excitation from the on-center (Figure 9F).

Most vertebrates also have some retinal ganglion cells with the properties exactly reciprocal to those just described. These are called OFF-CENTER ON-SURROUND neurons. Such a cell can be excited by turning off, or dimming, the light within the off-center region of its receptive field. Simultaneously turning off or dimming the light within the on-surround region inhibits the cell. Off-center on-surround cells in mammals can also be excited by increasing the illumination anywhere in the on-surround, as by projecting small spots of light in this region. In amphibians, however, it is generally the receptive field center that has the dominant effect on the cell's response (Werblin and Dowling, 1969; Ewert and Hock, 1972; Grüsser and Grüsser-Cornehls, 1976). Thus, for an amphibian off-center on-surround cell, the maximally effective stimulus is one that dims the entire center but not any of the surround. This dimming can be achieved by placing a black circle of just the right size in just the proper location to completely fill the off-center region of the receptive field for the particular cell in question (Figure 9G).

To complicate matters somewhat, some retinal ganglion cells in amphibians are excited by both brightening and dimming of the receptive field center, which can therefore be called a mixed on/off-center. In such cells, brightening or dimming the surround counteracts the effect of central brightening or dimming. These cells, then, can be called on/off-center on/off-surround cells. For simplicity, in the remainder of this chapter we will designate receptive fields of retinal ganglion cells by the shorthand labels *on-center* cells, *off-center* cells, or *on/off-center* cells. It should be understood that each of these has an antagonistic receptive field surround.

The responses of vertebrate retinal ganglion cells to small stationary spots of light, as just described, are consistent with the responses to moving wormlike, antiworm, and square stimuli that we have seen in the toad's ganglion cells. This can be seen, for instance, in the case of the toad's R3 cells (Figure 8C). Because these are on/off-center cells (Ewert and Hock, 1972), they are excited by either a brightening or a dimming of the receptive field center. Such a brightening occurs when a white object moves into this center region, and a dimming occurs when a black object moves into this region. Because the receptive field centers of the R3 cells have a diameter of about 8° visual angle (Ewert and Hock, 1972), an optimal moving stimulus would be either a white or a black object that sweeps through the receptive field center, covering the entire 8° diameter. As it sweeps, however, the stimulus object would first enter the receptive field surround, then the center, and then again the surround on the far side (Figure 10). It is not known in detail how this prior and subsequent visual stimulation of parts of the surround affect the cell's response to the interposed stimulation of the center (Grüsser and Grüsser-Cornehls, 1976). However, it is clear that a moving square or antiworm stimulus of 8° dimensions would just fully cover the receptive field center as it sweeps through, whereas smaller squares or antiworms would incompletely cover the center, and larger ones would cover an excess of the antagonistic surround (Figure 10). Moreover, a wormlike stimulus, no matter what its length, would only partially fill the receptive field center. These considerations would lead to the prediction that 8° squares or antiworms would maximally excite an R3 cell. As we have already seen, this prediction is fulfilled (Figure 8C). R2

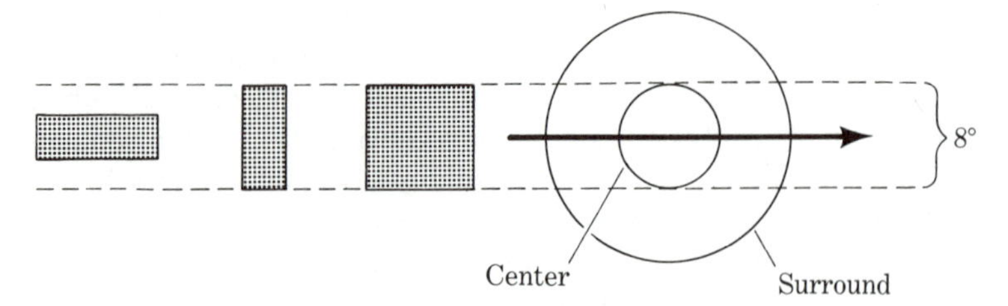

FIGURE 10. The optimal moving configurational stimuli for a cell with an 8° receptive field center are squares and antiworms that are 8° in length.

cells have primarily off-center receptive fields with center diamters of 4°, and R4 cells have primarily on-center receptive fields with center diameters of about 16°. One can thus predict that black squares and antiworms of 4° would be maximally excitatory for R2 cells and that white squares and antiworms of 16° would be maximally excitatory for R4 cells. Again, these predictions are borne out (Figure 8B and D).

How can one explain the antagonistic center–surround arrangement of the receptive fields of retinal ganglion cells, based upon what is known of the neuronal connections within the retina? As mentioned in Chapter 5 (see Figure 21), the retina consists of five major cell types; receptors (both rods and cones), horizontal cells, bipolar cells, amacrine cells, and ganglion cells. The synaptic interactions among these cells that appear to be responsible for an on-center off-surround receptive field of a retinal ganglion cell in an amphibian (and perhaps also in a mammal) are shown schematically in Figure 11. At the bottom of the figure is the ganglion cell in question, and above this is a bipolar cell that synapses upon it. The bipolar cell itself receives synaptic input directly from a small cluster of receptor cells (just two in the figure) that are responsible for producing the ganglion cell's excitatory reaction to light within its receptive field center. In response to illumination, these receptors excite the bipolar cell; and the bipolar cell in turn excites the ganglion cell. A surrounding group of receptor cells exerts not a direct but rather a disynaptic effect on the same bipolar cell, mediated through a horizontal cell.[7] This disynaptic pathway inhibits the bipolar cell. Thus, the excitatory response to light within the receptive field center is antagonized by light in the surround, both for the bipolar cell and for the ganglion cell that it drives. In this explanation, no role is offered for the amacrine cells, which in fact are especially numerous in amphibian retinae as compared with those of mammals and are often interposed between bipolars and ganglion cells. However, these amphibian amacrine cells may connect primarily to ganglion cells of a type other than those described in this chapter, which appear designed primarily for monitoring changes in illumination level rather than for detecting spatial patterns of light and dark (Dowling, 1976).

[7]As Figure 11 shows, the receptors of the receptive field center also have this indirect connection to the bipolar cell through a horizontal cell. However, because there are fewer receptors in the center than in the surround, the contribution of these central receptors to the indirect pathway is very small.

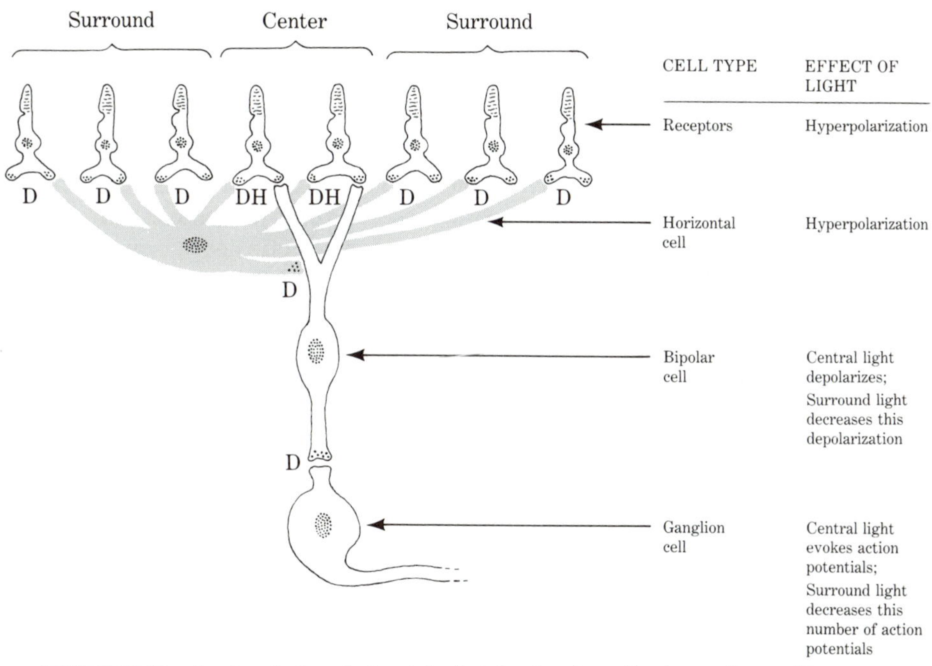

FIGURE 11. Anatomical and physiological basis of a retinal ganglion cell's on-center receptive field. See text for explanation.

The picture of retinal connections becomes a bit more complicated when one recalls that a vertebrate receptor cell is *hyperpolarized* by light and that this hyperpolarization *decreases* this cell's normally steady release of neurotransmitter (Chapter 5). Release of these neurotransmitter molecules appears to exert a hyperpolarizing effect (H in Figure 11) on the bipolar cell. Thus, illuminating the rods or cones of the receptive field center *decreases* the steady *hyperpolarization* that these receptors exert on the bipolar cell; that is, it *depolarizes* the bipolar cell. This same neurotransmitter of the receptor cells, however, appears to have a *depolarizing* effect on the horizontal cell (D in Figure 11). Thus, illuminating the rods and cones of the receptive field surround, by decreasing the release of their neurotransmitter, would *decrease* the ongoing *depolarization* of the horizontal cell; that is, it would hyperpolarize this cell. The neurotransmitter of the synapses from horizontal cells to bipolar cells is thought to have a depolarizing effect on the bipolar cells and may, like the transmitter of receptor cells, be released continuously in the dark. Thus, when the horizontal cell is hyperpolarized by illumination of the receptive field surround, this cell would decrease its release of depolarizing transmitter, thus *hyperpolarizing* the bipolar cell.

In summary, then, the bipolar cell is depolarized by direct synaptic action from illuminated central receptors and hyperpolarized by a disynaptic connection from illuminated surrounding receptors. Because the bipolar cell, when excited, releases a depolarizing neurotransmitter onto the ganglion cell, it passes on to this cell its own antagonistic on-center off-surround response.

RESPONSES OF NEURONS IN A TOAD'S BRAIN TO MOVING STIMULI

We return now to the search for prey-recognition neurons in a toad's brain. The axons of retinal ganglion cells, coursing through the optic nerve, project to at least five different locations within the amphibian brain (Ingle and Sprague, 1975; Fite and Scalia, 1976). Each of these areas is called a NUCLEUS—a brain region consisting of closely packed cell bodies and their processes. Into and out of a nucleus project TRACTS, or bundles of axons. The optic nerve is one example; it is sometimes called the optic tract. Nuclei, along with some other parts of the vertebrate central nervous system, are referred to as GRAY MATTER, owing to the grayish appearance of the cell bodies in fresh tissue. By contrast, tracts, whose axons are wrapped in glistening myelin sheaths, are referred to as WHITE MATTER.

The search for recognition neurons in toads has centered upon two of the five nuclei to which the optic nerve projects (Ewert, 1980). First is the OPTIC TECTUM, which in amphibians and other lower vertebrates receives by far the greatest number of ganglion cell axons (Figure 12A). The word *tectum* is Latin for "roof," and this brain region forms the roof of the midbrain. In amphibians, the neuronal projection from eye to tectum, called the RETINOTECTAL PROJECTION, is entirely contralateral, the left optic nerve terminating in the right tectum and vice versa. The second brain region of interest, lying near the tectum, is a composite of two regions: the posterior thalamus and the pretectum, together called the THALAMIC PRETECTAL, or TP AREA.

Within the TP area, many different categories of visually responsive neurons have been found (Ewert, 1976, 1980). Of these, one category, called TP3 cells, responds to moving visual stimuli. Each TP3 cell has a small receptive field. However, different TP3 cells have different receptive field locations, so that the whole visual world is represented in this group of neurons. This is important because the toad's behavior demonstrates an ability to recognize wormlike stimuli presented in any part of the visual field. These properties make the TP3 cells worth exploring further, to see whether they show the key property expected of recognition neurons: a profile of response strengths to moving worms, antiworms, and squares that matches the response profiles of the behavior. It was found, in fact, that the response profiles of the TP3 cells do not match at all that of the behavioral responses (Figure 13B; compare with 13A); these cells were least strongly excited by moving wormlike stimuli, the reverse of the behavioral response profile.

Within the optic tectum there are again several physiological categories of visually responsive neuron. The neurons of one category, called T5 cells, were investigated in detail because, like the TP3 cells, they gave strong responses to moving stimuli, and different T5 cells had their receptive fields in different positions, distributed throughout all regions of visual space (Grüsser and Grüsser-Cornehls, 1976). The T5 cells are all located in one layer of the tectum (Figure 12B); this was determined by first recording from a neuron, physiologically identifying it as a T5 cell (by criteria to be explained below), then marking the location of the electrode tip by passing a strong current through the electrode that burned a small hole in the tectal tissue near the electrode tip. This hole was then found

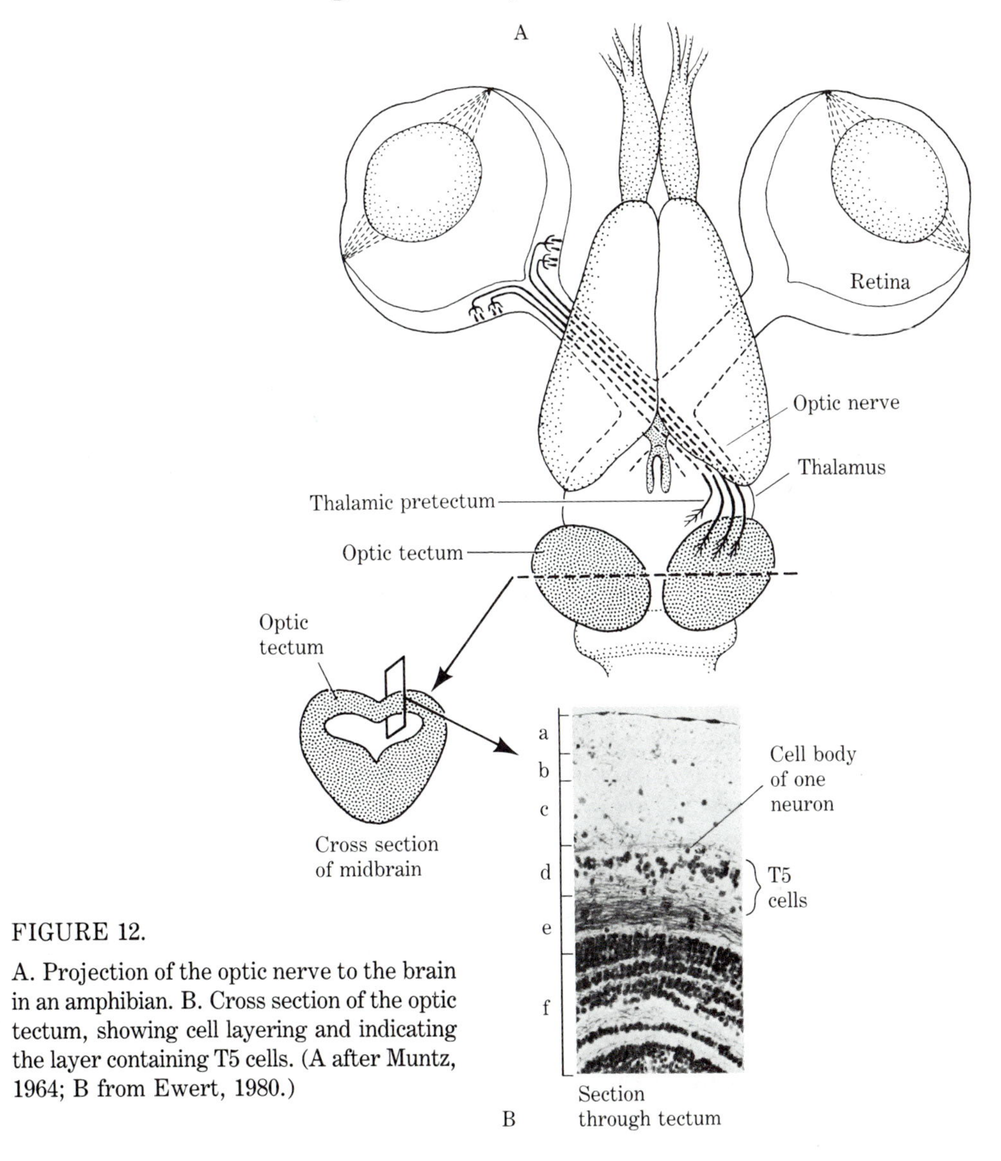

FIGURE 12.

A. Projection of the optic nerve to the brain in an amphibian. B. Cross section of the optic tectum, showing cell layering and indicating the layer containing T5 cells. (A after Muntz, 1964; B from Ewert, 1980.)

in histological sections of the tectum. It is within this T5 layer, then, that our search for recognition neurons now proceeds.

Recordings from T5 neurons reveal two subpopulations, called T5(1) and T5(2) cells (Grušser and Grušser-Cornehls, 1976; Ewert and Hock, 1972; Ewert, 1980). T5(1) cells show a response profile that is clearly different from that of the toad's behavior (Figure 13C). However, the response profile of T5(2) cells resembles rather closely that of the toad's behavior (Figure 13D; compare to 13A). For both the behavior and the T5(2) cells, the response strength is greatest to wormlike stimuli and least to antiworms. Moreover, the behavior and the T5(2) cells have response curves of nearly identical shape. This is clearest for antiworms

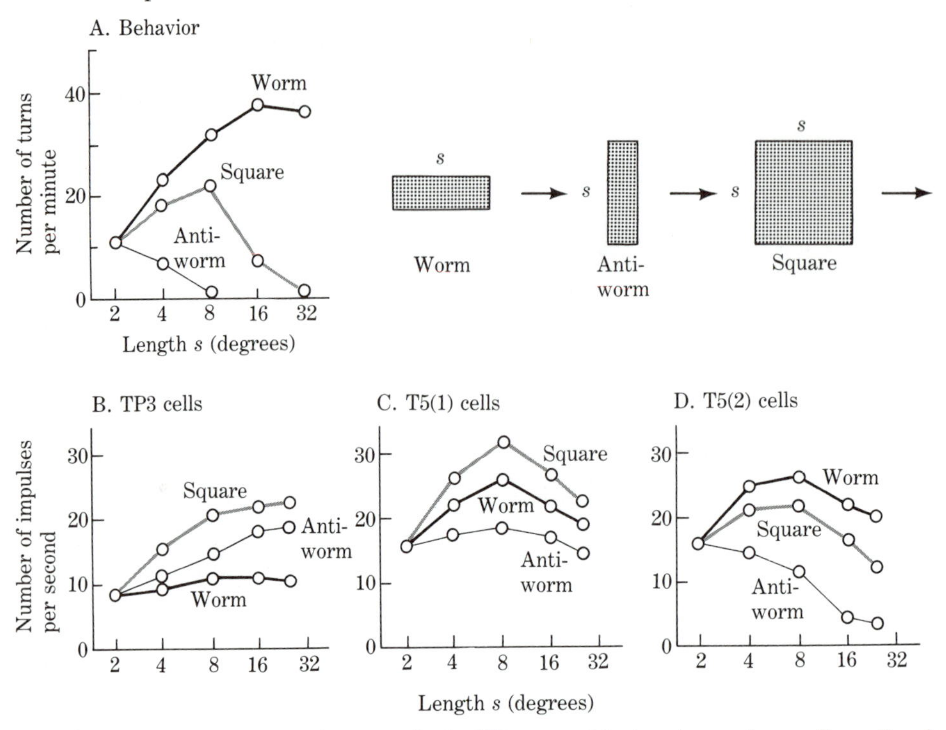

FIGURE 13. Responses of neurons in the TP area and tectum to moving configurational stimuli. A. Summary of behavioral responses (from Figure 5). B, C, and D. Responses of three classes of brain neurons. Only the T5(2) cells (D) have a response profile resembling that of the behavior. (After Ewert, 1980.)

and squares. As for wormlike stimuli, both the behavioral and the T5(2) responses increase for initially increasing worm length. But this increase stops at 16° for the behavior and at only 8° for the T5(2) cells. This is the only consistent difference between the response profiles of the behavior and the T5(2) cells. In fact, the close correlation between the behavioral and the T5(2) response profiles has received further support from a careful statistical analysis (Ewert *et al.*, 1978). Moreover, like the behavior, the T5(2) cell responses show invariance, consistently giving their strongest response to worms and weakest to antiworms under a variety of stimulus conditions: different directions of stimulus movement (Figure 6A and B), variations in contrast (black on white, white on black, black or white on checkered background), variations in movement velocity and in distance of the stimulus from the animal, and variations in the form of stimulus movement, including smooth and jerky displacement. Therefore, except for the reduced responses of the T5(2) cells to wormlike stimuli of 16° length, their properties closely match those of the toad's behavior. No other neurons have yet been found whose responses show such a close match. For these reasons, the toad's T5(2) neurons are good candidates for prey-recognition neurons. Further tests of this candidacy will be presented in the next section.

FURTHER TESTS OF T5(2) CELLS AS RECOGNITION NEURONS

The fact that the T5(2) response is closely *correlated* with prey recognition does not show that it is *causal* for this recognition. Indeed, it is implicit in the definition of recognition neurons that they must not only respond selectively to the particular object being recognized (in this case a moving wormlike stimulus), but they must actually be cells that the brain uses in the process of recognizing this object. In general, when one wishes to demonstrate that some component process [T5(2) activity] is responsible for a larger process (prey recognition), an important strategy is to show that the first process is both *necessary* and *sufficient* to account for the second.

We will first consider tests designed to show whether the T5(2) cells are *necessary* for prey recognition. To do this, one might like somehow to silence all of the T5(2) cells but not any other neurons; then, if a moving prey stimulus evokes no behavioral response, this would indicate that the T5(2) cells are necessary for a visually elicited behavioral response toward prey. This could mean that the toad is simply blinded by the T5(2) silencing rather than being unable specifically to recognize prey; but this objection could be overcome if behavioral responses to other visual stimuli persisted. Alternatively, it could be that the silencing of the T5(2) cells does not block recognition but rather blocks the ability to carry out the particular *motor* act involved in turning toward an object; but this objection could be eliminated if the same motor act could be evoked by some other, perhaps nonvisual, stimulus. With these objections overcome, the test for a *necessary* role of the T5(2) cells in prey-recognition would be completed.

Unfortunately, it is not possible with present techniques to silence just the T5(2) cells and spare all other cells, because other tectal cells are distributed among the T5(2) cells. However, as a much more crude experiment, one can ablate all, or a large part of, the optic tectum. Following ablation of the tectum on one side of the brain, visual stimuli moving in the opposite half of the visual world evoke no behavior (Ewert, 1980). (Recall that the left tectum analyzes the right visual field and vice versa; Figure 12A.) Tectal ablation does not cause blindness, however, because frogs with such ablations can still localize stationary objects in their visual field (Ingle, 1973, 1976). (This test has not been carried out as yet for toads.) Also, the failure to orient toward prey on the side of the visual world served by the ablated half-tectum probably does not result from a motor incapacity because frogs with tectal ablations show entirely normal prey orientation and tongue flicks in response to tactile stimulation of the legs or body (Comer and Grobstein, 1981a, b). These observations, then, indicate that *some* tectal cells [perhaps the T5(2) cells, though this is not proved] are necessary for visual prey recognition.

To test the idea that the T5(2) cells are *sufficient* for visual recognition of prey, one can activate small groups of these cells with electric currents from a stimulating electrode in the absence of visual stimuli to see whether the toad responds by giving an orientation behavior (Ewert, 1980). These experiments have been carried out on toads that are free to move about and that wear an electrode permanently implanted in a small region of the tectum. Before turning on the stimulating current, this same electrode was first used to record the activity of the several cells right near the electrode's tip, in order to determine their receptive field locations. It was found that for any given electrode location, all the cells recorded had their receptive fields in one region of space. (The broader significance of this point will be explained shortly.) Now, with the electrode's position unchanged, it was used to deliver the stimulating current. The toad responded by orienting toward the region of the environment containing the receptive fields of the cells at the electrode's tip (Figure 14). It was as though the sight of a moving wormlike stimulus at a particular location in space had triggered this behavioral response. This indicates that *some* tectal neurons that were excited by the electrode are sufficient for inducing the prey-orienting response. One must note, however, that even with the electrode tip located in the tectal layer containing the T5(2) cells, other cell types also located in this layer or in nearby layers would also be stimulated. Therefore, a mixed population of tectal neurons is excited by such extracellular stimulation, making impossible any definitive conclusions as to which cells are responsible for the evoked behavior. (This same difficulty arises with all extracellular brain stimulation experiments. Further examples will be presented in Chapter 8.) Nevertheless, these experiments provide useful *supportive* evidence that the T5(2) cells may be sufficient for prey recognition because these are almost surely among the neurons stimulated by the electrode.

These tests for a necessary and sufficient role of T5(2) cells in prey recognition, though necessarily somewhat crude, are suggestive. (A more rigorous and powerful test for a necessary and sufficient role of specific neurons in behavior will be described in Chapter 8.) However, additional correlations between the prey-orienting behavior and the responses of T5(2) cells further contribute to

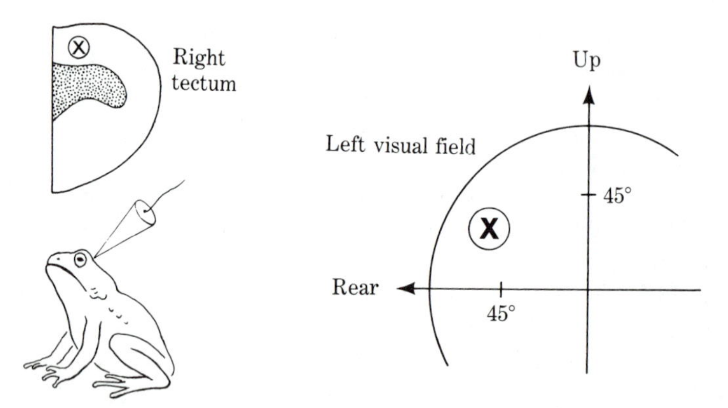

FIGURE 14. Orientation behavior evoked by electrical stimulation of small groups of neurons in the tectum. For the particular electrode position used (X on the drawing of the tectum) the orientation response was toward the upper left quadrant of visual space (X on the map of the visual field). This area of visual space contained the receptive fields of cells recorded from this electrode position. (After Ewert, 1980.)

implicate these neurons in prey recognition (Ewert and Wietersheim, 1974; Ewert, 1980). For instance, if one destroys the TP region or if one separates the TP from the tectum by a small knife cut, the response profile of the T5(2) cells to moving stimuli immediately takes on a different shape (Figure 15B). The behavioral response profile also changes and resembles almost exactly the altered T5(2) profile (Figure 15A). An additional correspondence between the behavioral prey-orienting responses and T5(2) responses is that both are decreased in well-fed satiated toads, as compared to hungry toads (Ewert, 1980).

Although the toad's T5(2) cells pass many of these tests that implicate them as recognition neurons for prey, there is one test they do not pass. If moving square stimuli of different sizes are presented to toads from a given distance, there is always one size of square that evokes a prey-orienting response of maximal

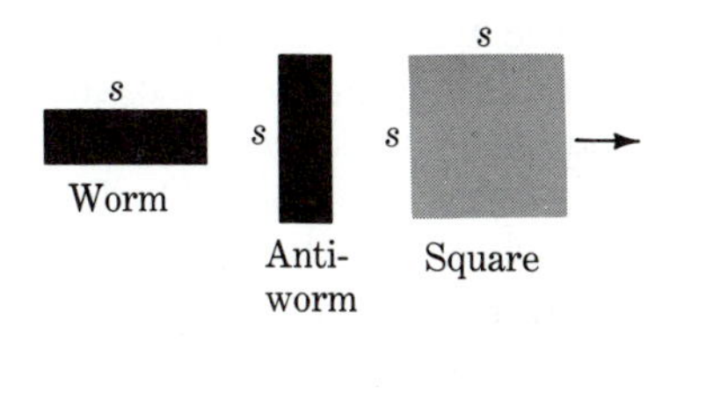

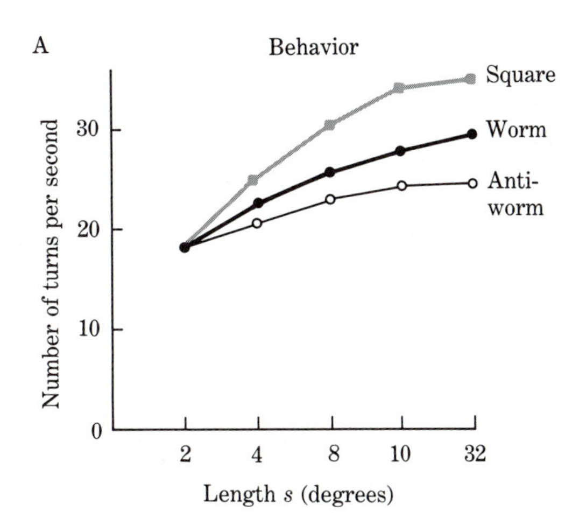

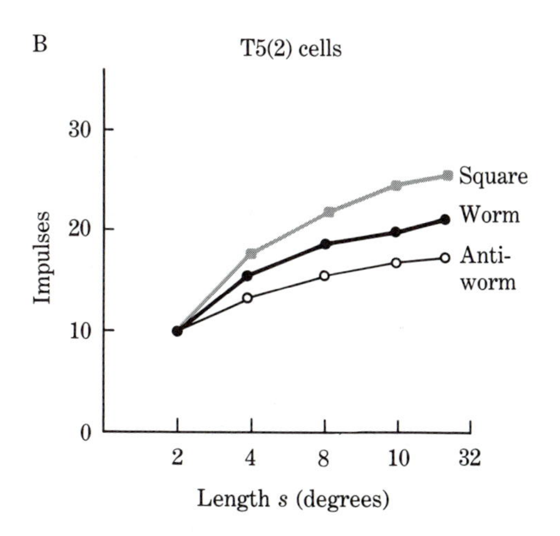

FIGURE 15.
Toad's orienting behavior (A) and responses of T5(2) cells (B) after TP lesions. (From Ewert, 1980.)

strength (Figure 5C). If now the same squares are presented at different distances from the toad, it is found that the size of the maximally effective square varies with distance in a characteristic way; the greater the viewing distance, the larger the maximally effective square. Do the T5(2) cells respond in a parallel manner? Actually, all the physiological recordings from T5(2) cells have been carried out with the stimulus at just one fixed distance—20 centimeters from the toad (Grüsser and Grüsser-Cornehls, 1976). At this 20-centimeter distance, the maximally effective square in the behavioral tests had a length of about 1.5° visual angle. For the T5(2) cells, the maximally effective square at 20 centimeters was five times this size. This mismatch has yet to be properly explained.

Before concluding this discussion of the T5(2) cells, it should be noted that under some conditions, a toad's prey-orienting behavior can be more complex than would be predicted on the basis of the T5(2) cells alone. For instance, if a toad or a frog views a prey stimulus moving beyond a barrier, often its response is first to sidestep the barrier and only then to orient toward the prey on the other side (Ingle, 1970). This sidestepping, which is directed toward the nearer of the two ends of the barrier, requires the toad to orient initially *away* from, not toward, the prey stimulus (Figure 16). The prey apparently is properly recognized, but a separate brain process, probably involving neurons of the thalamus (Ingle, 1973, 1976), intervenes to redirect the initial movement. Thus, even though the prey-orienting behavior as usually tested in the laboratory is reasonably stereotyped, this more complex perceptual situation reveals considerable flexibility, at least within the early, orienting phase of the prey-catching behavior.

In summary, we have seen several lines of evidence supporting the role of tectal T5(2) cells as visual recognition neurons for prey: (1) the response profiles of these cells are very similar (though not perfectly identical) to that of the prey orienting behavior (Figure 13); (2) unilateral tectal ablation eliminates all behavioral responses to moving visual stimuli in that half of the visual field normally subserved by the ablated tectal regions, while not causing blindness to stationary

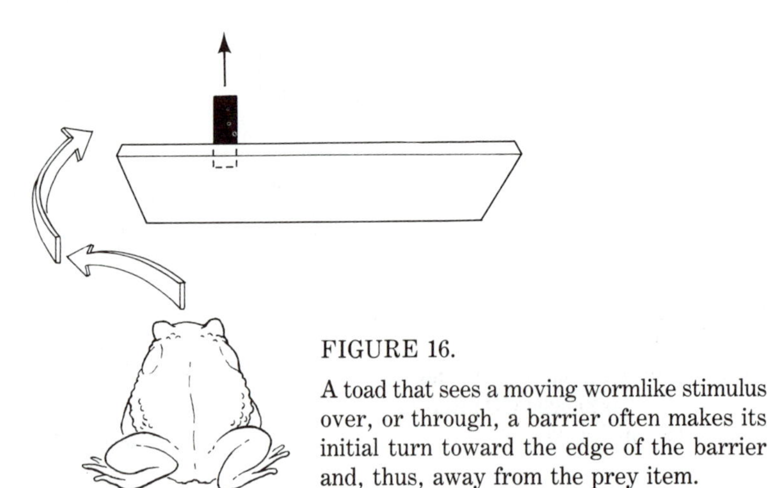

FIGURE 16.

A toad that sees a moving wormlike stimulus over, or through, a barrier often makes its initial turn toward the edge of the barrier and, thus, away from the prey item.

objects, and not preventing the orienting and snapping responses to tactile stimuli; (3) lesions of the TP area lead to similar alterations of the response profile for the T5(2) cells as for the behavior (Figure 15). There are also parallel changes in the responsiveness of the behavior and the T5(2) cells with changes in hunger state; (4) electrical stimulation of small populations of T5(2) cells, along with other tectal cells, evoke behavioral orientation toward the receptive field locations of these cells. Arguing against the role of T5(2) cells as recognition neurons is the observation that the size of a moving square that is optimal for exciting these cells does not match the size predicted from behavioral experiments carried out with stimuli at different distances from the toad. The weight of evidence, though, appears to point toward T5(2) cells as subserving a function quite close to that of recognition neurons, as defined in the beginning of this chapter. In fact, this evidence is probably as strong for these toad neurons as for any cells yet studied in any animal. However, contrary to our earlier presumption that our theoretical recognition neuron for a chair should be excited by a chair in any and all *locations* within visual space, wormlike stimuli presented in different regions of a toad's visual space excite not the same but different T5(2) cells. This important spatial aspect of prey recognition will be explored further in the next section.

STIMULUS LOCALIZATION AND TECTAL MAPS

Apart from recognizing a suitable prey item, a toad or other predator must be able to determine the location of its prey. That toads can do so is indicated by their properly oriented turning responses and tongue flicks. To some extent each T5(2) cell is able to serve for both prey recognition and prey localization because each has a fairly restricted receptive field, averaging 27° in diameter. Activity of a T5(2) cell, then, signals that a moving wormlike object is located within these 27° of visual space. In fact, because the receptive fields of different T5(2) cells partially overlap one another, the brain may be able to obtain much finer positional resolution by taking account of which particular ones of these T5(2) cells are and which are not excited at a given moment by this stimulus. As mentioned in the last section, it is neighboring T5(2) cells that have neighboring, and partly overlapping, receptive fields. This raises the question of the overall relationship between cell location and receptive field position in the tectum.

The relationship between cell position and receptive field location was explored by making recordings from a linearly arranged series of T5(2) cells. For each cell, the investigator determined the location of its receptive field and then marked the location of the cell itself by passing a strong electrical current through the electrode to produce a lesion that could be found later in brain sections viewed under the microscope (Ewert, 1980). Such experiments reveal a systematic point-to-point projection of receptive field location on the tectal surface, or a TOPOGRAPHIC MAP of the visual world (Sperry, 1943; Gaze, 1980; Constantine–Paton and Law, 1982). This is also called a RETINOTOPIC MAP (Figure 17). The spatial order of the map comes about as a two-stage process. The first stage does not involve nerve connections at all but rather the geometry of the eyeball. As mentioned in Chapter 5 (Figure 6), this geometry produces a two-dimensional map of space on the retina, such that a particular retinal ganglion cell encodes infor-

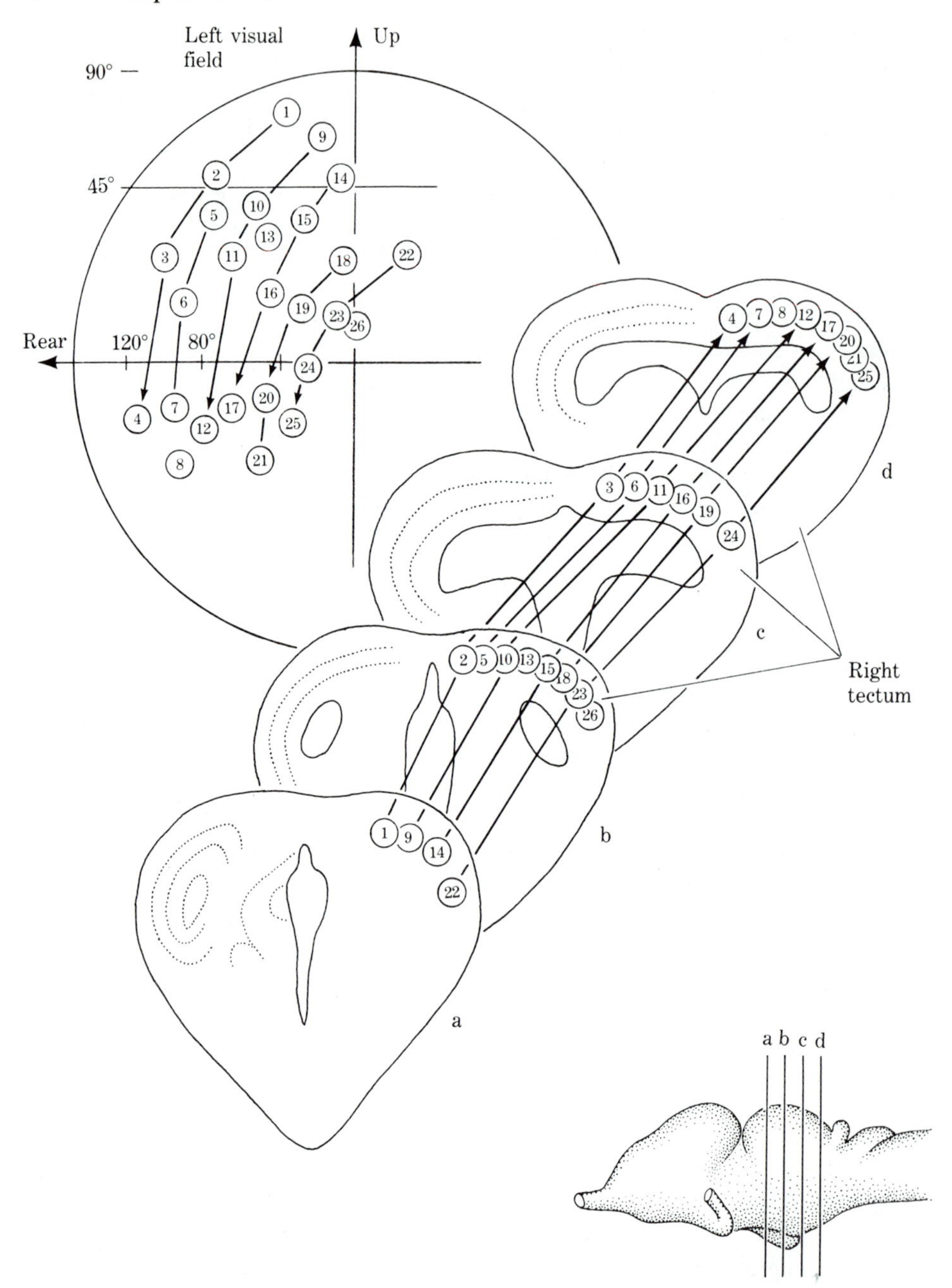

FIGURE 17. The toad's retinotectal map. Each circled number on the right tectum represents a region sampled by the recording electrode. The corresponding numbers on the map of the toad's left visual field indicate the receptive field locations for the cells recorded. Four cross sections (a through d) through the tectum are shown. The positions of these sections are indicated on the diagram of the whole brain, seen in lateral view in the lower right part of the figure. (After Ewert, 1980.)

mation about only one region of the visual world. The second (neural) stage consists of an ordered coursing of the axon of each retinal ganglion cell to terminate on the tectal surface in a position correlated with the retinal location of its cell body. Such a highly ordered coursing of the optic nerve into a brain region is often called a RETINOTOPIC PROJECTION.

The topographic organization of the tectum has an important functional consequence. Imagine two tectal cells, A and B, which have overlapping receptive fields. Because they are both involved in analyzing some of the same parts of the visual world, A and B would probably need to connect synaptically to the same postsynaptic cell, C, within the tectum. Because A and B lie close to each other, C can lie close to both of them. This permits A and B both to have short intratectal axons, providing a saving in terms of the energy cost of building and maintaining these axons. In addition, A and B may both need to send long axons to the same postsynaptic cell, D, outside the tectum. Because A and B are neighbors, their axons can travel the same course during development in finding their common distant target, simplifying the enormously complex problem of constructing the brain during development. Given these advantages, it should come as no surprise that not only toads but also several other species that are distributed among all vertebrate classes are known to have retinotopic tectal maps (Ingle and Sprague, 1975; Wurtz and Albano, 1980). In all these vertebrates, the optic tectum (or, as it is commonly called in mammals, the SUPERIOR COLLICULUS) appears to be the major brain region mediating a turn of the eyes, ear pinnae, head, or body toward small, moving, visual targets (Schiller and Stryker, 1972; Ingle and Sprague, 1975; Wurtz and Albano, 1980). Thus, it appears that the basic outline of tectal structure and function emerged early in vertebrate evolution and has been conserved ever since with relatively little fundamental change.

The tectum of most vertebrates, sometimes together with the underlying brain tissue, contains a topographic map not only for vision but also for other sensory modalities. For instance, tactile stimulation of a given region of the body surface excites cells within a given tectal or subtectal region, and changing the site of body stimulation correspondingly changes the location where tectal cells are excited. Just as with vision, tactile stimuli on the left side of the body activate tectal or subtectal cells on the right side and vice versa. Within each half-tectum, the cutaneously activated cells are topographically organized, forming a map of the body surface, or a SOMATOTOPIC MAP. In fact, in several species, this somatotopic map has been found to lie in a cell layer just underneath, and *in register with*, the retinotopic map (Drager and Hubel, 1975; Stein *et al.*, 1976; Gaither and Stein, 1979). Thus, for instance, cells excited by touching the left foreleg lie just below those excited by visual stimuli near this foreleg. This presumably facilitates the neural analysis of a stimulus by both visual and tactile sensory modalities. A topographic map for localization of infrared signals in rattlesnakes (Hartline *et al.*, 1978; Newman and Hartline, 1982) also lies underneath, and roughly in register with, the tectal retinotopic map.

The maps of many animals are distorted, with areas of frontal vision and corresponding facial tactile regions taking up more space than other parts of the map. This reflects the fact that more cells are devoted to processing stimuli in this important frontal position than elsewhere. This, in turn, reflects the situation in the retina of some vertebrates, which have a fovea containing an excess of cells for high resolution viewing of frontal space (Chapter 5). Such map distortions are expressed quantitatively by the term MAGNIFICATION FACTOR—defined as the distance across a given region of the brain's topographic map that corresponds to a fixed distance in visual space. In general, the magnification factor is highest in the frontal region, where a large distance across the brain's map corresponds to only a small distance in space.

To summarize, three common features of the tectum in vertebrates are the presence of spatial maps, spatial registry between maps of different sensory modalities, and variations of magnification factor along the maps. As we shall see, other brain regions also exhibit spatial maps. Moreover, such maps have been found in invertebrate nervous systems as well (Murphey *et al.*, 1980; Murphey, 1981).

The Special Case of Auditory Receptive Fields and Spatial Maps

As we saw in the last section, retinotopic and somatotopic maps are common in the brain. The retina constitutes a two-dimensional map of visual space. In like manner, the sensory receptors in the skin surface constitute a map of the outside world. Thus, ordered projections of axons from the receptor surface of the retina or the skin to the brain produce the topographic order of the maps. A different situation is found in the auditory system, however. Within the array of hair cells in the cochlea, no sensory map of external space exists; the position of a given hair cell along the basilar membrane is correlated with that cell's response, not to a particular spatial location of the sound source, but rather to a particular sound frequency (Chapter 6). Therefore, if the axons of the auditory nerve were to project and connect in an ordered manner to their postsynaptic cells in the brain, what should result is a brain map not of sound location but of sound frequency—a TONOTOPIC MAP like that in the cochlea itself. In fact, such tonotopic maps have been found in various auditory processing areas of many vertebrate brains (Yost and Nielsen, 1977).

How, then, can an animal determine the location of a sound source? As mentioned in Chapter 6, sound from one side of an animal produces binaural differences in the time of arrival, intensity, and phase of the sound wave. One theory of sound localization is that the brain neurons to which the left and right auditory nerves connect constitute something akin to a spatiotemporal pattern recognizer (Figure 2C), used in this case not for object recognition but for sound localization (van Bergeijk, 1967). By this theory, sound from 90° left would evoke activity in the left auditory nerve (and the brain cells excited by it) preceding that in the right auditory nerve (and the different group of brain cells excited by it). The left cells would also give more action potentials than the right cells. Sound from, say, 45° left would produce less of a left–right disparity, and sound from directly in front would produce no binaural disparity at all. By this scheme, then, each auditory neuron in the brain would be excited by sounds from all locations; none would have a restricted receptive field. It would be the relative timing and amount of activity among the different neurons that the brain would interpret as a particular sound location.

More recently, however, auditory neurons with restricted receptive fields have been found within a midbrain nucleus of barn owls (Knudsen and Konishi, 1978a, b, c; Knudsen, 1981, 1982; Moiseff and Konishi, 1981). The nucleus is the mesencephalicus lateralis dorsalis, or MLD, which is the avian homolog of the mammalian inferior colliculus. This is a major auditory nucleus just below the superior colliculus (optic tectum). The MLD appears to play an important role in sound-mediated prey capture by owls, a behavior (Figure 18) that involves auditory localization of a sound source with an accuracy of just a few degrees (Payne, 1971; Konishi, 1973; Knudsen and Konishi, 1979). In fact, just like retinal ganglion cells, these auditory neurons in owls have on-center off-surround receptive fields: sounds from within the center region (mean diameter 25°) excite the cell, and sounds from the surround antagonize this excitatory effect (Figure 19A). Moreover, the MLD neurons are arrayed to form a spatial auditory map (Figure 19B). Just like the retinotopic map in the tectum, the magnification factor of the auditory map is greatest for frontal stimuli.

How is it possible for an auditory neuron in the brain to have a restricted receptive field when each auditory receptor in the ear responds to sounds from all directions? A given MLD cell, whose receptive field center lies exactly in front of the animal, responds only when the intensity and arrival time of the sound is equal at the two ears (Moiseff and Konishi, 1981). An imbalance of these parameters at the two ears, indicating sounds not from straight ahead but rather from the surround, inhibits the cell. A different cell whose receptive field's center is at 30° to the left is activated when the left ear receives sounds earlier and louder than the right ear—earlier and louder by just that amount corresponding to the 30°-left position. The mechanism by which a given cell is excited by a particular amount of binaural difference in sound intensity and arrival time and is inhibited by other amounts of binaural difference is unknown. Whatever the mechanism, though, it is fundamentally different from that by which cells in the visual and somatosensory parts of the brain derive their receptive fields, namely, by ordered projection of axons from receptor surfaces whose cells derive their receptive fields from the geometry of the sensory organ. Rather, auditory receptive fields are derived de novo from a neuronal calculation of differences in the signals encoded in the two auditory nerves. The particular amount of binaural difference to which a given MLD neuron selectively responds varies from cell to cell across the MLD surface; the values for these differences are used to create the spatial auditory map. The fact that the auditory system bothers, so to speak, to construct de novo receptive fields and a spatial map suggests that these properties must be of major importance in the brain's handling of spatial information about the outside world.

Another instance of de novo construction of response properties is found in the auditory cortex of bats (O'Neill and Suga, 1982). Here cells are found that respond selectively to particular time intervals between an emitted pulse and its returning echo. Because this interval correlates with the distance to the echo-producing target (Chapter 6), these cells encode target distance. Moreover, this parameter of target distance is mapped on the bat's cortex.

FIGURE 18. Cine sequences, filmed in only infrared light, of an owl capturing in its feet a running mouse. Owls apparently are blind to infrared light, so the prey capture sequence shown was not visually mediated. Observations such as this, as well as numerous control experiments (Konishi, 1973; Knudsen, 1981) have shown that owls can localize their prey by auditory cues alone. (Courtesy of M. Konishi.)

axis. Number sequences show the correspondence between cell location and receptive field position in the dorsal–ventral axis. The MLD, then, contains a two-dimensional representation of auditory space. (After Knudsen and Konishi, 1978a).

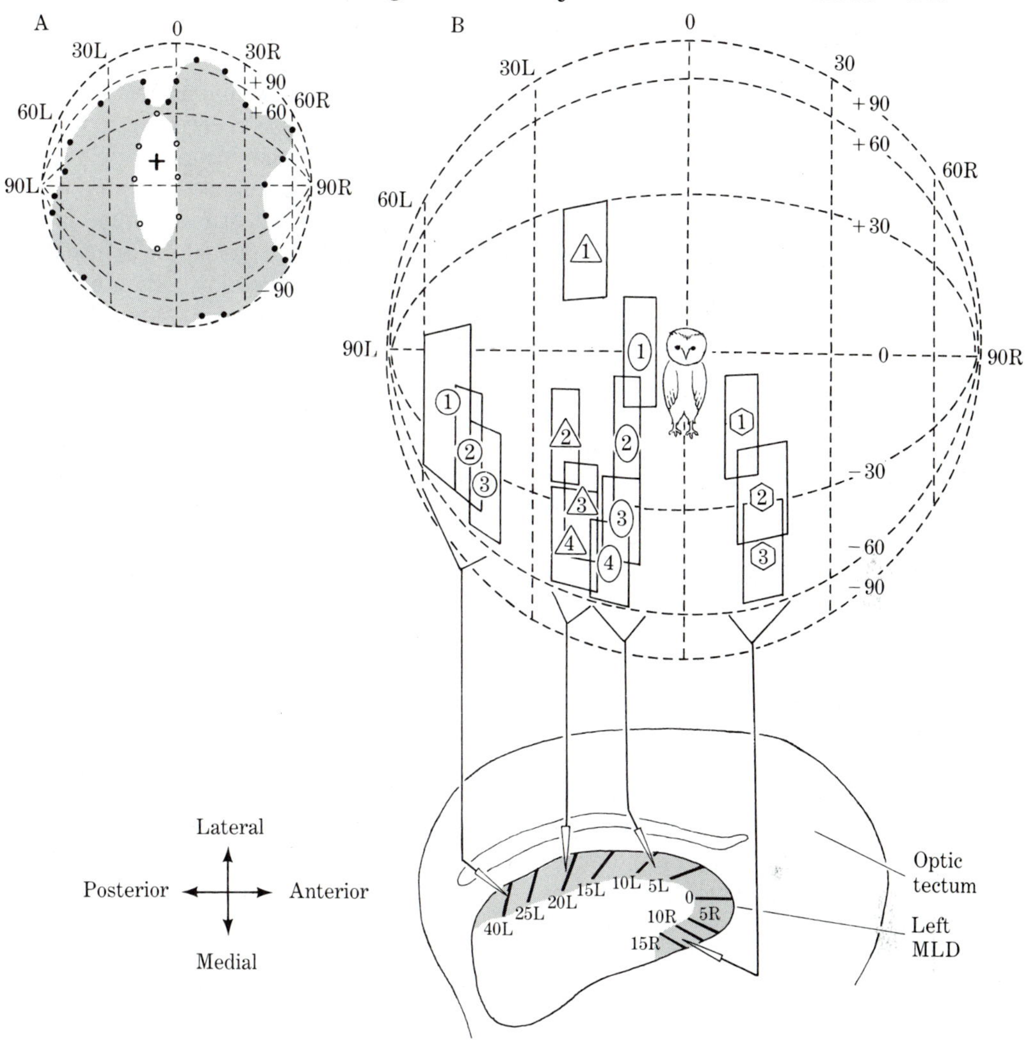

FIGURE 19. Auditory neurons in the owl brain. A. On-center off-surround auditory receptive field of a neuron in the MLD nucleus of a barn owl. The receptive field is plotted on an imaginary hemisphere of the space in front of the owl. The white area with the plus sign is the on-center, and the large shaded region is the off-surround. B. Spatial auditory map in a barn owl's MLD nucleus. The upper part of the figure shows the hemisphere of space in front of the owl. The position of the receptive field on-centers of 14 different neurons are shown by the numbered rectangles. Rectangles whose numbers are backed by the same symbols (circles, triangles, etc.) represent neurons recorded in the same passage of the electrode through the MLD; the numbers indicate the sequence in which these neurons were recorded as the electrode passed from the surface into the depth of the MLD. Below is a horizontal section through the midbrain, showing the left MLD nucleus. The shaded portion contains the neurons recorded from in these experiments. The four lines drawn from the hemisphere of space down to the MLD show the correspondence between the cell location and receptive field position in the left–right

Mammalian Vision and the Cerebral Cortex

In the lower vertebrates, as we have seen, the major retinal projection goes to the tectum, although numerous other brain regions also receive optic nerve axons. As was mentioned, different visual areas of amphibian brains serve somewhat different functions—prey recognition by the tectum, recognition of predators and of stationary boundaries by cells in the thalamus. As with these amphibians, mammals also show some functional segregation of different visual tasks into different brain areas. For instance, after ablation of the superior colliculus (the mammalian optic tectum), hamsters fail to make head orientations toward small food items presented within the visual field. However, these same individuals learn as well as normal hamsters to discriminate between different stationary visual patterns. If, instead of the colliculus, the visual cortex is ablated, the reciprocal deficits are found; these hamsters orient normally toward food items presented to them but show severely impaired learning of visual pattern discrimination (Schneider, 1969; Mort *et al.*, 1980). In spite of this dichotomy of visual functions, however, many mammals have a pronounced neuronal pathway from visual cortex to superior colliculus—the corticocollicular pathway. Disruption of this pathway can impair normal physiological responses from the colliculus (Schiller *et al.*, 1974). Thus, the separation of cortical from collicular function is by no means complete in all mammals.

The evidence just cited for hamsters, as well as numerous studies on other mammals (Hubel and Wiesel, 1962; Hubel, 1979) indicate that the mammalian cerebral cortex plays a key role in the analysis of pattern vision. This brain region is therefore an obvious candidate within which to search for visual recognition neurons. However, before describing this search, we will make a brief digression to consider the structure and evolution of the cerebral cortex—that region of the brain that most clearly distinguishes mammals, and in particular, the human species, from all other vertebrates.

EVOLUTION AND STRUCTURE OF THE CEREBRAL CORTEX

In the early stages of vertebrate evolution, as in early embryological development of most modern vertebrates, one can distinguish three major brain regions (Figure 20)—the FOREBRAIN (or PROSENCEPHALON), the MIDBRAIN (or MESENCEPHALON), and the HINDBRAIN (or RHOMBENCEPHALON). As the embryo develops, the forebrain subdivides into an anterior TELENCEPHALON and a more posterior DIENCEPHALON. The telencephalon includes the bilaterally paired cerebral hemispheres, as well as the olfactory bulbs. The diencephalon remains unitary rather than bilaterally paired in its gross structure. It contains the thalamus and the hypothalamus, some of whose functions will be discussed in Chapter 8. The midbrain, or mesencephalon, also persists as a unitary structure. It contains, among other subdivisions, the tectum (or, in mammals, colliculus). The hindbrain becomes subdivided into a more anterior METENCEPHALON and a posterior MYELENCEPHALON. These have such vital roles as the control of respiration, heart beat, and eye movements.

It is the expansion of the telencephalic cerebral hemispheres that epitomizes

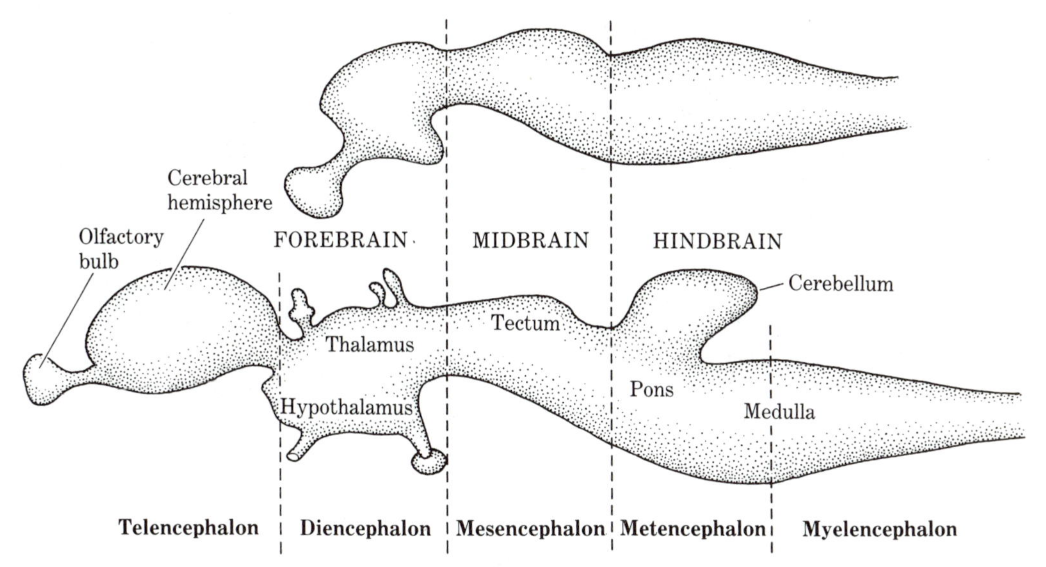

FIGURE 20. Development of the mammalian brain. Lateral views of an embryonic *(top)* and adult brain of a phyletically early mammal. The major brain divisions are labeled. (After Romer, 1962.)

the evolution of the vertebrate brain. In mammalian evolution, it is the CEREBRAL CORTEX in particular whose expansion totally outstrips all other areas (Figure 21). A cortex in the brain is a thin outer rind of gray matter (ranging from about 1 to 5 millimeters in thickness) that covers either the cerebral hemispheres or certain other brain regions. Cells within a cortex are arranged in several layers, as are cells of the tectum. However, unlike the tectum, with its visually responsive neurons overlying neurons for the other sensory modalities, the cerebral cortex has no such overlapping of different sensory modalities; rather, all cells throughout all the layers, in any given piece of cortex, are generally devoted to processing the same type of information. Anatomically, one finds that the cells of a cortex generally have vertically oriented dendrites that synapse upon cells of different cortical layers. By contrast, the lateral spread of these dendrites is rather limited (Chow and Leiman, 1970; Nauta and Karten, 1970; Shepherd, 1974).

Even in nonmammalian vertebrates, although the cerebral hemispheres are not cloaked by a cortex, small regions of simplified telencephalic cortex can be found. These regions have a smooth surface rather than the corrugated texture characteristic of the cerebral cortex in most mammals; and they possess a limited number of geometrical cell types and of cell layers. The earliest telencephalic cortex to evolve was the olfactory cortex, receiving input from the olfactory bulbs. Owing to its primitive origin, olfactory cortex is often called PALEOCORTEX. It is especially prominent in the telencephalons of phyletically older vertebrates. The second telencephalic cortex to evolve was that of the hippocampus, a region deep within the forebrain, whose several functions are currently much debated (Olton *et al.*, 1979). The hippocampal cortex is often called ARCHICORTEX. Also

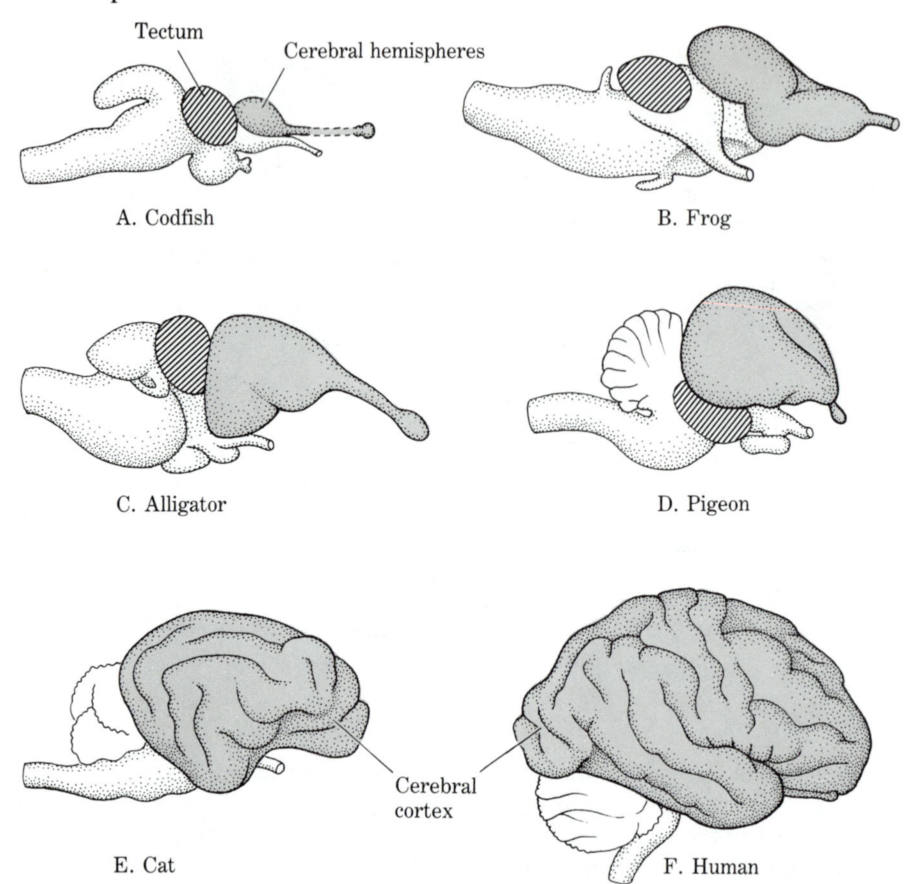

FIGURE 21. Lateral view of the brains of a series of vertebrate animals. All are drawn as though of roughly equal size. The cerebral hemispheres, cloaked by the corrugated cerebral cortex in mammals (E and F), became progressively larger through vertebrate evolution. By contrast, the tectum maintained a fairly constant size, relative to the whole brain, throughout evolution. It is covered by the cerebral hemispheres in E and F. (After Nauta and Karten, 1970.)

present on the lateral surface of the telencephalon of early mammals is a GENERAL CORTEX of simple design (Nauta and Karten, 1970).

As mammals evolved, paleocortex and archicortex remained, but general cortex was lost, being replaced by a new, corrugated form called NEOCORTEX (Nauta and Karten, 1970). It is this tissue that forms practically all of the cerebral cortex of higher mammals. It has up to six cell layers, and its cells make highly complex patterns of synaptic connections, both within the cortex itself and with other brain regions. The neocortex of higher mammals has so expanded that most of the remainder of the brain is hidden beneath it (Figure 21).

Within the mammalian cerebral cortex there are three PRIMARY SENSORY PROJECTION AREAS (Figure 22)—regions where ascending tracts of axons of in-

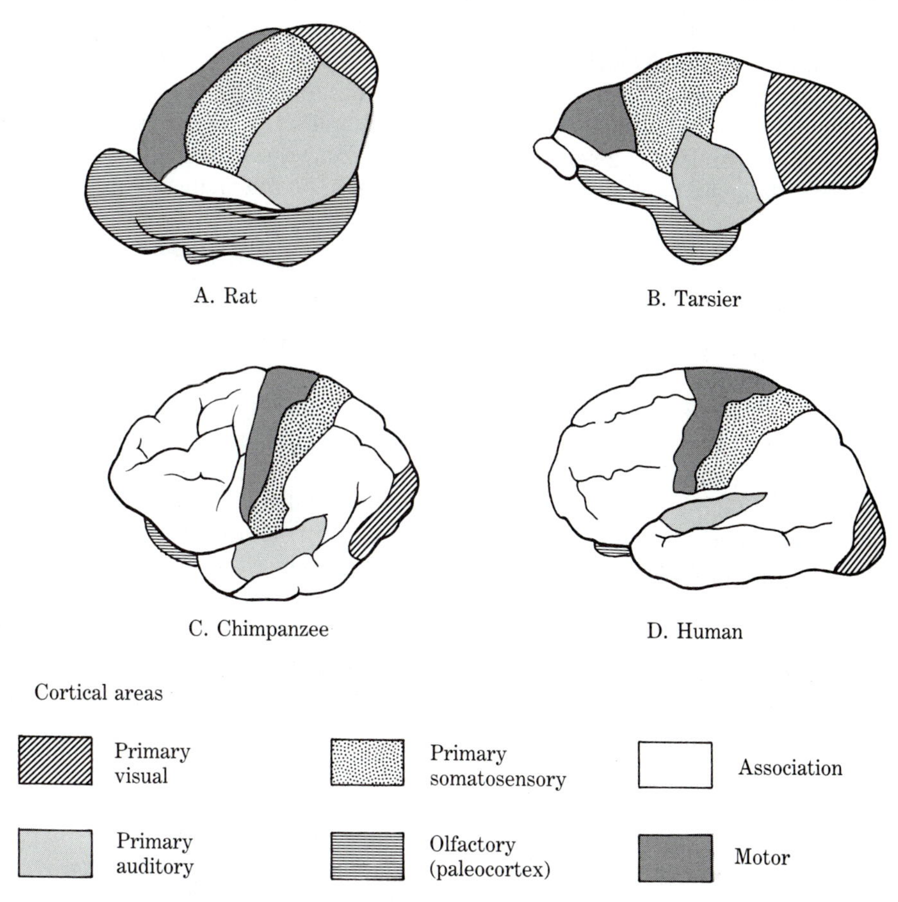

FIGURE 22. Functional subdivisions of the cerebral cortex in four animals of different phyletic ages, from rat through human.

dividual sensory modalities terminate. One of these, located most posteriorally, is the primary visual cortex, mentioned in the first section of this chapter. This region is also called STRIATE CORTEX (so named because of a clear striation formed by the ascending tract of axons), or area 17, or visual area I. The other two primary projection areas are the primary auditory cortex and the primary somatosensory cortex. In addition, there is a region of MOTOR CORTEX, whose axons descend toward motor centers and are involved in the production of movement.

Aside from these areas, there are other cortical regions, especially prominent in higher mammals, that receive only limited ascending input from noncortical structures; rather, they receive their main input through axons entering from one or more of the nearby primary sensory areas (Van Essen, 1979). These axons that run from one cortical region to another are called ASSOCIATION AXONS, and the cortical regions to which they carry information are called ASSOCIATION CORTEX. Thus, unlike the tectum, in which different sensory modalities are represented in layers overlying one another in topographic congruence, intermodality interactions within the cortex require fairly long axons from one cortical region to another. In the evolution of higher mammals, association cortex has expanded far more than any other brain region, and it occupies 80% of the human cerebral cortex (Figure 22). This expansion of association cortex, which carries out higher order processing of the sensory information from the primary projection areas, reflects the apparently increasing sophistication of neural analysis within the brains of some phyletically more recent mammals.

SKETCH OF THE MAMMALIAN VISUAL PATHWAY

Based upon the foregoing outline of mammalian neuroanatomy, we can now trace the major visual pathways within the mammalian brain. As in lower vertebrates, the mammalian optic nerve projects directly to the optic tectum, or superior colliculus. But in mammals, the optic nerve sends a much more massive projection to the thalamic LATERAL GENICULATE NUCLEUS, or LGN (Figure 23). Axons of LGN cells project onward as a tract, called the OPTIC RADIATION, to the striate cortex. Association axons from cells in striate cortex then project to neighboring association cortex. The neurons at each of the sequential locations—retina, LGN, striate cortex, and association cortex—carry out a further analysis of the visual information they receive. However, the visual pathway is not strictly linear because there are reciprocal connections between visual cortex and subcortical centers such as the superior colliculus, as well as many other complex connections (Van Essen, 1979).

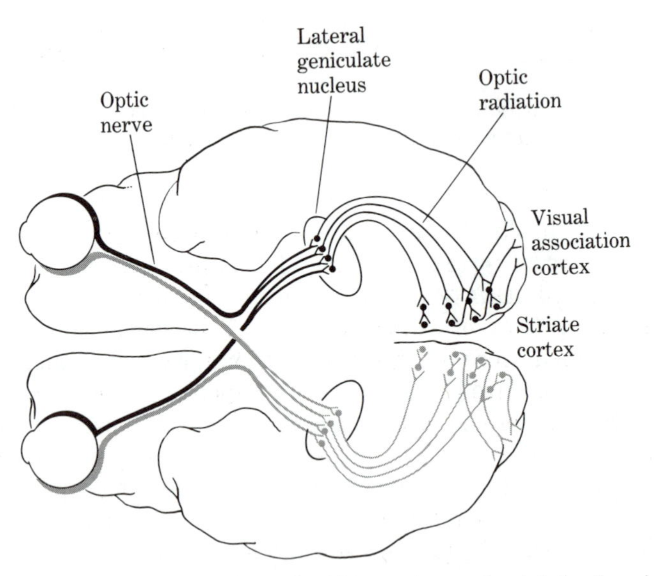

FIGURE 23. Outline of the major visual pathways in a primate's brain, viewed from below. This scheme is greatly oversimplified, omitting several brain nuclei that have involvement in visual processing. (After Kuffler and Nicholls, 1976.)

As is the case with the tectum, the LGN, striate cortex, and visual association cortex are retinotopically mapped. In fact, in some primates there are at least eight separate such retinotopic maps plotted onto different small regions of association cortex (Van Essen, 1979). Also, the primary somatosensory cortex is somototopically mapped. This map is congruent with a motor map in the adjacent motor cortex. That is, tactile stimulation of a given area of the body excites cells in a particular region of the somatosensory cortex; and just adjacent to these cells, in the motor cortex, are found neurons that control movements of this same area of the body.

DOES THE MAMMALIAN VISUAL SYSTEM CONTAIN RECOGNITION NEURONS?

In the beginning of this chapter, we discussed feature-detecting neurons of the mammalian visual cortex that are sensitive to edges or stripes of light or dark, oriented at particular angles (Figure 1D). These cells, which are located in striate cortex and immediately adjacent visual association cortex, have very small receptive fields. In fact, in the area of frontal vision, most receptive fields are smaller than 1° (Hubel and Wiesel, 1974). As we have seen, because of both this small size and the specificity for only very simple edge or stripe stimuli, these neurons do not exhibit the properties of recognition neurons but rather of feature detectors, as defined in this chapter. These feature detectors are probably important in the process of object recognition and could provide the input to any recognition neurons that may lie in later processing stages of the visual system.

Association axons from the striate cortex project to several areas of association cortex. It is in these areas that the search for recognition neurons has been carried on. One such region studied in monkeys is the inferotemporal cortex (Gross *et al.*, 1972, 1974). This region receives input not directly from striate cortex but rather indirectly from visual association areas surrounding it, as well as from other brain regions.

Cells of the inferotemporal cortex respond only to visual stimuli. Their receptive fields all include the frontal region of space. But unlike cells of the primary visual cortex, these receptive fields are fairly large, with diameters averaging about 20°. Many of these cells respond strongly to simple edge or stripe stimuli. However, a limited number of them respond especially vigorously to highly specific, complex stimuli. For instance, one such cell that was recorded responded apparently only to the sight of a monkey's hand. Although tests for stimulus specificity were not carried out quantitatively in these exploratory studies, more recent work has confirmed the presence of at least a few cells that respond only to the sight of a hand. These more recent studies were carried out in a region of temporal cortex (the region called the superior temporal sulcus) that received input from the inferotemporal cortex (Perrett *et al.*, 1982). The responses of these cells show invariance with angle of rotation of the hand and with different viewing distances (and therefore different image sizes). There were no responses to partial or distorted models of hands.

In the same region of the temporal cortex in monkeys, numerous cells have been found that appear specialized for responding to faces (Perrett *et al.*, 1982).

Between 50 and 100 of the 300 visually responsive neurons tested in this area gave significantly stronger responses to the sight of a monkey or human face than to any of a large number of other objects tested. The responses of many of these cells were invariant with changes in rotation, color, size, and viewing distance. Of these cells, some responded almost as strongly when the monkey viewed just the eyes, and other cells almost as strongly when he viewed just the mouth, as when viewing the whole face. Some other cells responded somewhat to each of several separate facial features, so that the slightly greater response to the whole face represented stimulus summation. A few cells, however, responded hardly at all to the sight of an eye, hair, mouth, or other complex facial components, but did respond strongly to the entire face. Though further tests remain to be carried out on this last group of neurons, these may be examples of gestalt encoders of faces.

These temporal cortex cells, then, are good candidates for facial recognition neurons. In this regard, it is interesting that local brain damage to this area in human patients suffering accidents can sometimes damage the recognition of faces without noticeably affecting other aspects of perception.

Are There Recognition Neurons in the Auditory System?

The search for recognition neurons has not been limited to the visual system. Because many animal species exhibit rich vocal communication, their auditory systems can be explored for cells each of which might respond specifically to just one biologically meaningful vocalization. Such a recognition neuron for a given communication sound, if it exists, would be termed a CALL DETECTOR.

A schematic drawing of the major components of the mammalian auditory system is shown in Figure 24, together with the major visual pathways that run somewhat parallel to it. The right auditory nerve terminates, within the right side of the hindbrain, in the COCHLEAR NUCLEUS. Cells from this nucleus project bilaterally to the hindbrain's left and right SUPERIOR OLIVARY NUCLEI, and cells in all three of these nuclei send their axons to the INFERIOR COLLICULUS. Projections from this midbrain nucleus go to the MEDIAL GENICULATE NUCLEUS (MGN) on the left side of the thalamus. Left MGN cells project to the left primary auditory cortex. Just as visual processing in the brain was carried out in lower vertebrates primarily by the midbrain and in mammals more so by the forebrain, likewise auditory processing was at first handled mainly by the hindbrain and through evolution came to be dominated by the forebrain (Nauta and Karten, 1970). Nevertheless, in all vertebrates including mammals, much of the primitive posterior circuitry remains, such that the optic nerve still enters the midbrain and the auditory nerve still enters the hindbrain. This evolutionary trend, whereby more anterior brain regions come to have a prominent integrative role, is called CEPHALIZATION.

Most explorations in search of call detector cells have been carried out within the primary auditory cortex. For instance, in squirrel monkeys recordings were made in this area in response to species-specific vocalizations, as well as to definable

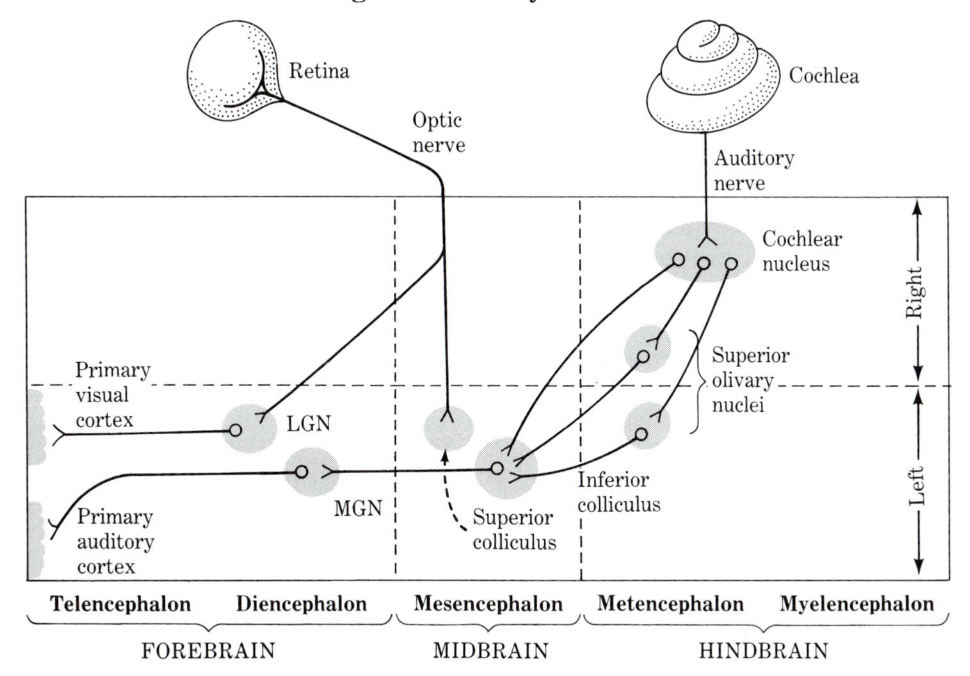

FIGURE 24. Schematic representation of the major auditory pathways and visual pathways of a mammal. Inputs only from the right ear and eye are shown. LGN, lateral geniculate nucleus; MGN, medial geniculate nucleus.

subfragments of these vocalizations. (Such a subfragment would be roughly comparable to a syllable of human speech, and a vocalization to a word or phrase.) Neurons were found that were selectively responsive to particular subfragments; but none were found that were selective for whole vocalizations (Symmes *et al.*, 1979). Thus, the primary auditory cortex of this animal appears functionally analogous to the primary visual cortex, with cells acting as feature detectors, selective for small components of an overall stimulus but not for the entire stimulus. In view of the recent findings of hand and face detectors in association cortex, the auditory association areas should now be explored more fully, looking for call detector cells.

Some workers recording from primary auditory and association cortex of the squirrel monkey have found cells that respond selectively not to one particular vocalization but to some group of vocalizations out of several that were tested. Surprisingly, many of these cells appear, over time, to change spontaneously their selectivity, becoming responsive to a *different* group of vocalizations (Manley and Mueller-Preuss, 1978; Glass and Wollberg, 1979). This observation raises serious difficulties for an understanding of auditory recognition because almost any theory of recognition, including those described at the beginning of this chapter, requires that any given cell used to recognize a stimulus must respond in a consistent manner. It will be important now to test visual and other neurons for the temporal permanence of their response specificities.

In bats, neurons have been found in the auditory cortex that respond selectively to specific components of an echolocating pulse, followed after definite intervals, by specific components of the echo (Suga *et al.*, 1978; O'Neill and Suga, 1979, 1982). These neurons, which are selective for fairly specific stimulus properties and yet are not used for a complete identification of a target, are again roughly comparable to feature detectors in the mammalian visual system.

In long CF–FM bats, the CF and the FM components are analyzed by different areas of the cerebral cortex. As we have seen in Chapter 6, the CF component of the pulse and echo permits the bat to determine several important target parameters. Feature detectors for some of these parameters have been found in the CF processing area of the cortex. The FM component of the pulse and echo enable the bat to determine target distance; feature detectors for this parameter have been found in the FM processing area of the cortex. This separation of the cortex into two distinct areas that process different aspects of pulse and echo information provides a particularly clear example of PARALLEL PROCESSING in the brain.

Is There a Future for the Recognition Neuron Theory?

If toads have recognition neurons for prey and if monkeys perhaps have such neurons for faces, should we guess that animals may identify most objects by means of such recognition neurons? This would seem to be a bad guess. Prey items, faces, and a handful of other critically important stimulus objects may merit individual coding by such highly specialized neurons as these. But it seems unreasonable to expect the brain of a monkey, ape, or human, which perhaps can recognize thousands to millions of different objects, to have one or more special neurons for each of these objects. Rather, what seems to be called for is a combination of compactness and flexibility, such as that offered by the theory of a spatiotemporal pattern recognizer, as discussed earlier in this chapter. Searching for and analyzing such spatiotemporal pattern recognizers (or other meticellular forms of coding that could exist) may be very difficult because one would need to make simultaneous recordings from the several cells that would constitute such a recognizer. And such cells need not be neatly clustered together but could be scattered among neurons subserving other functions. For these reasons, studies on the neural basis of object recognition are likely to become a good deal more difficult as research proceeds.

Beyond the problem of neurally encoding a given stimulus object that needs to be recognized for behavior, there remains the question, for the human brain and perhaps for other brains, of how *conscious* recognition comes about. That is, how do you not only act behaviorally as though you see your friend, but know *subjectively* that it is your friend that you are seeing? There has been some progress in determining *where* in the brain the information is encoded that is used by our consciousness-producing mechanism (Sperry, 1975). But there has been no real progress in understanding the mechanism itself. One possible future source of insight on this problem lies in direct communication between experimenters and their animal subjects (Griffin, 1981). Chimpanzees, for instance, can be taught to communicate with humans by means of sign language or by pressing the keys on a computer terminal (Gardner and Gardner, 1975; Ristau and Robbins, 1981). If these animals are indeed capable of subjective experience (which we do not know because we have not asked them, and it is not yet clear just how one asks such a thing of a chimp), it may be that by using a combination of communicating with and carrying out physiological studies upon chimps or other animals one can in the future obtain some useful insights. But behind all such thoughts lurks a major uncertainty: Are there perhaps some brain processes that are so complex as to be not amenable to scientific understanding? Whether our brains will indeed be capable of understanding themselves is one of the major scientific and philosophical questions of the present age.

Summary

Among the most useful theories of object recognition are (1) the theory of lists of feature-detecting cells; (2) the theory of recognition neurons; and (3) the theory of a spatiotemporal pattern recognizer. Behavioral tests of prey recognition in toads show that gestalt, or configurational, stimulus properties are used in recognition. A search for recognition neurons underlying this process has focused upon neurons called T5(2) cells, found in the toad's tectum. Though the evidence is incomplete, these cells may be both necessary and sufficient for visual prey recognition. In was also found that the tectum contains a retinotopic map that is used to localize visually detected objects in space. The tectum of most vertebrates is used similarly for localization of visually detected moving objects.

Most visually responsive neurons in vertebrates have limited receptive fields. Some cells, such as retinal ganglion cells, have receptive fields subdivided into antagonistic excitatory and inhibitory center and surround regions. Cells in the mammalian striate cortex show straight-edged borders between excitatory and inhibitory regions of their receptive fields. A few regions of visual association cortex in monkeys appear to contain neurons that respond selectively to highly specific visual images, such as faces.

In the auditory system, there is less evidence pointing toward recognition neurons such as hypothetical call detectors. However, there is good evidence for spatial maps, such as that in the midbrain of the barn owl. The cells of the map have antagonistic center–surround receptive fields like those found for retinal ganglion cells. However, the mechanisms for deriving auditory receptive fields and spatial maps in the auditory system are fundamentally different from those in the visual system. Auditory receptive fields and maps are derived by a neuronal calculation of intensity and time differences between the signal detected at each ear rather than by an ordered projection from a topographically organized sensory surface. The ability of some animals, such as primates, to communicate with experimenters through sign language may offer a future means to gain insight into the perceptual processes of these animals.

Questions for Thought and Discussion

1. Devise at least one coherent theory of object recognition by the visual system that is different from any presented in this chapter.
2. In what ways do the responses to the stimuli of a toad's T5(2) cells (Figure 6A, B) differ from the responses to these same stimuli by cells of the primary visual cortex in a cat or a monkey?
3. Suppose that it had been shown that the neurotransmitter synthesized and liberated by the T5(2) neurons in a toad's tectum is different from the neurotransmitter of any other neuron in the brain. (This is not actually the case.) How might you be able to use such a finding in carrying out studies to test whether the T5(2) cells are recognition neurons for prey? Outline your experimental program, complete with controls.
4. Given that the neurons of a mammal's primary visual cortex respond optimally to straight edges or stripes, how might such an animal's visual system encode curved shapes? Specifically, how does your cortex encode the sight of your dinner plate?
5. The existence of a retinotopic map in the tectum implies that during development individual optic nerve axons, growing into the tectum, somehow find their proper tectal termination points. Such target localization by a growing axon resembles sensory localization of an object by an entire animal that orients toward that object by virtue of some sensory cue. List the possible sensory cues by which the direction of the axonal growth into the tectum might be influenced, during the establishment of the retinotopic map. For each cue on the list, outline an experiment that could help confirm or refute whether this cue is actually used by the growing optic nerve axons.
6. Give examples of conservation and novelty in the evolution of the mammalian visual and auditory systems. Suggest what advantages there may have been for the evolution of new brain structures, as opposed to the redesign of old structures, in the phyletic history of sensory systems.
7. Design an experiment that could provide useful information on the mechanisms by which conscious experience is generated. If you think that no such experiment is possible, even in principle, state explicitly why.

Recommended Readings

BOOKS

Haber, R.N. and Hershenson, M. (1980) *The Psychology of Visual Perception.* Holt, Rinehart & Winston, New York.
This book outlines well several theories of object perception in human vision and covers many related topics.

Yost, W.A. and Nielsen, D.W. (1977) *Fundamentals of Hearing: An Introduction.* Holt, Rinehart & Winston, New York.
An excellent elementary treatment of human auditory processing, including material on the recognition and localization of sound sources.

Tinbergen, N. (1951) *The Study of Instinct.* Oxford University Press, New York.
Chapter 2 of this classic book summarizes much of the early ethological work on the visual recognition of biologically meaningful stimuli.

Ewert, J.P. (1980) *Neuroethology.* Springer-Verlag, New York.
This book contains a fairly complete review of Ewert's work on object recognition by toads.

Ingle, D. and Sprague, J.M. (eds). (1975) *Sensorimotor Function of the Midbrain Tectum. Neurosci. Res. Prog. Bul.* 13.
This compendium summarizes experiments and ideas from many leading researchers on tectal structure and function.

Shepherd, G.M. (1979) *The Synaptic Organization of the Brain,* 2nd Edition. Oxford University Press, New York.
This excellent introductory book presents the anatomy of the brain in terms of physiological function and cellular composition.

ARTICLES

Ingle, D. (1973) Two visual systems in the frog. *Science* 181: 1053–1055.
An excellent experimental report demonstrating functional compartmentalization of visual processing centers in the frog's brain.

Werblin, F.S. and Dowling, J.E. (1969) Organization of the retina of the mudpuppy, *Necturus maculosus.* II. Intracellular recording. *J. Neurophysiol.* 32: 339–355.
This classic article on the physiological organization of the retina remains well worth a careful reading.

Hubel, D.H. (1979) The visual cortex of normal and deprived monkeys. *Am. Sci.* 67: 532–543.
This article, by a 1981 Nobel laureate, summarizes much recent work on primate visual cortex.

van Essen, D.C. (1979) Visual areas of the mammalian cerebral cortex. *Annu. Rev. Neurosci.* 2: 227–263.
This excellent review presents recent findings on multiple visual processing areas in monkey cortex.

Nauta, W.J.H. and Karten, H.J. (1970) The general profile of the vertebrate brain, with sidelights on the ancestry of cerebral cortex. In *The Neurosciences: Second Study Program.* F.O. Schmidt (ed.). The Rockefeller University Press, New York.
The evolution of the cerebral cortex in the context of the overall structure of the brain is presented.

Knudsen, E.I. (1981) The hearing of the barn owl. *Sci. Am.* 245(6): 113–125.
An excellent treatment of the neuroethology of prey capture guided by auditory cues in owls.

PART III ORGANIZATION FOR ACTION

From many directions workers are tunneling hopefully into the mountain [of the brain], some with steam shovels and others with dental drills. Some travel blindly in a circle and come out close to their point of entrance; some connect, usually in a mismatched fashion, with the burrows of others; some have chosen to disregard the random activities of their fellows and have worked out in a small region an elegant system of interconnecting tunnels of their own.

(K. Roeder, 1963)

In the preceding three chapters, we have dealt with the acquisition and analysis by the nervous system of information about the outside world. Such information is obviously essential for eliciting behavioral acts that are well matched to an animal's surroundings. If this encoded sensory information is to elicit behavior, it must be passed on to those neurons responsible for bringing about coordinated movements. However, rather little is known about this sensory–motor interaction within the nervous system. In general, researchers studying sensory inputs and those studying motor outputs have worked as two largely independent groups, which have yet to meet up with one another in their excavations of the brain. However, in a few simpler nervous systems of invertebrates, this sensory–motor linkup has been successfully achieved. For these few examples, it has been possible to trace through, all the way from input to output, many of the individual neurons involved, their physiological responses to natural stimuli, and their synaptic interactions. This permits an in-depth analysis of the role played in behavior by specific neurons or groups of neurons within the central nervous system.

Interposed between a sensory input and a motor response, there can be a number of specific transformations of neuronal information. The cellular bases of some of these are explored in the three chapters of this section. Chapter 8 deals with the "decision" by an animal's nervous system whether or not to respond to a given sensory stimulus by producing a behavioral act. This chapter also discusses the mechanisms by which different parts of the body are integrated into a coordinated movement. Chapter 9 turns to the question of how the nervous system produces sequentially patterned behaviors. Finally, Chapter 10 considers feedback in behavior and the nervous system.

CHAPTER 8

Decisions and Commands for Behavioral Acts

In Chapter 2, fixed action patterns were defined partly in terms of their all-or-nothing property. That is, as the intensity of a releasing stimulus increases up to some critical threshold, little or no behavior is evoked; but for an intensity just at this threshold or higher, the behavior is usually evoked in full. This all-or-nothing property is one feature that distinguishes fixed action patterns from reflexes, which, instead, show a much more gradual increase in response strength with graded increases in stimulus intensity (Figure 1).

The all-or-nothing property of a fixed action pattern can be thought of as reflecting a *decision* carried out by the nervous system to execute the behavior. It should not be inferred, however, that there is necessarily anything conscious about this decision making; rather it can be regarded as a switchlike event that at some moment suddenly permits the activation of the motor circuits for a given

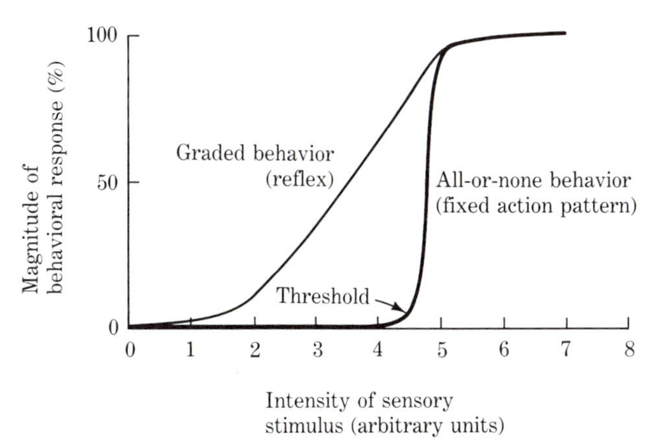

FIGURE 1. Stimulus–response graph of two hypothetical behaviors; an all-or-none response (fixed action pattern) and a graded response (reflex). (After Kandel, 1976.)

behavior. For most fixed action patterns, this switch is contained within the central nervous system because neither the sensory receptors nor the neuromuscular apparatus show the necessary all-or-nothing property.

The central neuronal switch for a given all-or-nothing behavior can be thought of as the watershed for the outflow of neural information that directs this particular behavior. The neural information that emanates from such a central decision switch and that activates a particular motor circuit is usually called a COMMAND. In terms of the overall organization of behavior, then, there must be many different central decision switches, each discharging a command for a different all-or-nothing behavior. This leads us to three questions concerning the control of fixed action patterns, which form the substance of this chapter: (1) What is the overall organization within the nervous system of the numerous decision switches and commands for different behaviors? (2) What is the cellular mechanism underlying a decision to perform a given behavioral act? (3) What are the cellular mechanisms by which commands direct properly coordinated sets of movements? This chapter will not consider the neural organization of reflexes, a topic that is discussed in Chapter 10.

Hierarchical Organization of Motor Systems

We begin by taking an overall view of the organization of decisions and commands in an animal's behavior. A conceptual tool that will be useful in this discussion is the idea of a HIERARCHY—a chain of command in which a highest authority instructs an intermediate authority, who in turn instructs a lower authority, and so on down to the lowest individual who obediently performs the required act. An example can be found in the army, where a given general commands his colonels, each of whom passes this information on to his majors, each of these to his captains, and so forth (Figure 2). A characteristic of a military hierarchy is that any individual is immediately responsible to only one person at the next higher level. For instance, a private has just one sergeant.

The sequence of neuronal events involved in the execution of some behavioral acts can also be represented as a hierarchy. Thus, one can speak of units of neural organization at higher, intermediate, and lower levels of organization. By this scheme, the neurons carrying out the switchlike decision to perform a given act would represent a unit of organization at a high level. Commands from these decision units may activate intermediate units (circuits involved in programming particular behaviors), which in turn activate the lowest units of organization (the motor neurons).

Such a scheme is illustrated in Figure 3, which shows the hierarchical control of the rhythmic movements of a single leg in the cockroach. The cockroach makes

FIGURE 3. Hierarchical organization of behaviors involving one leg of a cockroach. Both divergence and convergence of information flow are included. Explanation in text.

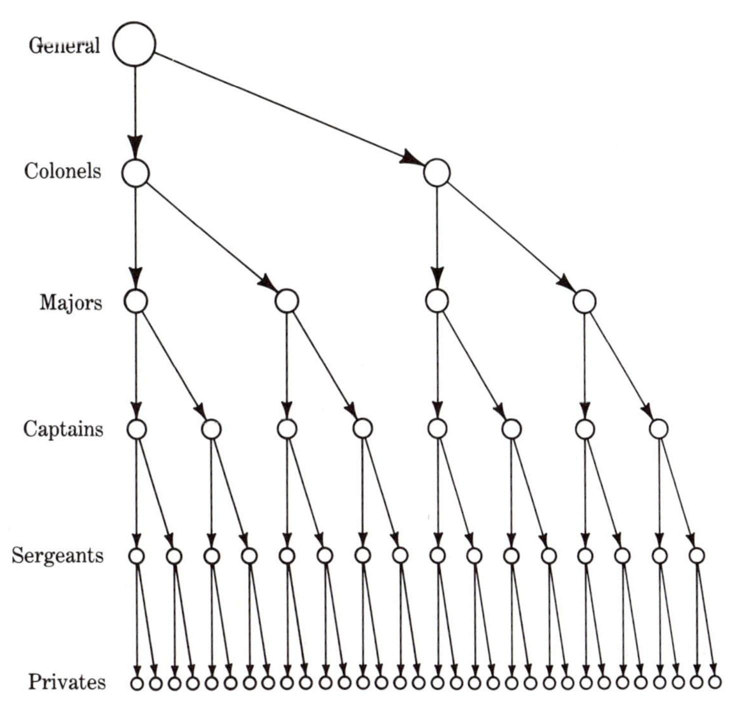

FIGURE 2. A military hierarchy, showing divergence but no convergence in the flow of information. That is, each person's major responsibility is to just one individual at the next higher level of command.

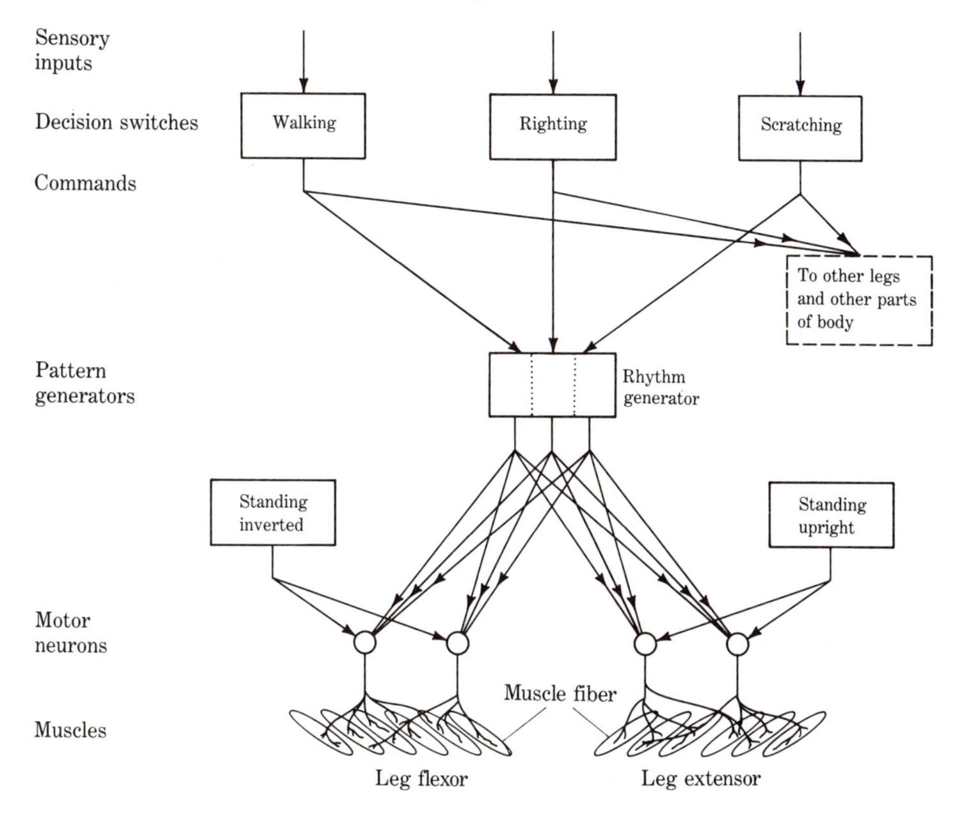

such rhythmic leg movements as a part of three different behaviors: walking; righting itself after having been turned on its back; and scratching a part of its body that has been irritated (Delcomyn, 1971; Pearson and Fourtner, 1975; Camhi, 1977; Reingold and Camhi, 1977; Sherman *et al.*, 1977). Different sensory stimuli elicit each of these behaviors, at least two of which (righting and scratching) occur as all-or-nothing acts. Therefore, these two, and possibly all three, behaviors must each have a separate central decision switch to turn it on (Camhi, 1977; Sherman *et al.*, 1977). From each of these switches there must emanate a command to turn on the rhythm for the leg's movement. (In fact, because the leg movements are slightly different for each of the three behaviors, the rhythm generator must produce three different varieties of output signal; alternatively, there could be three different rhythm generators, as indicated by the dotted lines in Figure 3.) The outputs from the rhythm generator must go to two groups of motor neurons—those to muscles that flex the leg (leg flexor motor neurons) and those to the opposite muscles (leg extensor motor neurons). As the figure shows, these motor neurons receive input from other sources as well. For instance, when a cockroach is standing upright, it resists the force of gravity by exciting the leg extensor motor neurons and thus contracting the leg extensor muscles. If, instead, the cockroach stands inverted on a ceiling, it resists gravity's pull by exciting the leg flexor motor neurons (Pearson, 1972).

In summary, Figure 3 shows that the overall control of leg movements in the cockroach can be viewed as a hierarchy, in which higher units of organization (decisions for walking, righting, and scratching) send out commands that control intermediate units (rhythm generators), which in turn control still lower units (the motor neurons). However, unlike a military hierarchy in which any individual has just one immediate superior, a unit of organization in this neurobehavioral hierarchy can be controlled by more than one unit at the next higher level. For instance, information from both antigravity neurons and the rhythm generator for all three leg rhythms converge to influence the flexor motor neurons. Thus, whereas the army exhibits only DIVERGENCE of information flow (Figure 2), nervous systems exhibit both divergence and CONVERGENCE (Figure 3). This situation gives rise to a latticelike crossing of diagonal lines on hierarchical diagrams such as that of Figure 3. Because of this, such behavioral hierarchies have been aptly referred to as *lattice hierarchies*, as distinct from *partition hierarchies* like that of the army (Gallistel, 1980).

One limitation of this hierarchical representation of behavior is that it implies that the flow of information is strictly unidirectional, from higher to lower levels of organization. This is rarely strictly the case because feedback pathways usually carry information upward in a hierarchy as well. We will defer the discussion of feedback, however, until Chapter 10.

The Machinery of Motor Systems

Before proceeding to a consideration of the mechanisms of decisions and commands, it will be useful to consider briefly the organization of muscles and motor neurons, which constitute the output machinery upon which central circuits act. As is shown in Figure 3, a given muscle fiber (that is, a muscle cell) of an invertebrate is generally innervated by more than one motor neuron. Usually these different motor neurons produce different effects upon the muscle fiber (Atwood, 1977; Usherwood, 1977). For instance, some motor neurons produce only tiny excitatory junctional potentials (EJPs) that must temporally summate in order to result in contraction. Others produce large EJPs that require little or no summation. Still others are inhibitory, producing inhibitory junctional potentials (IJPs). Some muscle fibers of invertebrates are unable to generate action potentials and contract only slowly and in a graded manner in response to different amounts of membrane depolarization. These are called SLOW, or TONIC, FIBERS. Others, in response to sufficiently large EJPs, produce action potentials, each of which is followed by a twitch contraction. These are called FAST, or PHASIC, or TWITCH FIBERS.

In the vertebrates, nearly all muscle fibers are twitch fibers, though they differ in the speed of contraction. Moreover, each is generally innervated by just one motor neuron, which is always excitatory (Stein, 1980). Thus, a particular vertebrate motor neuron provides the only route by which the central nervous system can influence the contraction of a particular muscle fiber. For this reason, the vertebrate motor neuron is often referred to as the FINAL COMMON PATH of the nervous system (Sherrington, 1906). Such a motor neuron generally branches to excite a small group of muscle fibers that constitute its private domain. This vertebrate motor neuron, together with its own set of muscle fibers, is called a MOTOR UNIT. A motor unit, then, is the lowest level of hierarchical organization in the vertebrate nervous system. Within a vertebrate motor unit, in general, a single action potential in the motor neuron evokes a large EJP, followed by an action potential and then a twitch in each of its muscle fibers. Larger contractions of these same muscle fibers can be evoked only by a rapid succession of action potentials such that each twitch begins before the previous twitch has fully terminated; this results in an addition of mechanical force. Larger contractions of the muscle as a whole can then occur only by activating additional motor units, a process called RECRUITMENT.

Motor neurons that activate fibers of the same muscle (or of neighboring muscles that produce very similar movements) are often excited together by a common set of presynaptic neurons. Such a group of motor neurons of common function and synchronous action is called a MOTOR NEURON POOL. The co-working muscles that they innervate are called SYNERGISTS. When one group of synergists is activated, muscles that work in mechanical opposition to these, called their ANTAGONISTS, are generally relaxed. This relaxation usually involves inhibition by the central nervous system of the motor neurons to these antagonists. The situation is symmetrical; when the second muscle group contracts, the first group is relaxed through inhibition of its motor neurons. This excitation of synergists occurring conjointly with inhibition of their antagonists is called RECIPROCAL INNERVATION. Such an arrangement assures a "single-minded" action, so that at any moment antagonistic muscles do not mechanically oppose one another. Not only individual muscular contractions, but also the behavior of an animal as a whole is generally characterized by such "single-minded" action; entire behavioral patterns usually occur not simultaneously with others but rather one at a time. As we shall see later in this chapter, a pattern of synaptic inhibition resembling reciprocal innervation can be important in producing this single-minded action.

The Escape Behavior of the Crayfish

One of the motor acts that is best understood, in terms of the cellular basis of decisions and commands, is the crayfish escape behavior. If one taps upon or directs a jet of water at a crayfish's abdomen (the posterior region of the body), the animal responds with a rapid abdominal downstroke, or flexion. This causes the whole body to move upward and often somewhat foreward, carrying it away from the initiating stimulus (Figure 4A). Immediately following this initial flexion, there is usually a swimming sequence consisting of alternating abdominal extensions (upstrokes) and flexions that cause continued movement away from the source of stimulation (Wine and Krasne, 1972).

Our discussion will focus upon the initial abdominal flexion, which has been studied most extensively. This flexion has three properties of particular interest (Wine and Krasne, 1982). First, it is all-or-nothing, or decisionlike in character.[1] Second, the movement begins very quickly (usually in less than 20 milliseconds after the stimulus) and involves a synchronous flexion of several abdominal segments. This implies that a command somehow initiates movements that are well coordinated in time and space along the body. Finally, at the onset of the escape behavior, the crayfish appears to cancel other ongoing behaviors that could interfere with a single-minded evasive maneuver. We will now consider in some detail the neural circuit for the initial abdominal flexion in an attempt to understand the neuronal bases of these three behavioral properties.

THE NEURONAL CIRCUIT FOR ESCAPE

The study of the neuronal circuit underlying a given invertebrate behavior often begins with a search for individually recognizable sensory cells, interneurons, and motor neurons that mediate this behavior. Recognizing as individuals the component neurons enables one to carry out repeated experiments on the same cell in different animals and, thus, to verify experimental results through such repetition. It also permits one to discover the contributions to particular behaviors of each of the individually identified cells. The most common experimental strategy makes use of electrophysiological techniques, both to search for any specialized physiological properties that any of the individual cells might display and to de-

[1]Although an abrupt or rapidly accelerating mechanical stimulus evokes this all-or-nothing behavior, more gradual stimuli evoke similar, though less stereotyped, swimming movements, which do not appear to occur in an all-or-nothing manner (Wine and Krasne, 1972). These less stereotyped movements have a largely separate neural mechanism, which will not be discussed.

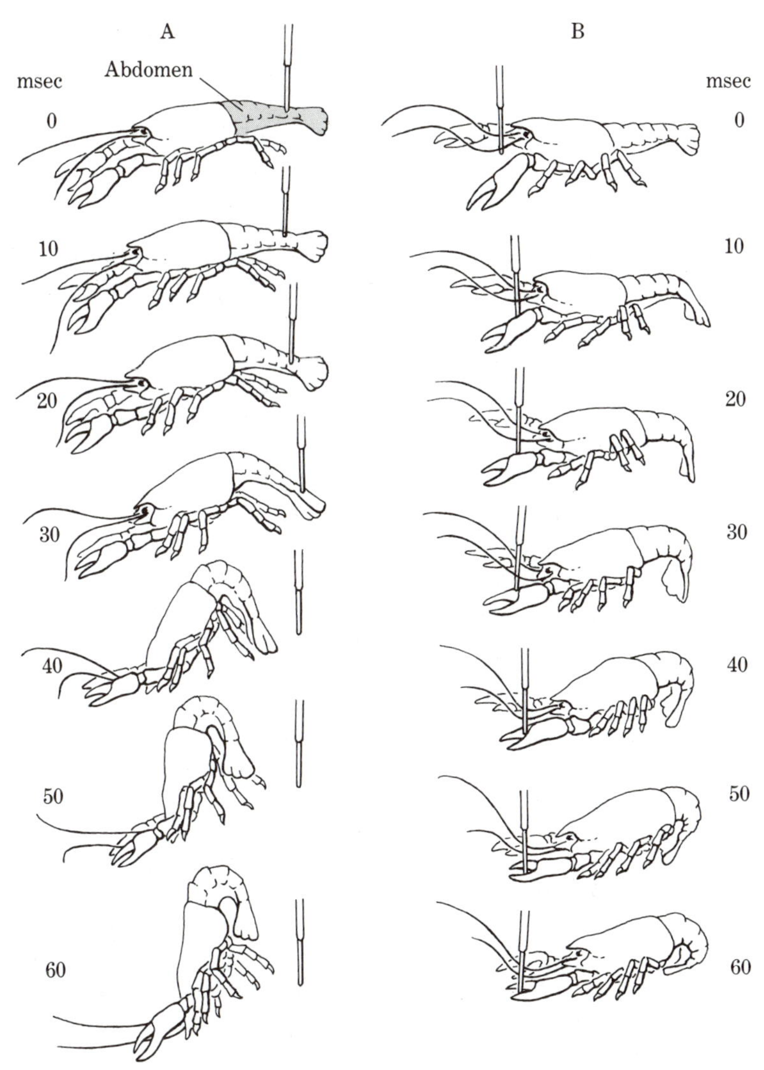

FIGURE 4. Responses of the crayfish *Procambarus clarkii* to tactile stimuli. A. Stimulation of the abdomen (shown in gray in the top left figure) evokes an abdominal flexion that moves the crayfish upward and forward. This response is mediated by the lateral giant interneurons (LGIs). B. Anterior stimulation (in this case, at the antennae) evokes an abdominal flexion of a different form that propels the animal backward. This response is mediated by the medial giant interneurons (MGIs). In both cases the movement is away from the source of stimulation. (After Wine and Krasne, 1972.)

termine which of the cells connect synaptically to which others. One then characterizes each such connection as either excitatory or inhibitory, chemical or electrical, strong or weak, monosynaptic or polysynaptic, as described in Chapter 3. This information then enables one to draw a detailed diagram of the neuronal circuit for this behavior. Much like an engineer's wiring diagram of a radio, this circuit diagram shows each cell and its connections with other cells and describes the properties of each cell and of each connection. Based upon this circuit diagram, one can sometimes identify particular cells, groups of cells, or synaptic connections as likely candidates to account for specific attributes of the behavior. Finally, one can attempt to verify this putative role of a given candidate cell or synapse by experimentally altering the candidate's physiological activity, or by totally silencing it, and determining whether the behavior is thereby altered. Although this program of research is simple to state, it can be extremely difficult and time-consuming to carry out. Therefore, there are very few nervous systems for which such detailed information is yet available.

Figure 5A shows examples of some of the neurons known to constitute the crayfish escape circuit, drawn in their actual positions within the nerve cord (Wiersma, 1947; Kennedy and Takeda, 1965a; Zucker, 1972; Wine and Krasne, 1982). As we saw in Chapter 4, an invertebrate nerve cord (which together with the brain constitutes the central nervous system) is composed of GANGLIA and CONNECTIVES (Figure 5A). Ganglia, analogous to the nuclei of vertebrate central nervous systems (Chapter 7), contain neuronal cell bodies, dendrites, and axons. Connectives, analogous to central nerve tracts in vertebrates, contain only axons. (In contrast to vertebrates, in which nerve tracts intermingle with nuclei, in many invertebrates the connectives and ganglion are spatially separated, forming a ladderlike arrangement.) PERIPHERAL NERVES, or ROOTS, many containing mixtures of sensory and motor axons, communicate between the central nervous system and sense organs or muscles. (Because invertebrate axons generally contain no myelin but only a loose wrapping of glial cells, there is no white matter and no white/gray distinction, as one finds in vertebrate central nervous systems.)

In Figure 5A, one TACTILE SENSORY CELL (TSC) is shown connecting to a SENSORY INTERNEURON (SI). (There are over 1000 TSCs, none of which are individually recognized. There are over 20 individually recognizable SIs that are part of the escape system.) The SI is shown connecting to a LATERAL GIANT INTERNEURON (LGI), individually recognizable by its axon of unusually great diameter, about 100 μm. This LGI synapses in the next anterior ganglion with another LGI, this with yet another, and so forth all the way up the nerve cord. Each LGI is shown synapsing on a motor neuron called the MOTOR GIANT (MoG), whose axon innervates a FAST FLEXOR MUSCLE (FFM). These muscles, found in every abdominal segment, are responsible for the abdominal flexion movement of the escape behavior.

Additional component neurons and the synaptic connections among them are shown in the circuit diagram of Figure 5B. For simplicity, this figure only shows the cells of one side of one abdominal segment. As can be seen, some TSCs, in addition to connecting to SIs, also have weak excitatory synapses onto the LGI. At least some of the SIs excite one another. LGI has a strong excitatory connection not only to MoG but also onto a cell called the SEGMENTAL GIANT (SG), which in turn strongly excites a group of four to eight motor neurons called the FAST FLEXORS (FF). In addition, weak excitatory connections are shown from both the SIs and LGI to FF and from SG to cells called CDIs, which in turn weakly excite the FFs.

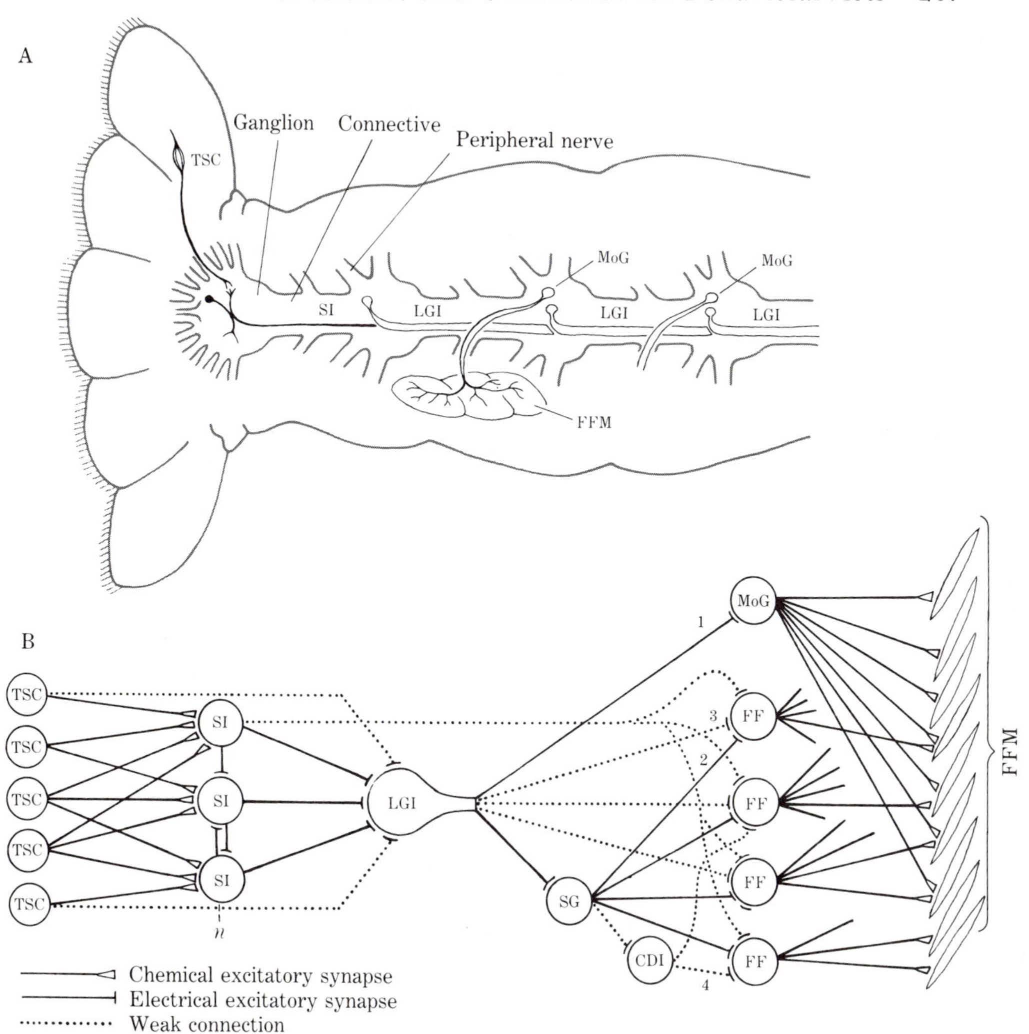

FIGURE 5. The nerve circuit for a crayfish's escape from a tactile stimulus to its abdomen. A. The abdomen of a dissected crayfish, showing the arrangement of some of the cells of the escape circuit. Only the four most posterior of the five abdominal ganglia, and the connectives between them, are shown. TSC, Tactile sensory cell; SI, sensory interneuron; LGI, lateral giant interneuron; MoG, giant motor neuron; FFM, fast flexor muscle. B. The escape circuit shown in detail. Only that part of the circuit involving one side of a single abdominal ganglion is shown. Three additional cell types are included: the fast flexor motor neurons (FF), the segmental giant cell (SG), and one of a group of cells called CDIs. The strongest synaptic connections, which are indicated by the solid lines, are as follows: the TSCs excite the SIs, which in turn excite LGI. LGI excites MoG, and also SG, which in turn excites the FFs. Most synapses in the circuit are electrical. Exceptions are the neuromuscular junctions (which are chemical in all known cases in all animals), and the TSC-to-SI connections. (A after Wine and Krasne, 1982.)

Although Figure 5B shows only the cells of one side of one ganglion, basically the same cellular complements and nearly identical synaptic connections occur in each half of each abdominal ganglion. Because each LGI synapses with the next more anterior LGI of the chain (Figure 5A), information is propagated along the whole chain. This propagation is assured by the fact that the LGI-to-LGI synapses are electrical relay synapses; that is, a single presynaptic action potential produces a single postsynaptic action potential (Chapter 3). Similar electrical relay synapses are found between the left and the right LGI in each ganglion (Payton *et al.*, 1969). Thus, a single action potential in any LGI is relayed through all LGIs of both chains.[2] The pair of LGI chains is therefore functionally analogous to a single neuron in that each action potential is transmitted uninterruptedly throughout the entirety of the two chains.

It is not clear why the crayfish escape circuit needs to be so complex, in view of the apparently simple nature of the initial abdominal flexion movement that this circuit controls. (For instance, it is not yet known why there are four separate pathways by which an LGI can excite the fast flexor motor neurons; numbers 1 through 4 on Figure 5B.[3]) Nevertheless, given this detailed understanding of the neuronal circuit for the initial abdominal flexion, we can now search for correspondences between the properties of this circuit and properties of the behavior.

To begin, one can see from Figure 5B that there is extensive neuronal convergence onto, and divergence from, LGI. Only one weak pathway (SI to FF) is known to bypass this cell. Owing to this pivotal position of LGI, in which it receives a wide array of sensory information and can in turn instruct all the fast flexor muscles to contract, this cell's axon is ideally situated for carrying the escape command from a prior decision switch to the hierarchically lower interneurons and motor neurons that induce the behavior. We will now consider the evidence for this command role of LGI. Following this, we will carry out a search for the decision switch.

THE COMMAND FOR ESCAPE

It may appear fully obvious from LGI's pivotal position on the circuit diagram that this cell's axon carries the command for escape. Indeed, in support of this idea, LGI has two special properties well suited for conveying quickly an escape command to the many hierarchically lower neurons onto which it synapses. First, because of its large diameter, the LGI axon conducts action potentials very rapidly (Chapter 3). (We have seen this same property in the GIs that mediate the escape

[2]These electrical synapses are of a variety referred to as nonrectifying, meaning that transmission is equally effective in either direction across the synapse (Kusano and Grundfest, 1965).

[3]It is known, however, that upon repeated stimulation, the neuromuscular junctions from MoG to the fast flexor muscle fibers defacilitate rapidly, whereas those from FF to these same muscle fibers facilitate (Kennedy and Takeda, 1965a). Therefore, when the crayfish is stimulated repeatedly, there may be a shift from primarily MoG to primarily FF control of the muscle. This may help conserve neurotransmitter molecules in the terminals of MoG, thus retaining this fast, most direct LGI-to-MoG-to-muscle pathway (number 1 in Figure 5B) for only occasional, emergency situations.

behavior of cockroaches: Chapter 4). Second, every known synapse from an LGI onto other cells of the circuit is electrical (Figure 5B), and electrical synaptic transmission always occurs more rapidly than chemical transmission (Chapter 3).

None of this, however, *proves* the command function of the LGIs. Although no alternative command neurons are revealed in the circuit diagram of Figure 5B (except for the weak pathway from SI to FF), one cannot rule out the possibility that neurons not yet identified carry out this command function. Such neurons may have axons parallel to LGI and may activate the MoG and FF motor neurons. Thus, although drawing the circuit diagram of identified cells, as in Figure 5B, is a crucial step in studying cellular functions underlying animal behavior, it cannot be the final step.

How can one test the hypothesis that the LGIs alone, and not any other cells (even cells yet to be discovered), carry the command to escape? As discussed in Chapter 7, one would want to determine both whether the action potentials of the LGI are *sufficient* to give rise to a fully normal escape behavior and whether action potentials in the LGI are *necessary* for the behavior. We shall now examine experiments that test both these points. In practice, for some of the experiments to be described, it was not the escape behavior per se that was observed, because the violent abdominal flexion movement would break the delicate recording microelectrodes. Rather, the peripheral nerves from the abdominal ganglia were cut, preventing the axons of motor neurons from activating the muscles. Action potentials in the MoG and FF axons were recorded proximal to the cut to indicate the animal's attempt to carry out an escape behavior.

Just a single action potential induced by a stimulating electrode in any LGI cell is indeed *sufficient* to evoke a fast abdominal flexion movement (Wiersma, 1947). This flexion is nearly identical to that evoked by tactile stimulation of a freely mobile crayfish. Very slight differences may, however, exist between the flexion produced by electrical stimulation of LGI and that produced by tactile stimulation. These could result from several factors: first, more than one LGI action potential often is evoked by tactile stimulation of the body, whereas only one action potential was evoked by electrical stimulation of LGI; second, the animal is freely mobile during tactile stimulation but is held by the thorax for electrical stimulation; and third, the weak SI to FF pathway (Figure 5B) is bypassed by the electrical, but not by the tactile, stimulus.

Whether LGI is also *necessary* for escape was tested using the arrangement of electrodes shown in Figure 6A. Sensory cells were stimulated (St) by means of electric shocks of constant voltage to a peripheral nerve containing only sensory axons. The response of the local LGI to this stimulus was recorded with an intracellular microelectrode (R_{LGI}). A second microelectrode in the same LGI was used to pass sufficient hyperpolarizing current to prevent the excitatory postsynaptic potentials (EPSPs) of this cell from reaching threshold and, thus, from inducing an action potential. A pair of extracellular wire electrodes placed more posteriorly on the nerve cord (R_{NC}) was used to verify that LGIs of other ganglia had not somehow been excited by the sensory stimulus presented during hyperpolarization. The presence or absence of an intended escape response was assayed by recording extracellularly from the entire proximal stump of the cut motor nerves (R_{MOT}). When the LGI was not blocked by hyperpolarization but rather

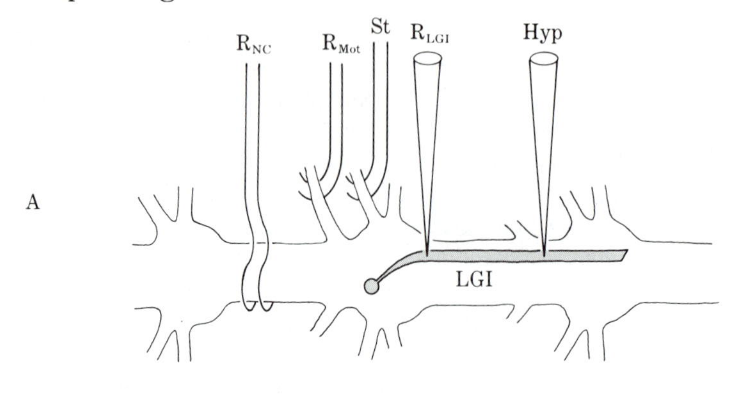

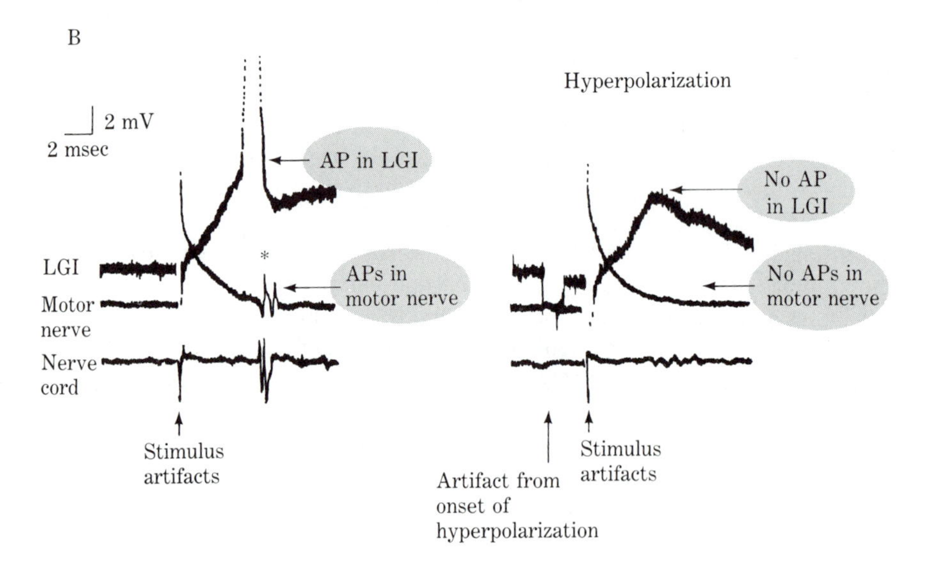

gave an action potential in response to the electrical sensory stimulation (Figure 6B, left panel, arrow), there were action potentials in the motor nerve as expected (Figure 6B, left panel, asterisk). However, when the action potentials of LGI were blocked, there were no action potentials in the motor nerve (Figure 6B, right panel). Thus, an action potential in LGI is necessary for the motor output of the escape behavior.

A single neuron that carries by itself the full command signal for a natural behavior, fulfilling both the necessary and sufficient criteria, is called a COMMAND NEURON[4] (Kupferman and Weiss, 1978). Whether a given behavioral act is mediated by such a command neuron is a question of general importance; it bears

[4]There has been much debate about the proper definition of this term and of the term *command system*, which will be discussed later in this chapter. The definitions presented here are rigorous and testable and are in accord with the ideas expressed in a recent insightful discussion of the subject (Kupferman and Weiss, 1978).

◀ FIGURE 6. Experiments showing that an action potential in LGI is necessary for the motor output of escape behavior. A. Experimental arrangement, described in the text. R_{NC}, Recording from nerve cord; R_{Mot}, recording from motor neuron; R_{LGI}, intracellular recording from LGI; St, electrical stimulus; Hyp, intracellular electrode used to hyperpolarize LGI by depositing negative ions into the cell; LGI, lateral giant interneuron. B. Recordings made while LGI *was not* hyperpolarized *(left panel)* and thus gave an action potential (AP); and while it *was* hyperpolarized *(right panel)* and thus gave no action potential. Action potentials in the motor nerve *(asterisk)* occurred only if there was an LGI action potential. This result was repeated many times. (After Olson and Krasne, 1981.)

upon the larger issue of the extent to which definable behavioral functions are relegated to individual neurons, as opposed to groups of neurons. (This question emerged in Chapter 7, in our discussion of recognition neurons, and it will reappear in Chapter 9 in relation to central pattern generators.)

Is a single LGI, then, a command neuron according to our definition? The particular LGI cell studied in the experiments of Figure 6 does fulfill the criteria of necessary and sufficient function and, thus, is a command neuron. But these criteria are fulfilled *only* under the conditions of the experiment, in which the sensory stimulus was confined to the same body segment as the hyperpolarized LGI. If, instead, the stimulus had been applied to the sensory nerve of a body segment *different* from that with the hyperpolarized LGI, the LGI of this stimulated segment would have been excited. This excited LGI would then have conveyed the command signal throughout both LGI chains and, thus, would have evoked an escape behavior. The particular LGI cell hyperpolarized in Figure 6, then—in fact, *any* one LGI cell—is not necessary by itself for commanding the escape; other LGIs of the chain can replace it. In this strict sense, then, LGI does not fully qualify as a command neuron; rather its command function is distributed among all the LGI cells of both chains. However, since the two chains function together as a single conducting unit, it is not unreasonable to think of the two LGI chains as closely resembling a single command neuron. In particular, it is clear that neurons *other* than the LGIs do not play a command function in the behavior because blocking action potentials only in LGI blocks tactually evoked behavior.

A DIGRESSION ON COMMAND NEURONS FOR OTHER SIMPLE BEHAVIORS

Are there single command neurons for behaviors other than the fast abdominal flexion of the crayfish? No other neurons have been tested as rigorously for a possible command function as have the crayfish LGIs. There are, however, cells with some properties that fit our definition of command neurons; these cells await further testing. Three well-studied examples will be mentioned briefly, each of which plays a role in a stereotyped escape behavior of some animal. (1) The MEDIAL GIANT INTERNEURON (MGI) of crayfish mediates an escape behavior slightly different from that commanded by LGI (Wine and Krasne, 1972, 1982). This behavior is evoked by visual stimuli or by tactile stimulation of the *anterior* region of the body. The initial fast abdominal flexion evoked has a more curled form than that evoked by LGI and directs the animal backward (away from the anterior stimulus), not forward as in LGI-mediated escape (Figure 4B). (2) One medial and two lateral giant interneurons in earthworms mediate, respectively, a forward and a backward escape behavior (Drewes, in press). (3) A pair of giant interneurons called the MAUTHNER NEURONS mediate the fast initial tail flick of the escape behavior in fish and larval amphibians (Diamond, 1971; Rock *et al.*, 1981; Eaton, in press). Incidentally, Mauthner neurons are among the very few individually recognizable neurons that are known in vertebrate animals, though new anatomical techniques are currently revealing many more in fish and amphibians.

The idea that some motor acts may be controlled by single command neurons has led some workers to suggest that animal behavior in general may be organized in a push-button manner. By this view, to produce a particular behavior the animal need only excite the appropriate command neuron within its central nervous system, and the behavior will follow in a robotlike fashion. On the surface, this one-command-neuron-for-one-behavior scheme is attractive for its simplicity. However, it is surely too simple an explanation for any but perhaps the most stereotyped of behaviors. As we have seen, even just a simple abdominal flexion of the crayfish is not, strictly speaking, mediated by only one command neuron but rather by two whole chains of LGIs. In fact, numerous studies on more complex behaviors, even those of invertebrates, indicate that neural command signals are generally conveyed by *groups* of neurons whose axons run more or less parallel to one another. Activity in several of these neurons appears to be required for the normal behavior to occur. Such a group of cells is often called a COMMAND SYSTEM (see Footnote 4 above). Two well-studied examples of command systems are found in crayfish: that controlling rhythmic swimmeret beating (Davis and Kennedy, 1972) and that controlling slow flexion and slow extension of the abdomen (Evoy and Kennedy, 1967).

THE CRAYFISH'S DECISION TO ESCAPE

We turn now to the question of how the crayfish carries out its all-or-nothing decision whether or not to escape. It might be thought that, having shown that a single action potential in any individual LGI induces a full abdominal flexion behavior, we have thereby shown that the LGI itself makes the decision to escape. The act of "deciding" could consist simply of producing the all-or-nothing action potential in LGI: stimuli too weak to evoke this action potential would result in a negative decision, and those strong enough to exceed the threshold of LGI and evoke an action potential would produce a positive decision. However, it is possible that there are some other all-or-nothing processes occurring in the escape circuit prior to the LGI. For instance, it could be that the SIs (Figure 5) respond to all sensory stimuli below some given intensity by evoking few or no action potentials but respond to any sensory stimulus above this intensity by evoking a vigorous barrage of action potentials. Such a barrage, then, would be all-or-nothing in character. This barrage might provide sufficient depolarization of LGI to evoke in it an action potential and would thus lead to an escape behavior. But by this scheme the SIs, and not LGI, would formulate the animal's decision to act.

The question of whether there are all-or-nothing processes prior to LGI that could account for the decision to escape has been studied through experiments using the same electrode arrangement as shown in Figure 6A (Olson and Krasne, 1981). The sensory nerve was stimulated (St) with shocks of different voltages, and the sizes of the resulting EPSPs in LGI were recorded. Shocks of low voltage evoked only a small EPSP in LGI, presumably because these stimuli excited only a few axons in the sensory nerve. As the stimulating voltage was increased, with more and more sensory axons being excited, the EPSP in LGI grew. At some critical voltage, this EPSP exceeded threshold for an action potential in LGI which, as usual, evoked the motor output for a full abdominal flexion. The key question in this experiment is, "As the strength of the shocks to the sensory nerve is increased up to the threshold for an LGI action potential, do the EPSPs in LGI show a sudden large increase in size or only a gradual increase?" A sudden increase could reflect an all-or-nothing process somewhere in the circuit prior to LGI. A gradual increase would mean that the only all-or-nothing process whose threshold equals the behavioral threshold is the generation of LGI's action potential. This action potential, then, would constitute the decision to act.

It was difficult to measure accurately the size of an EPSP that produced an action potential in LGI because the action potential itself obscured the peak of the EPSP. Therefore, the procedure adopted was first to give a stimulus shock of a particular voltage to the sensory nerve and observe whether the stimulus did or did not produce an action potential in LGI. Then the LGI was hyperpolarized by passing a fixed amount of current through a second intracellular electrode. Given this hyperpolarization, EPSPs no longer could bring the membrane potential to threshold, so no action potential could be evoked. During the hyperpolarization, the same shock to the sensory nerve was repeated; but now the height of the EPSP that it evoked in LGI could be measured, unobscured by an action potential. (This hyperpolarization was carried out whether or not the sensory stimulus had evoked an LGI action potential, so that all measured EPSPs occurred under identical conditions.) This entire procedure was then repeated using other stimulus voltages. Thus, the EPSP heights associated with different shock strengths were determined.

As the strength of shocks to the sensory nerve was increased up to and beyond the threshold for an LGI action potential, the resulting increase in EPSP size was clearly gradual (Figure 7). Thus, it is the generation of the all-or-nothing action potential in LGI that causes the all-or-nothing decision to escape. It is worth recalling that this all-or-nothing property of an action potential results from a positive feedback relationship within the cell membrane; depolarization increases the sodium conductance, which in turn increases depolarization, which in turn increases the sodium conductance, and so forth (Chapter 3). Any such positive feedback arrangement has an explosive, or all-or-nothing, character. Therefore, it is not surprising to find a positive feedback relationship as the basis for an animal's switchlike decision to act.

It is unlikely that such a simple form of all-or-nothing decision as the creation

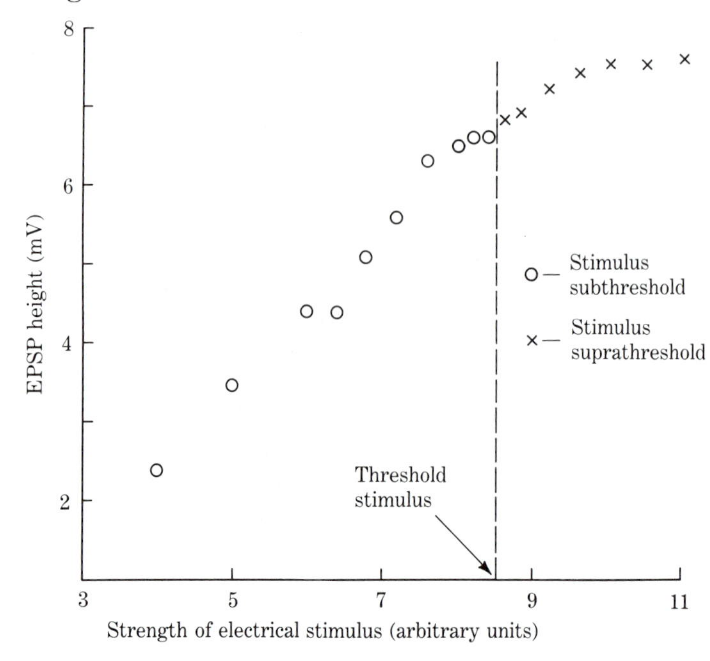

FIGURE 7. EPSP height in an LGI as a function of the strength of stimulation of the local sensory nerve. As the threshold for an LGI action potential (and thus for a motor output) is exceeded, there is only a gradual increase in EPSP height. This result was independent of the order of presentation of stimuli of different strengths. (After Olson and Krasne, 1981.)

of a single action potential in just one neuron is a common basis for decision-making in animal behavior. Rather, multicellular forms of positive feedback are likely to be quite common and have, in fact, been discovered underlying other switchlike decisions. One example is seen in the defensive release of ink from the ink gland in the mollusk *Aplysia* (Kandel, 1976; Carew and Kandel, 1977a, b). Three motor neurons innervate this gland. An intense barrage of action potentials in these three cells leads to the release of ink. These three cells generally show either very few action potentials or, when excited by presynaptic neurons, an intense barrage of action potentials; they rarely show an intermediate level of activity. The barrage, then, is roughly all-or-nothing in character.

This all-or-nothing property results in part from the fact that the three cells are electrically coupled to each other; that is, each has electrical synapses onto the other two. This constitutes a form of positive feedback; an action potential in any one of the three cells produces an electrical EPSP in the other two, and this increases the probability that each of these two cells will thereby be brought to its threshold and give an action potential. If either of these two cells *does* give an action potential, this then feeds back as an electrical EPSP onto the first cell, increasing in turn its own probability of giving a second action potential. Thus, each time any one of these three cells produces an action potential, it increases its own likelihood of giving yet another action potential. The overall effect is a rapid escalation into an intense barrage of action potentials. Similar escalating barrages in electrically coupled neurons have also been seen in the mollusk *Tritonia* (Getting and Willows, 1974; Getting, 1975).

THE SINGLE-MINDEDNESS OF THE CRAYFISH ESCAPE BEHAVIOR

As mentioned earlier, the performance of the crayfish escape behavior appears to cancel other ongoing behaviors. What is the neuronal basis of this single-mindedness of the escape action? A series of experiments has revealed that a single action potential in LGI not only commands the initial abdominal fast flexion but also produces widespread synaptic inhibition of circuits mediating behaviors antagonistic to fast flexion (Wine, 1977a, b; Wine and Mistick, 1977; Kuwada and Wine, 1979; Kuwada *et al.*, 1980). This can be thought of as an extension of the principle of reciprocal innervation, described earlier in this chapter, in which the excitation of one muscle, or synergistic group of muscles, is accompanied by inhibition of the motor neurons to the antagonistic muscles.

The inhibitory effects resulting from activity in LGI are generally mediated by polysynaptic pathways employing interneurons that, in most cases, have not yet been identified. Five such pathways from LGI are shown in Figure 8, labeled A through E. Pathway A inhibits a form of movement that is clearly antagonistic to fast abdominal flexion, namely, fast abdominal extension. This movement comes about by contraction of the fast extensor muscles, each of which is innervated by five excitatory motor neurons and one inhibitory motor neuron (Wine and Hagiwara, 1977). An action potential in LGI evokes inhibitory postsynaptic potentials (IPSPs) in the excitatory motor neurons and depolarizing PSPs in the inhibitory motor neuron (Figure 9). The depolarizing PSP appears to be an EPSP, because it sometimes give rise to an action potential (Wine, 1977a). The overall effect of pathway A, then, is a suppression of activity in the fast extensor muscles when the fast flexors are activated by LGI.

As mentioned briefly earlier in this chapter, the crayfish is capable of making not only fast abdominal flexions and extensions but also slow, postural extensions that are mediated by a command system. These movements are carried out by slow flexor and slow extensor muscles that are distinct from the fast muscles. The slow muscles have their own private motor neurons that do not branch to innervate the fast muscles. Each slow extensor or slow flexor muscle is innervated by five excitatory and one inhibitory motor neuron (Fields *et al.*, 1967; Kennedy and Takeda, 1965b). Contraction of the slow extensor muscles is clearly antagonistic to the initial fast flexion of the escape behavior. Contraction of the slow flexors, being prolonged, would last well into the subsequent fast extensions of the escape behavior and thus would antagonize these movements. Contractions of both these group of slow muscles are suppressed by inhibition during LGI-mediated escapes. For instance, an action potential in LGI excites the inhibitory motor neuron of the slow extensor (Figure 8, pathway B) and that of the slow flexor (Figure 8, pathway C). It also inhibits the slow flexor motor neurons (pathway C). Aside from these inhibitory effects on the slow abdominal motor system, pathways from LGI also inhibit the sensory receptor cells and sensory interneurons

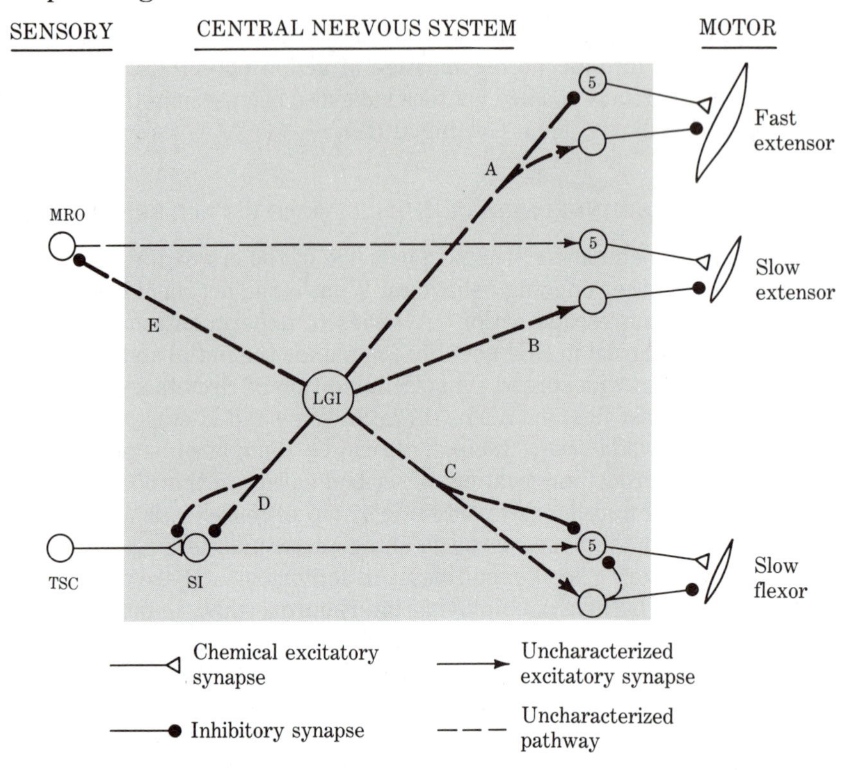

FIGURE 8. Inhibition deriving from LGI onto nerve circuits that would otherwise compete with the escape behavior. Effects shown include excitation of inhibitory motor neurons (pathways A, B, and C), inhibition of excitatory motor neurons (A and C), presynaptic and postsynaptic inhibition at a sensory-to-interneuron synapse (D), and inhibition of a sensory neuron at the periphery (E). MRO, Muscle receptor organ, a stretch-sensitive mechanoreceptor; TSC, tactile sensory cell; SI, sensory interneuron. The number 5 in each of the three excitatory motor neurons drawn indicates that there are five cells of each of these types.

that drive these two sets of muscles (Figure 8, pathways D and E). The overall effect of pathways B through E is a suppression of slow postural movements during the LGI-mediated escape behavior.

Thus, the LGI, in addition to deciding upon and commanding action by postsynaptic neurons of the escape circuit, also brings about the silence of neurons mediating competing behaviors. The inhibition is observed at all hierarchical levels, including receptor cells, interneurons, motor neurons, and muscle cells. In addition to the inhibition from LGI onto other circuits, there is also some inhibition of these circuits by branches of the fast flexor (FF) motor neurons (Wine, 1977). However, the dominant inhibiting influence stems from the LGIs. Therefore, these neurons carry the prime responsibility for *deciding upon*, *commanding*, and *preventing competition with* the escape behavior.

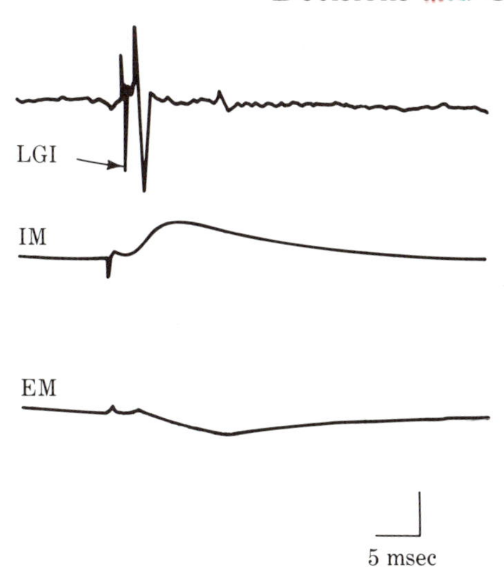

FIGURE 9.

Synaptic potentials recorded from cell bodies of fast extensor motor neurons in response to an action potential in LGI. *Top trace,* LGI action potential recorded with extracellular electrodes. (With the electrode position used, the action potential in LGI is followed by an action potential from a motor neuron, probably MoG.) *Middle trace,* The inhibitory motor neuron (IM) to the extensor muscles gives a depolarizing PSP in response to the action potential in LGI. *Bottom trace,* An excitatory motor neuron (EM) to the fast extensor muscles gives an IPSP in response to the LGI action potential. *Vertical scales:* 5 millivolts for IM; 2 millivolts for EM. (After Wine, 1977a.)

Commands and Brain Stimulation in Vertebrates

As was mentioned earlier in this chapter, most behaviors are probably mediated not by single command neurons but rather by command systems. This is especially likely to be true in vertebrates, whose behavior in general is more variable than that of invertebrates and whose nervous systems are endowed with far greater numbers of neurons. It is presently impossible to understand command systems, especially those of highly complex vertebrate brains, in the same detail with which the crayfish escape system is known. Nevertheless, it has been possible to identify, at least grossly, command systems controlling some vertebrate behaviors and to determine some features of the organization of these systems. In this section, we shall examine a few of the experiments that have led to the present understanding of these systems.

Three major experimental approaches have been employed in searching for and analyzing command systems in vertebrate nervous systems. One approach, developed most fully only in the last few years, is the recording from awake, freely behaving animals of the electrical activity of single neurons or small groups of neurons. Generally one attempts to correlate this activity with the animal's performance of specific behavioral acts. Cells whose activity is closely correlated become candidates for members of a command system (Evarts, 1966; Mountcastle, 1976). The second approach involves producing lesions by making small knife cuts, or by applying strong, damaging electric currents through strategically placed electrodes, or by directing laser beams or other forms of concentrated energy at specific neuronal targets. After making the lesion, one generally searches for deficits in the animal's natural behavior. If a specific behavior shows a deficit, the damaged neurons can be considered candidates for a part of a command system of that behavior (Valenstein, 1973). The third approach is to stimulate a local brain structure in an attempt to evoke some natural behavioral act. The stimulation can be electrical, as was described for the tectum of the toad (Chapter 7, Figure 14) or chemical, consisting of neurotransmitter molecules or their agonists, or hormones, or a variety of other pharmacological agents applied through small cannulae implanted into the brain (Valenstein, 1973). The procedures of implanting the stimulating electrodes or cannulae can be either ACUTE (that is, prepared for an immediate experiment that the animal is not expected to survive) or CHRONIC (the animal being intended for long-term survival and experimentation). The following discussion will focus on acute electrical and, to a lesser extent, chemical stimulation of brain structures to elicit behavior.

Before describing these results, it is important to recall a major limitation of such experiments, which we encountered in Chapter 7, namely, that one can never be certain precisely which cells have been excited by an extracellular electrode. Moreover, there is an additional limitation of brain stimulation experiments in which one wishes to ascribe particular behavioral functions to particular groups of cells; because one cannot work out at the single cell level a vertebrate nerve circuit, one is usually unsure what functional role is played by different parts of the circuit. To understand this problem, imagine that we were to place a gross extracellular stimulating electrode in the nerve cord of the crayfish (Figure 5). It should be clear that a sufficiently strong stimulus delivered not only to LGI but also to the TSCs or SIs could evoke a normal escape behavior. Thus, if the nerve circuit had not been elucidated at the cellular level and if the experiments on LGI reported earlier had not been carried out, we might have concluded on the basis of stimulating the TSCs or SIs that these were command neurons for the behavior. But, by our definition, a command neuron conveys the results of a decision made by the nervous system to carry out some behavioral act. So unless one knows where in the circuit this decision is made, one cannot determine whether or not a given cell, even though capable of inducing a behavior in an experimental situation, executes this decision and is thus a command neuron for that behavior.

As a further complication, an electrical stimulus within the brain may activate a behavior by way of neurons that, unlike the TSCs, SIs, and LGIs of the crayfish, do not lie within a direct sensory-to-motor circuit for the behavior; rather, the cells excited may somehow enhance the animal's *motivation* to perform the particular act. For instance, the electrical stimulation of some brain region may evoke drinking behavior not by exciting neurons that command specific movements of the lips, tongue, and throat but rather by somehow making the animal "thirsty" (whatever this may mean in physiological terms), so that it seeks out and drinks water. In fact, electrical stimulation of a certain region of a rat's hypothalamus causes this animal to lift its head and drink from a water tube placed just above it; or to lower its head to drink from a dish below it. If the rat instead had previously been trained such that it could obtain water only after running through a maze or only after pushing a certain lever, it would respond to the same electrical stimulus by running or pushing, accordingly (Jurgens, 1974). Thus, it is not a specific set of movements that is stimulated but rather the motivation to acquire a particular goal—in this case, water. When one evokes a behavioral act by brain stimulation, one must consider whether it is the particular motor acts constituting that behavior or the motivation to acquire a given goal that is being excited.

ELECTRICAL STIMULATION OF LOCOMOTION IN CATS

Specific regions of the brain and spinal cord of several vertebrate animals have been found that, if stimulated with repeated electric shocks, elicit apparently normal locomotory behavior (Grillner, 1975; Wetzel and Stuart, 1976). Because the animals proceed forward with automatonlike limb movements, rather than seeking out any particular goal, the stimulus apparently excites not a motivational system but rather some part of the circuit controlling locomotion. The most extensive work has been carried out on DECEREBRATE cats—animals whose brains have been transected at the upper midbrain region, as shown in Figure 10A. Such animals can survive for prolonged periods because most of the autonomic bodily functions, such as respiration and circulation, are controlled by nuclei in the hindbrain. Decerebrate animals, however, generally do not walk spontaneously, particularly during the first several weeks after the brain transection. But if one stimulates electrically a specific small region of the midbrain (Figure 10A) while the cat's body is positioned over a treadmill (Figure 10B), it responds by making normally coordinated walking movements (Figure 10C). With greater stimulating current strengths which presumably excite more cells and more action potentials in each cell, the cat walks faster and exerts greater force with its steps. As the current strength is raised still further, the cat switches from walking to trotting and then to galloping, each of these gaits showing its normal pattern of interlimb coordination (Shik *et al.*, 1966; Grillner, 1975).

Two factors suggest that the neurons stimulated electrically in these experiments may be closer to the motor side than the sensory side of the cat's neuronal circuitry and, thus, could in fact constitute a command system for locomotion. First, owing to the transection, the thalamic sensory areas, as well as the sensory areas of the cerebral cortex, have been disconnected from the rest of the central nervous system. Thus, if the electrode had excited some cells that would normally carry sensory information to these forebrain structures, these cells could not now be responsible for the evoked behavior. Second, the specific region stimulated appears to contain cells whose axons descend into the spinal cord toward motor circuits that control the legs (Steeves *et al.*, 1976). Thus, the group activity of the neurons stimulated appears to excite hierarchically lower levels of organization involved in locomotory behavior. Comparable command systems controlling locomotion have been found in turtles (Lennard and Stein, 1977; Stein, 1978) and fish (Kashin *et al.*, 1974; Grillner *et al.*, 1976).

BRAIN CENTERS VERSUS NEURAL NETWORKS: HYPOTHALAMIC STIMULATION AND MOTIVATED BEHAVIORS

An important question that emerges from studies employing brain stimulation is, "To what extent are neurons that mediate particular motor acts clustered together?" In Chapter 7 we saw that the optic tectum contains a topographic map of visual space and that stimulation of neighboring tectal areas produces similarly oriented turning behaviors. However, whereas the amphibian tectum is responsible for organizing visual sensory information, the actual *motor* organization that produces the coordinated turning movements may lie elsewhere than in the tectum.

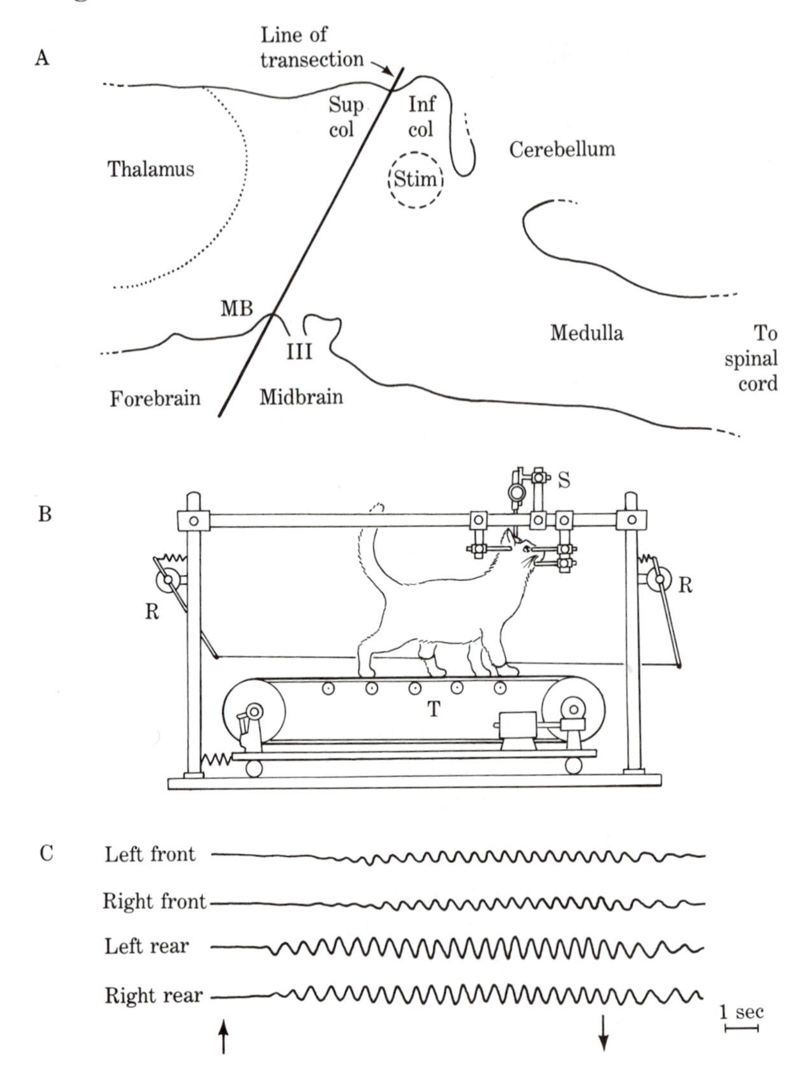

FIGURE 10. Electrical stimulation of locomotion in decerebrate cats. A. Location of the brain transection in the classic decerebrate procedure. Sup col, Superior colliculus; Inf col, inferior colliculus. MB, mammilary body; III, point of exit of third cranial nerve; Stim, region within which electrical stimulation evokes locomotion. B. Arrangement of decerebrate cat for electrical stimulation of the brain and recording of walking movements. S, Stereotactic instrument with which the head is held rigidly and the electrode is positioned in the brain. R, Position-recording devices normally attached to each leg. (Attachments to only two legs are shown.) T, Treadmill powered by a motor. The gear from the motor to the treadmill is loose, so that not only the motor, but also the walking cat, can control the speed of movement of the treadmill. C. Locomotory movements from all four legs evoked by stimulating the brain region shown in part A. Arrows show time of stimulus onset and termination. (A after Wetzel and Stuart, 1976; B and C after Shik *et al.*, 1966.)

This is suggested by studies on frogs whose tecta have been ablated; these individuals still perform normal turning and snapping movements when stimulated by tactile cues rather than by visual cues (Comer and Grobstein, 1981a). Thus, tectal topography tells us little about *motor* organization in the brain. In fact, relatively little is known in general about the spatial organization within vertebrate brains of motor circuits that organize different behaviors.

By contrast, more is known about the organization of circuits involved in the *motivation* of specific behaviors, such as drinking in rats. One question that was of particular interest to early investigators was whether the neurons controlling the motivation of a given behavioral function, such as drinking, eating, sex, or aggression, are clustered into separate BRAIN CENTERS—discrete, nonoverlapping regions, each devoted exclusively to a different behavior—or whether they are organized into more diffuse, overlapping NEURAL NETWORKS. This question has been studied most effectively by electrical stimulation of different hypothalamic regions (Valenstein, 1973; Jurgens, 1974). Figure 11 shows the positions of the tip of an electrode (determined by making subsequent markings in the brain by large, electrolytic currents) from which six different motivated behaviors could be evoked in rats. It can be seen that different hypothalamic regions produce different behaviors. However, the effective regions for some of the behaviors are so large, comprising perhaps tens of thousands of neurons, that it seems unlikely that all these neurons are devoted exclusively to just one behavior. Moreover, there is extensive overlap between the areas for different behaviors, especially for functionally related acts such as eating and drinking. Thus, although one can

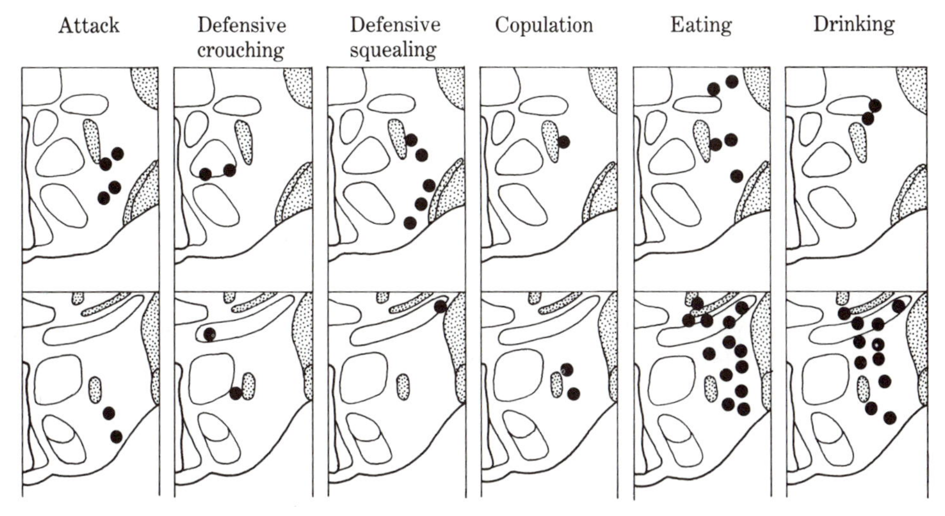

FIGURE 11. Hypothalamic sites of electrically evoked behaviors in rats. The top row of figures shows a frontal section through the anterior part of the hypothalamus; the bottom row shows similar sections through a more posterior part of the hypothalamus. Each symbol shows the location of the tip of the stimulating electrode that produced the particular behavior indicated. (After Jürgens, 1974.)

speak of general regions related to particular behaviors, there is no indication of organization by discrete, nonoverlapping brain centers but rather by more diffuse neural networks.

HIERARCHICAL ORGANIZATION OF ELECTRICALLY EVOKED BEHAVIOR

The hierarchical organization that can be observed for some behaviors (Figure 3) implies a hierarchy in the nervous system. Can one detect such a neuronal hierarchy by means of brain stimulation experiments? For instance, does a stimulating electrode placed in one location evoke a fully orchestrated behavior but in another location only a subcomponent of this larger behavior? This suggestion has been reasonably well borne out in a now classic set of experiments of the late 1950s on the thalamus and hypothalamus of domestic chickens (von Holst and von St. Paul, 1973). With the electrode in one position, a prolonged stimulus evoked the following sequence of behavior: (1) blinking of the left eye, then (2) shaking of the head, then (3) wiping the head repeatedly against the left shoulder, and finally (4) scratching the left side of the head repeatedly with the left foot. After a brief rest, a second, then a third, and each subsequent prolonged stimulus evoked the identical sequence. This behavior was very similar to that evoked naturally by topical irritation of the left side of the face.

With the electrode in other positions, many simpler behaviors could be evoked. Some of these appeared to be components of larger, more complex sequences. For instance, with some electrode positions, headshaking that looked identical to the second component of the sequence just described could be evoked independently of any other movements. Still other electrode positions led to this same head shaking; but the head shaking was now incorporated into a different complex behavior, as in the following sequence: (1) cessation of feeding, (2) salivation, tongue movement, and neck stretching, (3) head shaking. This sequence closely resembles the behavior shown by a normal chicken that has just eaten something distasteful.

It appears, then, that electrical stimulation can evoke hierarchically higher or lower behavioral components, depending upon electrode location. This provides independent support for the notion that hierarchical organization is one principle by which some behaviors are organized within the brain. Additional evidence has come from behavioral studies on a number of different animals (Tinbergen, 1951; Dawkins, 1976; Bentley and Konishi, 1978).

IS BEHAVIOR CHEMICALLY CODED?

Neurons can be excited not only by electrical stimulation but also by application of excitatory neurotransmitters or their agonists. These agents presumably exert this effect by binding to the receptor macromolecules on the postsynaptic membrane, an action that leads to the opening of ionic channels in the postsynaptic membrane. Because relatively few forms of neurotransmitter molecules have been discovered, one would expect that any given transmitter type would be employed by neurons mediating many different behaviors. Thus, application of a given transmitter, even fairly locally within the brain, should not be expected to evoke just one coherent behavioral response but rather should excite components of several different behaviors.

There is, however, some surprising evidence to the contrary. For instance, application of norepinephrine locally through a cannula implanted in that region of the rat's hypothalamus that subserves eating and drinking (Figure 11, last two columns) evokes only eating. This reponse is mimicked by agonists of norepinephrine and blocked by its antagonists. Application of acetylcholine (ACh) through the same cannula evokes only drinking, which likewise is mimicked by agonists and blocked by antagonists of ACh (Grossman, 1973). It appears, then, that there is some degree of coding by neurotransmitters of specific behaviors.

What is even more surprising is that intravenous injection rather than local application of some neurotransmitters or precursors can reproducibly evoke just one coherent behavior. For instance, SPINAL CATS—individuals whose spinal cords were transected in the neck region[5]—do not walk spontaneously but do usually walk if injected intravenously with an agent called L-DOPA (which is a precursor in the biosynthesis of the neurotransmitter dopamine), if the body is held over a moving treadmill (Grillner, 1969, 1975; Forssberg and Grillner, 1973; Steeves *et al.*, 1976). In spinal lamprey eels, application of L-DOPA to a saline solution bathing the spinal cord evokes apparently normal swimming undulations of the body (Cohen and Wallen, 1980; Poon, 1980). The higher the dose, the greater the rate of these swimming movements.

In lobsters and crayfish, injection of the neurotransmitter octopamine gives rise to a rigidly extended body posture, and injection of the neurotransmitter serotonin leads to a rigid body flexion (Figure 12). The slow flexor and slow extensor abdominal muscles (Figure 8) are responsible for these two behaviors (Livingstone *et al.*, 1980). By recording from the motor nerves to the slow extensor and slow flexor muscles, a consistent pattern of response to the injected octopamine or serotonin was found (Figure 13). Octopamine excited several excitatory motor neurons to each abdominal slow extensor muscle, as well as the one inhibitory motor neuron to each slow flexor. It also inhibited at least some of the excitor motor neurons to the flexors, as well as the inhibitor motor neuron to each extensor. Thus, octopamine produced exactly the expected set of reciprocal effects on the four sets of motor neurons to the slow muscles that is needed to produce a slow abdominal extension. Serotonin excited only a few of the excitatory motor neurons in the slow flexor muscles. However, it did excite the inhibitory motor neuron to each slow extensor and inhibited both the excitor motor neuron to the extensors and the inhibitory motor neuron to the flexors. Thus, serotonin produced motor effects that could largely account for the slow flexion.

These opposite sets of effects by octopamine and serotonin parallel exactly the pattern of motor excitation and inhibition that result from electrical stimulation of central neurons belonging to command systems for slow abdominal extension

[5]A spinal cat can survive for a substantial period because most of its basic autonomic functions, including respiration, circulation, and alimentation, are controlled by cranial nerves that are not severed by the spinal transection. Thus, the brain, although disconnected from the spinal cord, continues to function and to direct autonomic activities through the descending cranial nerves.

FIGURE 12. Two lobsters, one injected with the neurotransmitter serotonin (left) and the other with octopamine. Serotonin induces a general flexion of the body and limbs, whereas octopamine induces a general extension. (Courtesy of R. Harris-Warrick.)

and slow abdominal flexion (Evoy and Kennedy, 1967). That is, stimulation of interneurons of the slow extensor command system produces the same excitatory and inhibitory effects on the motor neurons as does octopamine; and stimulation of interneurons of the slow flexor command system produces the same excitatory and inhibitory effects as does serotonin. Therefore, it is possible that neurons of the slow extensor command system release octopamine and that those of the slow flexor command system release serotonin as their natural neurotransmitters. This possibility is presently under investigation.

There has been a major explosion of knowledge concerning neuropharmacology within the last several years (e.g., Cooper *et al.*, 1978). If the idea of chemical or pharmacological coding of specific behaviors is borne out in future research, this could offer important opportunities to apply a whole range of sophisticated neuropharmacological techniques to the solution of neuroethological problems. Thus, we may see in the near future a wide range of new approaches made available to the neuroethologist.

Summary

Some behaviors, such as those involving the legs of the cockroach, appear to be hierarchically organized. That is, higher units of organization control intermediate ones, which in turn control lower ones, to produce the behavior. The hierarchy is latticelike, a given unit of organization being controlled by more than one unit at the next higher level. Additional principles of motor organization are: the final common path and the motor unit (both of which apply primarily to vertebrates, in which muscle fibers are singly innervated); motor neuron pools; synergistic and antagonistic muscle groups; and reciprocal innervation.

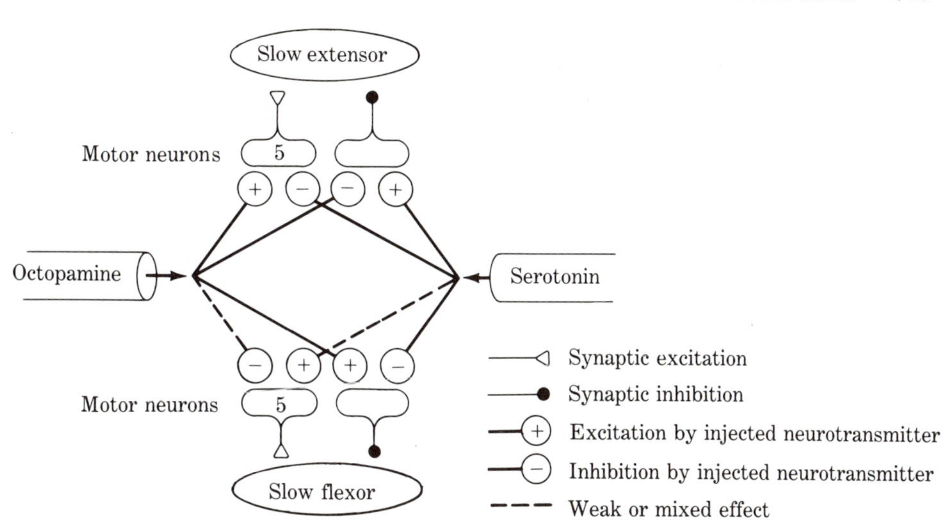

FIGURE 13. Summary of effects of injected octopamine and serotonin on motor neurons to the slow extensor and slow flexor muscles in the lobster. Only one of the five excitatory motor neurons to each muscle is shown.

An important first step in analyzing the neuronal bases of behavior in simpler, invertebrate systems is the working out of the detailed neuronal circuit for the behavior. As applied to the crayfish escape system, this has permitted an analysis of the mechanisms underlying several behavioral properties. First, the all-or-nothing decisionlike property results from the fact that a single all-or-nothing action potential in any lateral giant interneuron (LGI) evokes the fast flexion behavior. Second, the coherent distribution to all parts of the body of a unitary command results from the fact that LGI functions as a command neuron; it is both necessary and sufficient for carrying out the behavior. An LGI projects widely to the numerous motor neurons employed in escape. The command is executed quickly owing to the deployment of rapidly conducting giant axons and rapidly transmitting electrical synapses. Finally, the single-mindedness of the escape behavior results, in large part, from the widespread inhibition by LGI of other neuronal circuits. Thus, the burden of several major neurobehavioral functions rests largely or exclusively with LGI. It should be recalled, however, that LGI is not a single neuron but rather a bilateral pair of electrically coupled neuronal chains cooperating to form a single functional unit. Nevertheless, the delegation of such heavy responsibilities to such single functional units is probably very rare and should be expected only in the cases of extremely stereotyped behaviors. Other behaviors involve more complex, multiunit decision systems (as in the decision by *Aplysia* to release ink) and multiunit command systems (as in the control of swimmeret movements or slow abdominal flexion and extension in the crayfish or lobster, as well as in most vertebrate behaviors).

In the vertebrates, stimulation of selected brain regions can evoke well-coordinated behaviors. This stimulus may excite a part of the circuit for a fixed behavior (as apparently occurs with locomotion in the decerebrate cat), or it may affect the animal's motivation to acquire a particular goal (as in the drinking evoked by hypothalamic stimulation in rats). The sites within the hypothalamus from which a given motivated behavior can be evoked electrically, though reasonably circumscribed, overlap extensively with sites for other behaviors. Some stimuli evoke complex behavioral sequences, whereas others apparently evoke subcomponents of larger sequences—results that provide independent evidence for the hierarchical organization of behavior. Application of particular neurotransmitters to the nervous system can evoke coherent behaviors, a few of which apparently are specific for the particular transmitter applied. This finding suggests some degree of chemical coding of behavior.

Questions for Thought and Discussion

1. One possible approach to the neuronal basis of behavior, rather than laboriously working out the neuronal circuit for a given motor act, is to try to determine strictly theoretically what types of neuronal connections could possibly lead to a given behavior. This theoretical work could be aided by a computer and/or by a mathematical analysis. Do you feel that this approach is a fruitful one? Why? List a few specific questions that you would attack using this approach.
2. What alternatives to the hierarchical organization of behavior are imaginable? What selective advantages might be associated with a hierarchical neurobehavioral organization? What disadvantages?
3. The all-or-nothing properties of both the crayfish escape behavior and inking in *Aplysia* appear to result from positive feedback mechanisms. What problems do you envision arising in positive feedback systems in general? How might these problems be resolved in the decision systems of these two behaviors and among decision systems in general?
4. It has not been demonstrated whether the LGI-mediated escape behavior of the crayfish enhances this animal's survival in encounters with predators. Outline how you would test this question experimentally. What control experiments would be necessary?
5. Some crayfish live on the bottoms of lakes, and others live in rapid streams. What advantages, and what difficulties, might the physical parameters of each of these two environments confer upon the escape circuit as a means of evading predators? What physical measurements would be useful to confirm your answers? What behavioral and neuronal strategies might crayfish employ to take advantage of each of these two environments and to minimize the problems raised in each environment?
6. Design an experiment on the crayfish to determine the relative excitatory synaptic strengths of the LGI-to-FF connection (number 3 in Figure 5B) *vs* the SG-to-FF connection (number 2 in Figure 5B). What kinds of electrodes would you use, where would you place them, and how would you employ them? What controls are necessary? How would you define "excitatory synaptic strength"?

7. Suppose you had one set of stimulating hook electrodes over a sensory nerve of a crayfish's abdomen (St in Figure 6A) and one microelectrode in the nearest LGI, near to its trigger zone (R_{LGI} in Figure 6A). How could you use this stimulating and recording situation to gain confirming evidence for the following aspects of the nerve circuit shown in Figure 5B: (1) the TSCs excite SIs strongly and LGI weakly, whereas SIs excite LGI strongly; (2) the TSC-to-SI synapses are chemical, and both the TSC-to-LGI and the SI-to-LGI synapses are electrical?

Recommended Readings

BOOKS

Kandel, E.R. (1976) *Cellular Basis of Behavior.* W.H. Freeman and Company, San Francisco.
An excellent, clearly written exposition of the neuronal basis of behavior, primarily in the mollusk *Aplysia,* by an outstanding contemporary researcher. The cellular basis of reflexes and fixed action patterns are presented in Chapter 9.

Hoyle, G. (ed.). (1977) *Identified Neurons and Behavior of Arthropods.* Plenum Press, New York.
This volume contains a diversity of review articles presented at a symposium. The book is dedicated to C.A.G. Wiersma, who initiated research on the crayfish escape system.

Herman, R.M., Grillner, S., Stein, P.S.G. and Stuart, D.G. (1976) *Neural Control* of *Locomotion.* Plenum Press, New York.
This symposium report contains a rich diversity of contributions from workers on both invertebrate and vertebrate locomotion.

Gallistel, C.R. (1980) *The Organization of Action: A New Synthesis.* Lawrence Erlbaum Associates, Hillsdale, New Jersey.
An interesting account of the neuronal basis of behavior, encompassing the points of view of both physiological psychology and neurobiology.

Valenstein, E.S. (1973) *Brain Stimulation and Motivation: Research and Commentary.* Scott, Foresman & Company, Glenview, Illinois.
A good summary of techniques, interpretations, and problems in this field.

Sherrington, C. (1961) *The Integrative Action of the Nervous System.* Yale University Press, New Haven, Connecticut. (Reprint of the original 1906 edition.)
A landmark book in the history of integrative neurobiology, recommended for its historical interest.

ARTICLES

Kupferman, I. and Weiss, K.R. (1978) The command neuron concept. *Behav. Brain Sci.* 1: 3–10.
A highly thought-provoking article, followed by commentaries by numerous researchers in the field.

Wine, J.J. and Krasne, F.B. (1982) The cellular organization of crayfish escape behavior. In *The Biology of Crustacea.* Vol. IV. *Neural Integration.* D.E. Bliss, H. Atwood and D. Sandeman (eds.). Academic Press, New York.

An excellent, detailed review of this important work.

Olson, G.C. and Krasne, F.B. (1981) The crayfish lateral giants are command neurons for escape behavior. *Brain Res.* 214: 89–100.

This fine article demonstrates that the LGIs are both necessary and sufficient for escape behavior.

CHAPTER 9

Central Neuronal Control of Temporally Patterned Behaviors

In the last chapter we discussed the mechanisms by which a crayfish coordinates the simultaneous movement of many parts of its body during a single episode of behavior—the initial abdominal downstroke of the escape response. Most animal behaviors, however, consist not of single, isolated events but rather of sequences of movements. Indeed, even in the crayfish escape behavior, the initial downstroke is only the prelude to a more prolonged swimming sequence. In this chapter we will consider how an animal's nervous system, particularly its central nervous system, produces coordinated sequences of motor acts. We will begin with a behavioral description of these sequences.

Behavioral Sequences

Often a complex behavioral act can be divided into several distinct subunits that occur in succession. The particular sequence in which these various subunits are enacted may be (1) totally random, (2) strictly fixed, or (3) probabilistic (or, as it is sometimes called, stochastic). (In a probabilistic sequence, a particular subunit A would usually be followed by another specific subunit B but could instead occasionally be followed by a different subunit.) Totally random sequences are rare, though examples can be found in the trial-and-error searching behavior of protozoans, in which the sequence of left *vs* right turns is apparently random (Jennings, 1906). Strictly fixed sequences often occur in locomotory behaviors, as in the stepping sequence left–right–left–right of a bipedal animal. Such fixed sequences usually involve few subunits (just two in bipedal stepping) and the sequential immutability may result from some mechanical constraint, such as the requirement of maintaining balance. Probabilistically ordered sequences are found in a wide variety of behaviors other than locomotion, a few of which we shall examine in this section.

One can readily determine whether a given sequence of behavioral subunits is strictly fixed by simply recording the sequence and searching within it for a strictly repeating pattern. If the sequence is not perfectly fixed, determining whether it is probabilistic or completely random is more complicated and requires statistical analysis. One such analysis has been carried out on the courtship behavior of male guppies (Baerends *et al.*, 1955). This behavior is composed of 15 subunits. Two of these, called "following with fins folded" (F_f) and "sigmoid" (S), are shown in Figure 1A. After recording the sequence of all subunits for many courtship performances by males,the *transitions* from each subunit to the next were listed. As is common in such studies, a table was drawn showing the TRANSITION FREQUENCIES—the total number of times that any given behavioral subunit performed by the male (such as F_f) was followed by his performing each other subunit (such as S). To aid in visualizing these transitions, the data of the table can be represented in a number of different diagrammatic forms (Sustare, 1978)

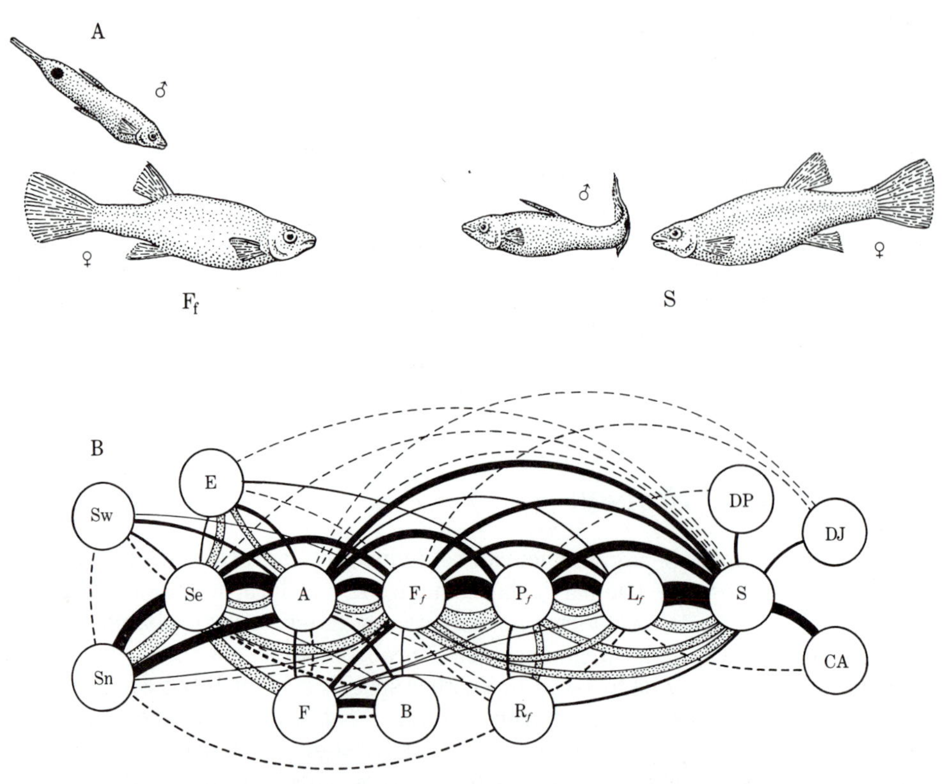

FIGURE 1. Courtship behavior of male guppies. A. Two subunits of courtship behavior: "following with fins folded" (F_f), and "Sigmoid" (S). B. Transition diagram. Each circle represents 1 of the 15 identified subunits of male courtship behavior. The thickness of a line between any two subunits represents the number of times a transition between these two subunits was observed. Black lines represent transitions to the right on the diagram; gray or dashed lines, transitions to the left on the diagram. (From Baerends *et al.*, 1955.)

such as that shown in Figure 1B. Here each behavioral subunit is indicated by a circle, and the transition frequency between any two subunits is represented by the thickness of the line running between their two circles. (Solid lines represent sequences proceeding to the right on the diagram, and gray or dashed lines represent sequences proceeding to the left.) The diagram suggests visually that although the sequencing is complex it is not random but probabilistic, because some transitions are extremely common and others are extremely rare or nonexistent. Statistical analysis verifies this.

During the male guppy's courtship behavior, the female usually remains passive and almost immobile. She does not appear to provide the male with any communication signals that might be responsible for his transitions from one subunit of courtship behavior to the next. Thus, the transitions among subunits appear to be produced by the male himself.[1] This suggests that for any two of the male's behavioral subunits that occur often in succession with each other, some neurons responsible for producing one of these subunits interact with neurons that produce the other subunit. This interaction would be such as to increase the probability that the second subunit will follow the first. For instance, neurons responsible for the subunit S_e presumably interact with those for A, which interact with those for F_f, and so forth (Figure 1B). Thus, even though the specific cells mediating these different subunits are not known either anatomically or physiologically, by observing in detail the sequence of behaviors, and thereby discovering rules of interaction among the behavioral subunits, one can infer underlying general rules of neural interaction.

What types of rules of probabilistic interaction are possible among behavioral subunits? Imagine that a behavioral act carried out by some animal consists of five subunits: A, B, C, D, and E. The simplest rule of transition would be one in which each subunit follows with high probability one particular other subunit. That is, B could follow A with high probability, and likewise C could follow B, D could follow C, and E could follow D. This rule of sequencing is called a MARKOV CHAIN. Specifically, it is a *first-order* Markov chain in that the occurrence of any given subunit depends *only* on the immediately preceding subunit.

A more complex rule is one where the occurrence of each subunit depends upon the *two* previously occurring subunits. This is a *second-order* Markov chain. For instance, subunit C could have its highest probability of occurrence following the sequential pair A–B. Subunit C would occur less often after other sequential pairs such as A–D, A–E, and B–A. This could result in a high probability of the sequential triad A–B–C. Likewise, D could occur with its highest probability immediately following the sequence B–C, and so forth. In a *third-order* Markov chain, the occurrence of each subunit depends upon the *three* preceeding subunits. Thus, D would occur with its highest probability following the sequential triad A–B–C, and so forth. One can test for the presence of a Markov chain and can distinguish among first-, second-, third-, and higher order chains by subjecting

[1] It remains possible that the female communicates with the male by chemical or other cues difficult for the investigator to detect. However, in sequential behaviors of other species, such as courtship singing in birds mentioned later in this section, the singing male performs his rhythmically patterned song far away from any courted female, ruling out female interaction in the establishment of the song's rhythm.

the transition frequencies among the subunits to Markovian statistical tests (Fagan and Young, 1978).

Markov chains have been demonstrated in many behavioral sequences, such as the sequential singing behavior of some birds. For instance, in different species of thrushes, the sequences of vocal components can be described by first-, second-, or third-order Markov chains (Lemon, 1977). A number of rules for sequential patterning other than Markov chains have also been described suggesting other, often complex, forms of behavioral and neural interaction (Dawkins, 1976; Fagan and Young, 1978; Sustare, 1978).

Most attempts to study directly the neurophysiological basis of sequential actions have turned to behaviors in which the sequence of subunits is strictly fixed. Locomotory rhythms have played a preeminent role in this work. One advantage of studying locomotion is that each cycle of movement, such as a single elevation of an insect's wings or a single step of a leg, represents an entire behavioral replication. Thus, repeated samplings of physiological data can be acquired very quickly. Moreover, because the sequence of events is fixed, one can expect the underlying physiological events to be relatively straightforward and comprehensible. A description of attempts to understand locomotory and other fixed rhythmic behaviors will occupy the bulk of this chapter. We will turn first to methods for describing such behavioral sequences. Next we will explore in depth the physiological basis of one locomotory rhythm: swimming in leeches. The remainder of the chapter will deal with some important related issues that have been explored by utilizing other rhythmic behaviors.

Describing Rhythmic Events

Figure 2 presents several standard graphic techniques for describing rhythmic behavior. In Figure 2A, the top trace is a schematic graph of the position of a single *hindwing* of a flying locust as a function of time. Because this graph fairly closely resembles a sine wave, the wing's movement can be described reasonably well by just two parameters: amplitude and frequency. For rhythmic behaviors, one often specifies not the frequency but rather 1/frequency, usually called CYCLE PERIOD. In Figure 2A, the cycle period is 60 milliseconds (corresponding to a frequency of 16.6 wingbeats/sec).

The movements of the ipsilateral *forewing* of the locust are shown by the second trace in Figure 2A. This wing has the same cycle period and about the

FIGURE 2. Graphical techniques for studying a rhythmic behavior, illustrated with an example from locust flight. A. The positions of a locust's hindwing and of a forewing, and impulses recorded from hindwing and forewing depressor muscles, as a function of time. The interval between hindwing and forewing and the cycle period of the hindwing are indicated. B. Interval histogram, hindwing to forewing. C. Phase histogram, forewing cycle with respect to hindwing cycle. D. Two sine waves that have no phase relationship to one another, and a phase histogram based upon such data. See text for further explanation. (A after Wilson, 1967.) ▶

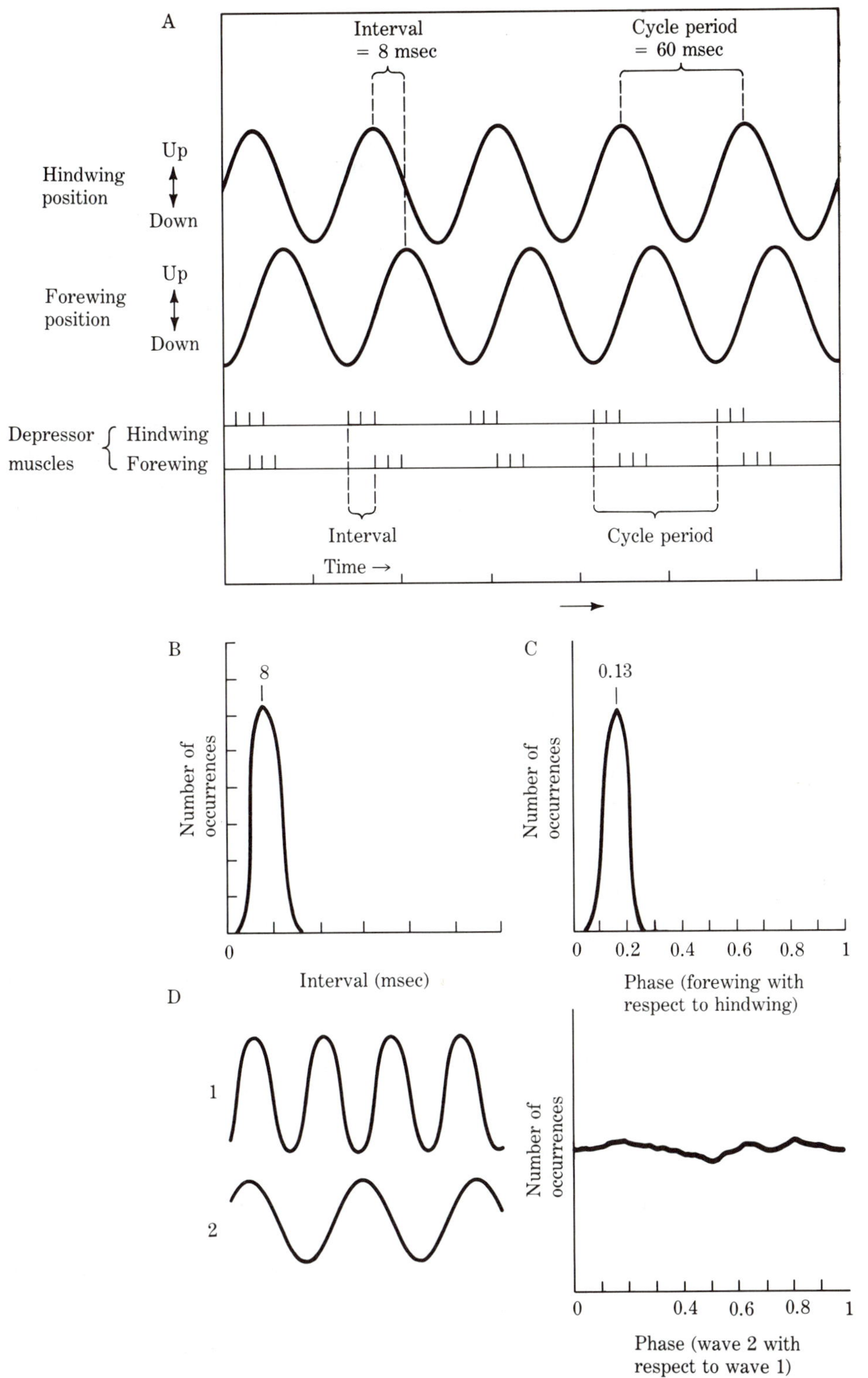
A
Interval
= 8 msec
Cycle period
= 60 msec
Up
Hindwing
position
Down
Up
Forewing
position
Down
Depressor muscles
Hindwing
Forewing
Interval
Cycle period
Time →
B
8
Number of occurrences
0
Interval (msec)
C
0.13
Number of occurrences
0
0.2
0.4
0.6
0.8
1
Phase (forewing with respect to hindwing)
D
1
2
Number of occurrences
0
0.4
0.6
0.8
1
Phase (wave 2 with respect to wave 1)

same amplitude as the hindwing. Its movements, however, are displaced in time relative to those of the hindwing. The temporal relationship of these two rhythms can be specified in either of two ways. The first of these, the *absolute* time difference between the two wings, can be determined by measuring the interval from the moment of maximal elevation on one cycle of the hindwing to the next occurring maximal elevation of the forewing. (Alternatively, some other point on the cycle, such as the moment of lowest wing position, could be used.) In Figure 2A this interval equals 8 milliseconds. To determine whether this same interval occurs repeatedly throughout flight and not just for the few wingbeats shown in the figure, one usually repeats the interval measurement for hundreds of successive cycles and then plots an INTERVAL HISTOGRAM (Figure 2B). The sharp peak of this histogram at 8 milliseconds, together with a statistical analysis showing that this peak almost certainly did not result by random chance, shows that the forewing's movement cycle consistently lags behind that of the hindwing.

The second way to specify the temporal relationship between two wings is to measure the *fractional* time difference between their rhythms. To do so, one first measures the interval between them as just described and then divides this interval by the cycle period. (Because cycle period usually varies slightly from cycle to cycle, it is important to measure the particular cycle period of the hindwing within which the measured interval occurs.) This fraction,

$$\frac{\text{interval (peak of hindwing to peak of forewing)}}{\text{cycle period (of hindwing)}}$$

is called the PHASE of the forewing cycle with respect to the hindwing cycle. It specifies the fraction of the hindwing cycle period that has elapsed at the moment when the next forewing cycle begins. The range of these possible fractions and thus the range of possible phase values extend from 0 to 1. In Figure 2A, the phase of the forewing cycle within the hindwing cycle is 8/60, or 0.13. As an alternative description, one could measure the phase of the hindwing cycle with respect to the forewing cycle. This phase value would be 52/60, or 0.87. To see whether phase is constant throughout the animal's flight, one would determine the phase for hundreds of successive wingbeats and then plot a PHASE HISTOGRAM (Figure 2C). Coupled with a statistical analysis showing the peak at 0.13 to be nonrandom, this histogram demonstrates a consistent phase lag of the forewing behind the hindwing.

Phase histograms offer a clear and simple way to describe the relationship between the two rhythms. For instance, if the two rhythms are exactly alternating in time (in which case they would be called RECIPROCAL), their phase histogram would have a peak of 0.5. If the two rhythms of fixed cycle period bear no temporal relationship to one another, their phase histograms will have no statisticaly significant peaks (Figure 2D).

One complexity in the representation of rhythmic behaviors occurs where the cycle period varies, as for instance when a person walks at different speeds. Whether one saunters with a very low cycle period, such as 0.2 seconds, or hurries at ten times this rate, each leg usually steps with a phase of just about 0.5 with respect to the cycle of the other leg. This is a simple necessity for keeping one's balance. A phase histogram of stepping would therefore show a sharp peak at 0.5. The rhythmic movements of the two legs, then, are said to be PHASE LOCKED to one another, or to show PHASE CONSTANCY. However, in order to achieve this phase constancy, as the cycle period of walking increases the *interval* from the onset of a step in one leg to a step in the other leg increases correspondingly. Thus, an *interval* histogram of these data would not show a sharp peak, incorrectly suggesting that the walking is not well coordinated. Other rhythmic behaviors show INTERVAL CONSTANCY, in which the peak of one rhythmically moving appendage follows the peak of another moving appendage with a fixed interval. To achieve this interval constancy, the phase varies as cycle period changes. Still other rhythmic behaviors show a compromise between phase constancy and interval constancy. Because of these complications introduced by variations in cycle period, such behaviors should be analyzed using both interval and phase histograms.

Interval and phase histograms are also useful in analyzing the *neuronal* activity underlying a rhythmic behavior. For instance, the action potentials from the locust's hindwing and forewing depressor muscles, shown on the lower two traces in Figure 2A, can be quantified in this way. One typically measures intervals and cycle periods beginning with the first action potential of a burst in each cell (as shown in the figure), or alternatively, with the middle action potential of a burst. Interval and phase histograms plotted from the physiological recordings of Figure 2A would resemble those of Figure 2B and 2C, respectively.

Neural Control of Swimming in Leeches

The bulk of this chapter consists of a description of just one neurobehavioral system: leech swimming. Through this example we will confront most of the experimental strategies, techniques, and problems of interpretation encountered by workers on a diversity of such rhythmic systems.

SWIMMING BEHAVIOR

The leech *Hirudo medicinalis* swims by means of dorsal and ventral undulations of its dorsoventrally flattened body (Figure 3). These undulations have the following characteristics (Kristan *et al.*, 1974). Any location along the body is thrown into alternating dorsal peaks and ventral troughs, as can be seen by following the movements of any of the four marks on the body in Figure 3. The cycle period varies between about 350 and 2000 milliseconds. The phase of dorsal peaks with respect to ventral troughs for any given body location is about 0.5. In addition, the peaks and troughs move posteriorly along the body in the form of waves, as shown by the short arrows. Such a progressive movement of a rhythmic activity along an animal's body is called a METACHRONAL WAVE. In the leech the metachronal wave progresses faster during faster swimming (shorter cycle periods) and ranges from 19 to 27 milliseconds per body segment. Given this brief description of the behavior, one can see that any satisfactory physiological explanation of leech swimming would have to account for at least three behavioral

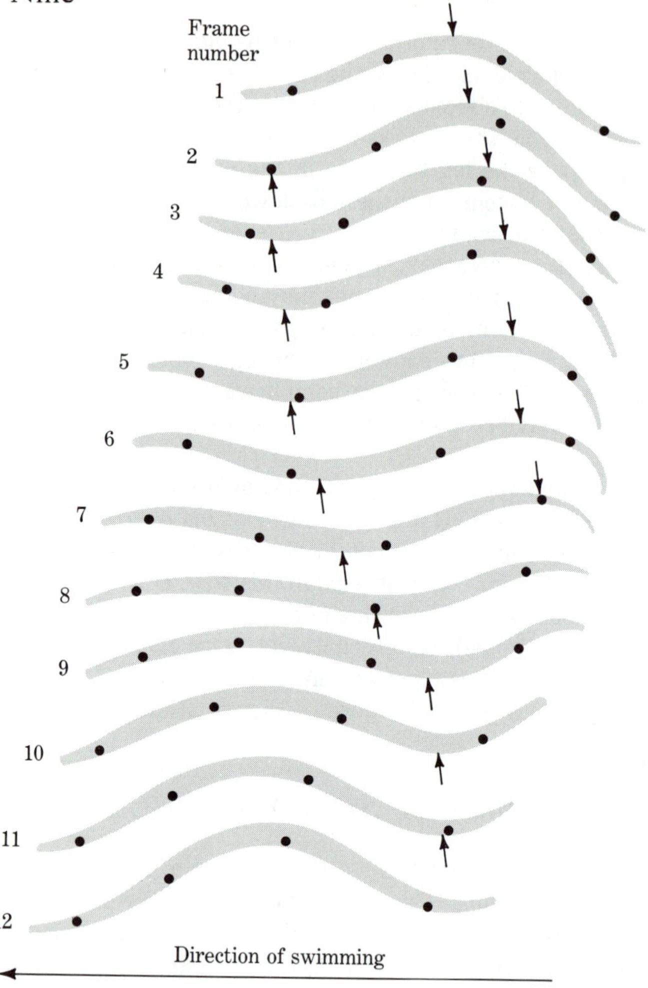

FIGURE 3. Swimming undulations of a leech, viewed from the side. Tracings of successive frames of motion picture film are arranged in sequence from top to bottom. The time occupied by this sequence is about 400 milliseconds, during which the leech moved to the left by about ½ body length. The leech's body had been marked with four reference spots as shown. A posteriorly propagating dorsal peak and ventral trough are indicated by the arrows. (After Kristan *et al.*, 1974.)

properties: the regular dorsoventral alternation at a given point on the body; the posteriorly moveing metachronal wave; and the measured durations of each of these activities.

CHAIN REFLEXES AND CENTRAL PATTERN GENERATORS

What type of neuronal mechanism might give rise to these rhythmic body movements? For leech swimming or for other temporally patterned behaviors, two general types of mechanism are possible. First, the rhythm could be derived from a CHAIN REFLEX—a sequence of sensory–motor interactions, in which each movement by the animal activates some sensory receptors which, in turn, induce the animal's next movement. In the case of a swimming leech, a chain reflex could, in theory, account for the dorsoventral rhythm at a given locus along the body in the following manner (see numbered sequence in Figure 4A). (1) The leech might begin the sequence by making an initial contraction of one local set of muscles, such as those of the ventral body wall. Contracting these muscles causes a local peak of the dorsal body surface. (2) This body position would stretch the dendrites of a stretch-receptor embedded in the dorsal body wall, thus activating this receptor neuron. The axon of this receptor might (3) excite motor neurons to the dorsal muscles and (4) by reciprocal innervation (Chapter 8) might also inhibit motor neurons to the ventral muscles. This would lead to (5) contraction of the dorsal muscles and relaxation of the ventral muscles, causing a local trough of the body (second panel). This body position would (6) activate a ventral stretch-receptor, which would then (7) excite ventral motor neurons and (8) inhibit dorsal motor neurons, resulting in a ventral contraction and thus a local peak of the body (third panel). Because this reestablishes the initial starting point, the entire sequence would repeat itself, over and over again. By this scheme, then, all that would be required to initiate a prolonged swimming sequence is an initial contraction of the ventral muscles. Alternatively, an initial contraction of the dorsal muscles would be just as effective.

The presence of a chain reflex of this type in leeches has been confirmed (Kristan and Stent, 1974). If one stretches a local region of the dorsal or ventral body wall, stretch-sensitive receptors are in fact excited, and these produce synaptic effects on the motor neurons that are functionally like those of Figure 4A. This does not prove, however, that these sensory-motor interactions are responsible, by themselves, for the swimming movements; in fact, as we shall see later in this chapter, they almost certainly are not; they are either too weak or improperly timed to account by themselves for swimming. Because chain reflexes represent a form of neuronal feedback, with movements exciting receptors that in turn induce movements, the further discussion of these reflexes is deferred until Chapter 10, which concerns the general question of feedback in behavior and the nervous system. We turn here instead to the second general type of mechanism capable of producing temporally patterned behavior.

The second type of mechanism is a CENTRAL PATTERN GENERATOR, or, as it is often called in the case of rhythmically repeating behaviors such as locomotion, a CENTRAL NEURONAL OSCILLATOR. This is a functional neural subsystem that is contained entirely within the central nervous system and that can produce rhythmically patterned motor outputs without the use of timing information from sensory receptors. It is thus like a circadian clock (Chapter 5) in that it somehow possesses an inherent timing capability, although it has a cycle period many orders of magnitude shorter than that of circadian clocks. In the case of leech swimming, such a central oscillator could produce locally alternating peaks and troughs of the body by the following hypothetical sequence of events (see numbered sequence in Figure 4B). Beginning with a leech in the resting position, the oscillator would (1) simultaneously excite the dorsal motor neurons and inhibit the ventral motor

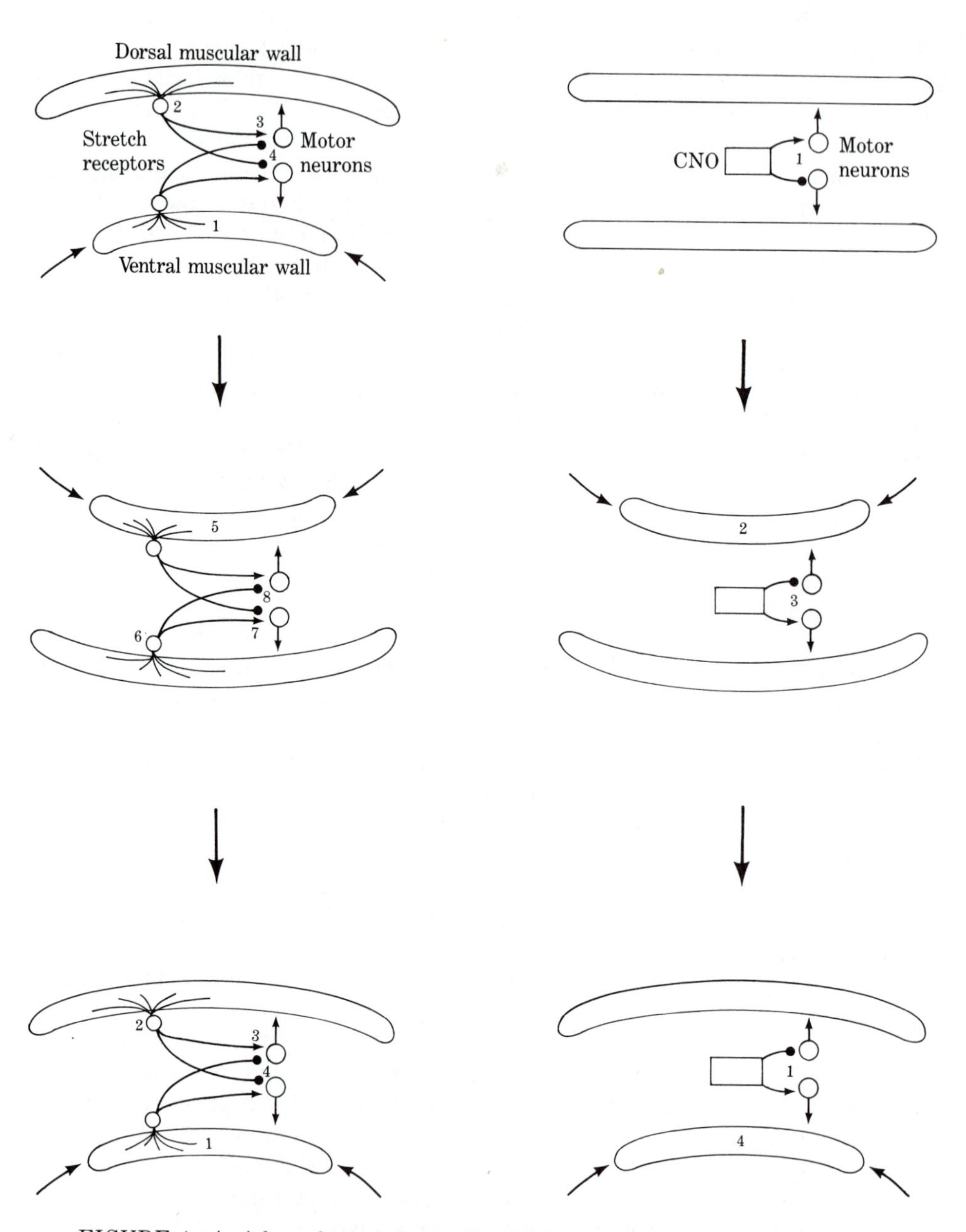

FIGURE 4. A. A hypothetical chain reflex for leech swimming. One possible sequence of events is indicated by the numbers. A short piece of dorsal and ventral muscular body wall is shown, viewed from the side. This piece of body wall alternates between a local peak and a local trough. B. A hypothetical central neuronal oscillator (CNO) for leech swimming. One possible sequence of events is indicated by the numbers. Explanation in text.

neurons. This would cause (2) local contraction of the dorsal muscles and thus a local trough of the body. Then, after an appropriate interval, the oscillator would (3) excite the ventral motor neurons and inhibit the dorsal motor neurons. This would cause (4) contraction of the ventral muscles and thus a local peak of the body. This entire sequence would be repeated over and over. Although this scheme says nothing about how the central neuronal oscillator would determine the appropriate interval between successive events in the sequence, the scheme does suggest a distinct alternative to rhythmicity based upon a chain reflex.

Does the leech have a central neuronal oscillator for swimming? One could demonstrate such an oscillator if one could observe swimming activity after ablating all of the animal's sensory receptors, a procedure called DEAFFERENTATION. This procedure would block chain reflexes, which depend upon sensory receptors but should not interfere with any central oscillator because it, by definition, would contain no sensory receptors (Figure 4B). In carrying out this search for a central oscillator, one must ablate all receptor types, not just those sensitive to stretch of the body wall, because it is at least conceivable that any type of receptor could detect some physical disturbance created by an animal's own movement and could thus trigger the next movement. Because the sensory axons of leeches, as in most invertebrates, run together with motor axons in mixed sensory–motor peripheral nerves, one must in fact cut *all* the peripheral nerves of the body, thereby both deafferenting and deefferenting the animal. This procedure leaves the central nervous system totally isolated from the periphery (Figure 5A). One would then record the neural activity of one or more peripheral nerves proximal to their transections to look for a temporal pattern of activity emerging from the central nervous system along the motor axon. (It may be necessary to stimulate the nerve cord briefly to get such a rhythm started.) If a rhythm closely resembling that for swimming could be observed in such recordings, this would suggest that the swimming leech has a central neuronal oscillator that may be used as part of its locomotory mechanism.

The experimental arrangement that has been used for this test is shown in Figure 5A (Kristan and Calabrese, 1976). A brief train of electric shocks delivered to a peripheral nerve, or to a connective of the isolated central nervous system, or to identified interneurons responsible for initiating swimming (Weeks and Kristan, 1978; Weeks, 1981) evoked a prolonged sequence of rhythmic motor outputs in the stumps of the leech's transected peripheral nerves (Figure 5B). Bursts of action potentials recorded from the stumps of the nerves to the dorsal muscles and those to the ventral muscles showed a regular alternation. The mean cycle period recorded was 385 milliseconds, and the phase of bursts in the dorsal nerve with respect to those in the ventral nerve was 0.51. Moreover, recordings from the nerves of different ganglia revealed the presence of a normal, posteriorly directed metachronal wave, traveling at 20 milliseconds/body segment. These parameters all fall within the normal range for an intact swimming leech. These results, then, show that the leech has a central neuronal oscillator that, matching as it does the normal swimming rhythm, probably plays an important role in producing this rhythm. It should be recalled, however, that a chain reflex probably also plays some role. In fact, a detailed comparison of the rhythmic motor outputs recorded from isolated nerve cords *vs* the motor outputs recorded from nearly

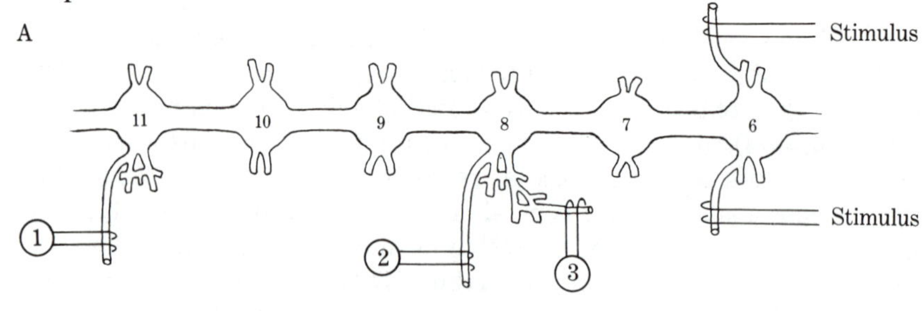

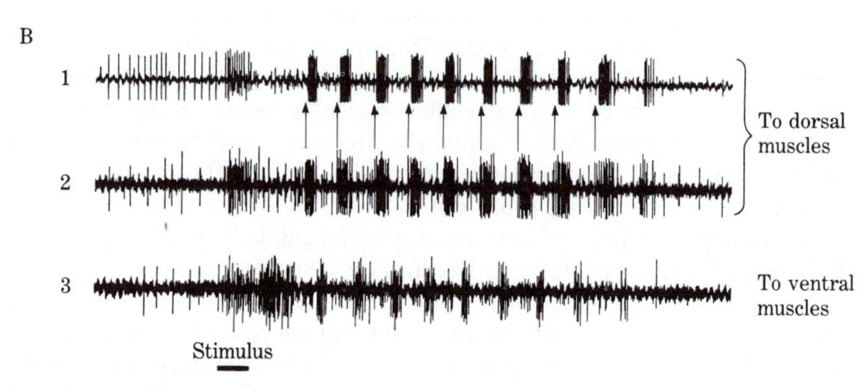

FIGURE 5. A. Part of the leech central nervous system, totally isolated from the periphery by cutting all peripheral nerves. Shown are the positions of two pairs of stimulating electrodes (stimulus) and three pairs of recording electrodes, (1, 2, 3), used for testing the presence of a central neuronal oscillator. A chain of ganglia much longer than that shown was actually used in this experiment. B. Recordings from peripheral nerves at positions 1, 2, and 3 of part A, in response to a brief train of electrical shocks delivered through both pairs of stimulating electrodes, at the time shown by the bar below. The bursts of action potentials in the peripheral nerve of ganglion 11 lagged behind those of the same nerve in ganglion 8 (arrows), showing that a normal, posteriorally directed metachronal wave is present. The bursts of large action potentials recorded in position 2 are known to be those of an excitatory motor neuron to a dorsal muscle, and those in position 3 belong to an excitatory motor neuron to the ventral muscles. The alternation between the bursts in these two recording positions reflects the normal dorsoventral alternation. (B after Kristan and Calabrese, 1976.)

intact, swimming leeches show some consistent, though minor, discrepancies (Kristan and Calabrese, 1976). These discrepancies could represent a role normally played by chain reflexes, or they could reflect undefined disturbances incurred by the central neuronal oscillator as a result of the trauma of isolating the nerve cord.

Almost every rhythmic behavior in invertebrates or vertebrates that has been investigated is mediated, at least in part, by a central neuronal oscillator. The list includes such behaviors as walking, flying, swimming, scratching, breathing, chewing, sound production, intestinal movements, and heartbeats (Delcomyn, 1980). It also includes a wide range of species, from phyletically ancient invertebrates to primates (Taub, 1976; Keele and Summers, 1976). Thus, the strategy of using a central neuronal timer as part of the mechanism for rhythmic behaviors was arrived at early in evolution and has been retained ever since.

METACHRONAL WAVES: MASTER OSCILLATOR OR COUPLED OSCILLATORS?

There are two different general ways in which central neuronal oscillators could produce metachronal waves such as those seen in leech swimming. One way would entail just a single central oscillator, perhaps located in the brain, plus a set of delaying mechanisms. These delays would cause the rhythmic signal from the single oscillator to arrive progressively later at more posterior points along the body. This single central oscillator would be called a MASTER OSCILLATOR because it would time the movements of the entire body. The second means of producing the metachronal wave would involve more than one oscillator, perhaps as many as one for each ganglion of the leech. Each oscillator would control the dorsoventral movements of only its local body region. These separate central oscillators would then need to be neuronally connected in such a way that the rhythm of each oscillator lags slightly behind its next anterior neighbor. The connected series would be called a system of COUPLED OSCILLATORS.[2]

If the leech uses coupled oscillators, it should be possible to record rhythmic motor outputs from at least two different subchains of ganglia from the nerve cord of a leech, separated from each other by cutting the connective between them. In fact, if one cuts a whole leech in half, both the front and the rear half can swim quite normally (Kristan *et al.*, 1974). However, because neither half has been deafferented, chain reflexes could be generating these rhythmic movements. Nevertheless, two other experiments on deafferented nerve cords strongly support the presence of coupled oscillators in the leech. In the first of these (Weeks, 1981), recordings were made from peripheral nerves of a totally isolated nerve cord in response to electrical stimulation of interneurons responsible for initiating swimming (Figure 6A). Normal rhythmic motor outputs for swimming occurred (Figure 6B) and the outputs of different ganglia were clearly phase locked to one another (Figure 6C). Then the connectives between two ganglia, numbers 11 and 12, were cut, leaving just a small nerve bundle between them that contains axons of the swim-initiating interneurons. Stimulation of these interneurons again excited rhythmic motor outputs for swimming from the ganglia both anterior and posterior to the cut connectives. However, the cycle periods were different in the anterior and the posterior ganglia (Figure 6E). This means that the rhythms anterior and posterior to the cut were produced by different central neuronal oscillators.

The second observation suggesting coupled oscillators is that adjacent pairs of ganglia, snipped out from an isolated nerve cord, produce a normal rhythm in

[2] Although the term *coupling*, when used by itself, often implies electrical synapses, the term *coupled oscillator* is not intended to imply any specific type of synaptic interaction within or between the oscillators.

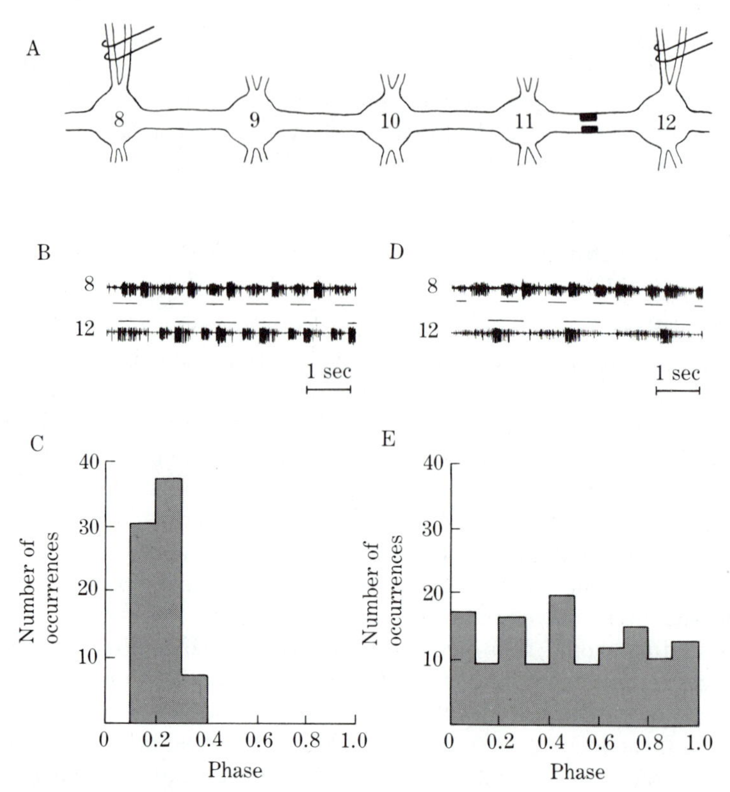

FIGURE 6. Motor outputs for leech swimming before and after cutting the interganglionic connectives, leaving only a small central bundle of axons intact. A. Recording arrangement. Recordings were made in segments 8 and 12 from peripheral nerves, whose motor axons go to both dorsal and ventral muscles. For some recordings, the connectives between ganglia 11 and 12 were cut, leaving only a small central bundle intact between these two ganglia. B. Recordings made before cutting the connectives between ganglia at 11 and 12. Bars demarcate bursts of action potentials known to be from motor neurons to the dorsal muscles. C. Phase histogram of bursts of active potentials in ganglion 12 with respect to those in ganglion 8, based upon data like that of part B. D. Recordings like those in B, but after cutting the connectives between ganglia 11 and 12, as shown. E. Phase histogram of ganglion 12 with respect to ganglion 8, based upon data like that of part D. (For some cycle periods of ganglion 8 in D, there was no burst in ganglion 12. For such cycles, no phase measurement could be made.) (From Weeks, 1981.)

response to stimulation of swim-initiating interneurons (Figure 7). Because different ganglionic pairs from a given leech can show this effect, several such oscillators must be present within the nerve cord (Weeks, 1981).

Does each ganglion have its own central neuronal oscillator for swimming? This was tested by cutting connectives in an isolated nerve cord, again leaving

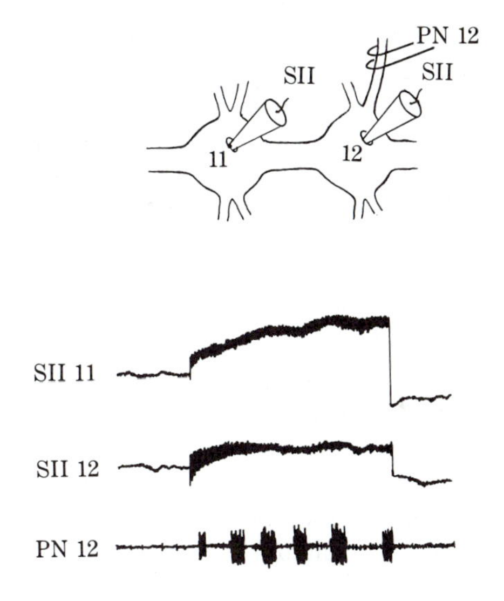

FIGURE 7.

Rhythmic bursts of action potentials produced in a peripheral nerve of ganglion 12 (PN12) of an isolated two-ganglion fragment (ganglia 11 and 12) of the leech nerve cord. The rhythm was initiated by a strong depolarization through two intracellular stimulating electrodes, in two swim-initiating neurons (SII), one in each ganglion. This stimulus induced trains of action potentials in each SII, which, however, are too small, as recorded in the cell bodies (the sites of SII impalement), to appear on the SII records. (From Weeks, 1981.)

intact just the small connecting nerve bundle containing axons of the swim-initiating interneurons, as in Figure 6. But now these cuts were made both anterior and posterior to a single ganglion (Weeks, 1981). This nearly isolated ganglion could therefore communicate with others only by way of the small swim-initiating bundles. Again, stimulating these interneurons evoked rhythmic motor outputs from all ganglia. The one nearly isolated ganglion was not phase locked to any other ganglia. Thus, this one ganglion must contain its own oscillator. Likewise, each other ganglion in the nerve cord is thought to have its own oscillator.

All these findings, then, point toward a coupled set of central neuronal oscillators, one in each ganglion, underlying leech swimming. As we shall see later in this chapter, the neuronal circuit that has been discovered for swimming also suggests a system of coupled oscillators rather than a master oscillator. Metachronal rhythms in the few other systems that have been studied (namely, the crayfish swimmerets and swimming in fish) also appear to employ coupled oscillators (Ikeda and Wiersma, 1964; Stein, 1971; Grillner, 1975).

ENDOGENOUS BURSTERS AND NETWORK OSCILLATORS

How does a central neuronal oscillator produce rhythmic bursts of action potentials? Two different general mechanisms are possible. In one mechanism, an *individual neuron* within the central nervous system would produce the rhythm all on its own, utilizing sequential openings and closings of specialized ionic channels within its cell membrane. (These would not be the same channels as are used in producing synaptic potentials or action potentials.) Suppose, for instance, that a set of such specialized channels permits certain positive ions to flow into the cell, or negative ions to flow out. Suppose, now, that such channels open for a duration of about 0.5 second. The cell would become depolarized during this interval. If the depolarization reaches threshold, a burst of action potentials occurs. Suppose that now a different set of specialized channels opens, for the next 0.5 second.

These channels permit certain positive ions to leave or negative ions to enter the cell. This would lead to a hyperpolarization and, thus, a termination of the burst of action potentials. This alternating sequence of opening and closing two sets of channels could continue, thus maintaining periodic bursts of action potentials, in this case at a rate of one burst per second. This overall mechanism, then, would not rely upon synaptic interactions with any other neurons for the production of the recurring bursts of action potentials. (Other neurons may indeed be required to start this rhythmic sequence of openings of the two specialized sets of channels but would not be involved in producing the rhythm itself.) A single rhythm-generating cell such as that just described would be called an ENDOGENOUS BURSTER. Such an endogenous burster could drive many postsynaptic neurons, causing these to become cyclically active; yet the *source* of the rhythm would be entirely confined to the endogenous burster cell itself.

One well-studied endogenous burster, the PB neuron of the mollusk *Aplysia*, achieves it rhythmicity in just this way—by opening first channels for sodium and calcium, whose entry depolarizes the cell, and then channels for potassium, whose exit hyperpolarizes the cell (Meech, 1979; Koester and Byrne, 1982). In response to each of these slow membrane depolarizations, the neuron's normal voltage-sensitive sodium and potassium channels open and close repetitively, giving rise in the usual manner to the individual action potentials of a burst.

The second possible mechanism for producing bursts of action potentials involves no one neuron whose cell membrane by itself possesses a rhythmic capability. Rather, the cyclic activity would be a NETWORK PROPERTY, or EMERGENT PROPERTY. That is, the rhythm would emerge as a result of the network of *synaptic interactions* among two or more neurons. We will see some examples of such network oscillators later in this chapter. (As a third possibility, a rhythm may be derived from endogenous burster cells that receive some assistance from rhythmic synaptic input produced by a network; Calabrese, 1979; Selverston, 1980.)

How can one determine whether a centrally produced rhythm is based upon endogenous bursters or network properties or both? To search for an endogenous burster, one might attempt somehow to isolate a given rhythmically active neuron from all sources of synaptic input and then determine whether its rhythmicity persists. One would repeat this process for each rhythmic cell. If the rhythm does persist in any cell that has been isolated, this cell, by definition, is an endogenous burster. If the rhythm does not persist, either the cell is not an endogenous burster, or perhaps it is one, but its bursting capability has been disrupted for some reason during the process of isolation. (It should be noted that this experimental strategy is the same as that of isolating the central nervous system from sensory inputs by means of deafferentation to determine whether it contains a central neuronal oscillator; Figure 5.)

The PB cell of *Aplysia* has a cell body large enough that one can dissect it out of its ganglion, effecting a total physical isolation from other neurons. When placed alone in a dish of saline, the PB cell continues to show its normal bursting rhythm of action potentials, demonstrating that this cell is an endogenous burster (Alving, 1968). However, many other rhythmically active neurons either have cell bodies too small to dissect out or else generate their rhythms not in their cell bodies but rather in their dendritic trees, which are structurally too complex to be dissected. As an alternate approach to studying such a neuron, one can wait for a moment when the behavioral rhythm stops and the cell to be studied becomes inactive. One can then verify by intracellular recording that the cell is now receiving no detectable rhythmic synaptic input. It is thus *functionally* isolated, at least from all the neurons whose synaptic potentials are large enough to be detected from the location of the microelectrode. Now if one depolarizes this cell by passing current with an intracellular microelectrode, one can determine whether the cell responds by producing rhythmic bursts, or only continuous trains, of action potentials. In practice, one usually increases the current gradually from subthreshold to well beyond threshold. Figure 8 shows such tests made on two neurons mediating the rhythmic movements of the stomach of a lobster (Selverston, 1976). Each was impaled with a microelectrode, and through both microelectrodes an increasing, and then a decreasing, amount of depolarizing current was injected, as shown by the bottom trace. The cell in the middle trace responded to the ramp of depolarizing current with a regular train of action potentials of increasing frequency but with no bursts of action potentials. Because no rhythmic bursts were produced, this cell presumably is not an endogenous burster. The cell in the top trace responds only with bursts of action potentials rather than with a continuous train. These bursts come at a higher frequency the greater the depolarization. This cell, then, may be an endogenous burster because its only mode of activity when stimulated directly, rather than through synaptic input, was by production of rhythmic bursts of action potentials.

There are, however, several uncertainties in interpreting the results of Figure 8. In the case of the rhythmically bursting cell, it is possible that some of the stimulating current passed through electrical synapses into other cells, or that the depolarization of this cell by the electrode activated chemical synapses onto other cells, or that the action potentials of this cell synaptically activated other cells. These other cells, then, may have participated in creating the bursts of impulses via synapses back onto the recorded cell. The bursts, then, could actually have resulted from a synaptic network. In the case of the cell that showed a continuous train of action potentials, it could be that this is in fact an endogenous

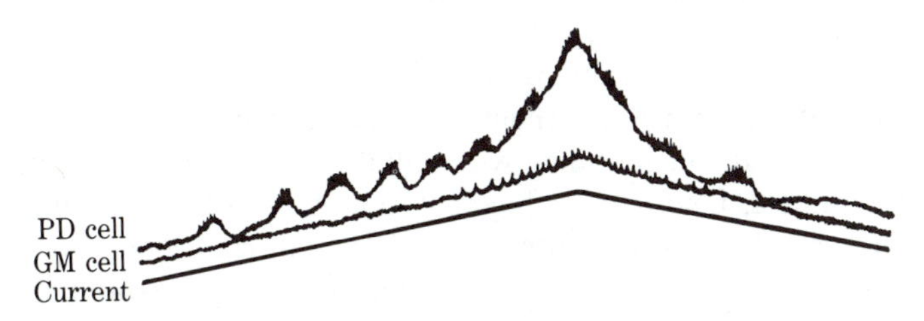

FIGURE 8. One test for endogenous bursting capability of a neuron. Two neurons of the stomatogastric ganglion of the lobster are impaled and simultaneously exposed to increasing, and then decreasing, depolarizing current. One of these cells, called a PD cell, gives only rhythmic bursts of action potentials; the other, called a GM cell, gives only continuous trains of action potentials. (From Selverston, 1976.)

burster but that the depolarizing current for some reason did not turn on its bursting mechanism. This mechanism might require, for instance, current in a different region of the neuron than the position of the microelectrodes; or it could require the presence of some neurotransmitter from a nonrhythmic presynaptic cell (Russell and Hartline, 1981). For these reasons, although the experiment of Figure 8 is suggestive, it is not conclusive. This should serve to indicate the great caution that is required in interpreting cellular physiological experiments. In fact, these problems of interpretation are particularly severe in the case of experiments on rhythm generation (Russell and Hartline, 1978; Hartline *et al.*, 1981).

One can apply additional experimental tests to help determine whether a given neuron is an endogenous burster. These tests involve further methods of functionally isolating the cell from all other cells that might influence it synaptically. For instance, if one uses a bathing solution in which the calcium has been replaced by magnesium, chemical synapses are blocked because calcium is required for the release of neurotransmitters at chemical synapses and magnesium antagonizes calcium's action (Chapter 3). (This synaptic blocking method generally does not block electrical synapses.) In using this technique, it is necessary to verify experimentally that the chemical synapses are in fact blocked because diffusion barriers may impede the free flow of calcium and other ions in the nervous system. This verification can be accomplished by stimulating a known presynaptic neuron that has a chemical synapse onto the cell of interest and noting the absence of the normally present postsynaptic potential. If normal rhythmic bursts of action potentials were to continue in the cell being studied, in spite of the successful blockage of chemical synapses, and if one knew through other experimental tests that only chemical and not electrical synapses connected to this cell, then one could conclude that the cell is an endogenous burster. An absence of bursting in a cell bathed in the calcium-free, magnesium-rich saline would be difficult to interpret, however. For, as we have seen, inward calcium currents are known to be important in generating the endogenous rhythmicity of the PB neuron of *Aplysia* and could likewise be important in other endogenous bursters. Such inward calcium currents would be all but abolished if the saline contained no calcium.

An apparently safer technique for functionally isolating a neuron has been used on the interneurons and motor neurons controlling the rhythmic beating of the leech's heart (Calabrese, 1979). All the detectable chemical synapses among these neurons are inhibitory (Thompson and Stent, 1976; Calabrese, 1979). Their IPSPs are produced by inward chloride currents (Nicholls and Wallace, 1978; Calabrese, 1979). Therefore, these synapses can be functionally blocked by replacing the chloride of the saline with sulfate ions, which do not pass through the chloride channels. Under these blocking conditions, the motor neurons show only continuous trains of action potentials, but the interneurons show bursts of action potentials. This suggests that the interneurons,but probably not the motor neurons, are endogenous bursters (Calabrese, 1979).

Returning now to the swimming system of the leech, specific neurons involved in producing the swimming rhythm have been identified, as will be described later in this chapter. These cells are interconnected almost exclusively by chemical inhibitory synapses (Friesen *et al.*, 1978). Each of these cells was tested for endogenous bursting capability by two of the techniques just described:

ramp depolarizations and synaptic blockade by substitution of sulfate for chloride ions in the saline. In both of these tests, only continuous trains of action potentials, and no bursts of action potentials, occurred (Friesen *et al.*, 1978). Therefore, the central neuronal oscillator for leech swimming appears not to rely upon endogenously bursting neurons, but rather it appears to be a network oscillator. It will be important, however, for workers to substantiate this claim further by additional experimental tests.

A DIGRESSION ON NEURONAL MODELING

Before proceeding with the description of the central neuronal circuit responsible for the alternating dorsal–ventral rhythm in leech swimming, it will be useful to consider some hypothetical circuit configurations that would be capable of producing such a rhythm. One reason this is useful is that the central oscillator circuit for leech swimming, as for several other rhythms, is very complex and difficult to understand without guiding principles. These principles have been developed through the theoretical study of hypothetical circuits (Wilson, 1966; Lewis, 1968). This approach is often called theoretical NEURONAL MODELING or CIRCUIT MODELING. This process, in its simplest form, involves simply drawing on paper a model consisting of a group of neurons that are interconnected by a particular configuration of excitatory and/or inhibitory synapses. One then attempts to follow through the sequence of neural activities expected to occur in each of the cells of the model, based upon their synaptic connections. In the present discussion, we will begin with this approach and will then consider briefly some more sophisticated modeling approaches employing electronic and computer techniques.

What kinds of model circuits could produce a rhythmic alternation of bursts of action potentials such as occur in the excitatory motor neurons to the dorsal and ventral muscles of the swimming leech? One such model consists of DELAYED EXCITATION between two neurons (Wilson, 1966). For instance, in Figure 9A, imagine that cell 1 receives, from a presynaptic neuron not shown, a brief train of EPSPs that produces in cell 1 a brief burst of action potentials. Once this initial event had occurred, rhythmically alternating bursts could be sustained by means of only the synaptic interactions shown between cells 1 and 2. The initial burst in cell 1, after a delay somehow built into the circuit, would cause in cell 2 a brief depolarization and a consequent brief burst of action potentials. This burst in turn would cause, after another delay, a depolarization and consequently a burst in cell 1, and so forth. The mechanism of the delay is not specified in this model but presumably could consist of a polysynaptic chain of neurons through which the excitatory signals between cells 1 and 2 pass. Such a system of mutual delayed excitation is thought to play a contributing role in controlling the rhythmic wingbeats in locust flight (Burrows, 1973) but has not yet been found to underlie any other rhythmic behavior.

The second model (Figure 9B) consists of two neurons interconnected not by excitation but by MUTUAL INHIBITION, or, as it is more often called, RECIPROCAL INHIBITION[3] (Wilson, 1966). In some forms of this model, each cell also

[3] This term should not be confused with reciprocal innervation (Chapter 8).

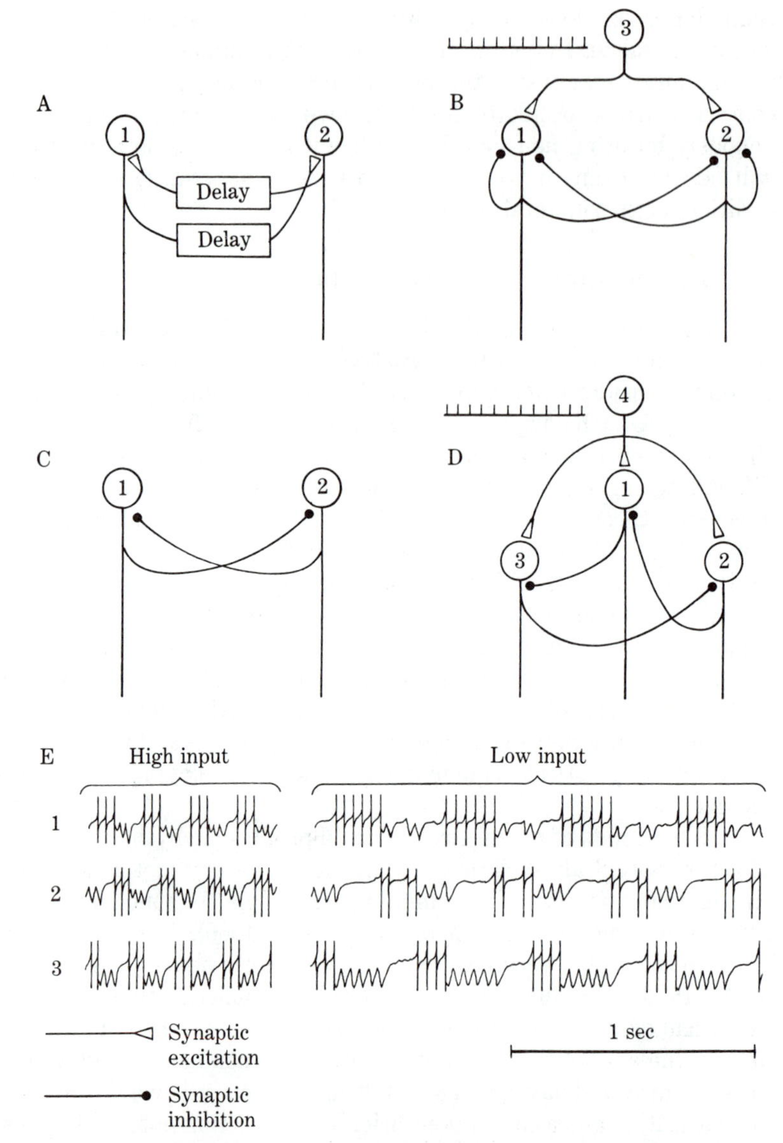

FIGURE 9. Theoretical modeling of rhythmic motor outputs. A. Delayed excitation. B. Reciprocal inhibition plus self-inhibition and shared excitation from a third neuron. C. Reciprocal inhibiton. D. Cyclic inhibition plus shared excitation from a fourth neuron. E. Impulses and simulated IPSPs recorded from three neuromimes connected according to the pattern of synapses among the neurons 1, 2, and 3 of D. When a high level of excitatory input was provided from a fourth neuron (simulating neuron 4 of D), the neuromimes produced short cycle periods. When a low level of excitatory input was provided, long cycle periods occurred. The phase was nearly constant, in spite of large changes in cycle period; phase of 2 with respect to 1 was 0.35 during short cycle periods and 0.45 during long cycle periods. (E after Friesen and Stent, 1977.)

inhibits itself, and both cells are excited by a third neuron (cell 3 in Figure 9B) that gives not brief bursts of action potentials but rather a continuous train. Imagine, for instance, that cell 3 of this model begins at some moment to give a regular train of action potentials, which then continues uninterruptedly. The onset of these action potentials would evoke a simultaneous depolarization of cells 1 and 2, each toward its threshold transmembrane voltage. By random chance, either cell 1 or cell 2 would reach its threshold before the other and at that moment would begin to give action potentials. Assume that cell 1 is the first to reach threshold. The synaptic inhibition from cell 1 to cell 2 would now prevent cell 2 from reaching its threshold and thus giving action potentials. Therefore, cell 1 alone gives action potentials. Soon, however, the inhibition by cell 1 upon itself would terminate its own train of action potentials. When this happens, cell 2 is thus freed from inhibition and can now become active. In fact, two separate factors could contribute to activate cell 2 at this moment. First, the continuing excitation from cell 3 is now unopposed by synaptic inhibition from cell 1. Second, as the inhibition from cell 1 ends, a phenomenon called postinhibitory rebound would contribute to the excitation in cell 2. A postinhibitory rebound, seen very commonly in neurons, is a brief burst of action potentials at the end of a period of inhibition.[4] When cell 2 begins to give its action potentials, these inhibit cell 1, maintaining its silence. But next, when cell 2 turns itself off by its self inhibition, cell 1 becomes active, and the whole cycle repeats itself.

It is also possible for two neurons connected by reciprocal inhibition to give alternating bursts of action potentials even if they do not have self inhibition or continuous input from a third neuron (Figure 9C). All that is required is that each neuron produce a brief postinhibitory rebound of substantial magnitude (Perkel and Mulloney, 1974). For instance, in Figure 9C, suppose that cell 1 is depolarized briefly (for instance, by an initial, brief excitatory input from another neuron not shown) so that it gives a brief burst of action potentials. During this burst, cell 2 would receive inhibition from cell 1. As soon as the burst in cell 1 terminates (which would occur when the initial brief excitatory input to it terminates), cell 2 would give its postinhibitory rebound. If this rebound were sufficiently large, it would produce in cell 2 a substantial burst of action potential. During this burst, cell 1 would be inhibited. But when the burst in cell 2 terminates (which occurs when its postinhibitory rebound terminates), cell 1 would give its own postinhibitory rebound and consequently a burst of action potentials. The whole cycle would then repeat itself.

As will be shown later in this chapter, not all central neuronal oscillators produce BIPHASIC rhythms—alternating bursts in just two groups of neurons. Rather some oscillators produce POLYPHASIC rhythms in which each of three or more neurons is activated at its own time within the cycle period. We consider

[4] Postinhibitory rebound is thought to result from the lowering of the threshold voltage of a cell in response to prolonged inhibition. That is, the neuron now needs only to be depolarized to a voltage very slightly above the resting potential in order to elicit action potentials. In fact, the threshold can actually become as low as the resting potential itself. In such a case, when the inhibiting hyperpolarization terminates and the transmembrane potential suddenly returns to the resting potential, action potentials result. Very soon thereafter, the threshold voltage returns to its normal value and the action potentials cease.

here one model capable of producing a TRIPHASIC rhythm, in which each of three neurons is activated in the sequence 1 - 2 - 3 - 1 - 2 - 3 - . . . (Friesen and Stent, 1977). This sequence of activity can be produced by three neurons, each of which inhibits the one *preceding* in the sequence (Figure 9D). This pattern of connectivity is called CYCLIC INHIBITION. (In the present model, a fourth neuron provides continuous, nonrhythmic excitatory input to all three cells, analogous to cell 3 of Figure 9B. This fourth neuron is not essential, however.) In this model, when cell 1 begins producing action potentials, it would inhibit cell 3, which would thus become silent. This inhibition of cell 3 would allow cell 2 to recover from the inhibition that it had just been receiving from cell 3, before cell 3 was turned off. Thus, cell 2, now free from inhibition, can respond to excitation from cell 4 and to its own postinhibitory rebound by producing a burst of action potentials. These action potentials in cell 2 would silence cell 1 through inhibition. But this silence of cell 1 now releases cell 3 from inhibition. Cell 3 would thus give a burst of action potentials. This entire sequence, then, would consist of a burst of action potentials in cell 1, then in cell 2, then cell 3. The cycle would continue in the same sequence 1-2-3-1-2-3- . . . .

To summarize, then, Figure 9 shows one model based upon delayed reciprocal excitation, two models based upon reciprocal inhibition, and one model based upon cyclic inhibition. In the next several sections of this chapter, we will consider the complex central neuronal circuit for leech swimming, with these theoretical circuits as our guide.

In this discussion of theoretical modeling, we simply examined visually the sequence of events expected of each of our paper-and-pencil models. A more powerful approach, commonly employed in research on neuronal modeling, is to verify the details of a model's activity pattern by means of either electronic or computer techniques. For instance, one can represent a single neuron of a model by an electronic component called a NEUROMIME (Lewis, 1968). A neuromime produces an output consisting of voltage impulses that resemble action potentials. These impulses are produced whenever the neuromime receives as an input a sufficiently high rate of small positive voltage signals that resemble summating EPSPs. The stimulating effect of these model EPSPs in producing output impulses can be counteracted by the simultaneous input to the neuromime of small negative voltage signals, resembling IPSPs. Each output action potential of a neuromime can be made to generate a single EPSP, or an IPSP, in another neuromime to which it is connected. Thus, by connecting a group of neuromimes in a pattern simulating a particular configuration of neural connections, one can determine the temporal pattern of impulses, or action potentials, that each neuron so connected would produce. Each neuromime can be independently manipulated by the experimenter to vary the sizes and time courses of the EPSPs and IPSPs, as well as the threshold voltage for producing impulses.

Model circuits with as many as eight neuromimes, representing eight different neurons, have been tested in this way (Friesen and Stent, 1977). In fact, the activity patterns of each model shown in Figure 9 have been studied in detail using neuromimes or computers (Wilson, 1966; Perkel and Mulloney, 1974; Dagan *et al.*, 1975; Friesen and Stent, 1977). For instance, as Figure 9E shows, the impulses and IPSPs of each of three neuromimes interconnected according to the cyclic inhibition model of Figure 9D reveal the expected triphasic rhythm. Moreover, by varying the amount of maintained excitatory input to each of the three neuromimes from a fourth neuromime (cell 4 in Figure 9D), additional useful information was obtained; the added excitation led to shorter cycle periods, while the phase of any one cell with respect to any other remained nearly constant. These details could not have been revealed simply by using paper-and-pencil models; rather, the use of neuromimes or a computer was essential.

MOTOR NEURONS AND THEIR SYNAPTIC CONNECTIONS IN LEECH SWIMMING

We will return now to the analysis of the neural circuit for leech swimming. Which neurons of the ganglia of a leech constitute its central oscillator for swimming? The search was begun with the motor neurons that innervate the dorsal and ventral muscles of the body wall, whose rhythmic contractions cause the swimming movements. Of course, these motor neurons were expected to show rhythmically alternating bursts of action potentials, which give rise to the swimming contractions. But is the rhythm of these neurons produced by their *own* synaptic interconnections, or are these cells merely *following* rhythmic signals from some presynaptic neurons? The initial strategy for answering this question was to determine the network of synaptic connections among the motor neurons and to see whether this network had a configuration that appeared capable by itself of producing rhythmic activity.

The cell bodies of many leech motor neurons, as well as some other cells, can be seen through a dissecting microscope in the living, translucent ganglia. Most of the cell bodies have constant positions relative to each other, a condition that permits individual neuronal identification (Figure 10). When making intracellular recordings from neurons of a ganglion, the first task is to determine whether a given impaled cell is a motor neuron, and if so, what muscles it innervates and what effect it has on these muscles. Figure 11 shows simultaneous recordings made intracellularly from a central cell body, extracellulary from a peripheral nerve, and intracellularly from a fiber (that is, a cell) of a dorsal muscle. The recording was made from a leech, part of whose body was pinned in place, but the rest of which was free to make rhythmic swimming movements. The impaled neuron shows rhythmic depolarizations and bursts of action potentials. Each of these action potentials is followed after a brief, fixed interval by an action potential in the peripheral nerve and then by a depolarizing, junctional potential in the muscle fiber. This is a slow muscle (Chapter 8), which is known (from other experiments) to contract in response to small depolarizations such as these. Thus, the physiological evidence of Figure 11 suggests that the cell body impaled is that of a motor neuron. Some cells physiologically identified in this manner have also been stained intracellularly and show an axon exiting the ganglion through a peripheral nerve, a finding that supports the physiological identification as a motor neuron (Muller and McMahan, 1976). By making repeated recordings of this type, investigators have found the cell bodies of many motor neurons, and the muscular targets of these neurons have been identified (Ort *et al.*, 1974). These neurons include four sets of motor neurons of interest to this discussion:

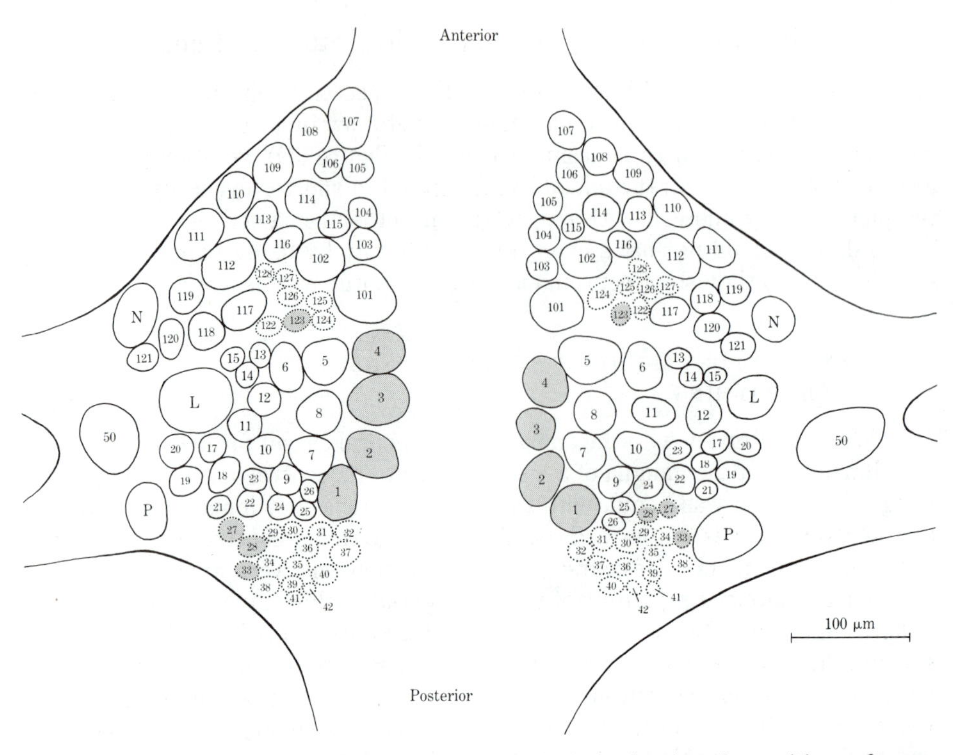

FIGURE 10. A ganglion of the leech, in dorsal view, showing the positions of many of its cell bodies. The four large, darkened cell bodies on either side belong to left and right motor neurons 1, 2, 3, and 4 discussed in the text. The four smaller darkened cell bodies on either side belong to the rhythm-producing interneurons discussed in the text. Throughout the ganglion, the larger cell bodies (with solid outlines) are more constant in their positions than are the smaller ones (in dashed outlines). (After Friesen *et al.*, 1978.)

excitatory and inhibitory motor neurons innervating the dorsal muscles and excitatory and inhibitory motor neurons innervating the ventral muscles. One cell body of each of these four types is darkened on either side of the ganglion in Figure 10. They are labeled cells 1, 2, 3, and 4.

Figure 12A shows simultaneous recordings from the cell bodies of two excitatory motor neurons, one to a dorsal and one to a ventral muscle, recorded while the pinned leech performed rhythmic swimming movements. These two neurons show reciprocal rhythms of depolarization and hyperpolarization. By recording also from an inhibitory neuron to the dorsal muscle and from one to the ventral muscle, a surprising complication was revealed—the rhythm is not biphasic, but *quadriphasic*. This four-part rhythm of the four motor neurons has the sequence 3-1-4-2-3-1-4-2- . . . (Figure 12B).

Can this quadriphasic rhythm of motor neurons be explained on the basis of the synaptic connections among these neurons themselves? The synaptic connections among these cells, shown in Figure 13, were elucidated by impaling pairs of the cells and stimulating each member of a pair while recording from the other,

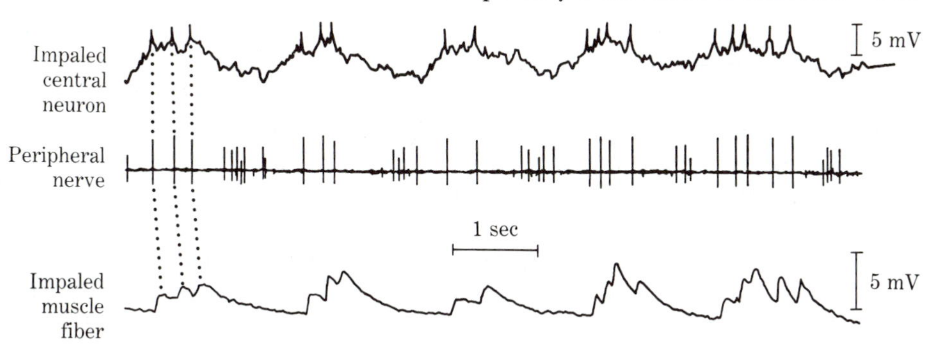

FIGURE 11. Experiment to determine whether or not a given impaled neuron is a motor neuron. *Top trace,* Recording of the impaled cell body. (Action potentials are small because the cell bodies of leech motor neurons are nonelectrogenic; thus, the action potentials decay electrotonically from distant electrogenic membrane.) Each action potential in the neuron is followed by an action potential in a peripheral nerve *(middle trace),* and then a depolarizing junctional potential in an impaled fiber of a muscle innervated by this nerve *(bottom trace).* (After Ort *et al.*, 1974.)

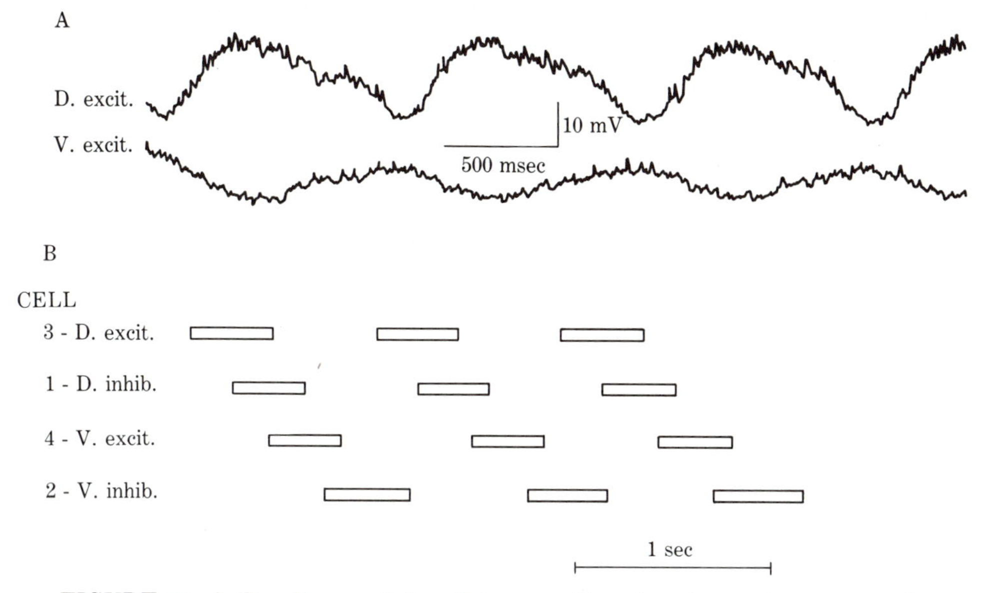

FIGURE 12. A. Simultaneous intracellular recordings from two motor neurons taken while the leech was making swimming movements. The action potentials in these particular recordings are very small and difficult or impossible to discern, having decayed electrotonically from distant electrogenic parts of the cell. B. Quadriphasic rhythm of motor neurons 1, 2, 3, and 4 from one side of one ganglion. The bars show the durations of the bursts of action potentials in each of the four cells. (A after Ort *et al.*, 1974; B after Poon *et al.*, 1978.)

as was described for the crayfish escape circuit (Chapter 8). Not shown in the figure are electrical synapses made by cells 1, 2, and 3, each to its cell of the same number on the opposite side of the ganglion (Ort *et al.*, 1974). The synaptic connections of Figure 13 do not exhibit reciprocal delayed excitation, reciprocal inhibition, cyclic inhibition, or any other circuit configuration that would seem capable of producing a quadriphasic rhythm.

Aside from simply inspecting the circuit diagram of Figure 13, a direct experimental test can be used to help determine whether any of these four motor neurons contributes to generating the rhythm. Such a test is important because one cannot be certain that all the synaptic connections among the four motor neurons have been found; for instance, there may be some synapses among these cells that are located too far away from the recording positions in the cell bodies to be detected by the microelectrode. The test consists of delivering a brief depolarizing or hyperpolarizing pulse of current to the neuron in question during its rhythmic activity. The rationale for this experiment is best understood by assuming first that the neuron tested *does* help to generate the rhythm. Suppose, for instance, that the cell belongs to a network oscillator that operates by virtue of some pattern of synaptic inhibition, such as that of Figure 9B, C, or D. If we depolarize the cell to prolong one of its bursts and increase its frequency of action potentials, this will prolong and increase the synaptic inhibition that this cell produces onto the other cells of the circuit. This would be expected to delay the next burst of the next cell in the cycle. Then, after our depolarizing pulse is terminated, the rhythm would return to normal. However, owing to the one prolonged interval,

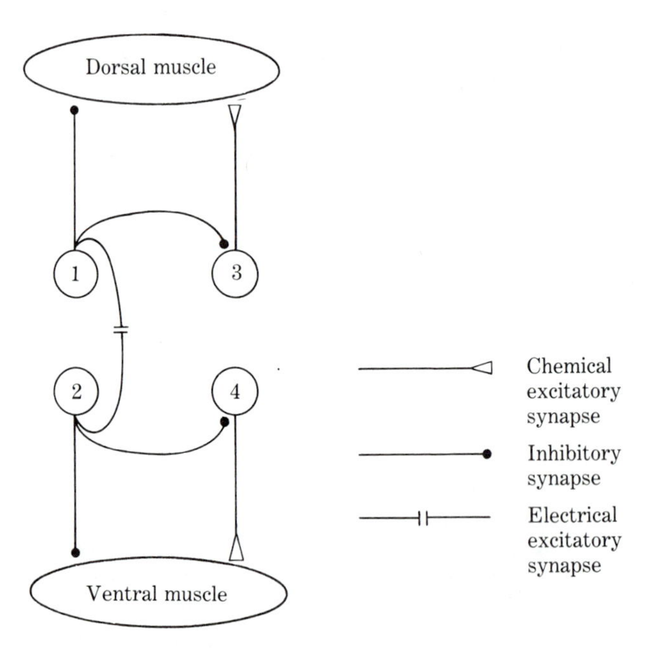

FIGURE 13. Synaptic interconnections among motor neurons 1, 2, 3, and 4 of one side of a leech ganglion.

the rhythm would have been *reset* relative to its original beat (Figure 14A). If the cell whose bursts we alter does not contribute to generating the rhythm but rather is only driven by an oscillator made up of other cells, we would expect our alteration of its bursts not to reset the rhythm (Figure 14B).

Thus, if altering the bursts of a rhythmically active neuron resets the rhythm of the circuit to which this neuron belongs, one can consider this cell to be part of the mechanism for generating the rhythm. (In this experiment, the duration and intensity of the altered bursts should fall within the naturally occurring range, in order to be certain that the effect of manipulating the cell on the rhythm is a natural one.) If the rhythm is not reset by altering a burst in one cell, this suggests that this cell does not contribute to generating the rhythm. However,it does not prove this; there could be some form of oscillatory mechanism that we are unaware of, that can somehow compensate for sudden alterations of bursts in one component neuron. Even with this restriction on interpreting negative results, however, the resetting experiment provides a very useful tool.

This experimental test has been applied to each of the four motor neurons of Figure 13. Altering a burst of motor neurons 2, 3, or 4 did not lead to a resetting of the swimming rhythm (Ort *et al.*, 1974). However, altering a burst of motor neuron 1 did lead to resetting (Kristan and Calabrese, 1976). Thus, motor neuron 1 does somehow contribute to generating the rhythm. However, further experiments suggest that this role of cell 1 is indirect, probably operating through the synaptic connections that this cell has onto a group of interneurons (discussed in the next section). As we shall see, it is these interneurons that appear to be more directly responsible for producing the rhythm (Poon *et al.*, 1978).

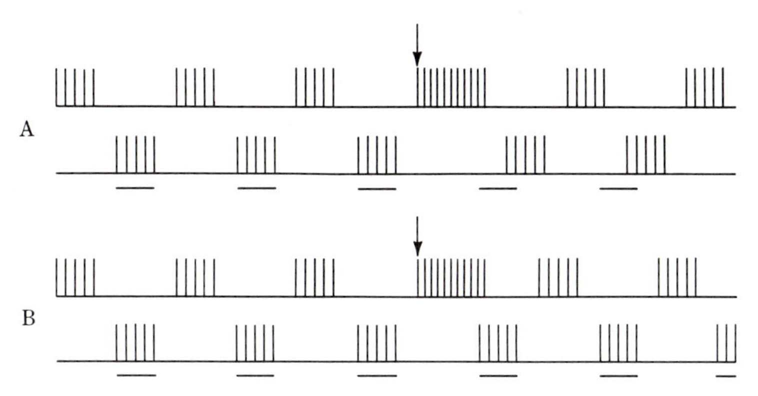

FIGURE 14. An experimental procedure used to help to determine whether a given rhythmically active neuron is part of the rhythm-generating mechanism. Hypothetical data are shown. A. Experimentally altering a burst of action potentials (arrow) in one neuron causes the time of occurrence of subsequent bursts in this and in other neurons to be reset. This demonstrates that the cell whose bursts were altered plays a role in generating the rhythm. B. Altering a burst does not lead to resetting; this finding suggests that the neuron whose burst was altered may not be involved in generating the rhythm.

An Oscillatory Network of Interneurons

Four interneurons on each side of each ganglion are at least partly responsible for generating the central neuronal rhythm for swimming (Friesen and Stent, 1977; Friesen *et al.*, 1978). The cell bodies of these four interneurons are darkened on each side of the ganglion in Figure 10; they are cells 27, 28, 33, and 123. Each of these cells is rhythmically active in synchrony with the animal's swimming movements. In fact, like the motor neurons, the rhythm of these four interneurons is essentially quadriphasic (Figure 15). Moreover, the swimming rhythm of the motor neurons can be reset by altering the bursts of any of these four interneurons (Figure 16).

What are the synaptic interconnections among a quartet of these interneurons? An example of the synaptic interactions between two of these four cells, numbers 27 and 28 from the same side of one ganglion, is shown in Figure 17. Suprathreshold depolarization of either of these cells leads to a hyperpolarization of the other. (It is not yet established whether these or other interactions among the quartet of interneurons are monosynaptic.) Thus, cells 27 and 28 are connected by reciprocal inhibition. As we have already seen, reciprocal inhibition can give rise to rhythmic alternation (Figure 9B and C). In fact, cells 27 and 28 do burst in reciprocal alternation (Figure 15), suggesting that their reciprocal inhibition may contribute to producing this aspect of the rhythm.

The full complement of synaptic interactions among the interneuronal quartets from one side of two neighboring ganglia is shown in Figure 18. In addition to the reciprocal inhibition between cells 27 and 28, there is also cyclic inhibition from cell 27 to 33 to 28 to 123.[5] In this cyclic inhibition, each cell inhibits the one preceding it in the activity cycle. As we have seen, this configuration can give rise to a polyphasic rhythm (Figure 9D and E).[6] Thus, cyclic inhibition could con-

[5]The inhibition of cell 33 by cell 27 may not result from the inhibitory synapse shown. This inhibition could instead arise indirectly through cell 28. That is, depolarizing cell 27 evokes a hyperpolarization in cell 28, and this hyperpolarization in turn would hyperpolarize cell 33 by virtue of the "rectifying," or one-way, electrotonic synapse from cell 33 to 28. Thus, although functional inhibition from cell 33 to 27 has been demonstrated, the inhibitory synapse itself has not. Several electrical synapses in different animals have been found to be rectifying, permitting current to spread in only one direction across the synapse. The mechanism of the unidirectionality is not clearly understood. Most electrical synapses studied are bidirectional, or nonrectifying.

[6]There are some theoretical complications with the generation of quadriphasic, as opposed to triphasic, rhythms by means of cyclic inhibition. These difficulties are actually reduced as one progresses to a five-phased rhythm (Friesen and Stent, 1977). This has led some workers in the field to consider the basic unit of rhythmicity to be a quintet, composed of the four interneurons on one side of one ganglion plus one interneuron from a neighboring ganglion. (This last interneuron then would also belong to another quintet, along with the three remaining interneurons of its own ganglion plus one from the next ganglion.) Considering a quintet the basic unit does not, however, seem to greatly illuminate the issue. The strategy of the present discussion instead is first to identify the neurons of the system and their interactions, then attempt to discern within the network patterns of connectivity known to be capable of producing a rhythm (Figure 9), and finally, as will be described later, to apply modeling techniques to help verify that the *totality* of the connections identified plays an important role.

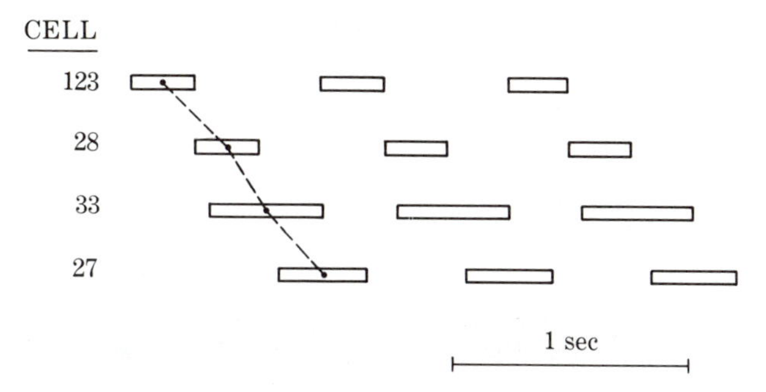

FIGURE 15. Quadriphasic rhythm of four interneurons on one side of one ganglion in a leech. The bars show the durations of the bursts of action potentials in each of the four cells. The dot in a bar shows the time of the middle action potential of the burst. Although the bursts in cells 28 and 33 begin almost simultaneously, the relative times of the middle action potential of each burst reveal a quadriphasic sequence. (After Friesen *et al.*, 1978.)

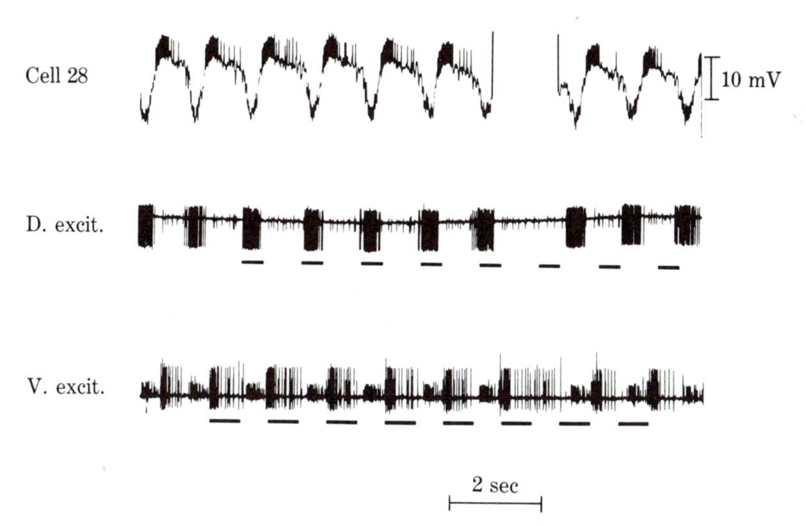

FIGURE 16. Test for a contribution by interneuron 28 in generating the swimming rhythm. The top trace is an intracellular recording from cell 28. A brief depolarizing pulse was delivered to this cell, between the upward and the downward arrows. (The recording from this cell went off scale for the duration of the pulse.) Extracellular recordings were made from two peripheral nerves. In one, the largest action potentials are from an excitatory motor neuron to the dorsal muscles (D. Excit.); in the other, the largest action potentials are from an excitatory motor neuron to the ventral muscles (V. Excit.). The depolarizing pulse to cell 28 reset both its own rhythm and that of the motor neurons. (After Friesen *et al.*, 1978.)

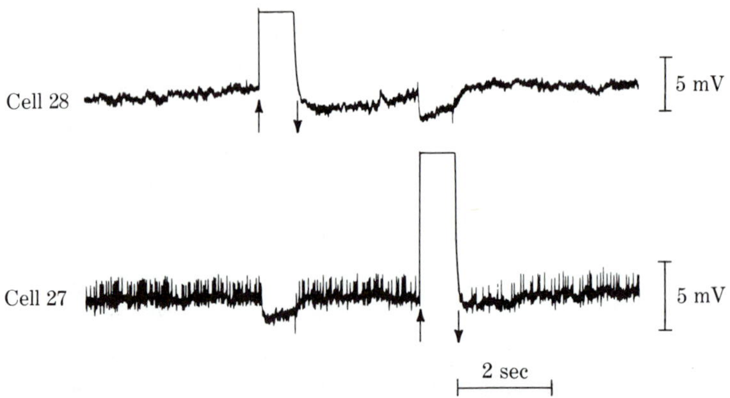

FIGURE 17. Synaptic connections between interneurons 27 and 28 from the same side of one ganglion. A brief depolarizing pulse delivered to either cell (arrows) results in a hyperpolarization of the other cell. Cell 27 had a continuous train of action potentials, whose frequency was greatly reduced when hyperpolarizing current was passed into cell 28. (After Friesen *et al.*, 1978.)

tribute to producing the rhythm among the quartet of interneurons. Only a single excitatory synapse is found in each quartet, namely, a "rectifying," or one-way, electrical synapse[5] from cell 33 to 28. Thus, reciprocal delayed excitation can be ruled out as a mechanism contributing to the rhythm. There are also intraganglionic inhibitory connections, as shown in Figure 18. These extend over a span of at least seven ganglia (not just the two ganglia shown in the figure) and may provide the necessary mechanism for coupling the leech's several central neuronal oscillators to achieve the metachronal wave of activity. Finally, though not shown in Figure 18, cells 28 and 33 of each ganglion connect by nonrectifying, or two-way, electrical synapses, each to its same-numbered cell across the ganglion. These cross-connections presumably help to synchronize the rhythms on opposite sides of the body.

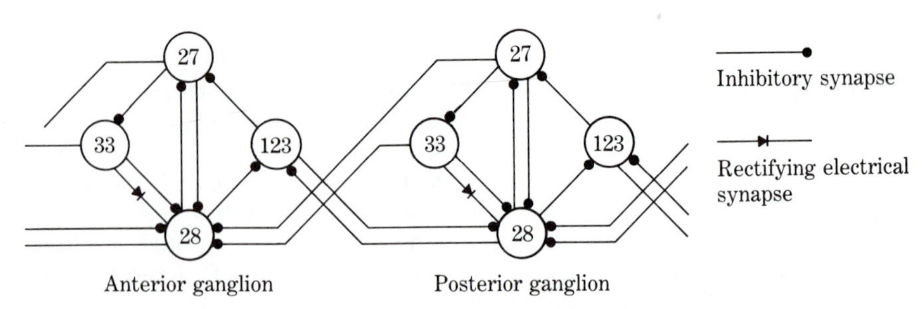

FIGURE 18. Synaptic interactions within and between quartets of interneurons from one side of two neighboring ganglia. Explanation in text. (After Friesen *et al.*, 1978.)

Although one can find reciprocal and cyclic inhibition in the interneuronal quartets of Figure 18, this does not prove that these synaptic interactions are adequate to explain the quadriphasic rhythm. Moreover, because the interganglionic connections among different quartets greatly complicate the network, mere visual inspection of the circuit diagram can provide at best only a hint as to how the network operates. To proceed further, one must use a more rigorous approach such as that available through neuromimes or a computer. The procedure followed here is formally identical to that described earlier for theoretical neuronal modeling. For instance, one would connect a group of neuromimes together in the configuration shown by the synaptic connections of Figure 18. One would adjust various properties of each neuromime to mimic known properties of the individual interneurons, such as the relative magnitudes and durations of the PSPs of each synaptic connection and the threshold voltage for each neuron. One would then determine whether the circuit of neuromimes produces a rhythm like that of the leech's real circuit. This procedure can be called *empirical* modeling.

Although empirical and theoretical modeling use the same neuromime or computer techniques, there is an important difference between them. In empirical modeling, the parameters, such as PSP size, designed into the circuit of neuromimes or the computer program are not imaginary. Rather they are known properties derived from physiological experiments. This enables one to test specific hypotheses, such as whether the circuit of Figure 18 is sufficient to account for the observed rhythm among these cells. For instance, if neuromimes connected so as to mimic the circuit of Figure 18 do *not* show a temporal pattern like that actually recorded from the real leech interneurons, then one knows for sure that additional cells or circuit properties must be sought through physiological experiments. (In fact, sometimes by examining the particular manner in which a circuit fails to reproduce the animal's real temporal pattern, one can discover which specific aspects of the circuit are insufficiently understood and then search directly within that part of the circuit for the missing information; Hartline, 1979.) By contrast, theoretical modeling enables one to elaborate several alternative hypotheses but not to determine whether any of them is an accurate representation of an actual neuronal oscillator. In fact, when one is not constrained by the actual physiological parameters of the particular system under study, it is always possible to elaborate several different theoretical models, all capable of producing the same rhythmic or other form of activity. The value of theoretical modeling, then, is to alert researchers to possible forms of neuronal circuitry that they may not have thought of. The value of empirical modeling lies in its ability to help establish concretely that a particular set of experimentally identified neuronal properties actually underlies a particular behavior. Such experimental modeling is currently beginning to play an increasingly valuable role in neuroethology.

How good is the rhythm produced by a set of neuromimes interconnected according to the circuit diagram of Figure 18? Figure 19 shows the temporal pattern of impulse activity in each of eight such interconnected neuromimes (Friesen and Stent, 1977). These eight include one full quartet of one ganglion plus two cells from a more anterior ganglion and two from a more posterior ganglion. (The interganglionic connections among the neuromimes incorporate delays equivalent to those not between immediately adjacent ganglia but rather between

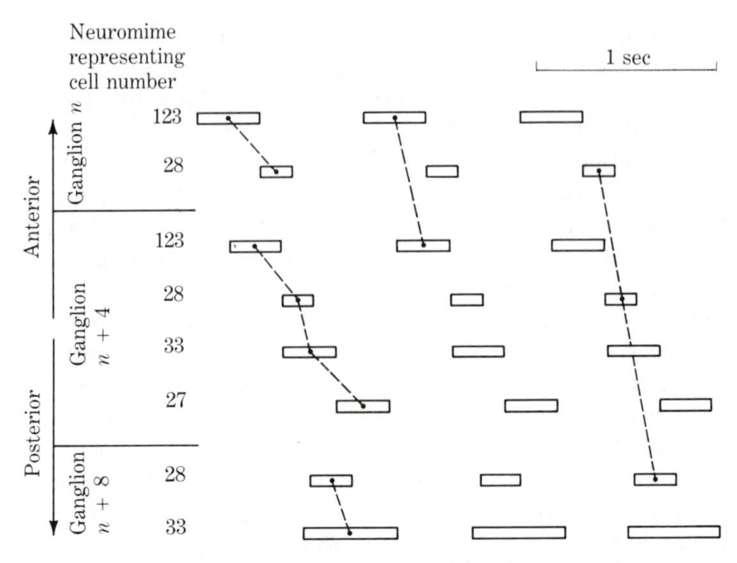

FIGURE 19. Temporal pattern of impulse activity produced by neuromimes connected so as to simulate the known synaptic interactions among rhythmic interneurons from ganglia n, $n + 4$, and $n + 8$ of a ganglionic chain. The bars show the durations of impulse bursts of each neuromime. Dashed lines connect the middle impulses of particular bursts. As shown by the dashed lines connecting the first burst for each cell in ganglion $n + 4$, a fairly normal quadriphasic rhythm occurs in the neuromimes. Also, as shown by the dashed lines connecting the bars of different ganglia for the cells 123 and those for cells 28, a posteriorly directed metachronal wave also occurs. (After Friesen *et al.*, 1978.)

ganglia four apart in the chain.) The four neuromimes that represent the quartet of one ganglion (ganglion $n + 4$ in the figure) show a rhythm with the normal sequence 123-28-33-27, though the bursts of neuromimes 28 and 33 overlap one another more than those of the real cells that they represent (Figure 15). This rhythm has a cycle period of about 840 milliseconds, which is within the naturally occurring range of cycle periods. Moreover, the bursts of impulses of neuromimes representing the same homologous cells in successive ganglia (for instance, cell 123 in ganglion n and $n + 4$, or cell 28 in ganglion n, $n + 4$, and $n + 8$) show a metachronal wave that proceeds posteriorly as in normal swimming. Thus, both a quadriphasic and a posteriorly directed metachronal rhythm having reasonably lifelike properties emerge from a group of neuromimes connected according to the experimentally derived synaptic pattern.

Does the similarity between the rhythm of the neuromimes and that of the real interneurons prove that the known connections among these neurons (as shown in Figure 18) are fully responsible for their quadriphasic and metachronal rhythms? It does not. The results from the neuromime study show only that the circuit of Figure 18 is *sufficient* to produce a nearly normal interneuronal rhythm.[7] It does not show that this circuit is *necessary* to produce this rhythm. That is, there could be other rhythmic neurons not yet identified that have strong synaptic interactions with the quartets of interneurons and that provide strong rhythmic activation of these, adding to the rhythmicity produced internally by the quartets themselves. In Chapter 8, when attempting to determine whether a particular neuron (LGI) was necessary for a particular event (the escape response of the crayfish), the technique of hyperpolarizing LGI to remove it functionally from the circuit provided the answer. This hyperpolarization eliminated escape response, showing that LGI was in fact necessary for the behavior. The comparable experiment in the leech, hyperpolarizing all four interneurons of a quartet (or all eight of a ganglion) would be technically impossible because these cells are sufficiently small that even impaling just two of them in the same ganglion is a substantial accomplishment. Therefore, the question of whether some neurons other than those of the quartet may play an important role in determining the rhythm of the quartet is still uncertain. This question must await specific experimental tests on additional neurons of the ganglia. Already, a few additional interneurons have been identified that appear to contribute to the rhythm (Weeks, 1982). At present, though, it is at least compelling to note that a circuit of interneurons has been uncovered that is sufficient to produce a reasonably normal replica of the leech's swimming rhythm, even though it may be only a part of a larger system of interconnected interneurons that constitute the actual network oscillator.

CONNECTIONS FROM RHYTHMIC INTERNEURONS TO RHYTHMIC MOTOR NEURONS

One crucial piece of information remains to be provided. Do these quartets of interneurons connect synaptically with motor neurons 1, 2, 3, and 4 in a manner that could account for the quadriphasic rhythm of these motor neurons? The pattern of synaptic connections that has been found between the interneuron quartets and the motor neurons is every bit as complex as that among the quartets themselves (Poon *et al.*, 1978). It includes both excitatory and inhibitory connections from interneurons of a given quartet to motor neurons, both within the same ganglion and in several nearby ganglia. By using a network of neuromimes mimicking these known connections, a quite normal (though not perfect) quadriphasic rhythm was obtained in the neuromimes representing the four motor neurons of

[7]In order to achieve a more complete test of the sufficiency of the circuit of Figure 18 for generating the rhythm, much more information should be acquired about each neuron and each synapse in the circuit. A few of the additional properties of interest are the rates of facilitation and defacilitation at each synapse, possible changes in the threshold voltage of the cell, and how quickly one action potential is able to follow upon another in a given cell. Moreover, a neuron's threshold voltage and the absolute PSP size are only meaningful if measured at the trigger zone, which for the leech is not the location where the recordings have been made. This additional information will be difficult to obtain, partly because of the small diameters of the neurites that appear to contain the trigger zones of these cells.

one ganglion (Figure 20). Thus, the observed synaptic connections from interneurons onto motor neurons are sufficient to account for a fairly lifelike motor rhythm (Poon *et al.*, 1978).

SUMMARY AND PERSPECTIVES OF LEECH SWIMMING

The swimming behavior of the leech displays alternating contractions of the dorsal and ventral body muscles. Recordings from excitatory and inhibitory motor neurons to each muscle reveal an underlying quadriphasic rhythm of motor output. In addition, there is a posteriorly directed metachronal wave of activity. A chain of central ganglia totally isolated from the periphery can give fairly normal rhythmic motor outputs, demonstrating the presence of a central neuronal oscillator. (This does not, however, rule out a contribution to swimming from chain reflexes; Chapter 10.) The central oscillation appears to result from a series of coupled oscillators rather than from a single master oscillator. There appears to be one oscillator for each ganglion. Moreover, each oscillator appears to be a network rather than an endogenous oscillator, because tests for endogenous bursting (ramp depolarization and chloride substitution) proved negative. This point, however, requires further testing.

The synaptic connections among the motor neurons are not adequate by themselves to account for the quadriphasic rhythm of motor output. Apparently more important are quartets of rhythmically active interneurons. One such quartet is found on each side of each ganglion. Current pulses injected into any of these four cells reset the swimming rhythm. Most of the synaptic connections among the cells of each quartet and between those of different ganglia are inhibitory. In fact, such inhibitory interactions among cells of a central neuronal oscillator have been found in several invertebrate animals studied and appear to constitute a general rule (Calabrese, 1977; Getting *et al.*, 1981; Selverston *et al.*, 1976; Selverston, 1980). The complex synaptic network includes reciprocal inhibition and cyclic inhibition, both of which are known (from theoretical modeling studies) to be capable of contributing to rhythmicity. Neuromimes, when interconnected so

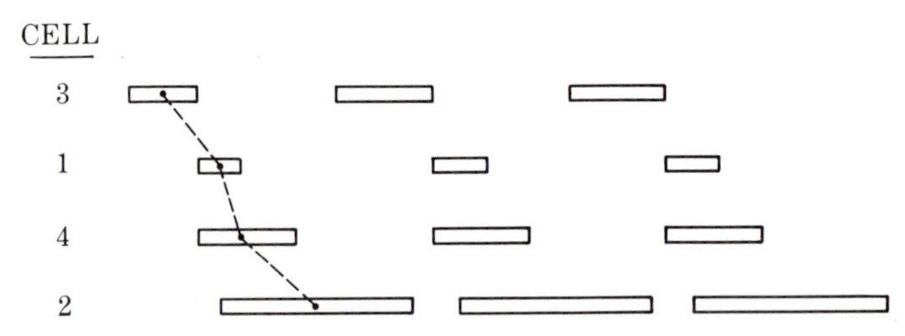

FIGURE 20. Temporal pattern of impulse activity produced by neuromimes interconnected according to the synaptic interactions from oscillatory interneurons to motor neurons and among the motor neurons themselves. The bars represent the durations of the impulse bursts of each motor neuron. The dashed lines connect the middle impulse of each burst in the four motor neurons. (After Poon *et al.*, 1978.)

as to represent the interneurons of quartets from several ganglia of a chain, produce nearly lifelike quadriphasic and metachronal rhythms. Also, the connections from interneurons to motor neurons, when represented by a network of neuromimes, produce fairly normal motor rhythmicity. These results suggest that the interneuronal quartets represent important elements in producing the central neuronal rhythm of swimming.

As a variation on a question raised in earlier chapters, one can ask to what degree the central nervous system of the leech invests responsibility for producing the swimming rhythm in a single cell rather than in a multicellular circuit. Apparently, in the leech swimming system, the responsibility is distributed among the several neurons of a circuit. However, even though each neuron of a quartet plays only a partial role, its role may be indispensable. Removal of almost any single connection from a circuit of neuromimes representing the oscillatory interneurons results in the instability of the rhythmic activity (Friesen and Stent, 1977).

It is sobering that an apparently simple behavior such as leech swimming should have a multicellular control system so complex that if one were to draw the complete circuit diagram, including all the connections among the quartets from several nearby ganglia, as well as the cross-ganglionic connections, it would appear nearly unintelligible. This leads one to wonder about the prospects for understanding on the cellular level temporally patterned behaviors of vertebrate animals, with their apparently more complex central nervous systems.

Ultimately, such complexities raise the philosophical question of what one means by *understanding* the cellular basis of behavior. Suppose, for instance, that a neuronal circuit diagram for some behavior were so complicated that, by just looking at it, we could derive no understanding at all. And yet, suppose that by means of empirical modeling using a computer, incorporating all the experimentally derived connections and cellular properties, one could recreate an exact replica of the temporal pattern in the real cells. Would this mean that we "understood" the neuronal basis of this behavior? Or are we simply unable to truly understand such complex phenomena? Acceptance of nonintuitive, or even counterintuitive, explanations based upon the use of modeling by computers or mathematics is commonplace in the physical sciences, but it is not yet familiar to most biologists. However, such procedures may soon increasingly enter the experimental realm of neuroethology (Selverston, 1980).

Complex Rhythmic Behaviors

Although leech swimming involves both dorsoventral rhythms and metachronal rhythms, both occurring over a range of cycle periods, these complex movements are coordinated into a unitary behavior that always has the same basic appearance (Kristan *et al.*, 1974). By contrast, some other rhythmic behaviors can occur in two or more discrete forms or can show other variations that imply still greater complexities in the underlying central oscillatory mechanisms. In this section we will consider a few of these added complexities.

PHASE TRANSITIONS

As was mentioned in Chapter 6, both walking and righting behavior in cockroaches involve rhythmic movements of the legs. The leg movements in these two behaviors are identical in most of their properties; they employ many of the same muscles and motor neurons, the same range of cycle periods, and the same phases of one leg's movement with respect to another (Sherman *et al.* 1977). Thus, the same central neuronal oscillator may produce both behaviors (though this is not yet known; Reingold and Camhi, 1977). However, the leg movements in walking and those in righting differ in one respect: the leg strokes in each behavior have a different direction. In walking, each leg moves in an anterior-to-posterior direction, whereas in righting it moves medial-to-lateral. Underlying this discrepancy is a difference in the phase of activation of one muscle, called CR, with respect to the activation of another muscle, called FE (Figure 21). Thus,

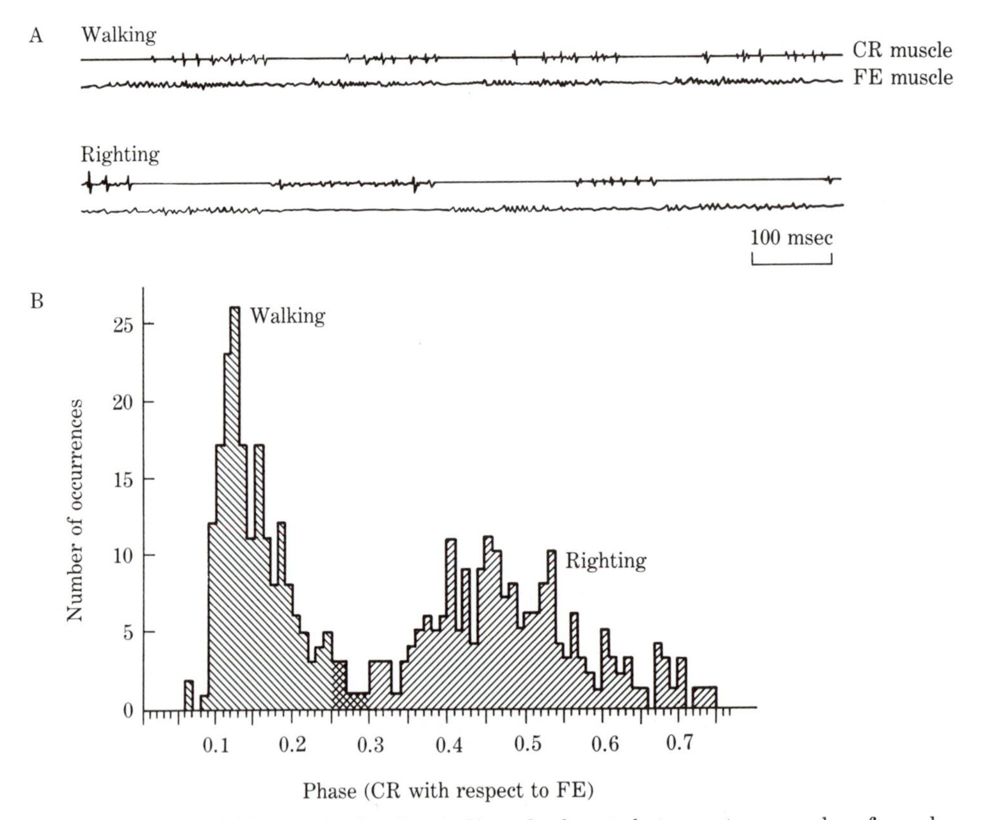

FIGURE 21. Difference in the phase of impulse bursts between two muscles of a cockroach's leg—muscles CR and FE—during walking and righting. A. Recording of the electromyogram (that is, using as electrodes a pair of wires implanted into a muscle) from CR and FE during the performance of walking and righting. B. Phase histogram of CR with respect to FE during walking and righting. (After Sherman *et al.*, 1977.)

the central oscillatory mechanisms for these rhythms appear to include a dual phase-producing capability. Similar phase changes have been observed in the flight systems of some insects, which vibrate their wings to warm the body using an intermuscle phase that is different from that used in flight (Kammer, 1970, 1971).

Phase transitions also occur in the locomotion of many mammals, which can utilize several different gaits such as walking, trotting, cantering, and galloping. Each of these gaits involves a different phase of one leg's movement with respect to that of another. There is evidence to suggest that in some mammals the rhythmic movements of each leg may be controlled by a separate central neuronal oscillator (Grillner, 1975). Thus, a change of gait may involve switching from one to another form of neural interaction among the four separate oscillators.

TRANSITIONS OF CYCLE PERIOD

Some complex rhythmic behaviors involve abrupt transitions from one cycle period to another, many times shorter or longer than the first. Such changes imply that one neuronal oscillatory mechanism suddenly gives way to another. One example is seen in the highly complex male courtship stridulation of the grasshopper *Stenobothrus* (Elsner, 1974). The stridulatory call begins with alternating up-and-down movements of the hindlegs at a frequency somewhere between 4 and 12 cycles/sec. On each upstroke and downstroke of a leg, its stridulatory file rubs against the stationary hindwings, producing a sound. Recordings during stridulation made from a hindleg levator muscle (muscle number one of Figure 22A) and from a depressor muscle (muscle number 2) reveal rhythmic bursts of impulses underlying the rhythmic leg movements. After several minutes, the stridulation changes, with many of the downstrokes of the leg now showing a different pattern. Each of these downstrokes begins normally but then suddenly switches to a series of brief downward jerking movements occurring at a fixed frequency of 65 per second and continuing till the end of the downstroke. Each of these jerking movements produces a brief pulse of sound. Muscle recordings show that at the moment of onset of these steps, the previously active muscle 2 turns off, and another leg depressor muscle, number 3, turns on (Figure 22B). This newly active muscle, unlike the previously active one, shows impulses at a highly fixed rate of 65 per sec. Muscle 3 extends from the thorax down into the legs, and its contractions can thus contribute to movements not only of the legs, as in Figure 22B, but also of the wings. This same muscle, then, is active during the third and final phase of stridulation, in which the wings flap at a frequency of 65 per second (though the animal does not become airborne). The wings create a sound as they rub together on each stroke. Again muscle 3 gives impulses at a rate of 65 per second, as do some other muscles that move only the wings (muscle 4, Figure 22C). Muscles 3 and 4 show a nearly identical pattern of impulses during flight.

This complex stridulatory pattern thus involves two rhythms, one slow and one fast, and thus probably two different central neuronal oscillators. It seems likely that the slow rhythm has been derived during evolution from a central oscillator for walking and the fast rhythm from a central oscillator for flight. The overall stridulation pattern, then, would come about by turning on and off each

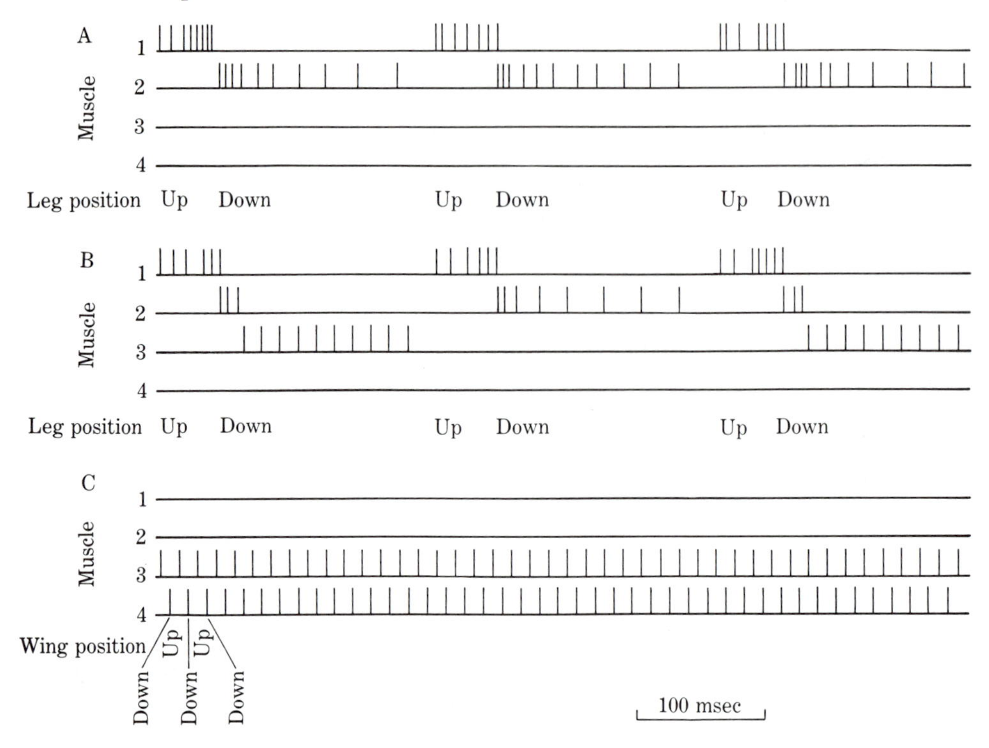

FIGURE 22. Electrical recordings from the muscles of the grasshopper *Stenobothrus* during three phases of stridulation. A. Alternating bursts of electrical activity in leg elevator and depressor muscles. B. These are now joined by a muscle of the wings and legs (muscle 3) that shows a characteristic impulse frequency of 65 per second. C. Now the leg muscles are inactive and a muscle used for the wings alone (muscle 4) joins the rhythm at 65 per second.

of these two oscillators at exactly the appropriate moments. Additional forms of complex interaction between two or more oscillators in producing rhythmic behaviors have been described for the movements of fins in several fish (von Holst, 1973a,b).

Summary

Most animal behavior consists of sequentially patterned motor acts. Sequences of behavior are often probabilistically organized. Though these may be difficult to study physiologically, careful behavioral and statistical analysis can suggest underlying rules of neural organization. Examples of such rules include interactions defined by first-, second-, and higher order Markov chains.

Rhythmic recurrence of sequentially fixed movements is easier to investigate physiologially. Important physiological concepts involving such behaviors include central neuronal oscillators and chain reflexes; a master oscillator *vs* a series of coupled oscillators; endogenous bursters and emergent network oscillators; delayed reciprocal excitation, reciprocal inhibition, and cyclic inhibition. Specific experimental tests can help to identify which of these various processes are operating within a given rhythmic neurobehavioral system. Most rhythmic behaviors studied to date involve central neuronal oscillators, though these are usually supplemented by chain reflexes. Coupled oscillators are probably more common than a master oscillator. Some central oscillators employ endogenous bursters and some may be entirely based upon emergent network properties. Synaptic inhibition is by far the most common form of neuronal interaction among the network oscillators so far studied. In order to demonstrate that a particular neuronal or network property is *sufficient* to account for a given rhythmic behavior, it may be necessary to employ electronic or computer modeling. Demonstrating that the same properties are also *necessary* for the rhythm often represents problems that are difficult to solve conclusively. Finally, central neuronal oscillations can exhibit more than one rhythmic pattern, and different oscillators can interact with one another in ways that produce complex behavioral rhythms.

Questions for Thought and Discussion

1. A dog barks repeatedly for a prolonged period. How would you determine whether there is any nonrandom temporal structure to its sequence of barks? What specific criteria would you use in making your determination? Now two dogs bark for a prolonged period. How would you determine whether there is any nonrandom temporal relationship, and of what form, between the barks of the two animals?
2. Suppose that a particular animal shows highly rhythmic locomotory behavior. Suppose now that you have totally deafferented this animal's central nervous system and found that no form of electrical stimulation that you try on the central nervous system evokes rhythmic motor outputs. What can you conclude about the neuronal basis of this animal's locomotion? How would you proceed with your investigation on the role of this animal's central nervous system in locomotory behavior?
   Now suppose that in a different species of animal, the central nervous system after deafferentation *can* be made to produce rhythmic motor outputs. You now dissect out from this animal's ganglia a single rhythmic neuron. You place this isolated cell in a dish of saline. Although the cell appears healthy and can give normal action potentials, it cannot be made to give rhythmic bursts of action potentials, no matter how you try to stimulate it with currents, with neurotransmitters, or in other ways. What can you conclude about this cell's role in the rhythmic behavior? How would you proceed with your investigation of this cell's role in the rhythmic behavior?
3. How likely do you think it is that the cellular mechanisms underlying rhythmic behaviors in vertebrate animals, including such phyletically recent species as our own, are the same as those of invertebrates? Why?
4. Suppose that in some species of animal, the males make highly rhythmic calls that the females must hear. At the location of a given female, however, the

background noise is just about as loud as the male's call. Now suppose that the females have a central neuronal oscillator just like that of the males, except that, unlike the male's oscillator, which is used to produce a motor output (calling), the female produces no such call. How might such a central oscillator help the female to detect the male's song? Outline in detail a strategy by which the female could use this oscillator to increase the signal-to-noise ratio. How would you demonstrate whether the female is actually using this strategy?

5. Which subjects treated in earlier chapters of this book do you think could profit from computer modeling? Would you suggest using theoretical or empirical modeling? What specific information would you suggest should be obtained in this way?
6. In some species of frog, males that are making courtship calls form aggregates, termed choruses. In some species, each male's call consists of a rhythmic sequence of short pulses, separated by silent, interpulse intervals. For some such species, each frog synchronizes its pulses and its silent, interpulse intervals with those of its neighbors; that is, all these frogs pulse together, then are silent together, then pulse together, and so forth. The synchrony may be nearly perfect. In order to bring about this synchrony, each frog must be listening to the pulses of others and adjusting the timing of its own pulses accordingly. One can regard this vocal–auditory interaction among the frogs as either excitatory (that is, a pulse by one frog promoting the occurrence of a pulse by its neighbors) or inhibitory (that is, a pulse of one frog retarding or preventing a pulse by its neighbors). Propose a specific hypothesis for such excitatory and/or inhibitory vocal–auditory interactions among frogs that could give rise to their rhythmic chorusing. Draw the proposed interactions as a circuit diagram, as though each frog were a single neuron, interacting with other neurons (frogs) through excitation and inhibition. Then propose a specific program of research, complete with controls, to be carried out on the frogs, that would help support or refute your circuit diagram.

Recommended Readings

BOOKS

Gallistel, C. R. (1980) *The Organization of Action: A New Synthesis.* Lawrence Erlbaum Associates, Hillsdale, New Jersey.

Chapters 4 and 5 present material relevant to concepts discussed in the present chapter. The author combines the approaches of psychology, integrative neurobiology, and animal behavior. Several classical concepts are well developed, and readings from classical papers are included.

Herman, R. M., Grillner, S., Stein, P.S.G. and Stuart, D. G. (1976) *Neural Control of Locomotion.* Plenum Press, New York.

This symposium report contains articles on the mechanisms underlying a variety of locomotory movements in vertebrates and invertebrates.

Berridge, M. J., Rapp, P. E. and Treherne, J. E. (1979) *Cellular Oscillators. J. Exp. Biol.* 81.

This symposium report contains several articles on the mechanisms of neural and other cellular oscillators.

Koester, J. and Byrne, J.H. (1982) *Molluscan Nerve Cells: From Biophysics To Behavior.* Cold Spring Harbor Labs, Cold Spring Harbor, New York.

A compendium of review articles by many of the leading workers on the cellular and biophysical basis of molluscan behavior. Considerable attention is devoted to neuronal oscillators.

Muller, K.J., Nicholls, J.G. and Stent, G.S. (1982) *Neurobiology Of The Leech.* Cold Spring Harbor Labs, Cold Spring Harbor, New York.

A compedium of review articles by many of the leading workers in the field.

von Holst, E. (1973) *The Behavioral Physiology of Animals and Man: Selected Papers.* Vol. I. University of Miami Press, Coral Gables, Florida.

The two behavioral articles reprinted in Part I of this book were first published in the late 1930s and strongly influenced neurobiologists to search within the central nervous system for the sources of pattern generation. Concepts developed in these two articles are still employed in neuroethological research and are recommended for their historical interest.

ARTICLES

Kristan, W.B. Jr. and Calabrese, R.L. (1976) Rhythmic swimming activity in neurones of the isolated nerve cord of the leech. *J. Exp. Biol.* 65: 643–668.

An important article demonstrating, among other things, that the leech swimming system contains a central neuronal oscillator.

Friesen, W.O., Poon, M. and Stent, G.S. 1978. Neuronal control of swimming in the medicinal leech. IV. Identification of a network of oscillatory interneurones. *J. Exp. Biol.* 75: 25–43.

This is an important article that presents the synaptic connections and neuromime modeling of the interneurons thought to produce the leech's swimming rhythm.

Weeks, J.C. (1981) Neuronal basis of leech swimming: Separation of swim initiation, pattern generation, and intersegmental coordination by selective lesions. *J. Neurophysiol.* 45: 698–723.

This fine article demonstrates the presence of multiple central neuronal oscillators underlying leech swimming.

Selverston, A.I. 1980. Are central pattern generators understandable? *Behav. Brain Sci* 3: 535–571.

A thoughtful article by a leader in the field, followed by commentaries from many other noted researchers.

CHAPTER 10

Feedback in Behavior and the Nervous System

In the last two chapters, and particularly in Chapter 8, there was an implicit assumption that the neuronal information underlying behavior progresses unidirectionally, from sensory cells to interneurons to motor neurons, and finally to muscles. Although such unidirectionality is sometimes found in nerve circuits, more commonly these circuits also contain return pathways (Figure 1A). The returning information is called FEEDBACK, and the closed circle formed by the

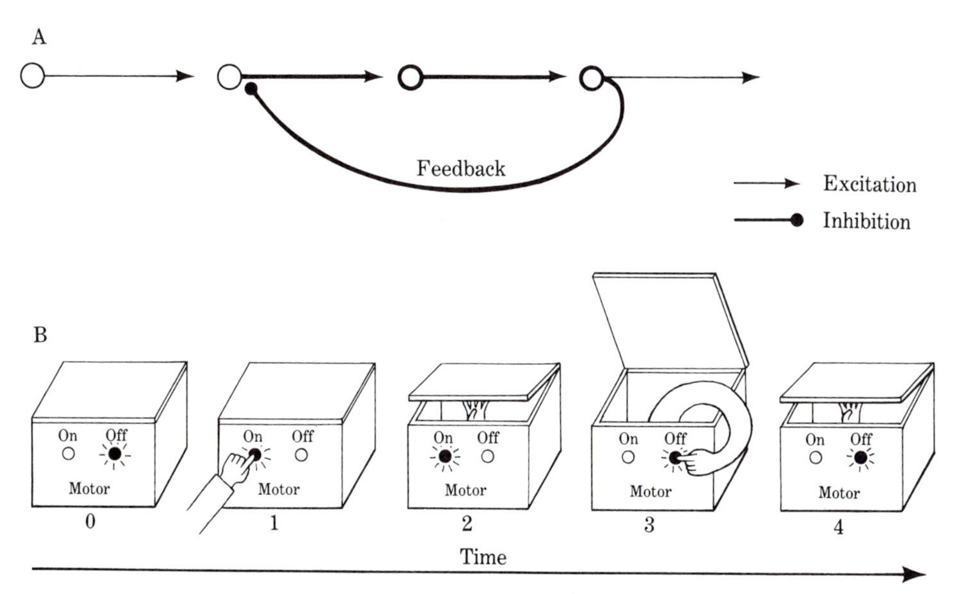

FIGURE 1. Two examples of negative feedback. A. Feedback in a hypothetical nerve circuit. B. Feedback in a hypothetical sequence of actions. The same "motor" is shown at five successive moments. At time 0, the motor is off. At time 1, somebody turns it on. This induces a sequence of events that result in the motor being shut off again.

return pathway is called a FEEDBACK LOOP. More generally, feedback occurs in any process where a succeeding event alters any of its causative preceding events (Figure 1B). Feedback that *inhibits* a preceding causative event, as in Figure 1, is called NEGATIVE FEEDBACK. As we shall see, negative feedback underlies many behavioral processes. Feedback that *excites* a preceding causative event, called POSITIVE FEEDBACK, appears to be less common in the nervous system. We have, however, seen two examples of positive feedback in behavior: the decision processes for crayfish escape behavior and for the release of ink by *Aplysia* (Chapter 8). A few other examples have also been reported (Camhi, 1971; Davis and Ayers, 1972; Getting and Willows, 1974; Gillette *et al.*, 1978; Ritzmann, 1974; Heitler and Burrows, 1976). Positive feedback will not be considered further in this chapter.

Open-Loop and Closed-Loop Behaviors

Feedback plays an important role in many goal-oriented behaviors. For instance, if you reach your hand out slowly to touch with one finger a preselected point on the table in front of you, during the act of reaching you may see that your hand is proceeding slightly off course. On the basis of this visual information, you may correct your movement so that your finger accurately hits the mark. This overall process contains a feedback loop because your *reaching* leads to a *visual evaluation*, which in turn controls your *reaching*. The fact that you actually do use visual feedback to guide your hand's motion can easily be demonstrated; simply look at the spot on the table, then close your eyes and reach for it. Now you will usually miss the mark. This method of testing has the disadvantage that it requires you to remember the location of the target while your eyes are closed. Thus, if you miss the mark, it could be because you have forgotten where it is rather than because you disengaged the feedback loop. To correct this difficulty, look at the spot on the table and reach for it while the sight of your hand's movement (but not the sight of the spot) is blocked by an opaque object, such as this book. Again you will usually miss the mark, demonstrating the importance of feedback.

This simple procedure, in which you removed the feedback normally incorporated in a behavior, is called OPENING THE FEEDBACK LOOP. It can be a useful procedure for testing the action of those parts of a control circuit that remain after the feedback loop has been disengaged. Opening a loop in this way transforms the behavior from one that is under CLOSED-LOOP CONTROL, or FEEDBACK CONTROL, to one under OPEN-LOOP CONTROL, that is, operating without feedback. A behavior under open-loop control is also called a BALLISTIC behavior—by analogy to the shooting of a bullet, which, having left the gun, is no longer subject to corrective aiming. It is important to note, however, that this open-loop behavior—reaching to touch a spot while your view of your hand is blocked—is open-loop only with respect to *visual* feedback. There could still be feedback from mechanoreceptors in the muscles and joints of the arm that is used to guide the movement. Moreover, there could also be feedback loops within the central nervous system's neural circuit that controls reaching movements of the arm.

NATURALLY OCCURRING OPEN-LOOP BEHAVIORS

A potential disadvantage of closed-loop control is that it takes time for the information, traveling around the feedback loop, to be integrated and to produce its effect. This presents no difficulty in slow reaching movements. However, where a behavior must be executed very quickly, the time occupied by feedback can be a liability. Among the quickest animal movements are the final strikes that many predators make in attacking their prey. Such strikes by several predators have been shown to operate by open-loop control, with no cues from the prey being used to steer the strike once it has been initiated. For example, in attacks on small crustacea by the molluscan cuttlefish, this predator ejects its two long tentacles with a rapid motion lasting only 30 milliseconds (Figure 2). Two observations demonstrate that this motion is performed by open-loop control. First, opening the feedback loop, by turning out the lights just after the tentacle ejection has begun, does not reduce the accuracy of the strike or the success in capturing prey. Second, if the crustacean target is tied to a string and is then pulled to one side just after the cuttlefish has begun to eject its tentacles, this pulling does not induce a corrective change in the strike's direction. Rather, the tentacles consistently miss their target, even though the lights have been left on throughout this strike (Messenger, 1968).

Such open-loop behaviors, lacking the capability of midcourse correction, require an animal to take particular care with its initial aiming. In fact, prior to its strike, the cuttlefish carefully lines up its body with the prey, using slow, closed-loop visual guidance. Many other animals that make quick predatory strikes exhibit similar prestrike behavior (Mittelstaedt, 1964; Ewert, 1980). This visual closed-loop property of the cuttlefish's initial aiming is shown by the following observations: (1) when a prey animal is introduced into the cuttlefish's environment, this predator slowly turns its eyes, head, and body toward it; (2) moving the prey around the cuttlefish causes it to turn to follow the prey; (3) this turning persists even if the prey is moved within an adjacent aquarium; and (4) these turning movements, once initiated, cease the moment that the lights are turned off or the prey is removed from the predator's sight (Messenger, 1968).

Open-loop behaviors also require that the animal's nervous system translate accurately the angle at which its sensory systems detect a target, into the angle at which its motor systems must produce a strike to hit the target. That is, motor systems must be accurately *calibrated* to sensory systems. Such internal accuracy

FIGURE 2. The predatory strike of the cuttlefish, *Sepia officianalis*, at a small crustacean. (After Wells, 1962.)

is less important in closed-loop behaviors, where errors of sensory-motor calibration can be corrected during the course of the behavior by means of feedback.

When a predator strikes, its potential victim often reacts with a very quick evasive movement. Like the predator, the escaping prey may also employ open-loop control, at least in its initial movements. For example, the initial abdominal downstroke of an escaping crayfish or lobster, lasting only about 30 milliseconds, can be evoked essentially in full by a single action potential in the lateral giant interneurons, or LGIs (Chapter 8). Stimuli received during these 30 milliseconds apparently have little or no effect on the form of the initial downstroke. After this downstroke, however, stimuli such as that produced by a continually attacking predator, can steer the crayfish's evasive swimming (Wine and Krasne, 1981). A similar sequence of open-loop and then closed-loop control in escape behavior appears to occur in cockroaches (Camhi and Tom, 1978; Camhi and Nolen, 1981) and possibly in fish (Eaton, 1982).

In summary, then, predator–prey interactions may often involve an overall sequence from closed-loop to open-loop control by the predator and open-loop to closed-loop control by the prey. At the moment of life and death interaction, both animals are operating in an open-loop manner, all efforts being directed toward speed at the expense of finely controlled accuracy.

CLOSED-LOOP CONTROL: THE OPTOMOTER RESPONSE OF THE FLY

Most animals use some form of feedback to compensate for unexpected changes in the position of the body or its parts. Such changes in body position could result from irregularities in the ground supporting the animal or from turbulence in the air or water surrounding it. To execute the proper corrective body movement, the animal requires information about the direction and magnitude of the unexpected change in its body angle. One possible means of acquiring this information is through mechanoreception. For instance, some mechanoreceptors, located on joints or muscles of the legs, by monitoring sudden imposed changes in leg position, can direct an animal's central nervous system to make the proper compensatory reaction (Matthews, 1972; Olivo and Jazak, 1980). Other mechanoreceptors, located in the equilibrium organs of the vertebrate inner ear or in comparable organs in some invertebrates, monitor changes in body angle with respect to gravity (Brodal and Pompeiano, 1972; Schone, 1961). Still other mechanoreceptors, such as the wind receptors of some insects, can detect changes in the wind angle that occur if the angle of the insect's body is suddenly shifted as a result of air turbulence (Camhi, 1970; Camhi and Hinkle, 1974). In addition to mechanoreception, however, vision is often used to direct behavioral reactions that compensate for imposed changes of body angle. The remainder of this section describes experiments on such visually induced feedback behaviors.

Any unexpected force on an animal that causes its head to move produces a slippage of the entire visual image across the retina of each eye. The direction and speed of this slipping image could be used to correct the body's posture. It is easy to demonstrate that vision is in fact used in this way. One can place an animal such as a housefly (*Eristalis*) on a platform inside a cylinder that has a striped pattern painted on its inner surface. (A great many invertebrate or vertebrate animals can be used equally well in this experiment.) When one rotates the platform holding the fly, the fly responds by turning its body in the opposite direction, as though it is attempting to stabilize the image of the cylinder's stripes on its retinae (Figure 3A). The fact that vision rather than mechanoreception or some other sensory cues mediates this turning response can be shown by holding the fly's platform fixed and instead rotating the cylinder around it. The fly responds

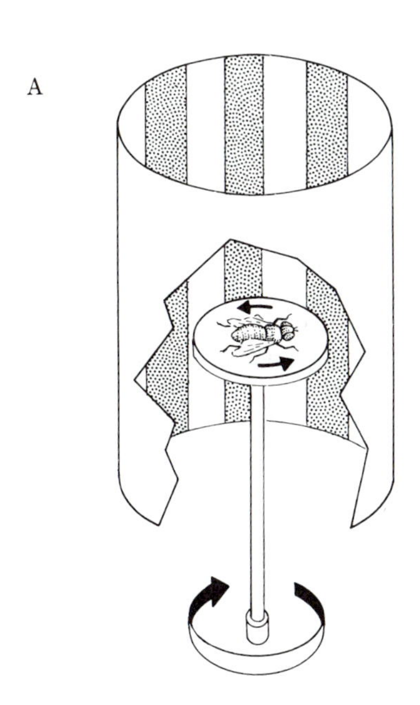

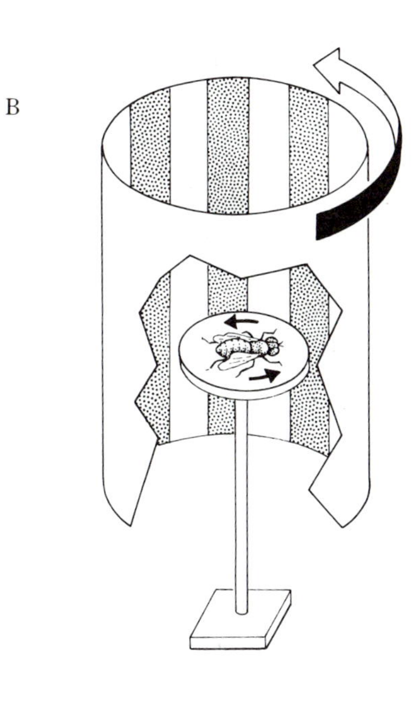

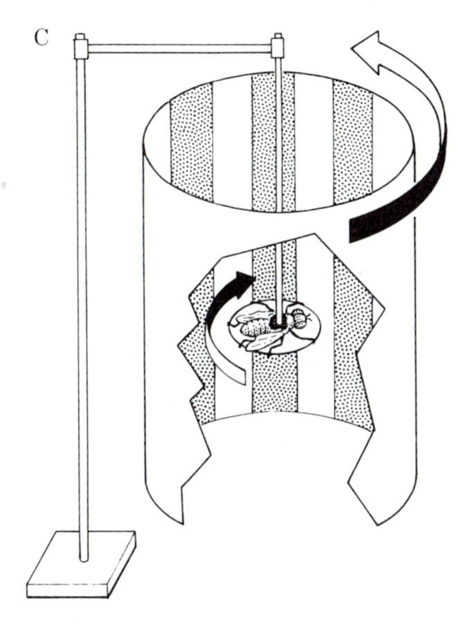

FIGURE 3.

Turning responses of the fly *Eristalis*. A. If one rotates clockwise the platform on which a fly stands *(lower arrows)*, the fly responds by turning counterclockwise *(upper arrows)*. B. The optomotor response: if one rotates counterclockwise the patterned cylinder surrounding the fly *(upper arrow)*, the fly responds by turning counterclockwise to follow the cylinder *(lower arrows)*. C. Open-loop turning response; the fly is tethered so that it cannot turn its body, but holds a card in its feet. If one rotates the cylinder counterclockwise *(upper arrow)*, the fly turns the card clockwise *(lower arrow)*. (After Mittelstaedt, 1964.)

by turning in the *same* direction as the cylinder, again helping to stabilize the image of the stripes on its retinae (Figure 3B). The faster the cylinder is rotated, the faster the fly turns. The visual signal from the cylinder is often called the visual EXAFFERENCE because it is an *afferent* signal coming from the animal's *external* world. The fly's turning reaction to this exafferent stimulus (that is, to the rotating cylinder) is called an OPTOMOTOR RESPONSE.

The sensory signal that evokes the fly's turning and that determines its direction and speed is not the cylinder rotation per se but rather the slippage over the retinae of the image cast by the rotating cylinder. The speed of this slippage is called the SLIP SPEED. The slip speed, of course, depends upon both the speed of the rotating cylinder and the speed with which the fly turns as it follows the cylinder's rotation. For instance, if one starts to rotate the cylinder around a stationary fly, the slip speed will be large at first, until the fly begins to turn in the same direction as the cylinder. The fly's turning will then reduce the slip speed toward zero. Thus, because *slip speed* evokes *turning* that itself reduces *slip speed,* the optomotor response is a closed-loop, negative feedback behavior (McFarland, 1971; Land, 1975).

Suppose now that one turns the cylinder at 20°/sec clockwise around the fly. This 20°/sec exafference induces an optomotor turning response of about 15°/sec clockwise (Mittelstaedt, 1964). The fly's turning behavior thus reduces the slip speed to $20 - 15 = 5$°/sec clockwise. The *change* in the slip speed of 15°/sec that is caused by the fly's own movement is called the REAFFERENCE because it is a *reaction* within an *afferent* system (namely, the eye) to the fly's own movement. The fly's reafference, then, closes the negative feedback loop, as is shown by the block diagram of Figure 4. (This figure employs diagrammatic conventions originally devised in cybernetic control theory, or systems analysis, and now commonly used for describing feedback in animal behavior and the nervous system.) Figure 4 emphasizes that the slip speed equals the *difference* between the exafference and the reafference. A more general term for such a difference between two parameters that serves as the activating agent of a behavior is a DIFFERENCE SIGNAL or ERROR SIGNAL.

One surprising feature of the optomotor feedback loop is that it is actually impossible for the fly to keep up perfectly with the cylinder's rotation; for to do so would be to create a reafference equal to the exafference, and this would create a slip speed equal to zero. But a slip speed of zero would produce zero behavior, so the fly's turning would, of necessity, stop. Given this circumstance, the best that the fly can do is to keep its turning speed very close, but not equal, to the cylinder's rotation speed. This it does, especially for slow cylinder rotations of up to about 20°/sec. (Another species of fly, during flight, keeps its turning speed very close to the cylinder's rotation speed, for rotations up to about 200°/sec; Collett, 1980.)

A useful term for expressing quantitatively the degree to which the turning fly keeps up with the rotating cylinder is the GAIN, or amplification, of the overall response. Beccuase this is a closed-loop behavior, the overall gain is commonly called the CLOSED-LOOP GAIN. The definition of gain for *any* biological or other response is the ratio

$$\frac{\text{magnitude of output}}{\text{magnitude of input}}$$

For the optomotor system, the input referred to is not the slip speed, which is the direct sensory input to the eyes, but rather the cylinder rotation, which is the input of the overall system. Thus, the closed-loop gain of the fly's optomotor response is

$$\frac{\text{fly's turning speed}}{\text{cylinder's rotation speed}}$$

Perfect following of the cylinder by the fly, which we have seen is impossible, would be represented by a closed-loop gain of 1. The actual closed-loop gain lies somewhere between 0 and 1, and accurate measurements show that for turning speeds up to 20°/sec it ranges between 0.75 and 0.95 (Figure 5).

With the closed-loop gain held close to 1, as it is for slow cylinder rotations, the slip speed remains fairly low. (For instance, when the cylinder speed is 20°/sec, as we have seen, the slip speed is only 5°/sec, owing to the 15°/sec reafference.) In fact, the slip speed is much lower than the turning speed of the fly, which this slip speed evokes. (For instance, with a cylinder speed of 20°/sec, the 5°/sec slip speed evokes a 15°/sec turning speed.) Therefore, the fly's optomotor circuitry considerably amplifies the slip speed signal to produce the turning speed.

A convenient way to measure the amplification of slip speed by the optomotor circuitry is shown in Figure 3C. The fly is waxed to a rod so that its body cannot turn to follow the cylinder's rotation. The fly is presented with a small circular card that it holds in its feet. Now when the cylinder is rotated, the fly, responding as usual by making turning movements with its legs, turns the card instead of its body. The fly's response therefore produces no visual reafference that can reduce the slip speed (Figure 4). We have thus opened the feedback loop so that,

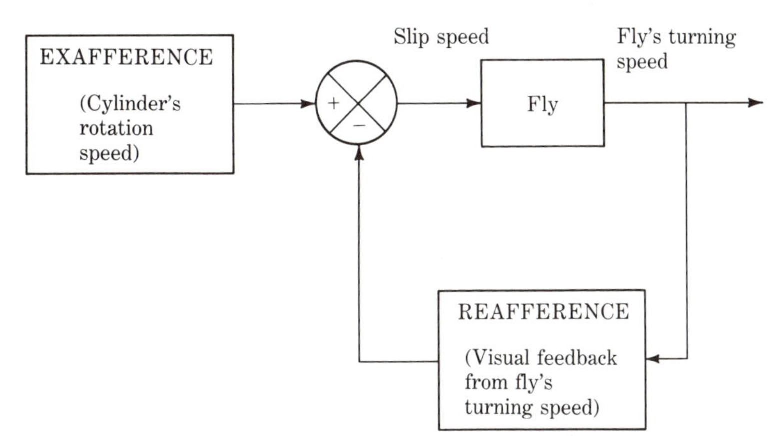

FIGURE 4. Block diagram of the fly's optomotor response. Lines with arrows indicate the flow of action or information. The circle is a point of convergence of flow. The negative sign in the circle indicates that the flow entering at this point (the reafference) is subtracted from that entering at the plus sign (the exafference). The figure should be read as follows: exafference minus reafference equals slip speed, which leads to the fly's turning speed, which causes reafference.

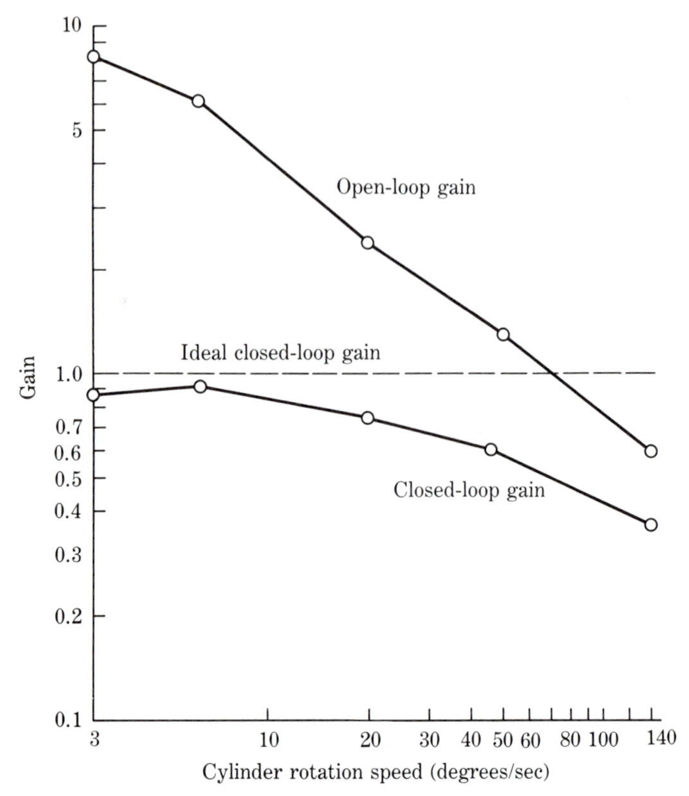

FIGURE 5. Closed-loop and open-loop gains of the optomotor response of the fly *Eristalis*, as a function of cylinder rotation speed. A closed-loop gain of 1 would be ideal, since the fly would then follow the cylinder perfectly. However, as explained in the text, this ideal value of closed-loop gain is impossible to achieve. See text for further explanation. (Gains calculated from Mittelstaedt, 1951.)

with reafference equal to zero, the slip speed equals the cylinder's rotation speed. In this situation, the ratio

$$\frac{\text{turning speed of the card}}{\text{cylinder's rotation speed}}$$

is called the OPEN-LOOP GAIN. It is a direct measurement of the optomotor circuit's amplification of slip speed.

The open-loop gain of the fly is much greater than 1 for all reasonably slow cylinder speeds (Figure 5). In some animals, the open-loop gain can approach 100 (Horridge, 1967) or even 1000 (Mittelstaedt, 1964) for very low cylinder speeds. The optomotor circuitry thus exercises a very strong amplifying effect on the slip speed signal that it receives. Clearly, then, with the reafferent feedback loop disengaged as in Figure 3C, the fly tries to turn much faster than a freely mobile fly ought to turn in order to stabilize its visual image of the moving cylinder. This indicates that the direct exafferent-to-motor interaction within the central nervous system is very poorly calibrated; it is only by virtue of the reafferent feedback that the overall optomotor behavior operates, as required, with a gain very close to the ideal value of 1.

It is also because of the reafferent feedback that the fly can gauge the success of its response to the rotating cylinder and, if necessary, speed up or slow down this response. For instance, if for some reason the turning movements of its legs were not producing sufficient rotation of its body to follow properly a rotating cylinder, then the slip speed would become large. As a result of this increased slip speed, the turning movements of its legs would increase. This is exactly what we have seen to occur in the extreme case where the fly's body is totally prevented from turning (Figure 3C).

It is important to recall that the natural role of the fly's optomotor response is not to compensate for rotations of cylinders produced by biologists; rather, it is to restore the body to its initial orientation when this orientation is disturbed by some outside force. Cylinder rotation is merely a convenient way to study the optomotor response.

In spite of the usefulness of optomotor responses in resisting unexpected changes of body position, this behavior raises a serious difficulty: How can a fly (or any other animal) ever *intentionally* turn its body or head? Such a turn would cause slippage of the retinal images, which should immediately induce a turn of the body or head back toward its starting position. We will now consider several possible solutions to this important problem.

IS THE OPTOMOTOR RESPONSE INHIBITED DURING INTENTIONAL TURNING?

One possible solution to the problem of optomotor interference with intentional movements would be for an animal to inhibit its optomotor neural circuitry whenever it intends actively to change its body angle. Such optomotor inhibition has been searched for in studies of fruitflies (Heisenberg and Wolf, 1979). When this and certain other fly species hover in a stationary visual environment, they make all their voluntary turns as quick, jerky movements lasting, in some species, only 40 milliseconds (Collett and Land, 1975). These quick turns are called BODY SACCADES, by analogy to the quick, jerky eye movements, called SACCADES, that most vertebrates use when visually scanning a stationary object (Dichtburn, 1973). The question, then, is whether a fly's vision is inhibited during its body saccades. Because body saccades are very brief, such SACCADIC SUPPRESSION, if it occurs, would interrupt only momentarily an animal's closed-loop visual control of body angle.

Saccadic suppression was demonstrated through experiments in which a fruitfly made normal flight movements with its wings, but its thorax was waxed to a rod. Thus, like the *Eristalis* fly in Figure 3C, it could not turn and, therefore, was operating in an open-loop manner. The force with which the fruitfly *tried* to turn, however, was measured by a torque meter attached to the rod. Because the torque meter responded quickly and accurately, it could record the precise turning force, or torque, of single body saccades. A body saccade to the left or to the right was found to be highly stereotyped in terms of the time course and magnitude of the torque produced. During such body saccades by the tethered fly, a tiny cylinder with a visual pattern painted on its inner surface was rotated by a motor. The motor was driven by the torque meter and could be made to restore exactly the normal reafferent feedback that had been disengaged by tethering the fly. For instance, when the fly intended a clockwise body saccade, the cylinder was made to move counterclockwise by just the amount that the fly would have turned if it were untethered. Thus, the cylinder's rotation reclosed the visual feedback loop. In this experimental situation, one could also choose to have the cylinder rotate too much or too little, producing an abnormal amount of reafference. In so doing, it was found that varying the amount and speed of the cylinder's rotation, from zero, to that needed for restoring normal reafference, to twice that amount, caused little or no alteration in the strength and time course of the torque in the fly's body saccade (Figure 6A, B, and C). Thus, when the fly carries out a body saccade, this saccade is not affected by slip speed of the visual environment over its retinae. That is, saccadic suppression has occurred.

In related experiments on human subjects, suppression of visual perception has been reported beginning about 15 milliseconds before the start of a saccade of the eyes. This suppression reaches a maximal strength at the time that the

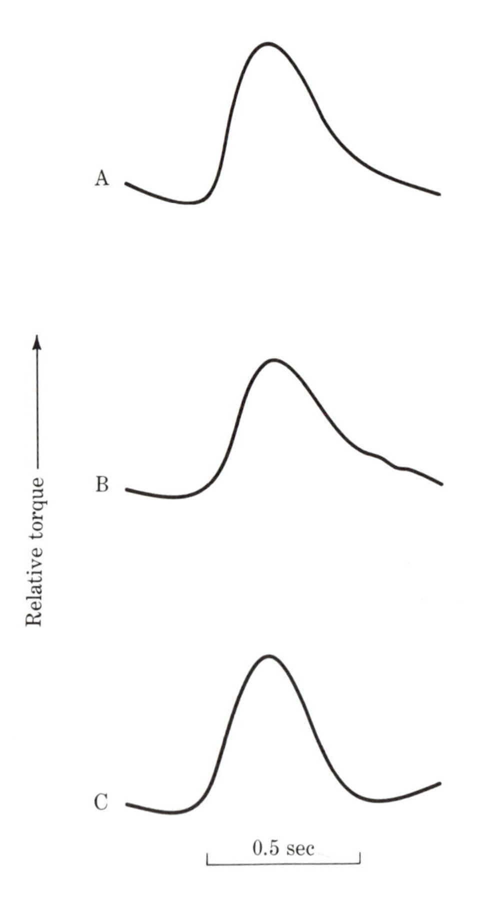

FIGURE 6.

Saccadic suppression in flies. Torque measurements of a tethered fly's turning effort under different conditions of visual feedback from a rotatable cylinder. A. The cylinder is stationary. B. The cylinder is moved so as to exactly close the visual feedback-loop that had been opened by tethering the fly. C. The cylinder is moved at twice the rate as in B. (After Heisenberg and Wolf, 1979.)

saccade begins and lasts for roughly the 45 milliseconds that the eye movements continue (Dichtburn, 1973). This saccadic suppression, then, helps to prevent a person from perceiving his visual environment as jerking about each time he makes a saccadic eye movement. Likewise in monkeys, neurons of the superior colliculus that respond to the movement of objects in the visual field are inhibited during saccades (Robinson and Wurtz, 1976; Richmond and Wurtz, 1980).

THE OPTOMOTOR RESPONSE IS NOT INHIBITED DURING SLOW TURNS

In addition to brief saccades, many animals can make slower movements of their eyes, head, or body. For instance, a walking housefly or a flying hoverfly often makes gradual turns lasting several seconds (Collett and Land, 1975; Collett, 1980). Is the fly's optomotor behavior inhibited during these prolonged turns? Apparently it is not. This was first demonstrated by the simple but elegant procedure of rotating a fly's head to alter its reafferent feedback. Because the fly's neck is soft and pliant, one can easily turn its head through 180° about the axis of the neck and wax it in place. Now the fly's left eye is on its right side and vice versa (Figure 7B). As a result, if this fly now makes an intentional turn, the

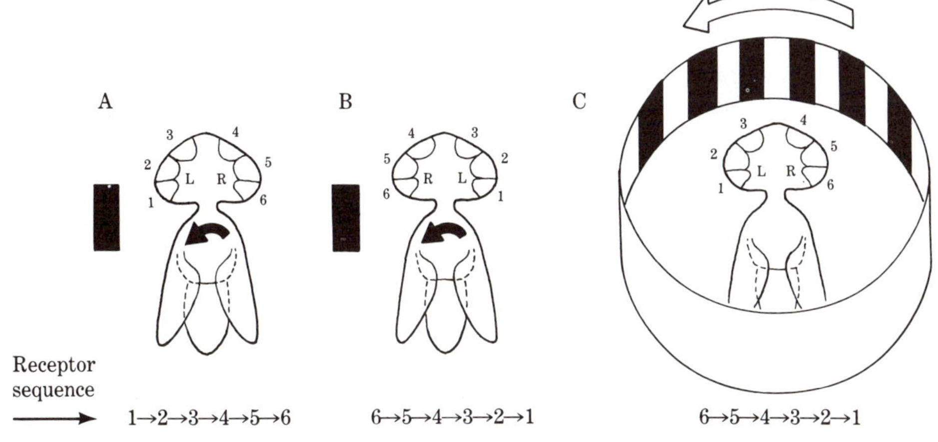

FIGURE 7. Test for suppression of optomotor responses during nonsaccadic turning in the fly *Eristalis*. A. Normally, when a fly turns counterclockwise (curved arrow on body), a visual cue, such as the black bar on the left side, excites the visual receptors shown in the sequence 1 through 6. B. The fly's head is rotated by 180° about the neck axis, interchanging the positions of the left and right eyes. Now when the fly turns counterclockwise (arrow), the same black bar excites receptors in the sequence 6 through 1. C. A stationary fly with its head not rotated; when the cylinder is rotated counterclockwise (arrow), each moving strip excites receptors in the sequence 6 through 1. Although only 6 receptors are shown, each eye actually has over a thousand. (After von Holst and Mittelstaedt, 1973.)

sequence in which the different visual receptors are stimulated by visual images slipping across the retinae is the reverse of normal. For instance, in the normal fly of Figure 7A, during a counterclockwise turn (arrow), the visual stimulus (black bar) would cause excitation of the six visual receptors shown in the sequence 1 through 6. By contrast, in the fly with its head rotated (Figure 7B), the same counterclockwise turn of the body would stimulate these same six receptors in the reversed sequence, 6 through 1.

This particular sequence of stimulation, 6 through 1, would also be produced in a normal fly (without its head rotated) if it stood still inside a painted cylinder that was rotated counterclockwise (Figure 7C). This rotating stimulus, as we have seen, would induce the fly to make a counterclockwise optomotor turn. Does this stimulation sequence, 6 through 1, also induce a counterclockwise optomotor turn in the fly that has its head rotated and that is already making a counterclockwise *intentional* turn (Figure 7B)? Or is this optomotor response inhibited during the fly's intentional turn? If the optomotor response were inhibited, then the visual sequence 6 through 1 that the fly experiences during its intentional turn would have no effect; the fly would simply complete its intentional turn and go about its business. But if the optomotor circuitry is not inhibited, the excitation of the receptor sequence 6 through 1 would cause a counterclockwise *optomotor* turn, superimposed upon the *intentional* turn. This would lead to a faster rate of turning of the body. This faster rate would speed up the slippage of the visual image across the retinae in the sequence 6 through 1. This increased slip speed would cause the fly to speed up further its optomotor turning, which would further increase the slip speed, and so forth. The fly would quickly enter a paroxysm of turning, spinning itself into utter exhaustion.

This is precisely what flies with rotated heads do (von Holst and Mittelstaedt, 1973). Thus, the optomotor response is not switched off during slow, nonsaccadic intentional turning but rather remains intact. The clever strategy for demonstrating this, rotating the fly's head, amounts to a transformation of the normal negative feedback organization of the optomotor response (in which reafference from the optomotor turning decreases slip speed) into positive feedback (in which this reafference increases slip speed).

A more quantitative demonstration of the persistence of optomotor responses during nonsaccadic turning has been carried out on hoverflies (Collett, 1980). Also, endless circling behavior such as that shown in houseflies with rotated heads occurs in frogs and fish whose eyes have been rotated by 180° and sewn into place (Sperry, 1950). Related to this, human subjects wearing spectacles that invert the visual image report a sense of the environment as constantly spinning (Teuber, 1960).

THE EFFERENCE COPY HYPOTHESIS

The problem that remains, then, is, "Because the optomotor response is not inhibited during intentional, nonsaccadic turns, why does it not block these turns?" There is as yet no conclusive answer. However, there is one especially elegant hypothesis, first suggested by von Helmholtz (1925), then formulated explicitly in the 1950s by von Holst and Mittelstaedt (1973) and reintroduced in recent studies (Land, 1975; Collett, 1980; Bell, 1982). In this section we will consider the merits of this idea.

The diagram of a fly's optomotor feedback loop (Figure 4) is reproduced in light gray in Figure 8. Suppose now that this fly makes a slow, nonsaccadic turn toward some specific target (perhaps another fly of the opposite sex) that has suddenly appeared in its peripheral field of view. This target constitutes for our fly an exafferent signal, that is, an external cue inducing a behavior (Figure 8, dark box). In the present hypothesis, this exafference is thought to lead to the activation within the fly's central nervous system of *two* separate neuronal signals (Figure 8, dark lines). One of these, called the EFFERENCE, is the signal that activates motor neurons to bring about the turn toward the target. (In the terminology used in Chapter 8, this signal would be called a command.) The second proposed signal is called the EFFERENCE COPY. This signal is thought of as being exactly proportionate in strength to the efference signal: the faster the efference instructs the fly's legs to turn, the stronger is the efference copy signal. This efference copy is thought of as projecting to a point near the input of the optomotor circuitry and there interacting with the reafference signal (Figure 8). At this point of interaction, the efference copy signal, being exactly proportionate to the efference signal (which causes the reafference) is used to exactly counterbalance the reafference.

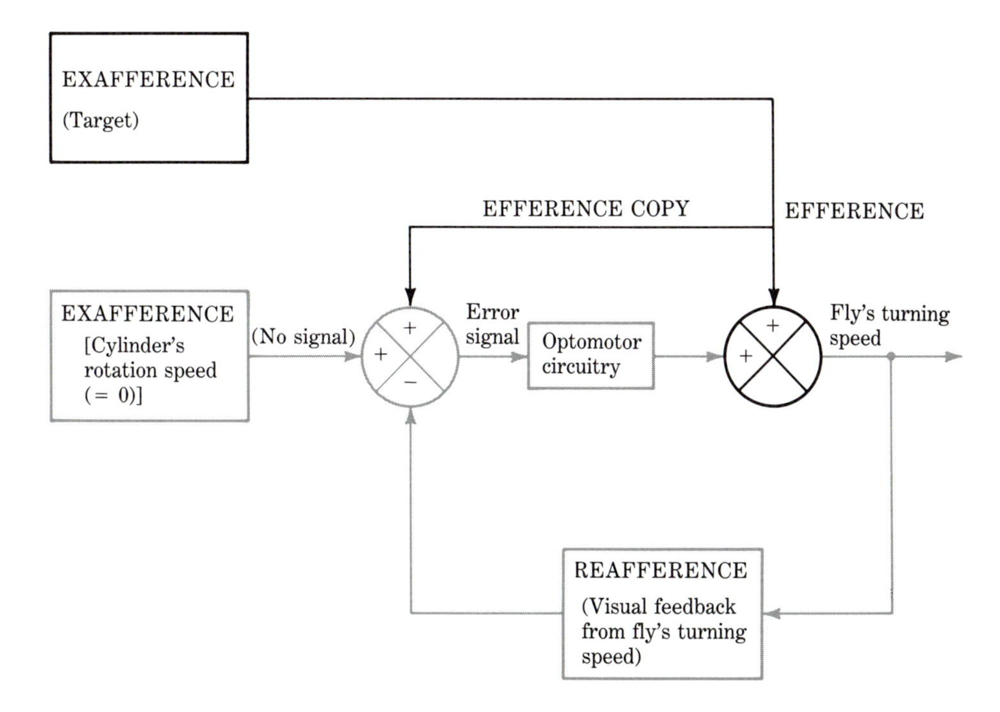

FIGURE 8. The efference copy hypothesis. The light gray part of the figure is basically the same as the diagram of Figure 4. Superimposed on this in black is an exafferent signal that leads to both an efference and an efference copy. Explanation in text.

As an example, suppose that a fly is turning toward its target clockwise at an angular rate of 20°/sec. The reafference created by this turn would be a slipping of the visual environment *counterclockwise* by 20°/sec. By our hypothesis, the efference copy signal would be of just such a strength to counteract this 20°/sec counterclockwise slip speed and would thus exactly balance out the reafference. With the reafference and efference copy thus equal and opposite (and with no exafferent input to the optomotor behavior because there is no cylinder rotation occurring), there would be an error signal of zero. Therefore, the optomotor response would not be activated and thus would not oppose the turn toward the target.

In summary, by this EFFERENCE COPY HYPOTHESIS, the optomotor circuitry, rather than being shut down as it is in saccadic suppression, lies poised with its excitatory and inhibitory inputs in a state of balance. The efference copy would represent the fly's *expectation* that a certain amount of visual reafference should occur during its turn toward the target; and because this reafference *does* occur, the efference copy exactly offsets it. However, if the amount of reafference during the turn *differs* from that which is expected (for instance, if the fly is standing on loose soil so that its moving legs do not produce the full turning speed expected), this would lead to some nonzero error signal and thus to an optomotor response superimposed upon the ongoing turning. This would bring the turning speed up to that which the fly had intended to make toward its target in the first place. Therefore, the persistence of the optomotor response during intentional, nonsaccadic turns can serve an important function—to compensate for disturbances that may alter the fly's expected rate of turning toward its target.

The efference-copy hypothesis provides an elegant possible solution to the question of how the optomotor response can remain operative during nonsaccadic turns and yet not block these turns. But is this hypothesis correct? There is, as yet, no answer. Other explanations are also possible (Collett, 1980). Unless physiological experiments can confirm the presence of an efference copy signal within the central nervous system and can demonstrate the effect of this signal in precisely offsetting reafference, the efference copy explanation of the fly's turning behavior will remain only a hypothesis. In another system, found in electric fish, just such a physiological demonstration has been accomplished, confirming the role of efference copy as originally suggested (Bell, 1981, 1982). There is also suggestive, but inconclusive, physiological evidence for an efference copy signal employed during optomotor turning in goldfish (Klinke, 1970). Moreover, as we shall see later in this chapter, strong physiological evidence has been obtained in one system—that involving the mammalian muscle receptors in the control of muscular contraction—for an organization similar (though not identical) to that proposed in the efference copy hypothesis.

One hallmark of an efference copy signal, as we have seen, is its *exact* proportionality to the efference signal that induces the behavior. In contrast, quite *inexact* copies of efference signals, originating within the central nervous system, have been recorded physiologically in various animals (Suga and Shimozawa, 1974; Russell, 1976; Wine and Mystick, 1977; Richmond and Wurtz, 1980). These inexact copies clearly could not be used for precise counterbalancing of reafference. Rather they are more commonly employed in shutting off the responses by various circuits to sensory input during the performance of certain behaviors. Some authors use the term *efference copy* to include even such inexact central signals. According to a more appropriate terminology, however, *any* neural signal that branches off centrally from an efference signal, whether or not it is exactly proportionate to this efference, is called a COROLLARY DISCHARGE.[1] Efference copy, then, is a special case of corollary discharge, in which the branched signal is strictly proportionate to the efference signal and is used to counterbalance reafference. One example of a corollary discharge that is *not* an efference copy is a central signal, associated with eye movements in monkeys, that is used to bring about saccadic suppression (Richmond and Wurtz, 1980).

Feedback in Locust Flight

Most of an animal's sensory receptors are designed to detect stimuli in the outside world. These visual, auditory, chemical, and other detectors are sometimes called EXTEROCEPTORS. Other sensors, designed for monitoring the internal conditions of the body, including parameters such as the CO_2 concentration of the blood, body temperature, and fullness of the stomach, are called INTEROCEPTORS. A largely separate class of sensory cells is a group called PROPRIOCEPTORS. These are mechanoreceptors whose normal function is to signal the movement or position of one body part relative to another. In this section we consider the role played in locust flight by feedback from proprioceptors of the wings. The next section will then discuss a more complex pattern of proprioceptive feedback: that involved in the control of mammalian movements.

The rhythmic beating of the wings of a locust (*Schistocerca gregaria*), like swimming in leeches (Chapter 9) or most other rhythmic behaviors, is controlled in part by a central neuronal oscillator (Wilson, 1961; Burrows, 1973, 1975). In addition, however, numerous proprioceptors on or at the base of each wing are excited at particular phases within each wingbeat (Burrows, 1976; Wendler, 1974). Some are active at the time of nearly maximal wing elevation and others at nearly maximal wing depression. Therefore, each of these receptors could, in principle, contribute through a chain reflex (Chapter 9) to the timing of the next movement of the wing.

The best understood of these wing proprioceptors is the WING HINGE STRETCH RECEPTOR, a cell whose dendrites are attached to a connective tissue strand at the wing base (Gettrup, 1962; Wilson and Gettrup, 1963; Burrows, 1975; Altman and Tyrer, 1977). There is one such receptor cell for each wing. When a wing is elevated, this receptor is stretched, causing it to be excited. During flight, this cell gives from one to a few action potentials on each wingbeat, just prior to or at the moment of maximal wing elevation (Figure 9). Given this timing, if the wing hinge stretch receptor were to excite synaptically the wing depressor motor

[1]Although efference copy was carefully defined by the workers who coined the term in the 1950s (von Holst and Mittelstaedt, 1973), the term *corollary discharge* has been used by different authors to mean somewhat different things (Sperry, 1950; Teuber, 1960; Bullock *et al.*, 1977; Heisenberg and Wolf, 1979; Gallistel, 1980). It is useful, however, to retain strict definitions, such as those employed here.

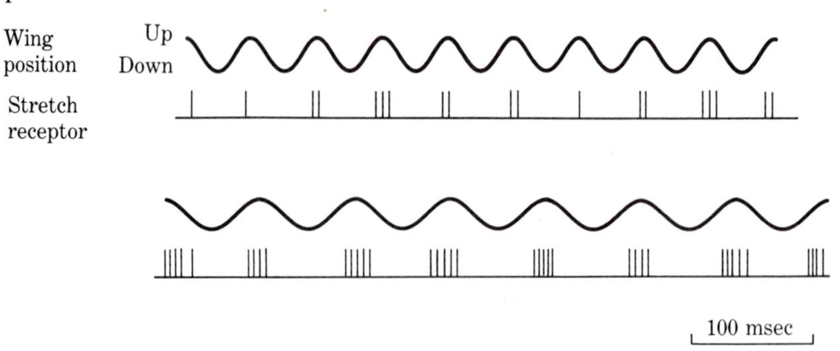

FIGURE 9. Responses of a locust's wing hinge stretch receptor cell to the wing's movements during flight. (Modified from Wilson and Gettrup, 1963.)

FIGURE 10. The three thoracic ganglia of the locust *Schistocerca gregaria*. Pro, Prothoracic; Meso, mesothoracic, and Meta, metathoracic ganglion. Filled circles show the positions of cell bodies of identified wing depressor motor neurons. The dotted lines show the axonal paths, determined physiologically, of two of the wing hinge stretch receptor cells—that of the right forewing and that of the left hindwing. Three sets of recordings used in this physiological determination are shown. A. Synchronous action potentials in the peripheral nerves to the prothoracic and the mesothoracic ganglia were evoked by elevating the right forewing. B. An action potential in the ipsilateral meso-to-metathoracic connective occurred at a fixed time after each action potential in A. The action potentials in the connective were too small to be seen directly against the background of activity from the many other axons in the connective. Therefore, the technique of computer summation was used; at a given moment following an action potential in trace A (for instance, 0.5 milliseconds after this action potential), the voltage registered by the electrode on the connective was entered into the computer. The voltage occurring at 0.5 milliseconds after each successive action potential in A was also entered into the computer, and the computer summed all these voltages. The same was done for each other moment following the action potentials in A, such as 0.6 milliseconds, 0.7 milliseconds, and so forth. The computer then plotted out a graph of these summed voltages as a function of time. By this method, all action potentials recorded by the connective electrode whose time of occurrence was fixed relative to an action potential in A added together to give a computer-summed action potential. Any other action potentials, occurring at times *not* fixed with respect to those in A, would be just as likely to occur at any one moment as at any other. Therefore, their net effect would be a straight line. The upper trace in the figure shows the action potential in A, and the lower trace the computer-summed action potential in the connective. C. Action potentials in the peripheral nerves to the mesothoracic and metathoracic ganglia, activated by elevating the hindwing. In no other extracellular recording position could action potentials be recorded that were synchronous with those shown, with or without computer summation. An electrode was placed in the body cavity and used as the second, or indifferent, electrode for all recordings in this figure. (After Burrows, 1975.) ▶

neurons, this receptor cell could contribute to producing the motor activity for each wing depression.

The anatomical distributions of the axon terminals from the wing hinge stretch receptor of each wing are consistent with the possibility that they may activate the depressor motor neurons. The main depressor motor neurons have their cell bodies and dendrites located in all three thoracic ganglia (Figure 10, filled circles). By recording in different locations the action potentials evoked by forced elevation of one wing, one can trace physiologically the axonal path of the receptor of that wing. Such recordings show that the axon of this receptor from a forewing projects ipsilaterally to all three thoracic ganglia (Figure 10A and B) and that the axon of this receptor from a hindwing projects to two of these three ganglia (Figure 10C). By staining each wing hinge stretch receptor with cobalt sulfate (Chapter 3), the distribution of the axons to different ganglia as shown in Figure 10 was confirmed, and the axon terminals were shown to branch very extensively within these ganglia (Figure 11). In fact, these terminals are intertwined with the dendrites of motor neurons to the wing depressor muscles (Altman

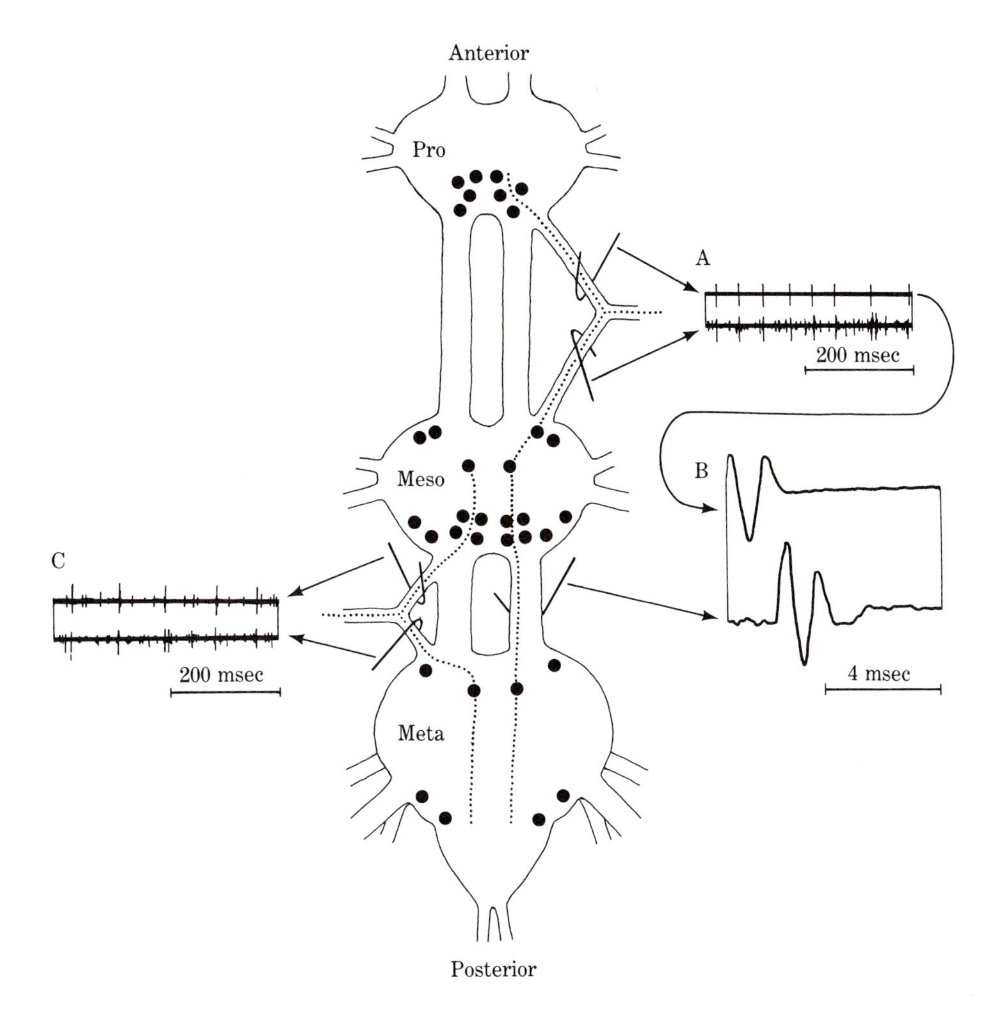

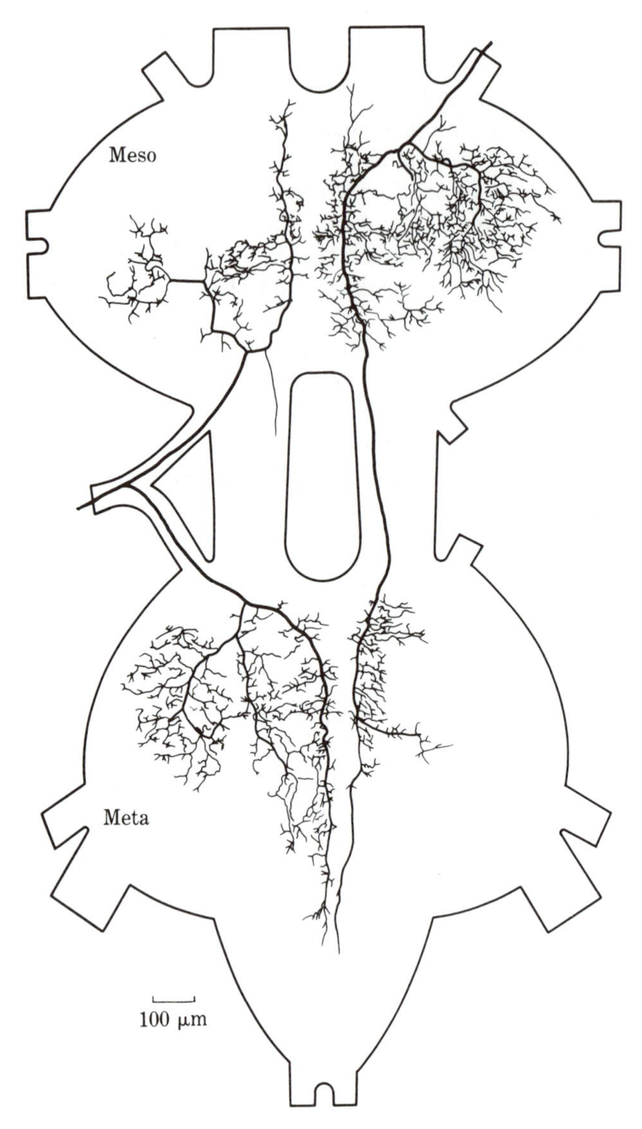

FIGURE 11. Axon terminals of the stretch receptor of the right forewing and of the left hindwing within the mesothoracic and metathoracic ganglia of a locust. (No similar study was made of the prothoracic ganglion.) The drawings were made from cobalt-filled axons. Many of the finest branches are not shown. The axonal pathways confirm those determined physiologically as shown in Figure 10. (From Burrows, 1975.)

and Tyrer, 1977). So far, though, no attempt has been made to identify with the electron microscope actual synapses that might connect a wing hinge stretch receptor axon to a wing depressor motor neuron.

Physiologically, however, such sensory-to-motor synapses have been discovered (Burrows, 1975). Each action potential in a wing hinge stretch receptor axon is followed by a depolarizing PSP, recorded intracellularly from the cell body of a wing depressor motor neuron (Figure 12A). This depolarizing PSP is an EPSP (Chapter 3, Appendix), because it can lead to an action potential if it occurs while one depolarizes the cell body with a steady injection of current through a second microelectrode (Figure 12B). The sensory-to-motor connection appears to be monosynaptic because it fulfills three of the necessary criteria for monosynaptic interaction (Chapter 3, Appendix): (1) there is a short and constant latency of 1–1.5 milliseconds, from the action potential recorded where it enters the ganglion to the start of the EPSP in the motor neuron; (2) the EPSP follows faithfully action potentials in the wing hinge stretch receptor occurring at a high frequencies, up to 125/sec; (3) replacing the calcium in the saline with magnesium leads to only a gradual, not a precipitous, drop in the EPSP size, and this drop is reversible (Figure 12C).

Not just 1 but at least 17 different depressor motor neurons of all three ganglia give an EPSP in response to each action potential of a wing hinge stretch receptor (Burrows, 1975). Only motor neurons ipsilateral to a given stretch receptor are excited by it. This pattern is consistent with the strictly ipsilateral distribution of the stretch receptor axon (Figure 11).

Many wing *elevator* motor neurons also respond to the action potentials of a wing hinge stretch receptor. But all such responses are hyperpolarizing IPSPs (Figure 12D). The latency from the action potential in the stretch receptor to the start of an IPSP was longer than for the EPSPs, about 4–6 milliseconds. However, like the EPSPs, these IPSPs showed some features expected of a monosynaptic connection. They could follow one-for-one the action potentials of the stretch receptor occurring at a fairly high frequency (50/sec). Moreover, they were decreased gradually, not precipitously, by replacement of calcium in the saline with magnesium. Thus, the IPSPs from a wing hinge stretch receptor to an elevator motor neuron could be monosynaptic, though this is not certain.

In total, then, elevation of a locust's four wings produces, through the central synaptic connections of all four wing hinge stretch receptors, depolarization of a great many wing depressor motor neurons and hyperpolarization of many wing elevator motor neurons. These connections presumably contribute to the rhythmic activation of wing depression during flight. In support of this idea, if one stimulates electrically just one stretch receptor cell in a temporal pattern that resembles its pattern of action potentials during flight, there results a flightlike rhythm of depolarizations in depressor motor neurons and hyperpolarizations in elevator motor neurons (Figure 12E). Although these depolarizations do not lead to action potentials, they undoubtedly would do so during flight, when they would sum with the depolarizations produced in the motor neurons by the central neuronal oscillator.

In addition to the wing hinge stretch receptor cell, which responds to wing elevation, each wing has many other proprioceptors that are excited whenever the wing is lowered (Burrows, 1976). When the axons of these depression-sensitive receptors are electrically stimulated as a group, they produce synaptic effects on the flight motor neurons that are of opposite polarity to those evoked by the elevation-sensitive wing hinge stretch receptor; that is, they hyperpolarize the

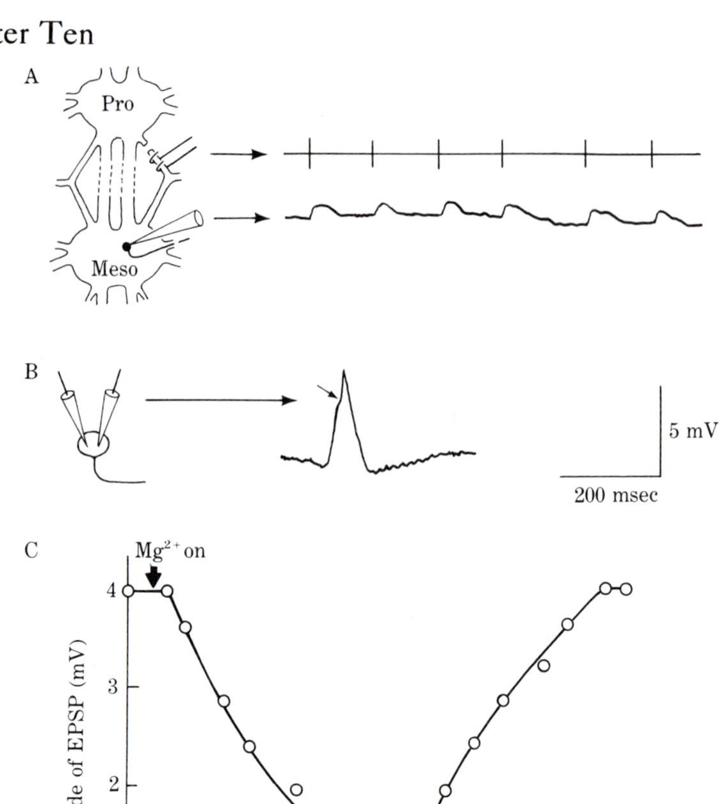
A
Pro
Meso
B
5 mV
200 msec

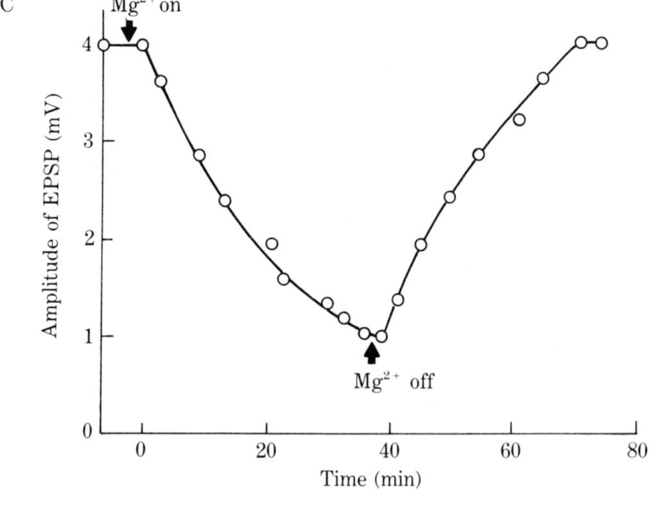
C
Mg^{2+} on
Mg^{2+} off
Amplitude of EPSP (mV)
Time (min)
0
1
2
3
4
0
20
40
60
80

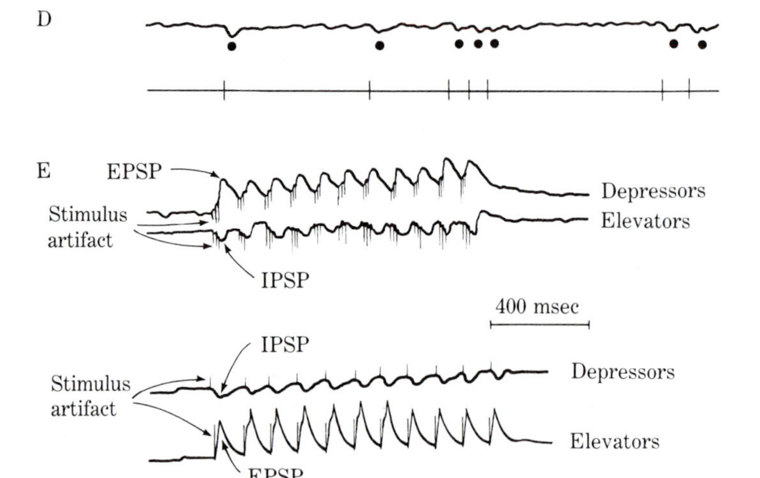
D
E
EPSP
Stimulus artifact
IPSP
Depressors
Elevators
400 msec
IPSP
Stimulus artifact
EPSP
Depressors
Elevators

◀ FIGURE 12. Synaptic connections from wing proprioceptors to wing motor neurons in a locust. A.–E. Connections from wing hinge stretch receptor. F. Connections from depression-sensitive receptors. A. Each action potential in the wing hinge stretch receptor is followed by a depolarizing PSP in a wing depressor motor neuron. B. With the motor neuron depolarized steadily by passing current through a second microelectrode, a depolarizing PSP now evokes an action potential (beginning at the arrow). The action potential is small because, with the motor neuron's cell body being nonelectrogenic, the action potential decays electrotonically as it travels from distant electrogenic regions of the cell to the cell body that contains the recording electrode. C. Replacing calcium in the saline with magnesium leads to a gradual, not a precipitous, decrease in EPSP amplitude. The effect is reversible. D. Action potentials in the stretch receptor evoke IPSPs (at the dots) in a wing elevator motor neuron. E. Electrical stimulation of a stretch receptor axon by rhythmic triplets of stimulus pulses evokes a flightlike rhythm of EPSPs in a wing depressor motor neuron *(top trace)* and IPSPs in a wing elevator motor neuron *(bottom trace)*. F. Repeated electrical stimulation of a group of depression-sensitive proprioceptors evokes a flightlike rhythm opposite in polarity to that in E. The second through the seventh depolarizing PSPs of the elevator motor neuron evoked action potentials, this evocation indicating that these are indeed EPSPs. (After Burrows, 1975, 1976.)

depressor motor neurons and depolarize the elevator motor neurons. When stimulated at or near flight frequency, these depression-sensitive receptors produce in the motor neurons a flightlike rhythm that is the inverse of that evoked by the elevation-sensitive stretch receptors (Figure 12F).

Of course, during flight, wing elevation (which activates the wing hinge stretch receptor) and wing depression (which activates the depression-sensitive receptors) occur in alternation with one another. Therefore, the proprioceptive feedback signals entering the central nervous system from these two sensory sources also alternate in time. It takes about 10 milliseconds for an action potential, initiated at the trigger zone of any of these receptors, to be conducted into a ganglion and to produce a PSP in a motor neuron. This timing is such that the PSPs produced by the wing hinge stretch receptor, whose net effect is to help depress the wing, occur right in the middle of wing depression (Figure 13). And the PSPs produced by the depression-sensitive receptors, which help to elevate the wing, come just during wing elevation (Figure 13). Thus, the two sets of receptors and their central connections constitute a chain reflex that would reinforce the effect of the central neuronal oscillator in producing the rhythmic wing movements.[2] Chain reflexes similar to these in locust flight have been observed in a number of rhythmic behaviors, including swimming in leeches (Kristan and Stent, 1974) and walking in both cockroaches and cats (Pearson *et al.*, 1976; Pearson and Duysens, 1976).

[2]Additional effects of proprioceptive feedback from the locust's wings include the control of wing beat frequency (Wilson and Gettrup, 1963) and the enhancement of coordination among the four wings (Wendler, 1974). These involve central synaptic connections not yet fully elucidated.

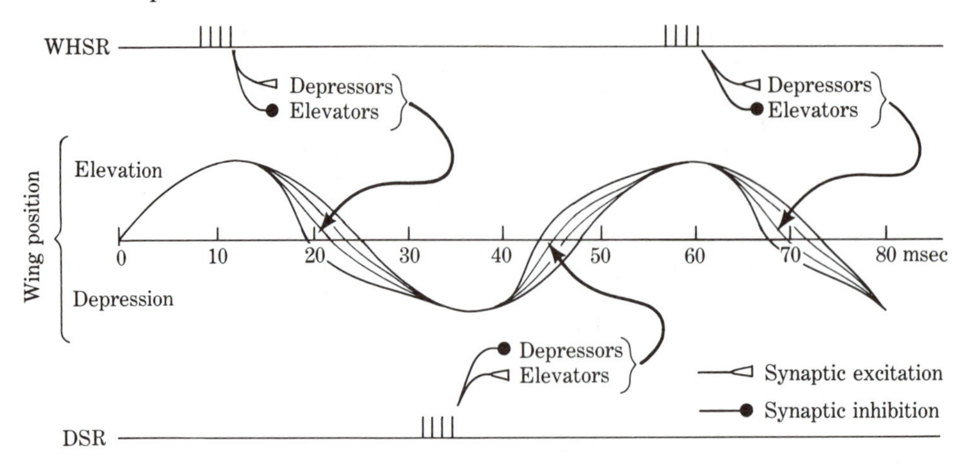

FIGURE 13. Summary of chain reflex action in locust flight. Trains of action potentials are shown in the wing hinge stretch receptor (WHSR) near the time of maximal wing elevation and in the depression-sensitive receptors (DSR) near the time of maximal wing depression. The effects of these receptors on the depressor and elevator motor neurons are shown. The influence of these motor neuronal effects on wing position is shown schematically by the arrows and the distortions of the sine wave.

A locust whose central nervous system has been totally deafferented is still capable of producing the normally patterned rhythmic motor outputs of flight (Wilson, 1961). Therefore, the locust has a central neuronal oscillator for flight, just as the leech has for swimming (Chapter 9). However, aside from any contribution by the central neuronal oscillator, the chain reflex involving the wing proprioceptors of a locust appears to be able, entirely on its own, to generate a normally patterned flight rhythm (Wendler, 1978). Why, then, has the locust evolved these two separate systems, each independently capable of producing rhythmic flight movements? Or, phrased somewhat differently, what special contribution might each of these two mechanisms make to the insect's flight?

When a locust flies (or, for that matter, when any animal produces a rhythmic behavior), its central neuronal oscillator produces a highly regular sequence of activity in the motor neurons. This high degree of regularity might be entirely suitable if the behavior were performed under perfectly constant environmental conditions. However, if a locust flies through turbulent air, as it probably does often, the forces exerted on the wings by the air would fluctuate. During any given wingbeat, any one of the wings might experience a sudden upward or downward push from the surrounding air. Also, if a locust's wing changes its weight, either by having a portion of its length lost to a predator or by virtue of the natural increase in weight that occurs during adult life, the highly regular motor output of the central oscillator will produce differing amounts of wing movement. The chain reflex of Figure 13 could help to overcome such uncontrolled variations in a wing's movement. For instance, if during elevation a wing were pushed too far upward by a local updraft, this would lead to enhanced activation of the wing hinge stretch receptor, which would induce more action potentials than normal in the wing depressor motor neurons on the immediate next downstroke. The wing, then, would push down harder and in so doing would counteract the effect of the updraft. By contrast, a local downdraft would reflexively induce an enhanced upward push by the locust on the immediate next upstroke. In confirmation of this role of the chain reflex, if one forcibly holds one of a locust's wings in an elevated position while the tethered insect is flying in place, the wing depressor motor neurons give an enhanced number of action potentials during each downstroke; also, the wing elevator motor neurons give a decreased number of action potentials on each upstroke. The opposite effects occur if one forcibly holds the wing depressed (Waldron, 1967).

But why has the locust evolved a central neuronal oscillator at all, rather than leaving the control of its flight entirely to its effective chain reflex? Without a central neuronal oscillator, flight performance would be left largely to the vagaries of environmental perturbations. If all four wings were bombarded by randomly occurring updrafts and downdrafts or if all four were of different weights, in compensating for these disturbances a locust might quickly lose all semblance of steady flight. But steady rhythmic movements, as in the case of an oscillating pendulum, are energetically the most efficient. Thus, it may be the combination of these two needs—a steady, energetically efficient, highly regular source of rhythm, along with the capability of responding to environmental uncertainties—that has led to the locust's blend of a central neuronal oscillator plus proprioceptive feedback. In fact, this central–peripheral combination appears to underlie almost all rhythmically patterned behaviors studied (Delcomyn, 1980), and therefore constitutes a major neuroethological principle.

Proprioceptive Feedback from Mammalian Muscle Spindles

Mammals have several types of proprioceptors that are excited in response to body movements. Stretch-sensitive proprioceptive organs of one type, called MUSCLE SPINDLES, are especially important in the feedback control of movement (Matthews, 1972). Tens to hundreds of these receptor organs are embedded directly within each muscle of a mammal's body.[3] Anatomically, muscle spindles are the most complex sensory structures of the body, aside from the eyes and ears. Each spindle (Figure 14) consists of (1) a small group of particularly short and thin muscle fibers called INTRAFUSAL FIBERS; (2) the axon terminals of GAMMA MOTOR NEURONS, which innervate these intrafusals; and (3) the receptive endings of two stretch-responsive sensory neurons attached directly to the intrafusal fibers. One of these sensory neurons, called a Ia AFFERENT, has its endings, called PRIMARY ENDINGS, wrapped spirally around the midportion of each intrafusal fiber

[3]Lower vertebrates also have muscle spindles. There are small but significant differences in their structure and physiology as compared with mammals (Ottoson, 1976). Because less is known about their role in behavior, this discussion is restricted to mammalian spindles.

A

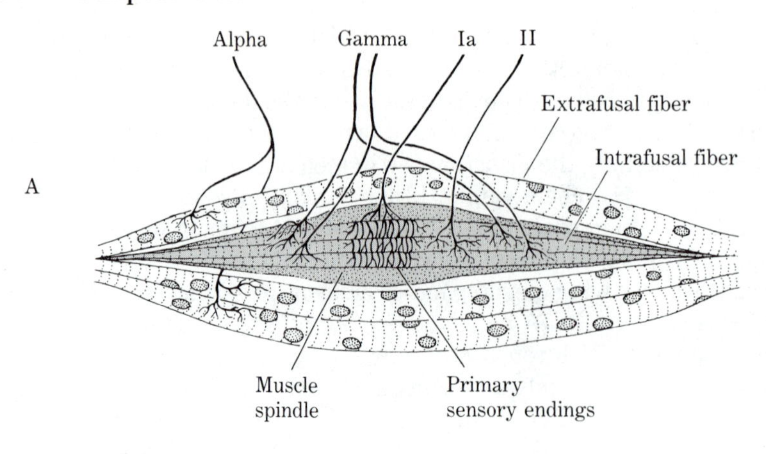

B

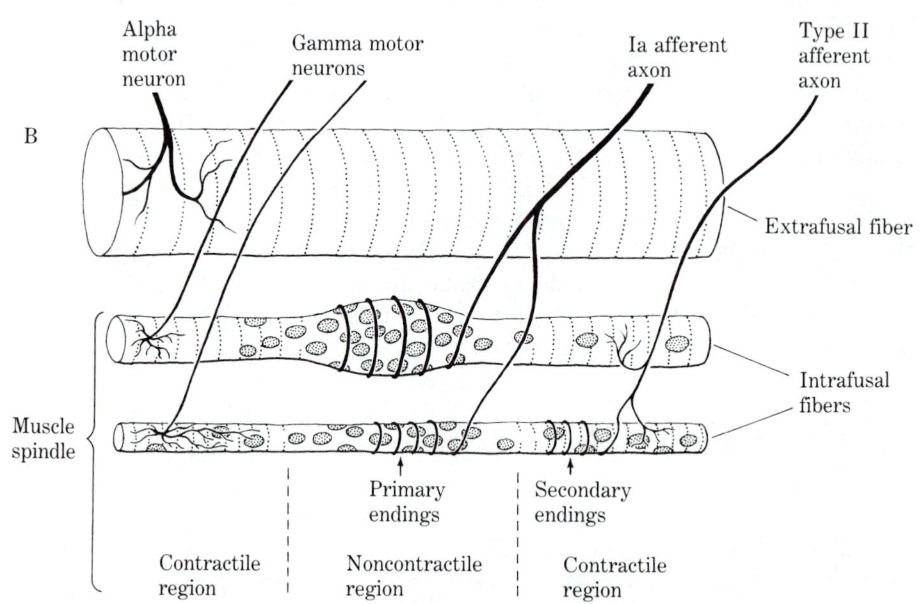

FIGURE 14. A. A mammalian muscle spindle (surrounded by gray sheath) embedded within a muscle. B. Enlargement of the central region of a spindle and of an extrafusal muscle fiber. Gamma motor neurons, shown innervating only the left contractile region of the intrafusal fibers, actually branch to both left and right contractile regions. The two intrafusal fibers shown differ in structure and in the rate at which they contract; most spindles have several intrafusals of each type. Explanation in text. (A after Eckert and Randall, 1978; B after Matthews, 1972)

in the spindle. This neuron responds with great sensitivity to a stretch of these spiral endings, as occurs if the whole muscle is stretched (Matthews, 1972). The second receptor, called a TYPE II AFFERENT, has its endings, called SECONDARY ENDINGS, lateral to the midportion of the intrafusal fibers. Though also responsive to stretch, the Type II cells are much less sensitive. They will not be considered further in this discussion.

The entire spindle apparatus sits alongside the muscle's thicker and larger EXTRAFUSAL FIBERS. These are innervated by ALPHA MOTOR NEURONS, whose axons are of greater diameter than those of the gamma motor neurons. All active movements of the body result from contractions of the extrafusal fibers of muscles. As we shall see somewhat later in this chapter, the intrafusal fibers and their gamma motor neurons play a complex and important role in regulating these movements. First, however, we shall consider the simpler role played by the Ia afferents in reacting to unexpected forces imposed on the body from the outside.

THE MYOTATIC REFLEX

If you hold your forearm out horizontally in front of you, in order to counteract the downward force of gravity, the biceps muscle of your upper arm must maintain a steady, partial contraction. It does so in response to a steady, low-frequency barrage of action potentials in its alpha motor neurons (Figure 15A, arrowheads on axon of alpha motor neuron). Now if a downward force is suddenly and unexpectedly added to your hand, this would lower your forearm and thus stretch your biceps muscle, along with its spindles. This spindle stretch would excite the Ia afferents[4] (Figure 15B, arrowheads on Ia axon).

What are the reflexive consequences of this Ia activity? Within the spinal cord, Ia afferents from the biceps muscle make apparently monosynaptic excitatory connections onto alpha motor neurons whose axons innervate the biceps (Figure 15B). [The full range of tests for the monosynaptic nature of these connections, described in the appendix of Chapter 3, has not yet been carried out, owing to the difficulties imposed by the large number of sensory and motor neurons and the deep spinal location of the synapses (Lloyd, 1943; Eccles *et al.*, 1957; Schmidt and Willis, 1963). Moreover, because the necessary physiological recordings are generally not attempted on human subjects, the monosynaptic nature of *human* myotatic reflexes is not fully certain, being based upon evidence from many other mammalian species.] These Ia afferents also make apparently disynaptic *inhibitory* connections onto alpha motor neurons of the antagonistic muscle, the triceps (Figure 15B). This inhibition decreases any low frequency of ongoing action potentials that might be occurring in these neurons (Matthews, 1972). These inhibitory connections constitute an example of reciprocal innervation (Chapter 8).

Owing to the central connections of the Ia afferents, a force that pushes the arm downward and thus stretches the spindles of the biceps muscle results in a reflexive biceps contraction and triceps relaxation (Figure 15C). This restores the arm toward its original position. The overall reflex involves a negative feedback loop. Figure 16 shows each of the events in this reflex and their feedback relationship. The downward force (box 1) is shown as having a subtractive effect on

[4]It is known from physiological recordings of numerous mammalian species that even while the limb is held in a static position there is some activity in the Ia afferents (Matthews, 1972). A sudden, unexpected downward push on the arm then increases this activity in the Ia afferents. For simplicity, in most of the discussion of this chapter we will assume that in the absence of a sudden unexpected downward force the Ia activity is zero.

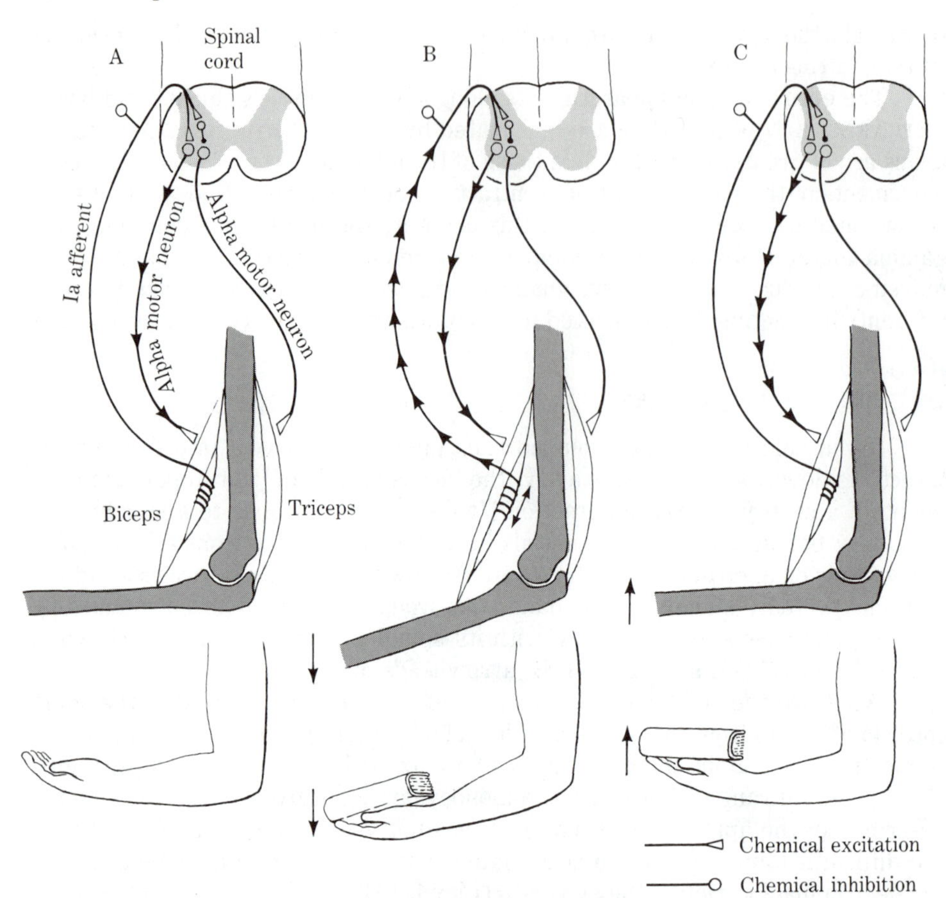

FIGURE 15. Myotatic reflex of human biceps muscle. A, B, and C. A single extrafusal fiber of the biceps and the triceps muscle is shown, along with their alpha motor neurons, as well as a biceps intrafusal fiber and its Ia afferent. A. The arm is held steadily up against gravity by a low frequency of action potentials in biceps alpha motor neurons *(arrowheads)*, driven by excitation from spinal interneurons that are not shown. B. A sudden downward force on the arm excites the Ia afferent of the biceps. C. This enhances the excitation of the biceps alpha motor neurons (and suppresses any competing activity that may occur in the triceps alpha motor neurons; no such competing activity is shown in the figure). This activity in the biceps alphas restores the forearm toward its initial position. (After Merton, 1972.)

the feedback circuit (negative sign in the junction circle) because it acts to decrease the next parameter, muscle shortness (box 2), that is, it causes a lengthening of the biceps muscle. This decrease in box 2 in turn leads to a decreased spindle shortening (box 3). *Subtracting* this *decrease* at the next junction circle leads to an *increased* stretch on the primary endings (box 4). As a result of the increase in box 4, the parameters of boxes 5, 6, and 7 all increase, leading to corrective muscle shortening (box 2). The latency of this overall sequence from the onset of the downward force on the arm to the onset of the arm's upward corrective movement is fairly brief—only about 40–50 milliseconds (Merton, 1972; Gottlieb and Agarwall, 1979). Because the reflex responsible for this quick onset of movement is confined to the spinal cord, the movement response begins before we become consciously aware that anything has happened. The general name for such short-latency, spindle-activated, force-compensating reflex is a STRETCH, or MYOTATIC, REFLEX. Functionally, it is analogous to the reflex evoked in a locust by depression or elevation of a wing (Figure 13). A familiar example of a human myotatic reflex is the knee-jerk reaction, used clinically to test for proper reflexive function. In this test a tap below the knee, by stretching a group of tendons, stretches the muscles that we use to extend the lower leg. The resultant activation of the Ia

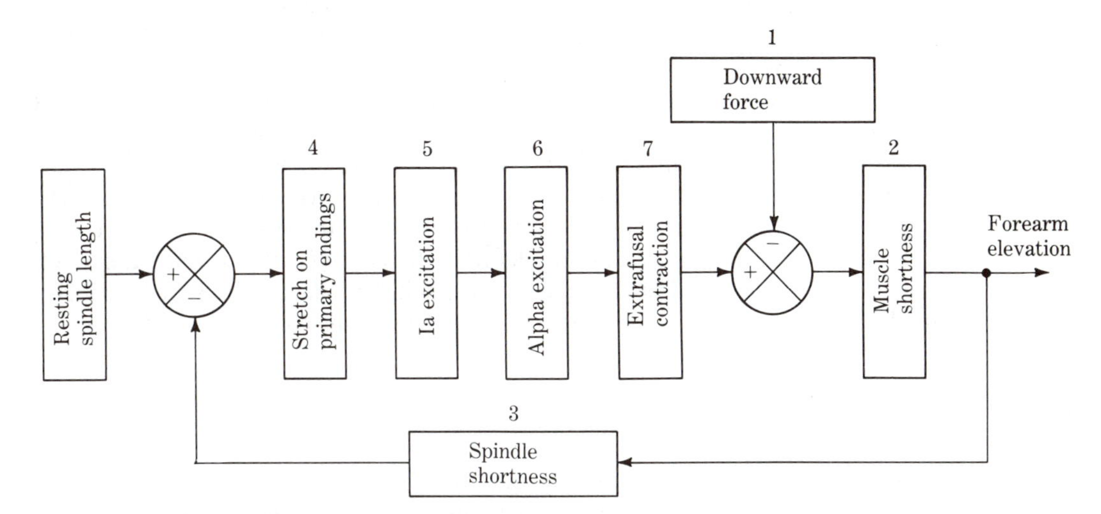

FIGURE 16. Block diagram of the myotatic reflex of the biceps muscle, showing this to be a negative feedback activity. A downward force on the arm (Box 1) causes a lengthening of the biceps muscle. Since a muscle's own natural activity is one of *shortening* rather than lengthening, the effect of the downward force is depicted as a *decrease* (negative sign in the junction circle) of muscle shortness (Box 2). There are two immediate consequences of this decrease in the parameter of Box 2: a decrease in forearm elevation (that is, the arm moves downward); and a decrease in the parameter of Box 3 (spindle shortness—that is, the spindles lengthen). The decrease of Box 3 has a negative effect (negative sign in left junction circle) and, thus, leads to an *increase* of the stretch on the primary endings (Box 4). That is, *subtracting* a *decreased* value of Box 3 amounts to *adding* an *increased* value. With Box 4 increased, there is an increase in 1a excitation (Box 5), in alpha excitation (Box 6), in extrafusal contraction (Box 7), in muscle shortness (Box 2), and, consequently, an increase in forearm elevation. This counteracts the original downward force that had lowered the arm. As the corrective increase in muscle shortness occurs (Box 2), there will be a consequent increase in spindle shortness (Box 3) and a decrease in the stretch on the primary endings (Box 4), thereby turning off the response.

afferents from these muscles leads to the reflexive contraction of these muscles and, thus, to leg extension.

In addition to the short-latency myotatic reflex, there are also spindle-activated responses of longer latency. These begin about 70 milliseconds after the onset of a downward force on the arm and last sometimes for hundreds of milliseconds. These later components, sometimes called postmyotatic responses, involve polysynaptic Ia-to-alpha pathways. (Some of the alpha motor neuronal activation probaby also derives from proprioceptors other than muscle spindles; Mathews, 1972; Janikowska, 1979; Lundberg, 1979.) Some of these polysynaptic pathways are restricted to the spinal cord, whereas others ascend to the brain and then descend back to the spinal cord to affect the motor neurons (Matthews, 1972; Gottlieb and Agarwall, 1979, 1980a,b). The pathways that ascend to the brain allow a person to become consciously aware of the ongoing movements of his forearm or other body regions and permit some degree of voluntary control over the postmyotatic response.

There is growing evidence that these postmyotatic responses may make a larger contribution than does the myotatic reflex to restoring a suddenly displaced limb to its original position (Gottlieb and Agarwall, 1980a,b). This evidence suggests that the earlier muscle contraction, resulting from the myotatic reflex, may serve primarily to increase muscle *stiffness* rather than to cause muscle shortening (Crago *et al.*, 1976; Nicholls and Hauk, 1976; Bizzi *et al.*, 1978). Such a sudden increase in stiffness would quickly present a substantial resistance to continued outside forces. Following this, the postmyotatic response could serve to restore the limb to its original position.

INTRAFUSAL CONTRACTION AND SET POINTS

The myotatic reflex raises the same problem that was created by the optomotor response of the fly: If the reflex always opposes movements of the arm or of other parts of the body, how can we ever move into new body positions? The solution to this problem lies in some special properties of the muscle spindles.

When an intrafusal muscle contracts, it does so only at its two end regions, not in its midregion, which is devoid of contractile filaments (Figure 14B). Because the contraction of the intrafusal fibers is too weak to pull the muscle to a shorter length, the two extreme ends of any intrafusal fiber remain fixed in place, bound to the surrounding muscle tissue, as the fiber contracts. Instead of the fiber's two ends being pulled inward, the central, noncontractile portion of the fiber is pulled outward (Figure 17). As this elastic midregion is stretched, so are the primary sensory endings that are wrapped about it. The surprising result is that contraction of an intrafusal fiber stretches and thus excites its own Ia afferent.

When a person holds his forearm out horizontally (as in Figure 15A), his central nervous system moderately excites not only the alpha motor neurons to the biceps muscle but also (not shown in the figure) the gamma motor neurons to the same muscle. The alpha activity, as we have said, maintains the partial contraction of the biceps extrafusal fibers that hold the arm up. In so contracting, of course, these fibers shorten the muscle. This shortening would slacken the biceps spindles, leaving their primary endings limp and thus relatively unre-

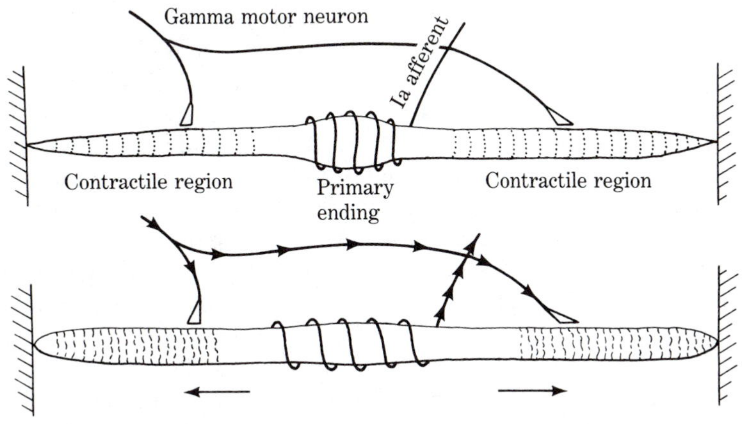

FIGURE 17. An intrafusal fiber at rest *(top)* and contracting in response to activity in a gamma motor neuron *(bottom)*. The contraction stretches the middle region of the fiber, thus stretching the primary sensory ending and activating the Ia afferent.

sponsive to any downward force on the arm. However, the moderate gamma-activated contraction of the biceps intrafusals, by applying a moderate stretch to the primary endings, restores partial tautness to these endings. Thus, the Ia afferents are kept poised just near threshold and therefore ready to respond to any stretches imposed by unexpected downward forces.

When a person intentionally lowers his forearm from an elevated position, the alpha motor neurons to the biceps muscle decrease their impulse rate (while the alpha motor neurons to the triceps generally increase their rate). The reason that this lowering is not opposed by the myotatic reflex is that the decrease in biceps alpha activity is accompanied by a decrease in the activity of the biceps gamma motor neurons (Vallbo, 1970; Phillips, 1969). Thus, during lowering of the forearm, although the primary endings of the biceps spindles are tending to be stretched owing to the greater length of the muscle, they are simultaneously tending to be slackened owing to the decrease of intrafusal contraction. In order for this intentional arm extension not to engage the myotatic reflex, the decrease in intrafusal contraction needs to be sufficient to counteract the stretching effect on the primary endings caused by the lengthening of the biceps muscle. However, this decrease in intrafusal contraction should not be *too* great or the primary endings would become so slackened as to be insensitive to any unexpected downward forces on the extended arm.

In fact, it appears that for each position in which one holds one's arm, the gamma motor neurons of the biceps (and those of each other arm muscle) are excited by the central nervous system with a characteristic frequency of action potentials, setting a characteristic amount of intrafusal contraction. This is presumably just the proper amount of intrafusal contraction needed to keep the primary endings just near their threshold tautness for that particular arm position.

The amount of excitation provided to the biceps gamma motor neurons by the central nervous system, then, constitutes the SET POINT or SOLLWERT (German for "should be value") for the control of arm position; that is, this gamma activity determines the forearm position that will be maintained by the feedback system against forces from the outside. To hold the limb in a different position, the central nervous system would select a different set point, that is, a different amount of gamma activation. This ability to establish a variable set point constitutes a major difference between muscle spindles and the stretch receptor of a locust's wing.[5] The locust receptor, lacking a variable set point, is always activated at the same wing position and thus always reflexively impels the wing away from that position. This rigidity is permissible in a locust, which uses its wings primarily for one behavior—flight (though some species also use the wings in stridulation and in righting behavior). By contrast, the human arm and the limbs of many other mammals perform a rich variety of skilled tasks, under a variety of environmental conditions that impose different forces on the body. The motor flexibility provided by variable set points is crucial for such behavior.

The set point of the spindle control system is analogous to the setting of a thermostat for a home furnace or air conditioner. For a given thermostat setting, the home's heating or cooling system maintains, by a negative feedback loop, a given room temperature, regardless of how the outside temperature may fluctuate. For instance, if the temperature of the room falls below the set point on the thermostat (creating an error signal, which is the difference between the set point and the room temperature), the furnace comes on to warm the room, and in the process decreases the error signal to zero. Now, with the error signal equal to zero, the furnace turns off. When one changes the thermostat to a different setting (that is, to a different set point), the system now maintains this different temperature against outside fluctuations.

SPINDLES AND THE CONTROL OF ACTION

We now consider how the muscle spindles are used in the control of *active* movements rather than just to oppose forces from the outside. When one makes an intentional movement, the alpha motor neurons that cause this movement obviously must be activated. In theory, the activation of these alpha motor neurons could come about in any of three different ways (Matthews, 1972; Keele and Summers, 1976); (1) indirectly, by the central nervous system exciting *gamma* motor neurons to the muscles (Figure 18A)—this would result in intrafusal contraction, which would lead to Ia excitation and thus, finally, to the activation of the alpha motor neurons that excite the extrafusal fibers; (2) directly, by central excitation of the *alpha* motor neurons to the muscles in question (Figure 18B); or (3) by a *combination* of central excitation to both alpha and gamma motor neurons (Figure 18C). We will consider each of these three hypotheses in sequence.

The first of these ideas, indirect control through the gammas, really constitutes a change of the *set point* of the muscle spindles. That is, a command to

[5]Only two stretch receptors that have variable set points have been discovered in invertebrates. One is the much-studied muscle receptor organ (MRO) of crayfish (Kuffler, 1960; Kennedy, 1976), and the other is found in adult moths (Finlayson and Lowenstein, 1958). By contrast, muscle spindles with variable set points are found in most vertebrates studied (Matthews, 1972).

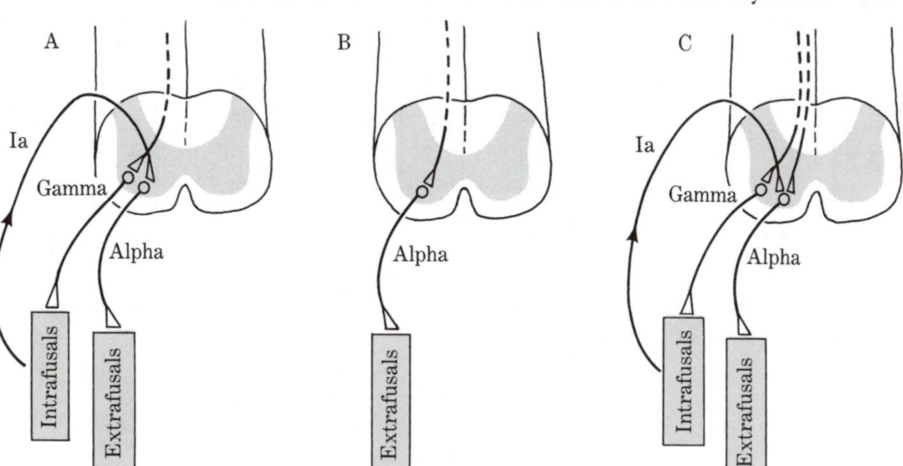

FIGURE 18. Three possible ways that central neuronal activity could excite alpha motor neurons to bring about contraction of extrafusal fibers and thus movement of the body. A. By exciting gamma motor neurons and thus utilizing the muscle spindles to activate the alphas. B. By exciting alpha motor neurons directly. C. By exciting both gamma and alpha motor neurons.

the gammas would fix a new, set amount of intrafusal contraction, which would bring about the new limb position by virtue of Ia activation (Figure 19A). This scheme of movement control is usually called FOLLOW-UP SERVO CONTROL. Figure 19A emphasizes that in this form of control, the movement would be initiated only after several stages of processing had occurred (boxes 1–7). The final position would be established and maintained through the agency of the feedback loop (box 8). That is, movement would stop when the spindles of the contracting muscles shorten (box 8) sufficiently to counteract the intrafusal contraction (box 2) that results from the gamma excitation by the central nervous system. With such a cumbersome control system, movements would come about relatively slowly, a disadvantage where rapid action may be needed. However, this follow-up servo system would offer some important advantages. For instance, a given strength of command to the gammas would result in the *same* final limb position, irrespective of the limb's starting position and irrespective of any outside forces on the limb (within the mechanical limits of the muscles). That is, once a fixed command to the gammas initiates and maintains a set amount of intrafusal contraction, the limb would continue to move for however long it takes to reduce the difference signal (prior to box 3) to zero and thus to establish the intended limb position. This fixed relationship between strength of command and final limb position would greatly simplify the brain's task of programming movements. For this reason, follow-up servo control has often been regarded as an attractive hypothesis.

In spite of its attractiveness, however, follow-up servo control is certainly not the only mechanism employed, as has been demonstrated by experiments on

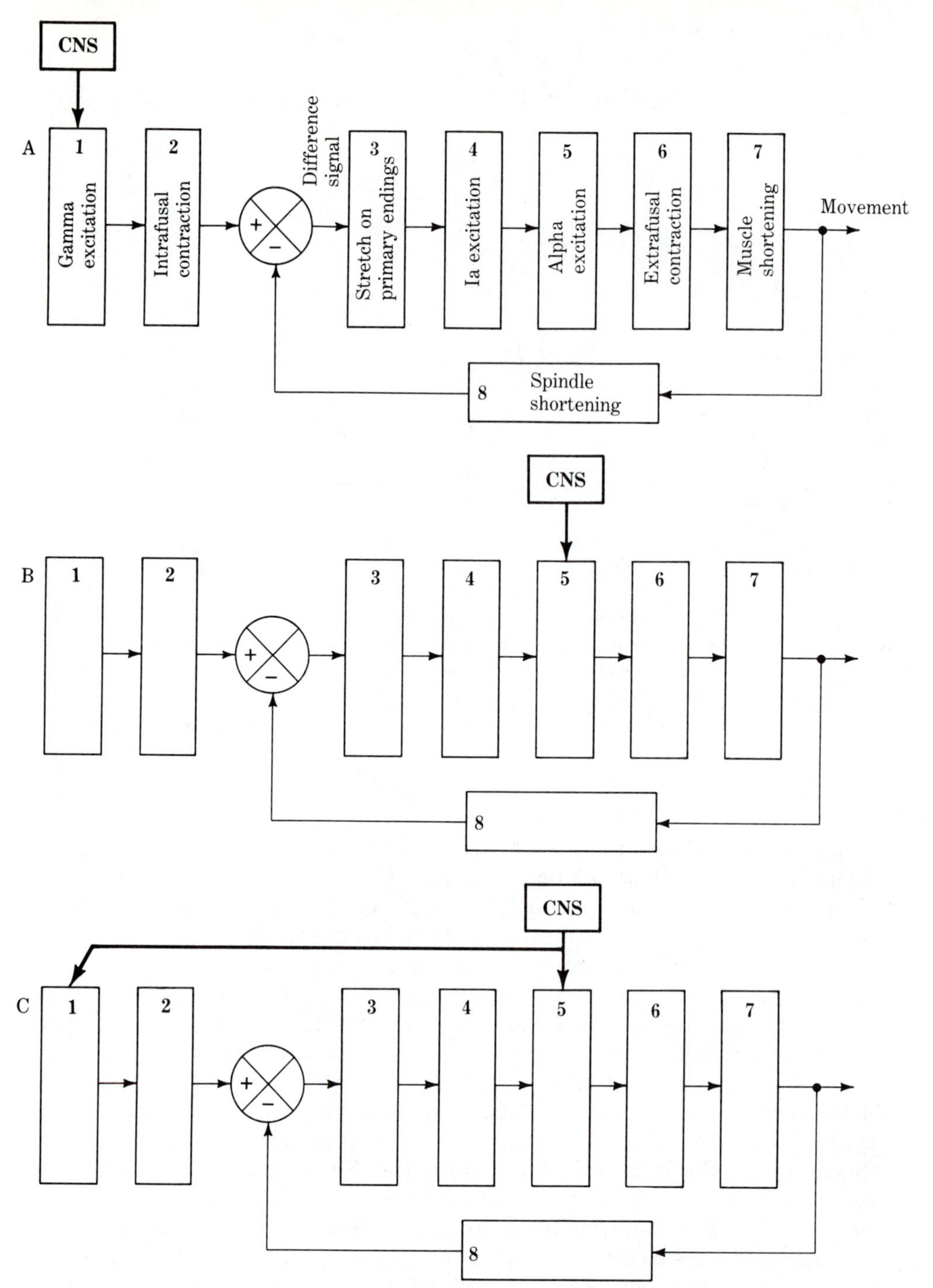

FIGURE 19. Three possible ways that the central nervous system could bring about a movement, corresponding to the three parts of Figure 18. Boxes 3 through 8 of each of these block diagrams are identical to the loop of Figure 16. A. Follow-up servo control, in which the central nervous system (CNS) acts by increasing the excitation of the gamma motor neurons (Box 1). B. Direct alpha control. C. Alpha–gamma coactivation. Boxes with the same numbers in A, B, and C represent the parameters specified in A. See text for explanation.

several mammalian species (Matthews, 1972; Taub, 1976). One type of experiment has involved deafferentation by dorsal rhizotomy (severing the dorsal roots that carry the sensory axons entering the spinal cord) of one or more limbs. Because this interrupts the Ia afferents, it is equivalent to severing the connections between boxes 4 and 5 in Figure 19A. By the follow-up servo scheme of Figures 19A, severing this connection should render the animal unable to make voluntary movements because the pathway between the central command to the gammas and the alpha motor neurons is interrupted. However, a nearly full range of different types of movement can be made by totally deafferented limbs of monkeys (Taub, 1976). These movements are carried out crudely, which suggests that the feedback control lost by deafferentation probably plays some modifying role. But the occurrence of fairly normal movements by these deafferented limbs proves that gamma excitation is not essential for voluntary movement. (This nearly normal performance of behavior after deafferentation recalls the nearly normal locomotory rhythms that we have seen can occur in deafferented leeches, locusts, and other animals. Both these sets of results point to a considerable role for the central nervous system in programming behavior.)

The second possible mechanism by which a mammal could execute movements, a direct command by the central nervous system to the alpha motor neurons, is illustrated in Figure 19B. Clearly such a command to the alphas could produce movements more quickly than the more indirect gamma command of Figure 19A. However, in spite of this advantage, direct alpha activation would have one negative consequence, which was referred to in a previous section of this chapter. Direct excitation of the alpha motor neurons (box 5), by causing muscle shortening (box 7), would lead to spindle shortening (box 8) that would decrease the stretch on the primary endings (box 3). Thus, the primary endings, lacking any excitatory input through a command to the gammas, would go totally slack and would be unavailable to respond to unexpected forces from the outside. In fact, physiological experiments suggest that central commands to the alphas do not constitute the sole mechanism by which mammals generally produce movements. The evidence against this is as follows: if any movement were produced in this way, one would expect there to be a decrease during the movement in the activity of Ia afferents from the contracting muscles (box 4). This is because the spindle shortening (box 8) that occurs during the movement would cause a decrease in the stretch on the primary endings (box 3) and this would decrease any ongoing Ia excitation (box 4). However, physiological recordings from numerous mammals show that during active movements, the Ia afferents from contracting muscles do not decrease their rate of action potentials (Sears, 1964; von Euler, 1966; Severin *et al.*, 1967; Matthews, 1972). Thus, mammals appear in general not to execute movements solely by means of direct commands to their alpha motor neurons.

The observation that the Ia afferents do not decrease their impulse frequency during contraction of their muscles can be explained only by excitation of the gamma motor neurons during the course of the movement. Having ruled out a central command to the gammas alone (Figure 19A) or to the alphas alone (Figure 19B), we are left with the mechanism of ALPHA–GAMMA COACTIVATION (Figure 19C). By this scheme, a direct command to the alphas (box 5) would induce the muscle contraction (box 7), whereas an exactly proportionate command to the gammas (box 1) would evoke just enough intrafusal contraction (box 2) to keep the primary endings (box 3) taut throughout the muscle contraction. The command to the gammas, then, would function in a manner analogous to the efference copy hypothesized in the fly optomotor system (Figure 8), in that it would exactly counterbalance the effects that intentional movements impose on the feedback loop. The advantage of this efference copy, both in the fly's optomotor system and in the mammalian spindle system, would be that reflexes used to oppose outside disturbances could function *during* the course of the intentional movement.

Is there quantitative evidence for this alpha–gamma coactivation mechanism? The exact balance of alpha and gamma activity implicit in this scheme predicts that, during an intentional movement, the impulse frequency of the Ia afferents (box 4) should remain exactly constant. This is because the intrafusal contraction (box 2) would be just exactly sufficient to hold constant the stretch on the primary endings (box 3) that would otherwise be reduced by the spindle shortening (box 8). Thus, the stretch on the primary endings would not decrease, so the Ia excitation (box 4) would not decrease during intentional movements. However, numerous recordings from Ia afferents actually show an *increase* in the frequency of their action potentials when their muscles contract in voluntary movements (Sears, 1964; von Euler, 1966; Severin *et al.*, 1967; Matthews, 1972). Therefore, the intrafusal contraction (box 2) is apparently *stronger* than that needed to just balance out the effect of spindle shortening (box 8). We can conclude that although alpha–gamma coactivation is occurring this does not constitute an efference copy mechanism as we have defined it, because such a mechanism would involve an *exact* counterbalancing of the feedback. The excess Ia excitation (box 4) that occurs contributes to drive the alphas (box 5) and thus contributes to cause the intentional movement. This mechanism, which does in fact appear to underlie many mammalian movements, is called SERVO-ASSISTED CONTROL. To reiterate, this mechanism involves (1) a direct command to the alphas; (2) a command to the gammas to counterbalance the feedback from the movement; and (3) excess command to the gammas to assist in producing the movement.

Servo-assisted control appears to offer a blend of those advantages inherent in the follow-up servo control scheme (Figure 19A) and those of the alpha command scheme (Figure 19B). As with follow-up servo control, because the primary sensory endings are not allowed to slacken during a movement, the system remains able, throughout the movement, to respond with a myotatic reflex to any unexpected external forces. Moreover, because there is a direct command to the alphas, movements could be initiated quickly. (There is, in fact, evidence that the first few milliseconds of human voluntary movements are controlled more by commands to the alpha than to the gamma motor neurons; Vallbo, 1970). Finally, to the extent that the excess gamma command contributes to the movement, a given strength of this excess command would determine the final position, irrespective of the limb's initial position.

CHANGES OF GAIN IN THE MYOTATIC REFLEX

The foregoing discussion of feedback control by muscle spindles has indicated that the myotatic reflex is functional both when a limb is stationary and when it makes a voluntary movement. Recent experiments on human subjects indicate, however, that the strength of this reflex varies to a considerable extent during the course of voluntary movements (Gottlieb and Agarwall, 1979, 1980a,b). Figure 20A shows an experimental arrangement in which a seated subject places his foot on a pedal that, like an automobile's accelerator pedal, can swivel upward or downward around a fulcrum beneath the heel. An opaque screen blocks the subject's view of the pedal and of his foot. The subject first is asked simply to hold the pedal still in a partially depressed position. While he is holding, an upward force is suddenly applied to the pedal by a motor, controlled by the experimenter. The time of occurrence of this upward force is unknown in advance to the subject. In response to the upward force, the subject's foot makes a downward myotatic reflex. Electrodes on the skin surface of the leg record the electrical responses of the underlying muscles that the subject uses to depress his foot. Also, force transducers connected to the pedal record the foot's active downward push on the pedal. The latency of the myotatic reflex, measured from the start of the motor's upward push on the pedal to the start of the foot's downward response, is about 40 milliseconds. This reflex lasts for about 30 milliseconds and then gives way to a postmyotatic response.

As a second task, the subject is asked to *move* the pedal, at a prearranged moment, downward by a fixed amount. The pedal is again presented with the same sudden upward force from the motor. This force is delivered either before, during, or after the subject has begun to move the pedal downward. (On some control trials, no upward force was given.) Again the subject does not know when (if at all) the upward force will be delivered. The magnitude of the myotatic reflex now depends upon when this upward force is delivered. If it is delivered within the first 50 milliseconds of the onset of the subject's voluntary downward push on the pedal, the myotatic reflex has a greater strength (larger muscle response and stronger downward push by the foot) than if delivered during prolonged maintenance of a fixed pedal position. But if the force is applied about 100 to 200 milliseconds after the foot's voluntary downward movement has begun, the reflex strength is greatly decreased (Figure 20B). The postmyotatic reflex shows similar changes late during the downward movement. These variations in strength of the foot's downward push represent changes in the closed-loop gain of the myotatic reflex. The mechanism by which the gain changes is unknown, but most likely it involves modulation by the central nervous system of the Ia-to-alpha synapses.

Interestingly, the extent of these changes of gain depends upon the speed with which the subject intentionally moves his foot downward. For instance, during voluntary pedal depressions as fast as 150°/sec, the gain late in the movement is reduced to zero (Figure 20B). That is, the myotatic reflex ceases to function so that, in the absence of this feedback, the foot's movement of the pedal to a predetermined final position is carried out in an open-loop manner. For slower downward movements of the foot, at 20°/sec, there is only a slight decrease in gain, so the feedback is still functional. We have seen (early in this chapter) other examples of slow movements employing closed-loop control and rapid movements carried out by open-loop control, namely, a predator's slow orientation toward its prey—a movement that uses visual feedback—followed by the rapid strikes in which visual feedback is not employed. This difference between the extent to which feedback is incorporated in slow *vs* fast movements is probably quite general.

A

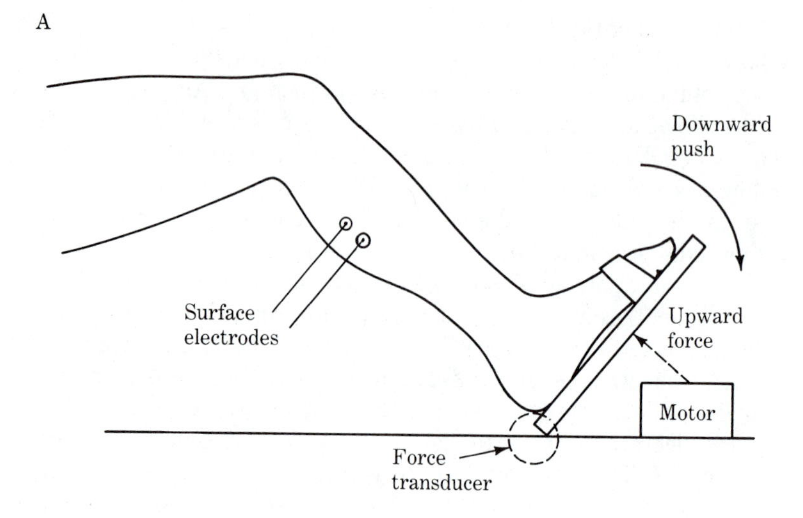

B

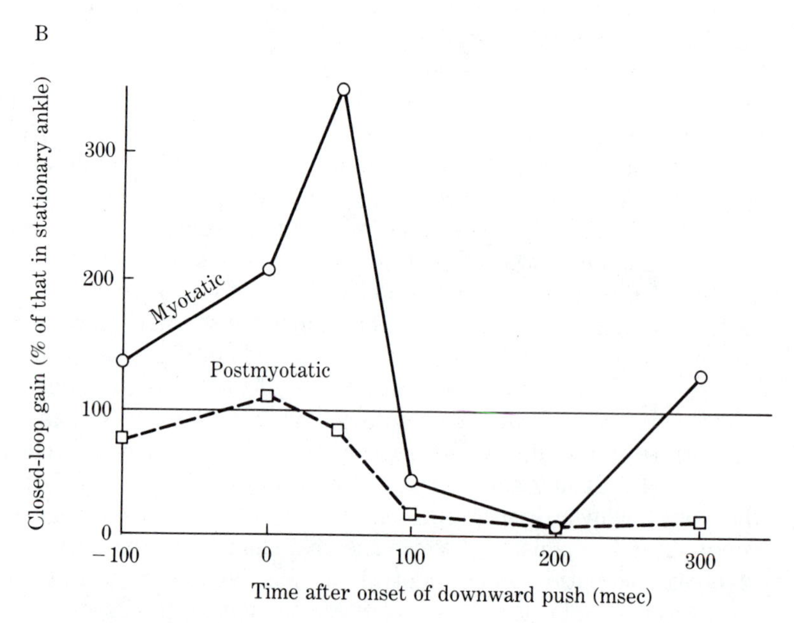

FIGURE 20. Variation in closed-loop gain during voluntary human movements. A. The subject's foot rests on a depressible pedal. While he holds the pedal fixed or pushes it downward, a motor delivers a sudden, unexpected upward force to the pedal. The strength of the subject's response to this force is monitored by surface electrodes and by a force transducer measuring his push on the pedal. B. The change in the closed-loop gain at different times during the subject's downward push on the pedal is plotted for the myotatic reflex and the postmyotatic response. A 100% response *(horizontal line)* is equal in strenght to the response given by the subject while holding the pedal in a steady, half-depressed position. (After Gottlieb and Agarwall, 1980a.)

Summary

Animals make some orientation movements, such as rapid predatory strikes, in an open-loop, or ballistic, manner—without reference to feedback from the target. Other behaviors, such as optomotor responses, are performed closed-loop, with the use of feedback from the target. The feedback loop is an integral component of the optomotor control system, giving this response its required closed-loop gain of close to one. The optomotor feedback system provides a means for correcting the body's position when it is unexpectedly disturbed by outside forces. However, there arises the potential problem of the optomotor response opposing *voluntary* movements and thus effectively immobilizing the animal. To overcome this problem, during rapid saccadic movements of the eyes, head, or body, some animals inhibit their optomotor responses, a process called saccadic suppression. During slower movements, however, optomotor responses are not suppressed. It has been suggested that flies may use the mechanism of efference copy to balance out the visual reafference that they receive during a slow turn. This would permit the turn to occur unopposed by the optomotor response and would permit the animal to respond to unintended disturbances of body angle that might occur during the turn.

Proprioceptive feedback from mechanoreceptors of a locust's wing is used to generate a chain reflex. This reflex can contribute to the activity of wing muscles during flight and can correct for wing displacements resulting from outside forces. Feedback from mammalian muscle spindles, used in the myotatic reflex, also permits corrective reactions to outside forces. In addition, however, because variations in the gamma-activated contractions of intrafusal muscle fibers can adjust the tautness of the primary sensory endings, this mammalian reflex remains sensitive to displacements at all muscle lengths and all limb positions. During active movements, both the gamma motor neurons to intrafusal fibers and the alpha motor neurons to the extrafusal fibers are excited. The gamma excitation is greater than that required to take up the slack on the primary sensory endings produced by the muscle shortening during the movement. Therefore, gamma-activation contributes to the excitation of the alphas, a mechanism called servo-assisted control. During some rapid voluntary movements, the closed-loop gain of the myotatic reflex decreases greatly, so that such movements are performed with the feedback loop opened.

Questions for Thought and Discussion

1. Whenever an animal carries out an intentional movement, it may be important for the brain to take account of the details of this action in order to plan the animal's next movement. What are the various sources of information, both within the central nervous system and in the periphery, by which the brain can be informed of the animal's ongoing movements? What are the relative advantages and disadvantages of the brain's utilizing each of these potential sources of information?

2. In this chapter, we have seen an example of *positive* feedback that presents an animal with a liability, namely, a fly with its head rotated (Figure 7B), which spins in endless circles. Because flies do not normally find their heads rotated in the manner described, the positive feedback produced presents no problem. However, here is an example of a naturally occurring positive feedback situation that could have adverse effects. When a crayfish is stimulated with a jet of water to give its escape response, the forceful movements of this response create violent water currents. These currents could re-excite the escape behavior, which would thus produce additional violent water currents, further re-exciting the escape behavior, and so forth in a never-ending cycle. This cycle is interrupted, however, because the lateral giant interneurons (LGIs) that mediate the escape response (Chapter 8) also inhibit neurons prior to themselves in the escape circuit (Kennedy *et al.*, 1974). That is, negative feedback within the central escape circuit interrupts the positive feedback involving the production of water currents by the behavior. In what ways is this example of positive feedback, and the mechanism of its prevention, similar to and in what ways different from the following two instances of positive feedback that have been mentioned in earlier chapters: (a) the positive feedback involved in generating an action potential by voltage-sensitive ion channels (Chapter 3) and (b) that involved in the decision by an *Aplysia,* using electrically coupled neurons that re-excite one another, to eject ink (Chapter 8)?
3. Draw the negative feedback circuit of a home heating system, including a thermostat with a variable set point, a furnace, and other items that you deem necessary. Use the diagrammatic method of boxes, junction circles, and flow lines that was used in Figures 4, 8, 16, and 19. Verify that the diagram is correct by showing that the furnace turns on when the room temperature drops or when a person raises the thermostat setting and that the furnace turns off when the temperature set on the thermostat is reached.
4. The figure here depicts the human optomotor response. This diagram resembles Figure 8, except that the eye motor neurons and eye muscles have been included and there is an added box that receives its input from the slip speed signal, labeled *perception of visual slip.* This box represents what one would actually see, in terms of the slippage of the visual world.

   A. Verify that the diagram does properly represent the events occurring in human vision, by considering the eye movements and the perception of visual slip that the diagram predicts when: (1) a striped cylinder is rotated around the person; or (2) the person makes voluntary eye movements; or (3) both occur simultaneously.

   B. Suppose now that you were to close one eye and, placing your finger on the lid of your *open* eye, you push your eyeball gently back and forth. Does this produce a perception of visual slip? How does your answer relate to the diagram? Specifically, where should your pushing of your eye be entered as a signal on the diagram; and by following the signal through the flow paths, does it produce the effect that you actually observed?

C. Suppose now that someone had injected your eye muscles with a local anesthetic so that, try as you may, you cannot move your eyes. (Don't try it!) What perceptual effect does the diagram suggest should result? Again, where on the diagram should the anesthetic be entered, and what does the flow of information through the diagram predict should be your perceptual experience?

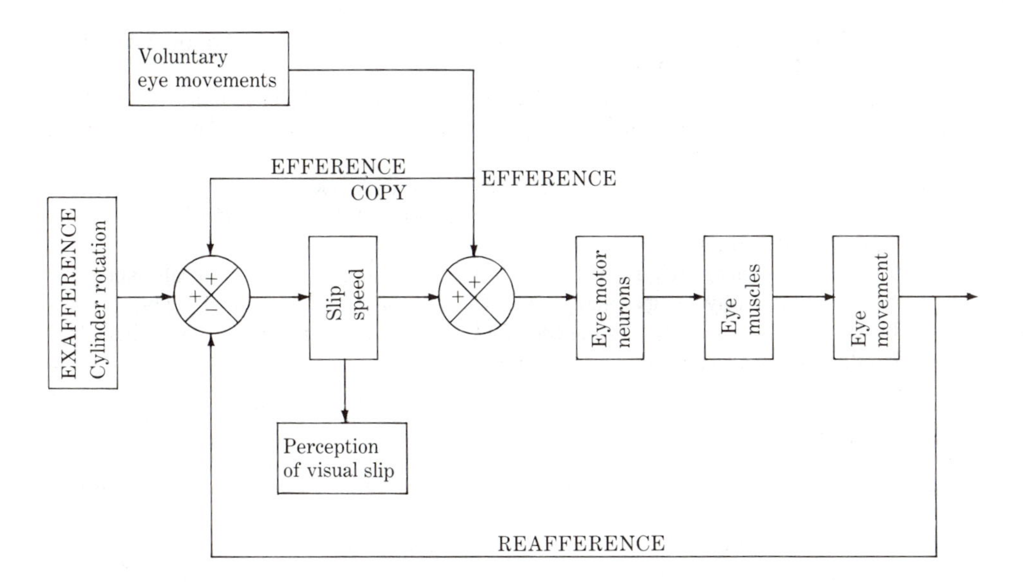

5. Imagine that you have found a species of fly with giant neurons, so that you can record the activity of any neuron you wish. Design an experiment to verify the efference copy hypothesis of optomotor behavior. State what results would convince you that this hypothesis is true. State all necessary controls.
6. Design an experiment to measure the influence of the wing hinge stretch receptor on the form of the wingbeat movements in locust fight. State all necessary controls.
7. Imagine that in an experiment on some mammal, electrical stimulation is delivered to each of the following three points, one at a time: (a) the axons of gamma motor neurons to the biceps muscle; (b) the axons of alpha motor neurons to the biceps muscle; (c) the axons of Ia afferents from the biceps muscle. For each of these three conditions, indicate what changes, if any, you would expect in each of the following: the activity of the Ia afferents; the activity of the alpha motor neurons; the activity of the gamma motor neurons; the length of the intrafusal fibers; the length of the extrafusal fibers.

Recommended Readings

BOOKS

Gallistel, C. R. (1980) *The Organization of Action: A New Synthesis.* Lawrence Erlbaum Associates, Hillsdale, New Jersey.
Chapters 6 and 7 of this book present a good introductory treatment of feedback in behavior.

Hinde, R. A. (1970) *Animal Behavior: A Synthesis of Ethology and Comparative Psychology.* McGraw-Hill, New York.
Chapter 7 contains a good introductory treatment of feedback in behavior.

McFarland, D. J. (1971) *Feedback Mechanisms in Animal Behavior.* Academic Press, London.
This book stresses the quantitative approach to feedback.

Matthews, P.B.C. (1972) *Mammalian Muscle Receptors and Their Central Actions.* Edward Arnold Ltd., London.
A superb historical account of research on muscle spindles covering all aspects of their anatomy, physiology, and role in the regulation of movement. This book is thoughtfully composed by a prime contributor to research on spindles. Though slightly dated, it still well repays a close reading.

ARTICLES

Miles, F. A. and Evarts, E. V. (1979) Concepts of motor organization. *Annu. Rev. Psychol.* 30:327–362.
An excellent review on the role of feedback in movement and related subjects.

Keele, S. W. and Summers, J. J. (1976) The structure of motor programs. In *Motor Control: Issues and Trends.* G. E. Stelnach (ed.). Academic Press, New York.
A thoughtful treatment of feedback in mammalian movement and related subjects.

von Holst, E. and Mittelstaedt, H. (1973) The reafference principle. In *The Behavioral Physiology of Animals and Man.* Vol. I. E. von Holst (ed.). University of Miami Press, Coral Gables, Florida.
This article contains a statement of the ideas of reafference and efference copy by two researchers especially prominent in developing these ideas. Originally written in 1950, this article is considerably dated but is still well worth reading.

Land, M. F. (1975) Similarities in the visual behavior of arthropods and man. In *Handbook of Psychobiology.* M. S. Gazzaniga and C. Blakemore (eds.). Academic Press, New York.
An excellent, thoughtfully written, general article on feedback in vision.

Collett, T. S. (1980) Angular tracking and the optomotor response: An analysis of visual reflex interactions in the hoverfly. *J. Comp. Physiol.* 140:145–158.
The possible role of efference copy during active flight movements is discussed.

Bell, C. (1981) An efference copy which is modified by reafferent input. *Science* 214:450–453.
Describes an excellent set of experiments revealing a true efference copy in electric fish.

Burrows, M. (1976) Neural control of flight in the locust. In *Neural Control of Locomotion.* R. M. Herman, S. Grillner, P.S.G. Stein and D. G. Stuart (eds.). Plenum Press, New York.

A fine review of feedback and other aspects of locust flight by a leading researcher in the field.

Wendler, G. (1974) The influence of proprioceptive feedback on locust flight coordination. *J. Comp. Physiol.* 88:173–200.
Sophisticated behavioral concepts and techniques establish specific roles for feedback in locust flight.

Epilog

It is necessary to avoid what I have come to call "simplicity filters.". . . One form of simplicity filter is the widespread tendency to think in terms of the *neuron,* the *mammal, or even* the *organism, with the implication that only a single system of each kind is significant. . . . It is desirable to strive for a happy mean between the belief that every subspecies is a totally different universe, and the highly focussed attitude of the molecular biologist who confidently expects in the near future to account for all the complexities of biology as neatly predictable reaction products.*

(D. Griffin, 1981)

In the early chapters of this book, we encountered numerous sensory and behavioral capabilities that, prior to their discoveries, would have been regarded as highly improbable. Who would have expected, for instance, to find animals that can see ultraviolet or infrared light; or that can read maps of plane polarized light in the sky; or that can hear ultrasound or infrasound; or that can sense the earth's magnetic field, as bees and a few vertebrate animals do; or that can communicate with one another directly by means of electric currents, as some fish do? And no doubt there are yet other sensory worlds of animals still to be discovered. It is safe to say that if there is useful information contained in some form of environmental energy, some animal has probably evolved a way to detect and utilize it. Evolution, then, has been highly opportunistic in exploiting environmental energies to meet behavioral needs. The discovery of this rich sensory diversity expands our understanding not only of the animals with which we cohabit the earth, but also of ourselves; for it shows that our own window on the world offers but one of many possible views.

We have also encountered in this book some surprising insights regarding the neural control of behavior. Who would have supposed, for instance, that the neural circuit underlying a behavior as simple as leech swimming would be so complex that, even after examining the circuit diagram of neurons and their connections, we are left wondering just how it all works? Moreover, though it has not been stressed in this book, individual nerve cells are now known to come in a surprising variety of physiological types. There are neurons that use action potentials and those that do not; those that use dendrites exclusively for input synapses and axons exclusively for output synapses, and others that have both types of synapse intermingled on many of their neurites; those whose synapses show various degrees of short- or long-term modifiability of transmission; and those whose plasma membranes contain many different types of ionic channels, giving rise to a rich variety of complex transmembrane potentials.

In view of these complexities at the cellular, network, and organismic levels, each animal species, each bit of behavior, and each neuron to be studied, should be approached with an open mind, realizing that almost anything is possible. In view of the diversity at all levels, the principles to be sought by neuroethologists are not global, but rather restrictive and conditional. That is, they do not state how *all* behaviors of *all* animals work. Rather, they take a form such as "under *these* conditions, animals having *these* properties are likely to use neurons of *this* type, organized in *this* manner." A major goal of neuroethology at present is to uncover as many such principles as possible. Whether these all will be subsumed eventually under some broader unifying principle remains to be seen.

The last 15 years or so have seen a tremendous growth in the technical preparedness of the neuroethologist to go out in search of the principles of his field. Prior to that time, for instance, it was not even clear whether particular neurons, with the exception of a few well-known giant cells, could be identified individually and reidentified in all members of a given species. It is now known that many neurons in the higher invertebrates, and a growing number in the lower vertebrates, can be so identified. This ability to recognize unique cells has provided a crucial underpinning for experiments such as those we have seen on cockroaches (Chapter 4), crayfish (Chapter 8), leeches (Chapter 9), and locusts (Chapter 10), where the role of individual neurons in behavior was investigated. Also, within the last 15 years, the methods have been developed that permit nerve cells to be filled with dyes such as Lucifer or Procion yellow or cobalt. This technique enables an investigator to identify the neuron from which recordings via a microelectrode have been made. It also enables the study of the three-dimensional shapes of nerve cells and greatly assists studies, using the electron microscope, that are designed to identify anatomically the synapses that connect two identified neurons.

The last 15 years have also seen the intensification of work at the cellular level on the role of the central nervous system in behavior. Previously, many workers had regarded the brain as a black box, whose sensory inputs and motor outputs were accessible for study but whose contents were a dark mystery. The techniques for using microelectrodes to study central neurons, now so common, were then in their infancy. Some workers even despaired of ever being able to unravel the brain's cellular processes; the cells seemed too numerous and their neurites too entangled ever to be sorted out.

Today we are armed with sharp microelectrodes that can be used to impale cell bodies or neurites larger than a few micrometers. A skilled worker can place up to 10 microelectrodes in different neurons of certain mollusks blessed with especially large cells and can stimulate, record, and fill these cells with dye. It is even sometimes possible in invertebrates to record intracellularly from a tethered animal while it carries out a complex behavioral act. And in some vertebrates, one can chronically implant an extracellular microelectrode in the brain and record for long periods from a single cell while the animal makes fairly normal movements. A variety of neurochemical and neuropharmacological tricks is also becoming available for the use of the neuroethologist.

Given all these recent technical advances, what principles of neuroethology have so far been discovered? A short list would include the following: (1) In social interactions among members of a species, where the signals evoking each behavioral interchange are highly predictable, the signals used are often simple, restricted sign stimuli. By contrast, when an animal orients through unfamiliar terrain and fluctuating climatic conditions during migration or homing, a rich variety of sensory cues of several modalities is often employed (Chapters 2 and 5). (2) Spatial information is often represented in the form of a spatial map in the brain even if, as in the auditory system, the receptor array does not contain a spatial map. Maps for different modalities are often aligned in register with one another (Chapter 7). (3) Behaviors that must be executed very quickly, such as escape behaviors, generally utilize neural circuits involving giant neurons, few synapses in series, and often some electrical synapses. The initial part of the behavior is often carried out in an open-loop manner (Chapters 4, 8, and 10). (4) Coordination of simultaneous movements by different parts of the body may be brought about by command neurons if the movement is extremely simple, or otherwise by command systems. In either case, these cells may have widespread synaptic divergence onto premotor or motor neurons subserving specific body regions (Chapter 8). (5) Rhythmic behaviors generally use both central neuronal oscillators and sensory feedback (Chapters 9 and 10). (6) Synaptic inhibition used by one neuron to partially or totally silence another, is common among cells of a network constituting a central pattern generator. Inhibition is also used to prevent movements that might compete with ongoing behavior (Chapters 8 and 9).

Among the most highly favored topics for neuroethological study have been spatial orientation and spatial maps in the brain, escape behaviors, and rhythmic behaviors. The reason for this is not hard to find. Maps impose order on the anatomically complex brain, escape behaviors involve large neurons that are relatively easy to impale with microelectrodes, and rhythms are, by definition, stereotyped and thus relatively easy to study. And yet, by focusing on neuroanatomical order, large neurons, and stereotypic movements, neuroethologists have tended toward behaviors that, to most ethologists, are just too simple to be really interesting.

Recognizing the restricted range of subjects previously studied, many neuroethologists are now turning their attention to behaviors that are both more complex and less rigidly fixed. This inevitably implies systems involving smaller neurons and more of them. Success in such studies will require the development of new techniques to solve several problems. New methods will be needed for exploring the details of cellular integration within individual small neurons, recording simultaneously from multicellular arrays, stimulating in a controlled manner selected individuals within these arrays, and analyzing the extremely complex data that will be acquired from these arrays. Inevitably, computers will become increasingly important as analytical tools in this work.

One useful guideline to the development of future neuroethological work emerges from considering the question, "what is the fundamental building block of the nervous system?" It has been almost axiomatic among many workers that this fundamental unit is the single neuron, which takes in synaptic information, processes it through cellular integration, and communicates the results of this integration to the next neurons in the chain (Chapter 3). Yet there are serious challenges to this point of view. For instance, a neuron that does not use action potentials may actually be functioning as several units. Each of several dendrites of such a cell may both receive and send synaptic information, and such dendrites may be several length constants away from one another, giving rise to functional isolation. In such a case, the fundamental unit would be, not the whole neuron, but the individual dendrite. On the other hand, a functionally related group of neurons, such as a command system, or a given region of a spatial map, or a spatiotemporal pattern recognizer (if such exists), can also be said to constitute a kind of building block. Thus, when investigating a particular neuroethological system, once the participating neurons have been identified, a next useful step is to try and determine what the fundamental units of the system are, be they subcellular, cellular, or multicellular. Next, one must study the interactions among these units. In the past, this type of analysis has been carried out most successfully at the single cell level. It is more difficult to study interactions among parts of a single cell or among large populations of cells. However, it is becoming increasingly important to proceed in both these directions—toward the more microscopic or subcellular and toward the more macroscopic or multicellular. Perhaps the most unique aspect of the discipline of neuroethology is its attempt to interrelate events at all levels of organization, from the subcellular to the cellular to the network to the organism and even to the community.

In Chapter 1, we saw that among the key concerns of the field of neuroethology are the *detection*, *recognition*, and *discrimination* of signals from the environment and the *decision* to carry out, on the basis of this information, *complex behavioral acts*. We also saw in Chapter 9 that a good deal of neural activity originates directly within the central nervous system, relatively independently of sensory inputs. Also in our own species, we know that the brain carries out relatively independently several types of functions that are as yet barely understood. Two examples are emotions and memory. As for emotions, one definition of mental health might be the ability to *detect* our own feelings, to *recognize* them accurately, and to *discriminate* among them, and on this basis to make *decisions* resulting in carrying out sometimes *complex acts*. The same set of capabilities might also apply to our ability to utilize information stored in memory. Human emotions and memory have never been accessible for electrophysiological exploration owing partly to the unwillingness of most people to serve as experimental subjects. It seems possible, however, that the mechanisms by which our brains retrieve and manipulate information from these deep storerooms of our inner lives are similar to those by which a brain acquires and organizes information from the external world. If so, we may be learning much more about ourselves than we realize in the process of exploring the brains of toads, bugs and bats.

References

THE NUMBERS THAT PRECEDE EACH ENTRY IDENTIFY THE CHAPTER(S) IN WHICH THE REFERENCE IS CITED.

5 Adler, J. 1976. The sensing of chemicals by bacteria. *Sci. Am.* 234(4): 40–47.

5 Adler, K. and Taylor, D.H. 1973. Extraocular perception of polarized light by orienting salamanders. *J. Comp. Physiol.* 87: 203–212.

2 Alcock, J. 1979. *Animal Behavior: An Evolutionary Approach*, 2nd Edition. Sinauer Associates, Sunderland, Massachusetts.

10 Altman, J.S. and Tyrer, N.M. 1977. The locust wing hinge stretch receptors. I. Primary sensory neurones with enormous central arborizations. *J. Comp. Neurol.* 172: 409–430.

9 Alving, B.O. 1968. Spontaneous activity in isolated somata of *Aplysia* pacemaker neurons. *J. Gen. Physiol.* 45: 29–45.

5 Aschoff, J. 1960. Exogeneous and endogenous components in circadian rhythms. *Cold Spring Harbor Symp. Quant. Biol.* 25: 11–28.

8 Atwood, H.L. 1977. Crustacean neuromuscular systems: Past, present and future. In *Identified Neurons and Behavior of Arthropods*. G. Hoyle (ed.). Plenum, New York.

5 Autrum, H. and Thomas, I. 1973. Comparative physiology of color vision in animals. In *Handbook of Sensory Physiology* Vol. VII 3A, *Central Processing of Visual Information*. R. Jung (ed.). Springer-Verlag, Berlin.

9 Baerends, G.P., Brouwer, R. and Waterbolk, H.T.J. 1955. Ethological studies on *Lebistes reticulatus* (Peters). I. Analysis of the male courtship pattern. *Behaviour* 8: 249–334.

2 Barlow, G.W. 1977. Modal action patterns. In *How Animals Communicate*. T.A. Sebeok (ed.). Indiana University Press, Bloomington, Indiana.

5 Barrett, R., Maderson, P.F.A. and Meszler, R.M. 1970. The pit organs of snakes. In *Biology of the Reptilia* Vol. 2. C. Gans and T.S. Parsons (eds.). Academic Press, London.

5 Baumann, F. 1968. Slow and spike potentials recorded from retinula cells of the honeybee drone in response to light. *J. Gen. Physiol.* 52: 855–875.

7 Baylor, D.A. and Fettiplace, R. 1977. Transmission from photoreceptors to ganglion cells in turtle retina. *J. Physiol.* 271: 391–424.

5 Baylor, D.A. and Fuortes, M.G.F. 1970. Electrical responses of single cones in the retina of the turtle. *J. Physiol.* 207: 77–92.

5 Baylor, D.A., Lamb, T.D. and Yau, K.W. 1979. Membrane currents of single rod outer segments. *J. Physiol.* 288: 589–611.

10 Bell, C.C. 1981. An efference copy which is modified by reafferent input. *Science* 214: 450–453.

10 Bell, C.C. 1982. Properties of a modifiable efference copy in an electric fish. *J. Neurophysiol.* 47: 1043–1056.

8 Bentley, D. and Konishi, M. 1978. Neural control of behavior. *Annu. Rev. Neurosci.* 1: 35–59.

5 Bernard, G.D. 1979. Red-absorbing visual pigment of butterflies. *Science* 203: 125–127.

9 Berridge, M.J., Rapp, P.E. and Treherne, J.E. 1979. *Cellular Oscillators. J. Exp. Biol.* 81.

10 Bizzi, E., Dev, P., Morasso, P. and Polit. A. 1978. Effect of load disturbances during centrally initiated movements. *J. Neurophysiol.* 41: 542–556.

5 Blakemore, R. 1975. Magnetotactic bacteria. *Science* 190: 377–379.

5 Bodznick, D. and Northcutt, R.G. 1981. Electroreception in lampreys: Evidence that the earliest vertebrates were electroreceptive. *Science* 212: 465–467.

5 Bowmaker, J.K. and Dartnall, H.J.A. 1980. Visual pigments of rods and cones in a human retina. *J. Physiol.* 298: 501–511.

6 Brenowitz, E.A. 1982. The active space of red-winged blackbird song. *J. Comp. Physiol.* 147: 511–522.

5 Brines, M.L. and Gould, J.L. 1979. Bees have rules. *Science* 206: 571–573.

10 Brodal, A. and Pompeiano, O. (eds.). 1972. *Basic Aspects of Central Vestibular Mechanisms.* Elsevier, Amsterdam.

2 Brower, J.V.Z. and Brower, L.P. 1962. Experimental studies of mimicry. 6. The reaction of toads *(Bufo terrestris)* to honeybees *(Apis mellifera)* and their dronefly mimics *(Eristalis vinetorum.) Am. Nat.* 96: 297–307.

2 Brower, L. P., Brower, J.V.Z. and Westcott, P.W. 1960. Experimental studies of mimicry. 5. The reactions of toads *(Bufo marinus)* to bumblebees *(Bombus americanurum)* and their rubberfly mimics *(Mallophora bomboides)* with a discussion of aggressive mimicry. *Am. Nat.* 94: 343–355.

5 Brown, P.K. and Wald, G. 1964. Visual pigments in single rods and cones of the human retina. *Science* 144: 45–51.

6 Bruns, V. 1976. Peripheral auditory tuning for fine frequency analysis by the CF–FM bat, *Rhinolophus ferrumequinum.* II. Frequency mapping in the cochlea. *J. Comp. Physiol.* 106: 87–97.

1 Bullock, T.H. 1977. *Recognition of Complex Acoustic Signals.* Dahlem Konferenzen, Berlin.

5 Bullock, T.H. and Diecke, P.J. 1956. Properties of an infrared receptor. *J. Physiol.* (Lond.) 134: 47–87.

5 Bullock, T.H. and Horridge, G.A. 1965. *Structure and Function in the Nervous System of Invertebrates.* Vol. 1. W.H. Freeman and Company, San Francisco.

10 Bullock, T.H., Orkand, R. and Grinnell, A. 1977. *Introduction to Nervous Systems.* W.H. Freeman and Company, San Francisco.

5 Bunning, E. 1973. *The Physiological Clock. Circadian Rhythms and Biological Chronometry.* Springer-Verlag, New York.

10 Burrows, M. 1973. The role of delayed excitation in the co-ordination of some metathoracic flight motor neurones of a locust. *J. Comp. Physiol.* 83: 135–164.

10 Burrows, M. 1975. Monosynaptic connexions between wing stretch receptors and flight motor neurons of the locust. *J. Exp. Biol.* 62: 189–219.

10 Burrows, M. 1976. Neural control of flight in the locust. In *Neural Control of Locomotion.* R.M. Herman, S. Grillner, P.S. G. Stein and D.G. Stuart (eds.). Plenum Press, New York.

6 Busnel, R.G. and Fish, J.F. 1980. *Animal Sonar Systems.* Plenum Press, New York.

2 Cade, W. 1979. The evolution of alternative male reproductive strategies in field crickets. In *Sexual Selection and Reproduction in Insects.* M.S. Blum and N.A. Blum (eds.). Academic Press, New York.

10 Calabrese, R.L. 1977. The neural control of alternate heart beat coordination states in the leech, *Hirudo medicinalis. J. Comp. Physiol.* 122: 111–143.

9 Calabrese, R.L. 1979. The roles of endogenous membrane properties and synaptic interaction in generating the heartbeat rhythm of the leech, *Hirudo medicinalis*. *J. Exp. Biol.* 82: 163–176.

10 Camhi, J.M. 1970. Yaw-correcting postural changes in locusts. *J. Exp. Biol.* 52: 519–531.

10 Camhi, J.M. 1971. Flight orientation in locusts. *Sci. Am.* 224(2): 74–81.

8 Camhi, J.M. 1977. Behavioral switching in cockroaches: Transformation of tactile reflexes during righting behavior. *J. Comp. Physiol.* 113: 283–301.

4 Camhi, J.M. 1980. The escape system of the cockroach. *Sci. Am.* 243(6): 158–172.

10 Camhi, J.M. and Hinkle, M. 1974. Response modification by the central flight oscillator of locusts. *J. Exp. Biol.* 60: 477–492.

4,10 Camhi, J.M. and Nolen, T.G. 1981. Properties of the escape system of cockroaches during walking. *J. Comp. Physiol.* 142: 339–346.

4,10 Camhi, J.M. and Tom, W. 1978. The escape behavior of the cockroach *Periplaneta americana*. I. Turning responses to wind puffs. *J. Comp. Physiol.* 128: 193–201.

4 Camhi, J.M., Tom, W. and Volman, S. 1978. The escape behavior of the cockroach *Periplaneta americana*. II. Detection of natural predators by air displacement. *J. Comp. Physiol.* 128: 203–212.

7 Campbell, F.W. 1974. The transmission of spatial information through the visual system. In *The Neurosciences: Third Study Program*. M.I.T. Press, Cambridge, Massachusetts.

6 Capranica. R.R. 1976. Morphology and physiology of the auditory system. In *Frog Neurobiology*. R. Llinas and W. Precht (eds.). Springer-Verlag, New York.

8 Carew. T.J. and Kandel, E.R. 1977a. Inking in *Aplysia californica*. I. Neural circuit for an all-or-none behavioral response. *J. Neurophysiol.* 40: 692–707.

8 Carew, T.J. and Kandel, E.R. 1977b. Inking in *Aplysia californica*. II. Central program for inking. *J. Neurophysiol.* 40: 708–720.

9 Carterette, E.C. and Friedman, M.P. 1973. *Handbook of Perception*. Vol. 3. *Biology of Perceptual Systems*. Academic Press, New York.

5 Carterette, E.C. and Friedman, M.P. 1978. *Handbook of Perception*. Vol. 6A. *Tasting and Smelling*. Academic Press, New York.

7 Chaplin, J.P. and Krawiec, T.S. 1974. *Systems and Theories of Psychology*. Holt, Rinehart & Winston, New York.

7 Chow, K.L. and Leiman, A.L. 1970. The structural and functional organization of the neocortex. *Neurosci. Res. Program. Bull.* 8:

5 Clayton, R.K. 1977. *Light and Living Matter*. Vol. 1. *The Physical Part*. Vol. 2. *The Biological Part*. R.E. Krieger Publishing Company, Huntington, New York.

8 Cohen, A.H. and Wallen, P. 1980. The neuronal correlate of locomotion in fish. "Fictive swimming" induced in an *in vitro* preparation of the lamprey spinal cord. *Exp. Brain Res.* 41: 11–18.

5 Cohen, A.I. 1972. Rods and cones. In *Handbook of Sensory Physiology*. Vol. VII-2. *Physiology of Photoreceptor Organs*. Springer-Verlag, Berlin.

1 Cole, K.C. 1975. Neuromembranes: Paths of ions. In *The Neurosciences: Paths of Discovery*. F.G. Worden, J.P. Swazey and G. Adelman (eds.) M.I.T. Press, Cambridge, Massachusetts.

6 Coles, R.B., Lewis, D.B., Hill, K.G., Hutchings, M.E. and Gawer, D.M. 1980. Directional hearing in the Japanese quail *(Coturnix eoturnix japonica)*. II. Cochlear physiology. *J. Exp. Biol.* 86: 153–170.

10 Collett, T.S. 1980. Angular tracking and the optomotor response. An analysis of visual reflex interactions in the hoverfly. *J. Comp. Physiol.* 140: 145–158.

10 Collett, T.S. and Land, M.F. 1975. Visual control of flight behavior in the hoverfly, *Syritta pipiens* L. *J. Comp. Physiol.* 99: 1–66.

7 Comer, C. and Grobstein, P. 1981a. Tactually elicited prey acquisition behavior in the frog. *Rana pipiens,* and a comparison with visually elicited behavior. *J. Comp. Physiol.* 142: 141–150.

7 Constantine-Paton, M. and Law, M.I. 1982. The development of maps and stripes in the brain. *Sci. Am.* 247(6): 54–62.

8 Cooper, J.R., Bloom, F.E. and Roth, R.H. 1978. *The Biochemical Basis of Neuropharmocology.* Oxford University Press, New York.

7 Corcoran, D.W.J. 1971. *Pattern Recognition.* Penguin Books Ltd., Harmondsworth, Middlesex, England.

2 Cott, H.B. 1957. *Adaptive Coloration in Animals.* Methuen, London.

10 Crago, P.E., Houk, J.C. and Hasan, Z. 1976. Regulatory actions of human stretch reflex. *J. Neurophysiol.* 39: 925–935.

5 Crane, J. 1955. Imaginal behavior of a Trinadad butterfly, *Heliconius erato hydara* Hewitson, with special reference to the social use of color. *Zoologica* (N.Y.) 40: 167–196.

2 Crane, J. 1957. Basic patterns of display in fiddler crabs. *Zoologica* (N.Y.) 42: 69–82.

5 Dacheux, R.F. and Miller, R.F. 1976. Photoreceptor-bipolar cell transmission in the perfused retina eyecup of the mudpuppy. *Science.* 191: 963–964.

4 Dagan, D. and Camhi, J.M. 1979. Responses to wind recorded from the cercal nerve of the cockroach *Periplaneta americana.* II. Directional selectivity of the sensory neurons innervating single columns of filiform hairs. *J. Comp. Physiol.* 133: 103–110.

9 Dagan, D., Vernon, L.H. and Hoyle, G. 1975. Neuromimes: Self-existing alternate firing pattern models. *Science* 188: 1035–1036.

4 Dagan, D. and Volman, S. 1982. Sensory basis for directional wind detection in first instar cockroaches, *Periplaneta americana. J. Comp. Physiol.* 147: 471–478.

4 Daley, D.L. 1982. Neural basis of wind-receptive fields of cockroach giant interneurons. *Brain Res.* 238: 211–216.

4 Daley, D.L. and Camhi, J.M. 1981. Neural basis of the directional selectivity of wind-receptive fields of cockroach giant interneurons. *Neurosci. Abst.* 7: 85.10.

4 Daley, D.L., Vardi, N., Appigrani, B. and Camhi, J.M. 1981. Morphology of the giant interneurons and cercal nerve projections of the American cockroach. *J. Comp. Neurol.* 196: 41–52.

2 Darwin, C. 1859. *On the Origin of Species by Means of Natural Selection.* Murray Ltd., London.

5 Daumer, K. 1956. Reizmetrische Untersuchung des Farbensehens der Beinen. *Z. vergl. Physiol.* 38: 413–478.

5 Daumer, K. 1958. Blumenfarben, wie Sie die Bienen sehen. *Z. vergl. Physiol.* 41: 49–110.

10 Davis, W.J. and Ayers, J.L. Jr. 1972. Locomotion: Control by positive-feedback optokinetic responses. *Science* 177: 183–185.

8 Davis, W.J. and Kennedy, D. 1972. Command interneurons controlling swimmeret movements in the lobster. I. Types of effects on motoneurons. *J. Neurophysiol.* 35: 1–12.

5 Davson, H. 1962. *The Eye.* Vol. 4. *The Visual Process.* Academic Press, New York.

8,9 Dawkins, R. 1976. Hierarchical organization: A candidate principle for ethology. In *Growing Points in Ethology.* P.P.G. Bateson and R.A. Hinde (eds.). Cambridge University Press, New York.

2 Dawkins, R. 1977. *The Selfish Gene.* Oxford University Press, New York.

7 Dean, J. 1980. Encounters between bombardier beetles and two species of toads. *(Bufo americanus, B. marinus):* Speed of prey-capture does not determine success. *J. Comp. Physiol.* 135: 41–50.

8 Delcomyn, F. 1971. The locomotion of the cockroach, *Periplaneta americana. J. Exp. Biol.* 54: 443–452.

9,10 Delcomyn, F. 1980. Neural basis of rhythmic behavior in animals. *Science* 210: 492–498.

2 Descartes, R. 1972. *Treatise of Man.* T.S. Hall (Translator). Harvard University Press, Cambridge, Massachusetts.

7 DeValois, R.L. and DeValois, K.K. 1980. Spatial vision. *Annu. Rev. Psychol.* 31: 309–341.

6 de Wilde, J. 1941. Contribution to the physiology of the Johnston organ and its part in the behavior of the *Gyrinus. Arch. Ne'er. Physiol.* 25: 381–400.

8 Diamond, J. 1971. The Mauthner cell. In *Fish Physiology.* Vol. 5. W.S. Hoar and D.S. Randall (eds.). Academic Press, New York.

10 Dichtburn, R.W. 1973. *Eye Movements and Visual Perception.* Clarendon Press, Oxford.

5 Dietz, M. 1972. Erdkroten konnen UV-lieht sehen. *Naturwissenschaften* 59: 316.

7 Dowling, J.E. 1976. Physiology and morphology of the retina. In *Frog Neurobiology.* R. Llinas and L. Precht (eds.). Springer-Verlag, Berlin.

7 Dowling, J.E. and Boycott, B.B. 1966. Organization of the primate retina: Electron microscopy. *Proc. R. Soc. Lond.* B 166: 80–111.

7 Drager, V.C. and Hubel, D.H. 1975. Responses to visual stimulation and relationship between visual, auditory and somatosensory inputs in mouse superior colliculus. *J. Neurophysiol.* 38: 690–713.

8 Drewes, C. In press. Escape response in earthworms and other annelids. In *Neuromechanisms in Startle Behavior.* R. Eaton, (ed.). Plenum Press, New York.

5 Eakin, R.M. 1968. Evolution of photoreceptors. *Evol. Biol.* 2: 194–242.

8 Eaton, R. (ed.) In press. *Neuromechanisms in Startle Behavior.* Plenum Press, New York.

2 Eberhard, W.G. 1977. Aggressive chemical mimicry by a bolas spider. *Science* 198: 1173–1175.

10 Eccles, J.C., Eccles, R.M. and Lundberg, A. 1957. The convergence of monosynaptic excitatory afferents on to many different species of alpha motoneurons. *J. Physiol.* 137: 22–50.

10 Eckert, R. and Randall, D. 1978. *Animal Physiology.* W.H. Freeman and Company, San Francisco.

2 Edmunds, M. 1974. *Defense in Animals.* Longman Group, Harlow, Essex.

5 Edrich, W. 1979. Honeybees: photoreceptors participating in orientation behavior to light and gravity. *J. Comp. Physiol.* 133: 111–116.

5 Edrich, W., Neumeyer, C. and von Helversen, O. 1979. Anti-sun orientation of bees with regard to a field of ultraviolet light. *J. Comp. Physiol.* 134: 151–157.

5 Eisner, T., Silberglied, R.E., Aneshansley, D., Carrel, J.E. and Howland, H.C. 1969. Ultraviolet video-viewing: The television camera as an insect eye. *Science* 166: 1172–1174.

6 Evans, E.F. and Wilson, J.P. 1975. Cochlear tuning properties: Concurrent basilar membrane and single nerve fiber measurements. *Science* 190: 1218–1221.

9 Elsner, N. 1974. Neural economy: bifunctional muscles and common central pattern elements in leg and wing stridulation of the grasshopper *Stenobothrus rubicundus* Germ. (Orthoptera: Acrididae). *J. Comp. Physiol.* 89: 227–236.

5 Emlen, S.T. 1967a. Migratory orientation in the indigo bunting, *Passerina cyanea*. Part I: Evidence for use of celestial cues. *Auk* 84: 309–342.

5 Emlen, S.T. 1967b. Migratory orientation in the indigo bunting, *Passerina cyanea*. Part II; Mechanisms of celestial orientation. *Auk* 84: 463–489.

5 Emlen, S.T. 1975. The stellar-orientation system of a migratory bird. *Sci. Am.* 223(2): 102–111.

6 Erulkar, S.D. 1972. Comparative aspects of spatial localization of sound. *Physiol. Rev.* 52: 237–360.

6 Evans, E.F. and Wilson, J.P. 1975. Cochlear tuning properties: Concurrent basilar membrane and single nerve fiber measurements. *Science* 190: 1218–1221.

8 Evarts, E.V. 1966. Methods for recording activity of individual neurons in moving animals. In *Methods in Medical Research*. Vol. II. R.F. Rushmer (ed.). Yearbook, Chicago.

8 Evoy, W.H. and Kennedy, D. 1967. The central nervous organization underlying control of antagonistic muscles in the crayfish. I. Types of command fibers. *J. Exp. Zool.* 165: 223–238.

7 Ewert, J.P. 1969. Quantitative analyse von Reiz-Reactionsbeziehungen bei visuellen Auslosen der Beutefang-Wendereaktion der Erdkrote (*Bufo bufo*) L.). *Pflungers Arch.* 308: 225–243.

7 Ewert, J.P. 1976. The visual system of the toad: Behavioral and physiological studies on a pattern recognition system. In *The Amphibian Visual System*. K.V. Fite (ed.). Academic Press, New York.

7,10 Ewert, J.P. 1980. *Neuroethology*. Springer-Verlag, Berlin.

7 Ewert, J.P., Borchers, H.W. and Wietersheim, A.v. 1978. Question of prey feature detectors in the toad's *Bufo bufo* (L.) visual system: A correlation analysis. *J. Comp. Physiol.* 126: 43–47.

4 Ewert, J.P., Capranica, R.R. and Ingle, D.S. (eds.). 1983. *Advances in Vertebrate Neuroethology*. Plenum Press, New York.

7 Ewert, J.P. and Hock, F. 1972. Movement-sensitive neurones in the toad's retina. *Exp. Brain Res.* 16: 41–59.

7 Ewert, J.P. and Wietersheim, A.v. 1974. Einfluss von Thalamus/Praetectum-Defekten auf die Antwort von Tectum-Neuronen gegenuber bewegten visuellen Mustern bie der Krote *Bufo bufo* (L.). *J. Comp. Physiol.* 92: 149–160.

9 Fagan, R.M. and Young, D.Y. 1978. Temporal patterns of behaviors: Durations, intervals, latencies and sequences. In *Quantitative Ethology*. P. Colgen (ed.). John Wiley & Sons, New York.

6 Feng, A.S., Gerhardt, H.C. and Capranica, R.R. 1976. Sound localization behavior of the green treefrog *(Hyla cinerea)* and the barking treefrog *(Hyla gratiosa)*. *J. Comp. Physiol.* 107: 241–252.

8 Fields, H.L., Evoy, W.H. and Kennedy, D. 1967. Reflex role played by efferent control of an invertebrate stretch receptor. *J. Neurophysiol.* 30: 859–874.

10 Finlayson, L.H. and Lowenstein, O. 1958. The structure and function of abdominal stretch receptors in insects. *Proc. R. Soc. Lond.* B 148: 433–449.

7 Fite, K.V. and Scalia, F. 1976. Central visual pathways in the frog. In *The Amphibian Visual System*. K.V. Fite (ed.). Academic Press, New York.

8 Forssberg, H. and Grillner, S. 1973. The locomotion of the acute spinal cat injected with clonidine i.v. *Brain Res.* 50: 184–186.

2 Fraenkel, G.S. and Gunn, D.L. 1961. *The Orientation of Animals: Kineses, Taxes and Compass Reactions*. Dover Publications, New York. (Reprint of the original 1940 edition.)

9 Friesen, W.O., Poon, M. and Stent, G.S. 1978. Neuronal control of swimming in the medicinal leech. IV. Identification of a network of oscillatory interneurones. *J. Exp. Biol.* 75: 25–43.

9 Friesen, W.O. and Stent, G.S. 1977. Generation of a locomotory rhythm by a neuronal network with recurrent cyclic inhibition. *Biol. Cybern.* 28: 27–40.

9 Friesen, W.O. and Stent, G.S. 1978. Neural circuits for generating rhythmic movements. *Annu. Rev. Biophys. Bioeng.* 7: 37–61.

5 Fuortes, M.G.F. and O'Bryan, P.M. 1972. Generator potentials in invertebrate photoreceptors. In *Handbook of Sensory Physiology.* Vol. 7. H. Autrum (ed.). Springer-Verlag, Berlin. pp. 279–319.

7 Gaither, N.S. and Stein, B.E. 1979. Reptiles and mammals use similar sensory organizations in the midbrain. *Science* 205: 595–597.

8,9,10 Gallistel, C.R. 1980. *The Organization of Action: A New Synthesis.* Lawrence Erlbaum Associates, Hillsdale, New Jersey.

7 Gardner, B.T. and Gardner, R.A. 1975. Evidence for sentence constituents in the early utterences of child and chimpanzee. *J. Exp. Psychol.* 104: 244–267.

7 Gaze, R.M. 1970. *The Formation of Nerve Connections.* Academic Press, New York.

8 Getting, P.A. 1975. *Tritonia* swimming: Triggering a fixed action pattern. *Brain Res.* 96: 128–133.

10 Getting, P.A., Lennard, P.R. and Hume, P.I. 1981. Central pattern generator mediating swimming in *Tritonia.* I. Identification and synaptic interaction. *J. Neurophysiol.* 44(1): 151–165.

8 Getting, P.A. and Willows, A.O.D. 1974. Modification of neuron properties by electrotonic synapses. II. Burst formation by electronic synapses. *J. Neurophysiol.* 37: 858–868.

10 Gettrup, E. 1962. Thoracic proprioceptors in the flight system of locusts. *Nature* 193: 498–499.

10 Gillette, R., Kovak, M.P. and Davis, W.J. 1978. Command neurons of *Pleurobranchaea* receive synaptic feedback from the motor network they excite. *Science* 199: 788–801.

7 Glass, I. and Wollberg, Z. 1979. Lability in the responses of cells in the auditory cortex of squirrel monkeys to species-specific vocalizations. *Exp. Brain Res.* 34: 489–498.

7 Gold, G.H. 1979. Photoreceptor coupling in retina of the toad, *Bufo marinus.* II. Physiology. *J. Neurophysiol.* 42: 311–328.

5 Goldsmith, T.H. 1973. Photoreception and vision. In *Comparative Animal Physiology.* C.L. Prosser (ed.). W.B. Saunders, Philadelphia.

5 Goldsmith, T.H. 1980. Hummingbirds see near ultraviolet light. *Science* 207: 786–788.

6 Goldstein, M.H. Jr. 1968. The auditory periphery. In *Medical Physiology.* Vol. 2. C.V. Mountcastle (ed.). Mosby, St. Louis.

10 Gottleib, G.L. and Agarwal, G.C. 1979. Response to sudden torques about ankle in man: I. Myotatic reflex. *J. Neurophysiol.* 42: 91–106.

10 Gottleib, G.L. and Agarwal, G.C. 1980a. Response to sudden torques about ankle in man. II. Postmyotatic reactions. *J. Neurophysiol.* 43: 86–101.

10 Gottleib, G.L. and Agarwall, G.C. 1980b. Response to sudden torques about ankle in man. III. Suppression of stretch-evoked responses during phasic contraction. *J. Neurophysiol.* 44: 233–246.

5 Gould, J.L. 1975. Honeybee recruitment: The dance language controversy. *Science* 189: 685–693.

10 Gould, J.L. 1980. Sun compensation by bees. *Science* 207: 545–547.

2 Gould, J.L. 1982. *Ethology: The Mechanisms and Evolution of Behavior.* W.W. Norton, New York.

5 Gould, J.L., Kirschvink, J.L. and Deffeyes, K.S. 1978. Bees have magnetic remanence. *Science* 201: 1026–1028.

6 Gourevitch, G. 1980. Directional hearing in terrestrial mammals. In *Comparative Studies of Hearing in Vertebrates.* A.N. Popper and R.R Fay (eds.). Springer-Verlag, New York.

9 Granzow, B., Freed, E. and Kristan, B. 1980. Long-term behavioral and neuronal changes induced by section of interganglionic connectives in leeches. *Neurosci. Abst.* 6: 235.1.

5 Gribakin, F.G. 1969. Cellular basis for colour vision in the honey bee. *Nature* 223: 639–641.

4,6 Griffin, D.R. 1958. *Listening in the Dark.* Yale University Press, New Haven, Connecticut.

6 Griffin, D.R. 1971. The importance of atmospheric attenuation for the echolocation of bats *(Chiroptera). Anim. Behav.* 19: 55–61.

2,7 Griffin, D.R. 1981. *The Question of Animal Awareness: Evolutionary Continuity of Mental Experience.* The Rockefeller University Press, New York.

6 Griffin, D.R., Friend, J.H. and Wendler, F.A. 1965. Target discrimination by the echolocation of bats. *J. Exp. Zool.* 158: 155–168.

8 Grillner, S. 1969. Supraspinal and segmental control of static and dynamic motoneurons in the cat. *Acta Physiol. Scand. [Suppl.] 327: 1–34.*

8,9 Grillner, S. 1975. Locomotion in vertebrates: Central mechanisms and reflex control. *Physiol. Rev.* 55: 247–306.

8 Grillner, S., Perret, C. and Zangger, P. 1976. Central generation of locomotion in the spinal dogfish. *Brain Res.* 109: 255–269.

7 Gross, C.G., Bender, D.B. and Rocha-Miranda, C.E. 1974. Inferotemporal cortex: A single unit analysis. In *The Neurosciences: Third Study Program.* M.I.T. Press, Cambridge, Massachusetts.

7 Gross, C.G., Rocha-Miranda, C.E. and Bender, D.B. 1972. Visual properties of neurons in inferotemporal cortex of the macaque. *J. Neurophysiol.* 35: 96–111.

8 Grossman, S.D. 1973. Direct adrenergic and cholinergic stimulation of hypothalamic mechanism. In *Brain Stimulation and Motivation.* E.S. Valenstein (ed.). Scott, Foresman & Company, Glenview, Illinois.

7 Grusser, U.J., and Grusser-Cornehls, U. 1976. Neurophysiology of the anuran visual system. In *Frog Neurobiology.* R. Llinas and W. Precht (eds.). Springer-Verlag, Berlin.

7 Haber, R.N. and Hershenson, M. 1980. *The Psychology of Visual Perception.* Holt, Rinehart & Winston, New York.

2 Hailman, J.P. 1967. The ontogeny of an instinct. *Behavior [Suppl.]* 15:1–159.

2 Hailman, J.P. 1969. How an instinct is learned. *Sci. Am.* 221(6): 98–108.

5 Hardie, R.C. 1979. Electrophysiological analysis of fly retina. I: Comparative properties of R1-6 and R7 and R8. *J. Comp. Physiol.* 129: 19–33.

5 Harris, W.A., Ready, D.F., Lipson, E.D., Hudspeth, A.J. and Stark, W.S. 1977. Vitamin A deprivation and *Drosophila* photopigments. *Nature* 266: 648–650.

9 Hartline, D.K. 1979. Pattern generation in the lobster *(Panulirus)* stomatogastric ganglion. II. Pyloric network simulation. *Biol. Cybern.* 33: 223–236.

9 Hartline, D.K., Gassie, D.V. Jr. and Sirchia, C.D. 1981. Perturbation responses of endogenous and network-derived bursting patterns. *Neurosci. Abst.* 7: 134.9.

7 Hartline, H.K., Wagner, H.G. and Ratliff, F. 1956. Inhibition in the eye of *Limulus. J. Gen. Physiol.* 39: 651–673.

5 Hartline, P.H. 1974. Thermoreception in snakes. In *Handbook of Sensory Physiology.* Vol. III/3: *Electroreceptors and Other Specialized Receptors in Lower Vertebrates.* A. Fessard (ed.). Springer-Verlag, Berlin.

5 Heiligenberg, W. 1977. *Principles of Electrolocation and Jamming Avoidance in Electric Fish.* Springer-Verlag, New York.

3 Heimer, L. and RoBards, M.J. 1981. *Neuroanatomical Tract-Tracing Methods.* Plenum Press, New York.

10 Heisenberg, M. and Wolf, R. 1979. On the fine structure of yaw torque in visual flight orientation of *Drosphila melanogaster. J. Comp. Physiol.* 130: 113–130.

10 Heitler, W.J. and Burrows, M. 1976. The locust jump. II. Neural circuits for the motor programme. *J. Exp. Biol.* 66: 221–241.

8,9 Herman, R.M., Grillner, S., Stein, P.S.G. and Stuart, D.G. 1976. *Neural Control of Locomotion.* Plenum Press, New York.

6 Hill, K.G. and Boyan, G.S. 1976. Directional hearing in crickets. *Nature* 262: 390–391.

6 Hill, K.G. and Boyan, G.S. 1977. Sensitivity to frequency and direction of sound in the auditory system of crickets *(Gryllidae). J. Comp. Physiol.* 121: 79–97.

6 Hill, K.G., Lewis, D.B., Hutchings, M.E. and Coles, R.B. 1980. Directional hearing in the Japanese quail *(Coturnix coturnix japonica).* I. Acoustic properties of the auditory system. *J. Exp. Biol.* 86: 135–151.

2 Hinde, R.A. 1970. *Animal Behavior: A Synthesis of Ethology and Comparative Psychology.* McGraw-Hill, New York.

5 Hoffmann, K. 1960. Experimental manipulation of the orientational clock in birds. *Cold Spring Harbor Symp. Quant. Biol.* 25: 379–387.

5 Hopkins, C.D. 1974. Electric communication in fish. *Am. Sci.* 62: 426–473.

5 Hopkins, C.D. 1980. Neuroethology of electric communication. *Trends in Neuroscience* 4: 4–6.

10 Horridge, G.A. 1967. The study of a system, as illustrated by the optokinetic response. In *Nervous and Hormonal Mechanisms of Integration. Symp. Soc. Exp. Biol.* 20: 179–198.

6 Hoy, R.R. 1978. Acoustic communication in crickets. A model system for the study of feature detection. *Fed. Proc.* 37: 2316–2323.

2 Hoy, R.R., Hahn, J. and Paul, R.C. 1977. Hybrid cricket auditory behavior: Evidence for genetic coupling in animal communication. *Science* 195: 82–84.

2 Hoy, R.R. and Paul, R.C. 1973. Genetic control of song specificity in crickets. *Science* 180: 82–83.

8 Hoyle, G. (ed.). 1977. *Identified Neurons and Behavior of Arthropods.* Plenum Press, New York.

7 Hubel, D.H. 1979. The visual cortex of normal and deprived monkeys. *Am. Sci.* 67: 532–543.

7 Hubel, D.H. and Wiesel, T.N. 1962. Receptive fields, binocular interaction and functional architecture in the cat's visual cortex. *J. Physiol.* 160: 106–154.

7 Hubel, D.H. and Wiesel, T.N. 1974. Uniformity of monkey striate cortex: A parallel relationship between field size, scatter and magnification factor. *J. Comp. Neurol.* 158: 295–306.

6 Huber, F. 1975. Sensory and neural mechanisms underlying acoustic communication in orthopteran insects. In *Sensory Physiology and Behavior.* R. Galun, P. Hillman, I. Parnas and R. Werman (eds.). Plenum Press, New York.

6 Hudspeth, A.J. 1982. Extracellular current flow and the site of transduction by vertebrate hair cells. *J. Neurosci.* 2: 1–10.

5 Hughes, A. 1977. The topography of vision in mammals of contrasting life style: Comparative optics and retinal organization. In *Handbook of Sensory Physiology.* Vol. VII-5 *The Visual System in Vertebrates.* Springer-Verlag, Berlin.

9 Ikeda, K. and Wiersma, C.A.G. 1964. Autogenic rhythmicity in the abdominal ganglia of the crayfish. The control of swimmeret movements. *Comp. Biochem. Physiol.* 12: 107–115.

2 Immelmann, K. 1980. *Introduction to Ethology* Plenum Press, New York.

7 Ingle, D. 1970. Visuomotor functions of the frog optic tectum. *Brain Behav. Evol.* 3: 57–71.

7 Ingle, D. 1973. Two visual systems in the frog. *Science* 181: 1053–1055.

7 Ingle, D. 1976. Spatial vision in anurans. In *The Amphibian Visual System: A Multidisciplinary Approach.* K.V. Fite (ed.). Academic Press, New York.

7 Ingle, D. and Sprague, J.M. (eds.). 1975. *Sensorimotor Function of the Midbrain Tectum. Neurosci. Res. Program. Bull.* 13.

10 Janikowska, E. 1979. New observations on neuronal organization of reflexes from tendon organ afferents and their relation to reflexes evoked from muscle spindle afferents. *Prog. Brain Res.* 50: 29–36.

2 Jeffords, M.R., Sternburg, J.G. and Waldbauer, G.P. 1979. Batesian mimicry: Field demonstration of the survival value of pipevine swallowtail and monarch color patterns. *Evolution* 33: 275–286.

9 Jennings, H.S. 1906. *Behavior of Lower Organisms.* Columbia University Press, New York.

8 Jurgens, V. 1974. The hypothalamus and behavioral patterns. *Prog. Brain Res.* 41: 445–464.

5 Kalmijn, A.J. 1978. Experimental evidence of geomagnetic orientation in elasmobranch fishes. In *Animal Migration, Navigation and Homing.* K. Schmidt-Koenig and W.T. Keeton (eds.). Springer-Verlag, New York.

9 Kammer, A.E. 1970. A comparative study of motor patterns during pre-flight warm-up in hawkmoths. *Z. vergl. Physiol.* 70: 45–56.

9 Kammer, A.E. 1971. The motor output during turning flight in a hawkmoth, *Manduca sexta. J. Insect Physiol.* 17: 1073–1086.

3,8 Kandel, E.R. 1976. *Cellular Basis of Behavior.* W.H. Freeman and Company, San Francisco.

3 Kandel, E.R. and Schwartz, J.H. (eds.). 1981. *Principles of Neural Science.* Elsevier/North-Holland, New York.

8 Kashin, S.M., Feldman, A.G. and Orlovski, G.N. 1974. Locomotion of fish evoked by electrical stimulation of the brain. *Brain Res.* 82: 41–47.

9,10 Keele, S.W. and Summers, J.J. 1976. The structure of motor programs. In *Motor Control: Issues and Trends.* G.E. Stelnach (ed.). Academic Press, New York.

5 Keeton, W.T. 1969. Orientation by pigeons: Is the sun necessary? *Science* 165: 922–928.

5 Keeton, W.T. 1971. Magnets interfere with pigeon homing. *Proc. Natl. Acad. Sci. USA* 68: 102–106.

5 Keeton, W.T. 1979a. Pigeon navigation. In *Neural Mechanisms of Behavior in the Pigeon.* A.M. Granda and J.H. Maxwell (eds.). Plenum Press, New York.

5 Keeton, W.T. 1979b. Avian orientation and navigation. *Annu. Rev. Physiol.* 41: 353–366.

5 Keeton, W.T. and Gobert, A. 1970. Orientation by untrained pigeons requires the sun. *Proc. Natl. Acad. Sci. USA* 65: 853–856.

2 Kennedy, D. 1976. Neural elements in relation to network function. In *Simpler Networks and Behavior.* J.C. Fentress (ed.). Sinauer Associates, Sunderland, Massachusetts.

10 Kennedy, D., Calabrese, R.L. and Wine, J.J. 1974. Presynaptic inhibition: Primary afferent depolarization in crayfish neurons. *Science* 186: 451–454.

8 Kennedy, D. and Takeda, K. 1965a. Reflex control of abdominal flexor muscles in the crayfish. I. The twitch system. *J. Exp. Biol.* 43: 211–227.

8 Kennedy, D. and Takeda, K. 1965b. Reflex control of abdominal flexor muscles in the crayfish. II. The tonic system. *J. Exp. Biol.* 43: 229–246.

6 Khanna, S.M. and Leonard, D.G.B. 1982. Basilar membrane tuning in the cat cochlea. *Science* 215: 305–306.

5 King-Smith, P.E. 1969. Absorption spectra and function of the coloured oil drops in the pigeon retina. *Vision Res.* 9: 1391–1399.

5 Kirschfeld, K., Lindauer, M. and Martin, H. 1975. Problems of menotactic orientation according to the polarized light of the sky.. *Z. Naturforsch.* 30: 88–90.

6 Kleindienst, H.-V., Koch, V.T. and Wohlers, D.W. 1981. Analysis of the cricket auditory system by acoustic stimulation using a closed sound field. *J. Comp. Physiol.* 141: 283–296.

10 Klinke, R. 1970. Efferent influence on the vestibular organ during active movements of the body. *Pflugers Arch.* 318: 325–332.

5 Knowles, A. and Dartnall, J.A. 1977b. The visual pigment in the receptor. In *The Eye*. Vol. 2B. *The Physiology of Vision.* H. Davson (ed.). Academic Press, New York.

6 Knudsen, E.I. 1980. Sound localization in birds. In *Comparative Studies of Hearing in Vertebrates.* A.N. Popper and R.R. Fay (eds.). Springer-Verlag, New York.

7 Knudsen, E.I. 1981. The hearing of the barn owl. *Sci. Am.* 245(6): 113–125.

7 Knudsen, E.I. 1982. Auditory and visual maps of space in the optic tectum of the owl. *J. Neurosci.* 2: 1177–1194.

7 Knudsen, E.I. and Konishi, M. 1978a. A neural map of auditory space in the owl. *Science* 200: 795–797.

7 Knudsen, E.I. and Konishi, M. 1978b. Center–surround organization of auditory receptive fields in the owl. *Science* 202: 778–780.

7 Knudsen, E.I. and Konishi, M. 1978c. Space and frequency are represented separately in the auditory midbrain of the owl. *J. Neurophysiol.* 41: 870–884.

7 Knudsen, E.I. and Konishi, M. 1979. Sound localization by the barn owl measured with the search coil technique. *J. Comp. Physiol.* 133: 1–11.

9 Koester, J. and Byrne, J.H. 1982. *Molluscan Nerve Cells: From Biophysics to Behavior.* Cold Spring Harbor Labs, Cold Spring Harbor, New York.

7 Koffka, K. 1935. *Principles of Gestalt Psychology.* Harcourt Brace, New York.

2 Konishi, M. 1965. The role of auditory feedback in the control of vocalization in the white-crowned sparrow. *Z. Tierpsychol.* 22: 770–783.

7 Konishi, M. 1973. How the owl tracks its prey. *Am. Sci.* 61: 414–424.

6 Konishi, M. 1977. Spatial localization of sound. In *Recognition of Complex Acoustic Signals.* T.H. Bullock (ed.). Dahlem Konferenzen, Berlin.

5 Korf, H.W. 1976. Histological, histochemical and electron microscopical studies of the nervous apparatus of the pineal organ in the tiger salamander, *Ambystoma tigrinum. Cell Tissue Res.* 174: 475–497.

5 Kramer, C. 1957. Experiments on bird orientation and their interpretation. *Ibis* 99: 196–227.

2 Krebs, J.R. and Davies, N.B. 1978. *Behavioural Ecology: An Evolutionary Approach.* Sinauer Associates, Sunderland, Massachusetts.

2 Krebs, J.R. and Davies, N.B. 1981. *An Introduction to Behavioural Ecology.* Sinauer Associates, Sunderland, Massachusetts.

5 Kreithen, M.L. 1978. Sensory mechanisms for animal orientation—Can any new ones be discovered? In *Animal Migration, Navigation and Homing.* K. Schmidt-Koenig and W.T. Keeton (eds.). Springer-Verlag, New York.

5 Kreithen, M.L. and Eisner, T. 1978. Ultraviolet light detection by the homing pigeon. *Nature* 272: 347–348.

5 Kreithen, M.L. and Keeton, W.T. 1974. Detection of polarized light by the homing pigeon, *Columba livia. J. Comp. Physiol.* 89: 83–92.

6 Kreithen, M.L. and Quine, D.B. 1979. Infrasound detection by the homing pigeon: A behavioral audiogram. *J. Comp. Physiol.* 129: 1–4.

9 Kristan, W.B. Jr. 1974. Neural control of swimming in the leech. *Am. Zool.* 14: 991–1001.

9 Kristan, W.B. Jr. and Calabrese, R.L. 1976. Rhythmic swimming activity in neurones of the isolated nerve cord of the leech. *J. Exp. Biol.* 65: 643–668.

9,10 Kristan, W.B. Jr. and Stent, G. 1974. Peripheral feedback in the leech swimming rhythm. *Cold Spring Harbor Symp. Quant. Biol.* 40: 663–674.

9 Kristan, W.B. Jr., Stent, G.S. and Ort, C.A. 1974. Neuronal control of swimming in the medicinal leech. I. Dynamics of the swimming rhythm. *J. Comp. Physiol.* 94; 97–119.

10 Kuffler, S.W. 1960. Excitation and inhibition in single nerve cells. In *The Harvey Lectures 1958–1959*. Academic Press, New York.

3,5,7 Kuffler, S.W. and Nicholls, J.G. 1976. *From Neuron to Brain*. Sinauer Associates, Sunderland, Massachusetts.

8 Kupferman, I. and Weiss, K.R. 1978. The command neuron concept. *Behav. Brain Sci.* 1: 3–10.

8 Kusano, K. and Grundfest, H. 1965. Circus reexcitation as a cause of repetitive activity in crayfish lateral giant axons. *J. Cell Comp. Physiol.* 65: 325–336.

8 Kuwada, J.Y., Hagiwara, G. and Wine, J.J. 1980. Postsynaptic inhibition of crayfish tonic flexor motor neurons by escape commands. *J. Exp. Biol.* 85: 343–347.

8 Kuwada, J.Y. and Wine, J.J. 1979. Crayfish escape behavior: Commands for fast movement inhibit postural tone and reflexes and prevent habituation of slow reflexes. *J. Exp. Biol.* 79: 205–224.

5 Land, E.H. 1977. The retinex theory of color vision. *Sci. Am.* 237(6): 108–129.

10 Land, M.F. 1975. Similarities in the visual behavior of arthropods and men. In *Handbook of Psychobiology*. M.S. Gazzaniga and C. Blakemore (eds.). Academic Press, New York.

6 Larsen, O.N. 1981. Mechanical time resolution in some insect ears. II. Impulse sound transmission in acoustic tracheal tubes. *J. Comp. Physiol.* 143: 297–304.

5 Laughlin, S.B., Menzel, R. and Snyder, A.W. 1975. Membranes, dichroism and receptor sensitivity. In *Photoreceptor Optics*. A.W. Snyder and R. Menzel (eds.). Springer-Verlag, New York. pp. 237–259.

9 Lemon, R.E. 1977. Bird song: An acoustic flag. *Bioscience* 27: 402–407.

8 Lennard, P.R. and Stein, P.S.G. 1977. Swimming movements elicited by electrical stimulation of turtle spinal cord. I. Low-spinal and intact preparations. *J. Neurophysiol.* 40: 768–778.

1 Leppelsack, H.-J. 1978. Unit responses to species-specific sounds in the auditory forebrain center of birds. *Fed. Proc.* 37: 2336–2341.

7 Lettvin, J.Y., Maturana, H.R., McCulloch, W.S. and Pitts, W.H. 1959. What the frog's eye tells the frog's brain. *Proc. Inst. Radio Eng.* 47: 1940–1951.

9 Lewis, E.R. 1968. Using electronic circuits to model simple neuroelectric interactions. *Proc. Inst. Elec. Electron Engrs.* 56: 931–949.

5 Liebman, P.A. 1962. *In situ* microspectrophotometric studies on the pigments of single retinal rods. *Biophys. J.* 2: 161–178.

6 Lim, D.J. 1980. Cochlear anatomy related to cochlear micromechanics. A review. *J. Acoust. Soc. Am.* 67: 1686–1695.

5 Lindauer, M. 1961. *Communication Among Social Bees*. Harvard University Press, Cambridge, Massachusetts.

8 Livingstone, M.S., Harris-Warwick, R.M. and Kravitz, E.A. 1980. Serotonin and octopamine produce opposite postures in lobsters. *Science* 208: 76–79.

10 Lloyd, D.P.C. 1943. Conduction and synaptic transmission of reflex response to stretch in spinal cats. *J. Neurophysiol.* 6: 317–326.

2 Loeb, J. 1964. *The Mechanistic Conception of Life.* Harvard University Press, Cambridge, Massachusetts. (Reprint of the original 1912 edition.)

1 Lorenz, K. 1960. Foreward to *The Herring Gull's World.* N. Tinbergen. Harper & Row, New York.

1 Lorenz, K. 1970. *Studies in Animal and Human Behavior.* Vols. I and II. Harvard University Press, Cambridge, Massachusetts.

2 Lorenz, K. 1981. *The Foundations of Ethology.* Springer-Verlag, New York.

2 Lorenz, K. and Tinbergen, N. 1938. Taxis und Instinkthandlung in der Eirollbewegung der Graugans. *Z. Tierpsychol.* 2: 1–29.

10 Lundberg, A. 1979. Multisensory control of spinal reflex pathways. *Prog. Brain. Res.* 50: 11–28.

5 Lythgoe, J.N. 1979. *The Ecology of Vision.* Clarendon Press, Oxford.

7 Manley, J.A. and Mueller-Preuss, P. 1978. Response variability in the mammalian auditory cortex: An objection to feature detection? *Fed. Proc.* 37: 2355–2359.

2 Manning, A. 1979. *Introduction to Animal Behavior.* Addison-Wesley, Reading, Massachusetts.

6 Markl, H. and Tautz, J. 1975. The sensitivity of hair receptors in caterpillars of *Barathra brassicae* L. (Lepidoptera, Noctuidae) to particle movement in a sound field. *J. Comp. Physiol.* 99: 79–87.

2 Marler, P. and Hamilton, W.J. III. 1966. *Mechanisms of Animal Behavior.* John Wiley & Sons, New York.

5 Martin, G.R. and Muntz, W.R.A. 1978. Spectral sensitivity of the red and yellow oil droplet fields of the pigeon *(Columba livia). Nature* 274: 620–621.

10 Matthews, P.B.C. 1972. *Mammalian Muscle Receptors and their Central Actions.* Edward Arnold Ltd., London.

5 Mautz, D. 1971. Der Kommunikationseffekt der Schwanzeltanze bei *Apis mellifera carnica* (polln.). *Z. vergl. Physiol.* 72: 197–220.

6 McCosker, J.E. 1977. Flashlight fishes. *Sci. Am.* 236(3): 106–113.

10 McFarland, D.J. 1971. *Feedback Mechanisms in Animal Behavior.* Academic Press, London.

9 Meech, R.W. 1979, Membrane potential oscillations in molluscan "burster" neurones. *J. Exp. Biol.* 81: 93–112.

5 Menzel, R. 1975. Polarization sensitivity in insect eyes with fused rhabdoms. In *Photoreceptor Optics.* A.W. Snyder and R. Menzel (eds.). Springer-Verlag, New York. pp. 372–387.

5 Menzel, R. 1979. Spectral sensitivity and color vision in invertebrates. In *Handbook of Sensory Physiology.* Vol. VII 6A. *Comparative Physiology and Evolution of Vision in Invertebrates: Invertebrate Photoreceptors.* H. Autrum (ed.). Springer-Verlag, Berlin.

5 Menzel, R. and Blakers, M. 1976. Color receptors in the bee eye—Morphology and spectral sensitivity. *J. Comp. Physiol.* 108: 11–13.

5 Menzel, R. and Snyder, A.W. 1974. Polarized light detection in the bee, *Apis mellifera. J. Comp. Physiol.* 88: 247–270.

10 Merton, P.A. 1972. How we control the contraction of our muscles. *Sci. Am.* 226(5): 30–37.

10 Messenger, J.B. 1968. The visual attack of the cuttlefish, *Sepia officinalis. Anim. Behav.* 16: 342–357.

6 Michelsen, A. 1979. Insect ears as mechanical systems. *Am. Sci.* 67: 696–706.

6 Michelsen, A. and Nocke, H. 1974. Biophysical aspects of sound communication in insects. *Adv. Insect Physiol.* 10: 247–296.

10 Miles, F.A. and Evarts, E.V. 1979. Concepts of motor organization. *Annu. Rev. Psychol.* 30: 327–362.

7 Miller, R.F. and Dacheux, R.F. 1976. Synaptic organization and ionic basis of on and off channels in mudpuppy retina. I. Intracellular analysis of chloride-sensitive electrogenic properties of receptors, horizontal cells, bipolar cells and amacrine cells. *J. Gen. Physiol.* 67: 639–659.

10 Mittelstaedt, H. 1951. Zur Analyze physiologischer Regelungssystems. *Verh. Dtsch. Ges. Zool.* (Wilhelmshaven) 150–157.

10 Mittelstaedt, H. 1964. Basic control patterns of orientation homeostasis. *Symp. Soc. Exp. Biol.* 18: 365–385.

7 Moiseff, A. and Konishi, M. 1981. Neuronal and behavioral sensitivity to binaural time differences in the owl. *J. Neurosci.* 1: 40–48.

6 Moiseff, A., Pollack, G.S. and Hoy, R.R. 1978. Steering responses of flying crickets to sound and ultrasound: Mate attraction and predator avoidance. *Proc. Natl. Acad. Sci. USA* 74: 4025–4056.

6 Morin, J.G., Harrington, A., Nealson, K., Kreiger, N., Blad, T.O. and Hastings, J.W. 1975. Light for all reasons: Versatility in the behavioral repertoire of the flashlight fish. *Science* 190: 74–76.

7 Mort, E., Finlay, B.L. and Cairns, F. 1980. The role of the superior colliculus in visually-guided locomotion and visual orienting in the hamster. *Physiol. Psychol.* 8: 20–28.

8 Mountcastle, V.B. 1976. The world around us: Neural command functions for selective attention. *Neurosci. Res. Program Bull.* 14. [Suppl.]: 1–47.

9 Muller, K.J. and McMahan, U.J. 1976. The shapes of sensory and motor neurons and the distribution of their synapses in ganglia of the leech: A study using intracellular injections of horseradish peroxidase. *Proc. R. Soc. Lond.* B 194: 481–499.

9 Muller, K.J., Nicholls, J.G. and Stent, G.S. 1982. *Neurobiology of the Leech.* Cold Spring Harbor Labs, Cold Spring Harbor, New York.

5 Muller-Schwarze, D. and Mozell, M.M. 1977. *Chemical Signals in Vertebrates.* Plenum Press, New York.

7 Muntz, W.R.A. 1964. Vision in frogs. *Sci. Am.* 210(3): 110–119.

7 Murphey, R.K. 1981. The structure and development of a somatotopic map in crickets: The cercal afferent projection. *Dev. Biol.* 88: 236–246.

7 Murphey, R.K., Jocklet, A. and Schuster, L. 1980. A typographic map of sensory cell terminal arborizations in the cricket CNS: Correlation with birthday and position in a sensory array. *J. Comp. Neurol.* 191: 53–64.

5 Naka, K.-I. and Eguchi, E. 1962. Spike potentials recorded from the insect photoreceptor. *J. Gen. Physiol.* 45: 663–680.

7 Nauta, W.J.H. and Karten, H.J. 1970. A general profile of the vertebrate brain, with sidelights on the ancestry of cerebral cortex. In *The Neurosciences: Second Study Program.* F.O. Schmidt (ed.). The Rockefeller University Press, New York.

6 Neuweiler, G. 1980. Auditory processing of echoes: Peripheral processing. In *Animal Sonar Systems.* R.G. Busnel and J.F. Fish (eds.). Plenum Press, New York.

3,5 Newman, E.A. and Hartline, P.H. 1982. The infrared "vision" of snakes. *Sci. Am.* 246(3): 116–124.

9 Nicholls, J.G. and Wallace, B. 1978. Modulation of transmitter release at an inhibitory synapse in the CNS of the leech. *J. Physiol.* (Lond.) 281: 157–170.

10 Nichols, T.R. and Hauk, J.C. 1976. Improvement in linearity and regulation of stiffness that results from actions of the stretch reflex. *J. Neurophysiol.* 39: 119–142.

4 Nicklaus, R. 1965. Die Erregung enzelner Fadenhaare von *Periplaneta americana* in Abhangigkeit van der Grosse und Richtung der Auslenkung. *Z. vergl. Physiol.* 50: 331–362.

5 Noble, G.K. and Schmidt, A. 1937. Structure and function of the facial and labial pits of snakes. *Proc. Am. Phil. Soc.* 77: 263–288.

2 Notobohm, F. 1970. Ontogeny of bird song. *Science* 167: 950–956.

5 Obara, Y. 1970. Studies on the mating behavior of the white cabbage butterfly, *Pieris rapae crucivora* Boisduval. III. Near-ultra-violet reflection as the signal of intraspecific communication. *Z. vergl. Physiol.* 69: 99–116.

10 Olivo, R.F and Jazak, M.M. 1980. Proprioception provides a major input to the horizontal oculomotor system of crayfish. *Vision Res.* 20: 349–353.

8 Olson, G.C. and Krasne, F.B. 1981. The crayfish lateral giants are command neurons for escape behavior. *Brain Res.* 214: 89–100.

7 Olton, D.S., Becker, J.T. and Handelmark, G.E. 1979. Hippocampus, space and memory. *Behav. Brain Sci.* 2: 313–365.

1,7 O'Neill, W.E. and Suga, N. 1979. Target range-sensitive neurons in the auditory cortex of the mustache bat. *Science* 203: 69–73.

1,2,6 O'Neill, W.E. and Suga, N. 1982. Encoding of target range and its representation in the auditory cortex of the mustached bat. *J. Neurosci.* 2: 17–31.

9 Ort, C.A., Kristan, W.B. Jr. and Stent, G.S. 1974. Neuronal control of swimming in the medicinal leech. II. Identification and connections of motor neurons. *J. Comp. Physiol.* 94: 121–154.

10 Ottoson, D. 1976. Morphology and physiology of muscle spindles. In *Frog Neurobiology.* R. Llinas and W. Precht (eds.). Springer-Verlag, Berlin.

5 Papi, F., Keeton, W.T., Brown, A.I. and Benvenuti, S. 1978. Do American and Italian pigeons rely on different homing mechanisms? *J. Comp. Physiol.* 128: 303–317.

2 Pavlov, I.P. 1960. *Conditioned Reflexes: An Investigation of the Physiological Activity of the Cerebral Cortex.* Dover Publications, New York. (Reprint of original 1927 edition.)

7 Payne, R.S. 1971. Acoustic location of prey by barn owls *(Tyto alba). J. Exp. Biol.* 54: 535–537.

8 Payton, B.W., Bennett, M.V.L. and Pappas, G.V. 1969. Permeability and structure of junctional membranes at an electrical synapse. *Science* 166: 1641–1643.

8 Pearson, K.G. 1972. Central programming and reflex control of walking in the cockroach. *J. Exp. Biol.* 56: 173–193.

10 Pearson, K.G. and Duysens, J. 1976. Function of segmental reflexes in the control of stepping in cockroaches and cats. In *Neural Control of Locomotion.* R.M. Herman, S. Grillner, P.S.G. Stein and D.G. Stuart (eds.). Plenum Press, New York.

8 Pearson, K.G. and Fourtner, C.R. 1975. Non-spiking interneurons in walking system of the cockroach. *J. Neurophysiol.* 38: 33–52.

10 Pearson, K., Wong, R. and Fourtner, C. 1976. Connexions between hair-plate afferents and motoneurones in the cockroach leg. *J. Exp. Biol.* 64: 251–266.

9 Perkel, D.H. and Mulloney, B. 1974. Motor pattern production in reciprocally inhibitory neurons exhibiting postinhibitory rebound. *Science* 185: 181–183.

7 Perrett, D.I., Rolls, E.T. and Caan, W. 1982. Visual neurones responsive to faces in the monkey temporal cortex. *Exp. Brain Res.* 47: 329–342.

10 Phillips, C.G. 1969. Motor apparatus of the baboon's hand. *Proc. R. Soc. Lond.* B 173: 141–174.

5 Phillips, J.B. 1977. Use of the earth's magnetic field by orienting cave salamanders *(Eurycea lucifuga). J. Comp. Physiol.* 121: 273–288.

6 Pierce, G.W. and Griffin, D.R. 1938. Experimental determination of supersonic notes emitted by bats. *J. Mammology* 19: 454–455.

2 Pietrewicz, A.T. and Kamil, A.C. 1977. Visual detection of cryptic prey by blue jays *(Cyanocitta cristata)*. *Science* 195: 580–582.

6 Platt, C. and Popper, A.N. 1981. Fine structure and function of the ear. In *Hearing and Sound Communication in Fish.* W.N. Tavolga, A.N. Popper and R.R. Fay (eds.). Springer-Verlag, New York.

4 Plummer, M.R. and Camhi, J.M. 1981. Discrimination of sensory signals from noise in the escape system of the cockroach: The role of wind acceleration. *J. Comp. Physiol.* 142: 347–357.

6 Pollak, G., Henson, O.W. and Novick, A. 1972. Cochlear microphonic audiograms in the "pure tone" bat *Chilonycteris parnellii parnellii. Science* 176: 66–68.

2 Pollak, G.S. and Hoy, R.R. 1979. Temporal pattern as a cue for species-specific calling song recognition in crickets. *Science* 204: 429–432.

8 Poon, M.L.T. 1980. Induction of swimming in lamprey by L-DOPA and amino acids. *J. Comp. Physiol.*136: 337–344.

9 Poon, M., Friesen, W.O. and Stent, G.S. 1978. Neuronal control of swimming in the medicinal leech. V. Connexions between the oscillatory interneurones and the motor neurones. *J. Exp. Biol.* 75: 45–63.

6 Popov, A.V. and Shuvalov, V.P. 1977. Phonotactic behavior of crickets. *J. Comp. Physiol.* 119: 111–126.

6 Popper, A.N. and Fay, R.R. 1980. *Comparative Studies of Hearing in Vertebrates.* Springer-Verlag, New York.

5 Prosser, C.L. 1973. *Comparative Animal Physiology.* W.B. Saunders, Philadelphia.

8,9 Reingold, S.C. and Camhi, J.M. 1977. A quantitative analysis of rhythmic leg movements during three different behaviors in the cockroach *Periplaneta americana. J. Insect Physiol.* 23: 1407–1420.

6 Rheinlaender, J., Gerhardt, H.C., Yager, D.D., and Capranica, R.R. 1979. Accuracy of phonotaxis by the green treefrog *(Hyla cinerea). J. Comp. Physiol.* 133: 247–255.

6 Rheinlaender, J. Walkowiek, W. and Gerhardt, H.C. 1981. Directional hearing in the green treefrog: A variable mechanism? *Naturwissenschaften* 67 S: 430.

5 Ribi, W.A. 1975. The first optic ganglion of the bee. I. Correlation between visual cell types and their terminals in the lamina and medulla. *Cell Tissue Res.* 165: 103–111.

5 Ribi, W.A. 1979. Do the rhabdomeric structures in bees and flies really twist? *J. Comp. Physiol.* 134: 109–112.

10 Richmond, B.J. and Wurtz, R.H. 1980. Vision during saccadic eye movements. II. A corollary discharge to monkey superior colliculus. *J. Neurophysiol.* 43: 1156–1167.

7 Ristau, C.A. and Robbins, D. 1981. Cognitive aspects of ape language experiments. In *Animal Mind—Human Mind.* D.R. Griffin (ed.). Academic Press, New York.

10 Ritzmann, R.E. 1974. Mechanisms for the snapping behavior of two Alpheid shrimp, *Alpheus californiensis* and *Alpheus heterochelis. J. Comp. Physiol.* 95: 217–236.

4 Ritzmann, R.E. 1981. Motor responses to paired stimulation of giant interneurons in the cockroach *Periplaneta americana.* II. The ventral interneurons. *J. Comp. Physiol.* 143: 71–80.

4 Ritzmann, R.E. and Camhi, J.M. 1978. Excitation of leg motor neurons by giant interneurons in the cockroach *Periplaneta americana. J. Comp. Physiol.* 125: 305–316.

4 Ritzmann, R.E. and Pollack, A.J. 1981. Motor responses to paired stimulation of giant interneurons in the cockroach. I. The dorsal interneurons. *J. Comp. Physiol.* 143: 61–70.

4 Ritzmann, R.E., Tobias, M.L. and Fourtner, C.R. 1980. Flight activity initiated via giant interneurons of the cockroach: Evidence for bifunctional trigger interneurons. *Science* 210: 443–445.

10 Robinson, D.L. and Wurtz, R.H. 1976. Use of an extraretinal signal by monkey superior colliculus neurons to distinguish real from self-induced stimulus movement. *J. Neurophysiol.* 39: 852–870.

7 Rock, I. 1973. *Orientation and Form.* Academic Press, New York.

8 Rock, M.K., Hackett, J.T. and Brown, D.L. 1981. Does the Mauthner cell conform to the criteria of the command neuron concept? *Brain Res.* 204: 21–23.

7 Rodieck, R.W. 1973. *The Vertebrate Retina: Principles of Structure and Function.* W.H. Freeman and Company, San Francisco.

4 Roeder, K.D. 1948. Organization of the ascending giant fiber system of the cockroach *(Periplaneta americana). J. Exp. Zool.* 108: 243–261.

4,8 Roeder, K.D. 1963. *Nerve Cells and Insect Behavior.* Harvard University Press, Cambridge, Massachusetts.

7 Romer, A.S. 1962. *The Vertebrate Body.* W.B. Saunders, Philadelphia.

5 Rossel, S., Wehner, R. and Lindauer, M. 1978. E-vector orientation in bees. *J. Comp. Physiol.* 125: 1–12.

4 Roth, L. and Willis, E. 1960. The biotic associations of cockroaches. *Smithsonian Misc. Coll.* 141.

2 Rovner, J.S. and Barth, F.G. 1981. Vibratory communication through living plants by a tropical wandering spider. *Science* 214: 464–466.

9 Russell, D.F. and Hartline, D.K. 1978. Bursting neural networks: A reexamination. *Science* 200: 453–456.

9 Russell, D.F. and Hartline, D.K. 1981. A multifunction synapse evoking both EPSPs and enhancement of endogenous bursting. *Brain Res.* 223: 19–38.

6,10 Russell, I.J. 1976. Amphibian lateral line receptors. In *Frog Neurobiology—A Handbook.* R. Llinas and W. Precht (eds.). Springer-Verlag, Berlin.

6 Russell, I.J. and Sellick, P.M. 1977. Tuning properties of cochlear hair cells. *Nature* 267: 858–860.

5 Rutowski, R. 1977. The use of visual cues in sexual and species discrimination by males of the small sulphur butterfly *Eurema lisa* (Lepidoptera, Pieridae). *J. Comp. Physiol.* 115: 61–74.

6 Saito, N. 1980. Structure and function of the avian ear. In *Comparative Studies of Hearing in Vertebrates.* A.N. Popper and R.R. Fay (eds.). Springer-Verlag, New York.

5 Salvini-Plawen, L.V. and Mayr, E. 1977. On the evolution of photoreceptors and eyes. *Evol. Biol.* 10: 207–263.

5 Sauer, E.G.F. and Sauer, E.M. 1960. Star navigation of nocturnal migrating birds. *Cold Spring Harbor Symp. Quant. Biol.* 25: 463–473.

4 Schal, C. 1982. Intraspecific vertical stratification as a mate-finding mechanism in tropical cockroaches. *Science* 215: 1405–1407.

7 Schiller, P.H. and Stryker, M. 1972. Single-unit recording and stimulating in superior colliculus of the alert rhesus monkey. *J. Neurophysiol.* 35: 915–924.

7 Schiller, P.H., Stryker, M., Cynader, M. and Berman, N. 1974. Response characteristics of single cells in the monkey superior colliculus following ablation or cooling of visual cortex. *J. Neurophysiol.* 37: 181–194.

3 Schmidt, F.O. and Worden, F.G. (eds.). 1979. *The Neurosciences: Fourth Study Program.* M.I.T. Press, Cambridge, Massachusetts.

10 Schmidt, R.F. and Willis, W.D. 1963. Intracellular recording from motoneurons of the cervical spinal cord of the cat. *J. Neurophysiol.* 26: 28–43.

5 Schmidt-Koenig, K. 1960. Internal clocks and homing. *Cold Spring Harbor Symp. Quant. Biol.* 25: 389–393.

5 Schmidt-Koenig, K. and Keeton, W.T. 1978. *Animal Migration, Navigation and Homing.* Springer-Verlag, New York.

2,5 Schneider, D. 1974. The sex-attractant receptor of moths. *Sci. Am.* 231(1): 28–35.

7 Schneider, G. 1969. Two visual systems: Brain mechanisms for localization and discrimination are dissociated by tectal and cortical lesions. *Science* 163: 895–902.

6 Schnitzler, H.U. 1968. Die Ultraschall-Ortungslaute der Hufeisen-Fledermause (Chiroptera-Rhinolophidae) in verschiedenen Orientierungssituationen. *Z. vergl. Physiol.* 57: 376–408.

6 Schnitzler, H.U. 1970. Echoortung bei der Fledermaus *Chilonycteris rubiginosa. Z. vergl. Physiol.* 68: 25–38.

6 Schnitzler, H.U. and Henson, O.D.W. Jr. 1980. Performance of airborne animal sonar systems. I. Microchiroptera. In *Animal Sonar Systems.* R.G. Busnel and J.F. Fish (eds.). Plenum Press, New York.

10 Schone, H. 1961. Complex behavior. In *The Physiology of Crustacea.* Vol. II. T.H. Waterman (ed.). Academic Press, New York.

6 Schuller, G. 1979. Coding of small sinusoidal frequency and amplitude modulations in the inferior colliculus of "CF–FM" bat, *Rhinolophus ferrumequinum. Exp. Brain Res.* 34: 117–132.

6 Schuller, G., Beuter, K. and Rubsamen, R. 1975. Dynamic properties of the compensation system for Doppler shifts in the bat, *Rhinolophus ferrumequinum. J. Comp. Physiol.* 97: 113–125.

3 Scientific American. 1979. *The Brain. Sci. Am.* 241(3).

10 Sears, T.A. 1964. Efferent discharges in alpha and fusimotor fibres of intercostal nerves of the cat. *J. Physiol.* 174: 295–315.

2 Sebeok, T.A. (ed.). *How Animals Communicate.* Indiana University Press, Bloomington, Indiana.

9 Selverston, A. 1976. A model system for the study of rhythmic behavior. In *Simpler Networks and Behavior.* J.C. Fentress (ed.). Sinauer Associates, Sunderland, Massachusetts.

9,10 Selverston, A.I. 1980. Are central pattern generators understandable? *Behav. Brain. Sci.* 3: 535–572.

9 Selverston, A.I., Russell, D.F., Miller, J.P. and King, D.G. 1976. The stomatogastric nervous system: Structure and function of a small neural network. *Prog. Neurbiol.* 6: 1–75.

10 Severin, F.V., Orlovskii, G.N. and Shik, M.L. 1967. Work of the muscle receptors during controlled locomotion. *Biophysics.* 12: 575–586. (Translation of *Biofizika* 12: 502–511).

7 Shepherd, G.M. 1979. *The Synaptic Organization of the Brain,* Second Edition. Oxford University Press, New York.

8,9 Sherman, E., Novotny, M. and Camhi, J.M. 1977. A modified walking rhythm employed during righting behavior in the cockroach *Gromphadorhina portentosa. J. Comp. Physiol.* 113: 303–316.

2,8 Sherrington, C. 1961. *The Integrative Action of the Nervous System.* Yale University Press, New Haven, Connecticut. (Reprint of the original 1906 edition.)

8 Shik, M.L., Severin, F.V. and Orlovski, G.N. 1966. Control of walking and running by means of electrical stimulation of the midbrain. *Biofizika* 11: 659–666.

5 Silberglied, R.E. 1979. Communication in the ultraviolet. *Annu. Rev. Ecol. Syst.* 10: 373–398.

2 Silberglied, R.E., Aiello, A. and Windsor, D.M. 1980. Disruptive coloration in butterflies: Lack of support in *Anartia fatima*. *Science* 209: 617–619.

2,6 Simmons, J.A. 1973. The resolution of target range by echolocating bats. *J. Acoust. Soc. Am.* 54: 157–173.

6 Simmons, J.A. 1974. Response of the Doppler echolocation system in the bat, *Rhinolophus ferrumequinum*. *J. Acoust. Soc. Am.* 56: 672–682.

6 Simmons, J.A. 1979. The perception of phase information in bat sonar. *Science* 204: 1336–1338.

6 Simmons, J.A., Fenton, B.M. and O'Farrell, M.J. 1979. Echolocation and pursuit of prey by bats. *Science* 203: 16–21.

6 Simmons, J.A., Howell, D.J. and Suga, N. 1975. Information content of bat sonar echoes. *Am. Sci.* 63: 204–215.

6 Simmons, J.A. and Vernon, J.A. 1971. Echolocation: Discrimination of targets by the bat, *Eptesicus fuscus*. *J. Exp. Zool.* 176: 315–328.

2 Smith, S.M. 1975. Innate recognition of coral snake pattern by a possible avian predator. *Science* 187: 759–760.

2 Smith, S.M. 1977. Coral snake pattern recognition and stimulus generation by naive great kiskadees (Aves: Tyrannidae). *Nature* 265: 535–536.

5 Synder, A.W. 1975. Optical properties of invertebrate physiology. In *The Compound Eye and Vision of Insects*. G.A. Horridge (ed.). Oxford University Press, London. pp. 179–236.

7 Sperry, R.W. 1943. Visuomotor coordination in the newt. *(Triturus viridescens)* after regeneration of the optic nerve. *J. Comp. Neurol.* 79: 33–55.

10 Sperry, R.W. 1950. Neural basis of the spontaneous optokinetic response produced by visual inversion. *J. Comp. Physiol. Psych.* 43: 482–489.

7 Sperry, R.W. 1975. In search of psyche. In *The Neurosciences: Paths of Discovery*. F.G. Worden, J.P. Swazey and G. Adelman (eds.). M.I.T. Press, Cambridge, Massachusetts.

4 Spira, M.E., Parnas, I. and Bergman, F. 1969. Organization of the giant axons of the cockroach *Periplaneta americana*. *J. Exp. Biol.* 50: 615–627.

8 Steeves, J.D., Jordan, L.M. and Lake, N. 1976. The close proximity of catecholamine-containing cells to the "mesencephalic locomotor region" (MLR). *Brain Res.* 100: 663–670.

7 Stein, B.E., Magalhaes-Castro, B. and Kruger, L. 1976. Relationship between visual and tactile representation in cat superior colliculus. *J. Neurophysiol.* 39: 401–419.

9 Stein, P.S.G. 1971. Intersegmental coordination of swimmeret motoneuron activity in crayfish. *J. Neurophysiol.* 34: 310–318.

8 Stein, P.S.G. 1978. Swimming movements elicited by electrical stimulation of the turtle spinal cord: The high spinal preparation. *J. Comp. Physiol.* 124: 203–210.

8 Stein, R.B. 1980. *Nerve and Muscle: Membranes, Cells, and Systems*. Plenum Press, New York.

2 Steinberg, J.B. 1977. Information theory as an ethological tool. In *Quantitative Methods in the Study of Animal Behavior*. B.A. Hazlett (ed.). Academic Press, New York.

2 Sternburg, J.G., Waldbauer, G.P. and Jeffords, M.R. 1977. Batesian mimicry: Selective advantage of color pattern. *Science* 195: 681–683.

5 Stork, D.G. and Levinson, J.Z. 1982. Receptive fields and the optimal stimulus. *Science* 216: 204–205.

6 Suga, N. 1965. Functional properties of auditory neurones in the cortex of echo-locating bats. *J. Physiol.* (Lond.) 181: 671–700.

6 Suga, N., Neuweiller, G. and Moller, J. 1976. Peripheral auditory tuning for fine frequency analysis by the CF–FM bat, *Rhinolophus ferrumequinum*. IV. Properties of peripheral auditory neurons. *J. Comp. Physiol.* 106: 111–125.

6 Suga, N. and O'Neill, W.E. 1979. Auditory processing of echoes: Representation of acoustic information from the environment in the bat cerebral cortex. In *Animal Sonar Systems.* R.-G. Busnel and J.F. Fish (eds.). Plenum Press, New York.

6 Suga, N. and O'Neill, W.E. 1980. Auditory processing of echoes: Representation of acoustic information from the environment in the bat cerebral cortex. In *Animal Sonar Systems.* R.-G. Busnel and J.F. Fish (eds.). Plenum Press, New York.

6 Suga, N., O'Neill, W.E. and Manabe, T. 1978. Cortical neurons sensitive to combinations of information-bearing elements of bisonar signals in the mustache bat. *Science* 200: 778–781.

10 Suga, N. and Shimozawa, T. 1974. Site of neural attenuation of responses to self-vocalized sounds in echolocating bats. *Science* 183: 1211–1213.

6 Suga, N., Simmons, J.A. and Jen, P.H.-S. 1975. Peripheral specialization for fine analysis of Doppler-shifted echoes in the auditory system of the "CF–FM" bat *Pteronotus parnellii. J. Exp. Biol.* 63: 161–192.

9 Sustare, D. 1978. Systems diagrams. In *Quantitative Ethology.* P. Colgen (ed.). Wiley, New York.

5 Swihart, C.A. 1971. Color discrimination by the butterfly *Heliconius charitonius* Linn. Amin. Behav. 19: 156–164.

7 Symmes, D., Newman, J.D. and Kojima, S. 1979. Evidence for feature detection in cortical auditory neurons of squirrel monkey. *Neurosci. Abst.* 5: 1601.

9,10 Taub, E. 1976. Motor behavior following deafferentation in the developing and motorically mature monkey. In *Neural Control of Locomotion.* R.M. Herman, S. Grillner, P.S.G. Stein and D.G. Stuart (eds.). Plenum Press, New York.

6 Tautz, J. 1979. Reception of particle oscillation in a medium—An unorthodox sensory capacity. *Naturwissenschaften* 66: 452–461.

6 Tautz, J. and Markl, H. 1978. Caterpillars detect flying wasps by hairs sensitive to airborne vibration. *Behav. Ecol. Sociobiol.* 4: 101–110.

5 Taylor, D.H. and Adler, K. 1973. Spatial orientation by salamanders using plane-polarized light. *Science* 181: 285–287.

5 Taylor, D.H. and Adler, K. 1978. The pineal body: Site of extraocular perception of celestial cues for orientation in the tiger salamander *(Ambystoma tigrinum). J. Comp. Physiol.* 124: 357–361.

10 Teuber, H.L. 1960. Perception. In *Handbook of Physiology, Neurophysiology.* Sect. 1, Vol. III. Am. Physiol. Soc., Washington, D.C.

9 Thompson, W.J. and Stent, G.S., 1976. Neuronal control of heartbeat in the medicinal leech. II. Intersegmental coordination of heart motor neuron activity by heart interneurons. *J. Comp. Physiol.* III. 281–307.

2 Thorndike, E.L. 1911. *Animal Intelligence.* Macmillan, New York.

2,5,7,9 Tinbergen, N. 1951. *The Study of Instinct.* Oxford University Press, New York.

5 Tomita, T. 1970. Electrical activity of vertebrate photoreceptors. *Q. Rev. Biophys.* 3: 179–222.

8 Usherwood, P.N.R. 1977. Neuromuscular transmission in insects. In *Neurons and Behavior of Arthropods.* G. Hoyle (ed.). Plenum Press, New York.

7 Uttal, W.R. 1978. *The Psychobiology of Mind.* Lawrence Erlbaum Associates, Hillsdale, New Jersey.

8 Valenstein, E.S. 1973. *Brain Stimulation and Motivation: Research and Commentary.* Scott, Foresman & Company, Glenview, Illinois.

10 Vallbo, A.B. 1970. Slowly adapting muscle receptors in man. *Acta Physiol. Scand.* 78: 315–333.

7 van Bergeijk, W.A. 1962. Variation on a theme of Békésy: A model of binaural interaction. *J. Acoust. Soc. Am.* 34: 1431–1437.

7 van Essen, D.C. 1979. Visual areas of the mammalian cerebral cortex. *Annu. Rev. Neurosci.* 2: 227–263.

4 Vardi, N. and Camhi, J.M. 1982a. Functional recovery from lesions in the escape system of the cockroach. I. Behavioral recovery. *J. Comp. Physiol.* 146: 291–298.

4 Vardi, N. and Camhi, J.M. 1982b. Functional recovery from lesions in the escape system of the cockroach. II. Physiological recovery of the giant interneurons. *J. Comp. Physiol.* 146: 299–309.

7 Vazulla, S. and Schmidt, J. 1976. Radioautographic localization of ^{125}I-bungarotoxin binding sites in the retinae of goldfish and turtle. *Vision Res.* 16: 878–881.

4 Volman, S.F., Camhi, J.M. and Vardi, N., 1980. Role of sensory activity in behavioral recovery of the directional escape response in the cockroach. *Neurosci. Abst.* 6: 232.11.

4 Volman, S.F. and Camhi, J.M. 1982. The role of activity in neuronal plasticity in the cockroach. *Neurosci. Abst.* 8: 9.12.

6 von Békésy, G. 1960. *Experiments in Hearing.* McGraw-Hill, New York.

10 von Euler, C. 1966. Proprioceptive control in respiration. In *Muscular Afferents and Motor Control.* R. Granit (ed.). Almqvist and Wiksell, Stockholm.

5 von Frisch, K. 1967. *The Dance Language and Orientation of Bees.* Harvard University Press, Cambridge, Massachusetts.

10 von Helmholtz, H. 1925. *Physiological Optics.* J.P. Southall (ed.). Optical Society of America, New York.

5 von Helversen, O. and Edrich, W. 1974. Der Polarisationsempfanger im Beinenauge: Ein Ultraviolettrezeptor. *J. Comp. Physiol.* 94: 33–47.

9 von Holst, E. 1973a. On the nature of order in the central nervous system. In *The Behavioral Physiology of Animals and Man: Selected Papers.* Vol. I. Univerisity of Miami Press, Coral Gables, Florida.

9 von Holst, E. 1973b. Relative coordination as a phenomenon and as a method of analysis of central nervous function. In *The Behavioral Physiology of Animals and Man: Selected Papers.* Vol. I. (Collected papers of E. von Holst, translated). University of Miami Press, Coral Gables, Florida.

10 von Holst, E. and Mittelstaedt, H. 1973. The reafference principle. In *The Behavioral Physiology of Animals and Man.* Vol. I. E. von Holst (ed.). University of Miami Press, Coral Gables, Florida.

8 von Holst, E. and von St. Paul, V. 1973. On the functional organization of drives. In *Behavioral Physiology of Animals and Man: The Collected Papers of Erich von Holst.* University of Miami Press, Coral Gables, Florida.

5 Walcott, C., Gould, J.L. and Kirschvink, J.L. 1979. Pigeons have magnets. *Science* 205: 1027–1029.

10 Waldron, I. 1967. Neural mechanisms by which controlling inputs influence motor output in the flying locust. *J. Exp. Biol.* 47: 213–228.

5 Walls, G.L. 1942. *The Vertebrate Eye and its Adaptive Radiation.* Hafner, New York.

7 Waloga, G. and Pak, W.L. 1978. Ionic mechanism for the generation of horizontal cell potentials in isolated axolotl retina. *J. Gen. Physiol.* 71: 69–92.

5 Waterman, T.H. 1975. The optics of polarization sensitivity. In *Photoreceptor Optics.* A.W. Snyder and R. Menzel (eds.). Springer-Verlag, New York. pp. 339–371.

5 Waterman, T. 1975b. Expectation and achievement in comparative physiology. *J. Exp. Zool.* 194: 309–344.

2 Watson, J.B. 1930. *Behaviorism.* W.W. Norton, New York.

6 Webster, F.A. and Griffin, D.R. 1962. The role of the flight membranes in insect capture by bats. *Anim. Behav.* 10: 332–340.

9 Weeks, J.C. 1981. Neuronal basis of leech swimming: Separation of swim initiation, pattern generation, and intersegmental coordination by selective lesions. *J. Neurophysiol.* 45: 698–723.

9 Weeks, J.C. 1982. Segmental specialization of a leech swim-initiating interneuron, cell 205. *J. Neurosci.* 2: 972–985.

9 Weeks, J.C. and Kristan, W.B. Jr. 1978. Initiation, maintenance and modulation of swimming in the medicinal leech by the activity of a single neurone. *J. Exp. Biol.* 77: 71–88.

5 Wehner, R. 1976. Polarized light navigation by insects. *Sci. Am.* 235(1): 106–115.

10 Wells, M.A. 1962. *Brain and Behavior in Cephalopods.* Heinemann, London.

10 Wendler, G. 1974. The influence of proprioceptive feedback on locust flight coordination. *J. Comp. Physiol.* 88: 173–200.

10 Wendler, G. 1978. The possible role of fast wing reflexes in locust flight. *Naturwissenschaften* 65: 65–66.

6 Wendler, G., Dambach, M., Schmitz, B. and Scharstein, H. 1980. Analysis of the acoustic orientation behavior in crickets (*Gryllus campestris* L.). *Naturwissenschaften* 67:, S99.

5,7 Werblin, F.S. and Dowling, J.E. 1969. Organization of the retina of the mudpuppy, *Necturus maculosus.* II. Intracellular recording. *J. Neurophysiol.* 32: 339–355.

4 Westin, J. 1979. Responses to wind recorded from the cercal nerve of the cockroach *Periplaneta americana.* I. Response properties of single sensory neurons. *J. Comp. Physiol.* 133: 97–102.

4 Westin, J., Langberg, J.J. and Camhi, J.M. 1977. Responses of giant interneurons of the cockroach *Periplaneta americana* to wind puffs of different directions and velocities. *J. Comp. Physiol.* 121: 307–324.

8 Wetzel, M.C. and Stuart, D.G. 1976. Ensemble characteristics of rat locomotion and its neural control. *Prog. Neurobiol.* 7: 1–98.

6 Wever, E.G. 1978. *The Reptile Ear.* Princeton University Press, Princeton, New Jersey.

2,5 Wickler, W. 1968. *Mimicry in Plants and Animals.* World University Library, McGraw-Hill, New York.

8 Wiersma, C.A.G. 1947. Giant nerve fiber system of the crayfish. A contribution to comparative physiology of synapse. *J. Neurophysiol.* 10: 23–38.

6 Wiley, R.H. and Richards, D.G. 1978. Physical constraints on acoustic communication in the atmosphere: Implications for the evolution of animal vocalizations. *Behav. Ecol. Sociobiol.* 3: 69–94.

2 Williams, G.C. 1966. *Adaptation and Natural Selection.* Princeton University Press, Princeton, New Jersey.

2 Willows, A.O.D. 1971. Giant brain cells in mollusks. *Sci. Am.* 244: 68–75.

10 Wilson, D.M. 1961. The central nervous control of flight in a locust. *J. Exp. Biol.* 38: 471–490.

9 Wilson, D.M. 1966. Central nervous mechanisms for the generation of rhythmic behavior in arthropods. *Symp. Soc. Exp. Biol.* 20: 199–228.

9 Wilson, D.M. 1967. An approach to the problem of control of rhythmic behavior. In *Invertebrate Nervous Systems.* C.A.G. Wiersma (ed.). University of Chicago Press, Chicago.

10 Wilson, D.M. and Gettrup, E. 1963. A stretch reflex controlling wingbeat frequency in grasshoppers. *J. Exp. Biol.* 40: 171–185.

2 Wilson, E.O. 1975. *Sociobiology: The New Synthesis.* Harvard University Press, Cambridge, Massachusetts.

8 Wine, J.J. 1977a. Crayfish escape behavior. II. Command-derived inhibition of abdominal extension. *J. Comp. Physiol.* 121: 173–186.

8 Wine, J.J. 1977b. Crayfish escape behavior. III. Monosynaptic and polysynaptic sensory pathways involved in phasic extension. *J. Comp. Physiol.* 121: 187–203.

8 Wine, J.J. and Hagiwara, G. 1977. Crayfish escape behavior. I. The structure of efferent and afferent neurons involved in abdominal extension. *J. Comp. Physiol.* 121: 145–172.

8,10 Wine, J.J. and Krasne, F.B. 1972. The organization of the escape behavior in the crayfish. *J. Exp. Biol.* 56: 1–18.

8,10 Wine, J.J. and Krasne, F.B. 1982. The cellular organization of crayfish escape behavior. In *The Biology of Crustacea*. D.E. Bliss, H. Atwood and D. Sandeman (eds.). Vol. IV. *Neural Integration*. H. Atwood and D. Sandeman (eds.). Academic Press, New York.

10 Wine, J.J. and Mistick, D.C. 1977. Temporal organization of crayfish escape behavior: Delayed recruitment of peripheral inhibition. *J. Neurophysiol.* 40: 904–925.

5 Wright, W. 1972. Color mixture. In *Visual Psychophysics*. D. Jameson and L. Hurvich (eds.). Springer-Verlag, Berlin.

7 Wurtz, R.H. and Albano, J.E. 1980. Visual-motor function of the primate superior colliculus. *Annu. Rev. Neurosci.* 3: 189–226.

7 Yost, W.A. and Neilsen, D.W. 1977. *Fundamentals of Hearing: An Introduction*. Holt, Rinehart & Winston, New York.

1 Young, J.Z. 1975. Sources of discovery in neuroscience. In *The Neurosciences: Paths of Discovery*. F.G. Worden, J.P. Swazey and G. Adelman (eds.). M.I.T. Press, Cambridge, Massachusetts.

8 Zucker, R.S. 1972. Crayfish escape behavior and central synapses. I. Neural circuit exciting lateral giant fiber. *J. Neurophysiol.* 35: 599–620.

6 Zwislocki, J.J. 1980. Five decades of research on cochlear mechanics. *J. Acoust. Soc. Am.* 67: 1679–1685.

Index